Lecture Notes in Mathematics

Edited by A. Dold, B. Eckmann and F. Takens

1423

Stanley O. Kochman

Stable Homotopy Groups of Spheres

A Computer-Assisted Approach

Springer-Verlag
Berlin Heidelberg New York London Paris Tokyo Hong Kong

Author
Stanley O. Kochman
Department of Mathematics, York University
4700 Keele Street, North York, Ontario M3J 1P3, Canada

Mathematics Subject Classification (1980): Primary: 55Q45
Secondary: 55T25, 55S30, 55Q50

ISBN 3-540-52468-1 Springer-Verlag Berlin Heidelberg New York
ISBN 0-387-52468-1 Springer-Verlag New York Berlin Heidelberg

Printed in Germany

Printing and binding: Druckhaus Beltz, Hemsbach/Bergstr.
2146/3140-543210 – Printed on acid-free paper

PREFACE

This work develops the theoretical basis for an efficient method for the inductive calculation of the stable homotopy groups of spheres, π_*^S. Most of the steps of this method are algorithmic and are done by computer. We will apply this method to compute the first 64 stable stems. This method is based upon the analysis of the Atiyah-Hirzebruch spectral sequence:

$$(*) \qquad E^2_{n,t} = H_n BP \otimes \pi_t^S \Longrightarrow \pi_{n+t} BP.$$

H_*BP and π_*BP are well known. Moreover, the Hurewicz homomorphism $h: \pi_*BP \longrightarrow H_*BP$ is a monomorphism. Therefore, $E^\infty_{n,t} = 0$ if $t \neq 0$, and $E^\infty_{n,0} = h(\pi_n BP)$ which is also well known. If π_t^S is known for $t < T$ then, with the exception of one step, it is algorithmic to deduce the composition series Image $[d^r: E^r_{r,T-r+1} \longrightarrow E^r_{0,T}]$, $2 \leq r \leq T+1$, of π_T^S. The determination of π_T^S from this composition series, the solution of the "additive extension problem", is accomplished using Toda brackets.

A distinctive feature of this method is that all the hard computations are done by computer. This includes the determination of differentials using Quillen operations and the computation of

$$E^{r+1}_{N,t} = \text{Kernel } [d^r: E^r_{N,t} \longrightarrow E^r_{N-r,t+r-1}] \ / \ \text{Image } [d^r: E^r_{N+r,t-r+1} \longrightarrow E^r_{N,t}].$$

On the other hand there are two key steps which require human intervention in the computation of each π_T^S:

(1) the matching of the list of "new" elements in degree T+1 which are hit by differentials with the list of "new" elements in degree T+2 on which nonzero differentials originate;

(2) the solution of the additive extension problems.

Chapter 1 is devoted to the exposition of the background of this computation and to a detailed description of the method we will use. Even the most experienced reader should read the exposition of our notation for elements of the stable stems at the end of that chapter. In Chapter 2 we develop the three and four-fold Toda bracket methods which are used to solve extension problems. In Chapter 3 we give a global computation of the spectral sequence in the first eight rows. In higher rows our computations are inductive and rarely achieve a global understanding of the rows beyond the range of our computations. In Chapter 4 we recall some facts about the Image of J and use them to compute all the differentials which originate on $E_r^{n,0}$ for $n \le 70$. Chapters 5 to 7 contain our calculations of the first 64 stable stems. In Chapter 8 we identify the elements $\theta_4 \in \pi_{30}^S$ and $\theta_5 \in \pi_{62}^S$ of Arf invariant one as well as the Mahowald elements $\eta_5 \in \pi_{32}^S$ and $\eta_6 \in \pi_{64}^S$. The new proof that θ_5 exists and has order two is based upon Mahowald's ideas [34A] and the computations of this paper. It is a rewording of a detailed proof which Mahowald sent to me. We also show that η_6 has order four. We conclude with Appendices 1 - 4, 7 which contain tables that summarize and give references for all the computations of this paper. In the fifth appendix, we discuss the Fortran computer programs which are used in this computation. A copy of the program listings is available from the author. The most important output of these programs is contained in the last sections of Chapters 4 - 7. The sixth appendix depicts the mod 2 Adams spectral sequence through degree 64.

We will work exclusively at the prime two. Our methods, however, apply at all primes. Of course, the computations at odd primes would be very different from these computations at the prime two. In addition the size of the numbers involved at the prime two reached 2^{32}, the limit of the computer, requiring the use of some multiprecision arithmetic. The computations at odd primes would involve much larger numbers.

I wish to thank The University of Western Ontario and York University for their support of this research as well as the University of Toronto for their hospitality during my sabbatical leave there. In addition, the Natural Sciences and Engineering Research Council of Canada supported this research through Operating Grants as well as an Equipment Grant which allowed the purchase of the IBM PC/AT computer on which the calculations were performed. Last, but not least, I am very grateful to Mark Mahowld for detecting errors in earlier versions of this paper, for his ideas on θ_5 and for his assistance in constructing the Adams spectral sequence tables in Appendix 6.

TABLE OF CONTENTS

CHAPTER 1: INTRODUCTION

1. History of the Problem

The calculation of the stable homotopy groups of spheres is one of the most central and intractable problems in algebraic topology. In the 1950s Serre [57] used his spectral sequence to study this problem. In 1962, Toda [60] used his triple brackets and the EHP sequence to calculate the first 19 stems. These methods were later extended by Mimura, Mori, Oda and Toda [44], [45], [46], [50] to compute the first 30 stems. In the late 1950s the study of the classical Adams spectral sequence was begun [1]. Computations in this spectral sequence are still being pursued using the May spectral sequence and the lambda algebra. The best published results are May's thesis [39] and the computation of the first 45 stable stems by Barratt, Mahowald, Tangora [10], [37] as corrected by Bruner [16]. The use of the BP Adams spectral sequence on this problem was initiated by Novikov [49] and Zahler [62]. Its most spectacular success has been at odd primes [42]. A recent detailed survey of the status of this computation and the methods that have been used has been written by Ravenel [55].

An exotic method for computing stable stems was developed in 1970 by Joel Cohen [19]. Recall [20] that for a generalized homology theory E_* and a spectrum X there is an Atiyah-Hirzebruch spectral sequence:

(1.1.1) $$E^2_{N,p} = H_N(X;E_p) \Longrightarrow E_{N+p}X.$$

Joel Cohen studied this spectral sequence with X an Eilenberg-MacLane spectrum and E equal to stable homotopy or mod p stable homotopy. His idea was to take advantage of the fact that in these cases the spectral sequence is converging to zero in positive degrees. Since the homology of the Eilenberg-MacLane spectra are known, one can inductively deduce the stable stems. This is analogous to the usual inductive computation of the cohomology of Eilenberg-MacLane spaces by the Serre spectral sequence [17]. In that example, however, all the work can be incorporated into the Kudo transgression theorem. Joel Cohen was able to compute a few low stems, but the computation became too complicated to continue. His method was discarded since the Adams spectral sequence computations seemed much more efficient. In 1972, however, Nigel Ray [56] used this spectral sequence with X = MSU and E = MSp. He took advantage of the fact that H_*MSU and MSp_*MSU are known to compute the first 19 homotopy groups of MSp. Again this method was discarded since David Segal had computed the first 31 homotopy groups of MSp by the Adams spectral sequence and his computations were extended to 100 stems in [31].

My interest in Atiyah-Hirzebruch spectral sequences began in 1978. In a joint paper with Snaith [32] we studied the case where X is BSp and E_* is stable homotopy. The methods we developed there, in particular the use of Landweber-Novikov operations to study differentials, were clearly applicable to a wide class of examples. In 1983, I observed that if Joel Cohen's method were applied to the case where X is BP and E_* is stable homotopy then the computations would be greatly simplified over Cohen's case because of the sparseness of H_*BP and because Quillen operations could be used to compute the differentials. So, I began computing at the prime two. I soon discovered that the computations became too complicated to do by hand, but since they were mostly algorithmic they could be done by a computer. Using an IBM PC/AT micro-computer I was able to compute the first 64 stable stems. This work is the account of that computation.

Kaoru Morisugi informed me that in 1972 he attempted to use this method to compute π_*^S at the prime three, but he became bogged down with technical problems.

2. The Brown-Peterson Spectrum and Quillen Operations

In this section we list some of the basic facts about the Brown-Peterson spectrum BP. The notation introduced here will be used throughout the computation.

Let MU denote the unitary Thom spectrum. By the Pontryagin-Thom isomorphism, π_*MU is isomorphic to Ω_*^U, the ring of bordism classes of compact smooth manifolds without boundary which have a complex structure on their stable normal bundles. Using the Adams spectral sequence, Milnor [43] computed π_*MU to be a polynomial algebra over Z with one generator in each even degree. Brown and Peterson [15] discovered that when the spectrum MU is localized at a prime p, it decomposes into a wedge of various suspensions of a spectrum BP. This spectrum defines a generalized homology theory BP_* and a generalized cohomology theory BP^*. We list several basic properties of BP at the prime two. The standard references are the expositions of Adams [7] and Wilson [61].

(1.2.1) There are $M_N \in H_*BP$ of degree $2(2^N-1)$ such that $M_0 = 1$ and

$$H_*BP = Z_{(2)}[M_1,\ldots M_N,\ldots].$$

(1.2.2) The Hurewicz homomorphism $h:\pi_*BP \longrightarrow H_*BP$ is a monomorphism. Henceforth we consider h as an inclusion.

(1.2.3) Define $V_N \in H_*BP$ of degree $2(2^N-1)$ recursively by $V_0 = 2$ and for $N\geq 1$:

$$V_N = 2M_N - \sum_{k=1}^{N-1} M_k \cdot V_{N-k}^{2^k}.$$

The $V_N/2$, $N \geq 1$, are polynomial generators for H_*BP. Moreover, all the V_N are in the image of h and $\pi_*BP = Z_{(2)}[V_1,\ldots,V_N,\ldots]$. The V_N are called the Hazewinkel generators [22], [23].

(1.2.4) BP^*BP is the algebra of BP-operations. These operations act on BP_*X for any spectrum X including $BP_*S = \pi_*BP$ and $BP_*KZ = H_*BP$. These operations are natural. In particular, they commute with the Hurewicz homomorphism h.

(1.2.5) $BP^*BP = \pi_*BP[[\, r_\omega \mid \omega$ is a finite sequence of nonnegative integers$]]$.

The r_ω are called the Quillen operations [54]. They have the following properties.

(a) The r_ω are $Z_{(2)}$-module homomorphisms.

(b) If $f: X \longrightarrow Y$ is a map of spectra then $f_* \circ r_\omega = r_\omega \circ f_*$. In particular, $h \circ r_\omega = r_\omega \circ h$.

(c) If X is a ring spectrum and $A, B \in BP_*X$ then we have the Cartan formula

$$r_\omega(A \cdot B) = \sum_{\omega=\omega'+\omega''} r_{\omega'}(A) \cdot r_{\omega''}(B).$$

In [32] we showed how Landweber-Novikov operations act on the Atiyah-Hirzebruch spectral sequences for $\pi_*^S BU$ and $\pi_*^S BSp$. The following theorem shows that the Quillen operations act on Atiyah-Hirzebruch spectral sequences for BP_*X.

THEOREM 1.2.6 Let F be a ring spectrum. Consider the Atiyah-Hirzebruch spectral sequence for F_*BP:

$$E^2_{N,t} = H_N BP \otimes F_t \Longrightarrow F_{N+t}$$

Then each Quillen operation r_ω of degree K induces a map of spectral sequences:

$$r_\omega : E^s_{N,t} \longrightarrow E^s_{N-K,t}.$$

These r_ω have the following properties:

(a) The r_ω are $Z_{(2)}$-module homomorphisms.

(b) The r_ω are natural with respect to maps of spectral sequences induced by maps of spectra.

(c) The r_ω satisfy the Cartan formula

$r_\omega(A \cdot B) = \sum_{\omega=\omega'+\omega''} r_{\omega'}(A) \cdot r_{\omega''}(B)$ for all $A, B \in E^s$.

(d) The action of r_ω on E^2 is given by $r_\omega \otimes 1$ where the latter r_ω is the usual Quillen operation on H_*BP.

(e) $d^s \circ r_\omega = r_\omega \circ d^s$ for all $s \geq 1$.

(f) The action of r_ω on $E^{s+1} = H_*(E^s, d^s)$ is induced by the action of r_ω on E^s.

(g) The action of r_ω on the E^s induce an action of r_ω on $E^\infty = \varinjlim E^s$.

(h) The action of r_ω on E^∞ defined by (g) agrees with the action of r_ω on E^∞ induced by the usual action of the Quillen operations on $F_*BP = BP_*F$.

PROOF. Since $r_\omega \in BP^kBP$, we can represent r_ω by a map of spectra $r_\omega : \Sigma^K BP \longrightarrow BP$. Since the Atiyah-Hirzebruch spectral sequence is natural we have an induced map of spectral sequences. All of the properties are immediate except for the Cartan formula (c). It follows from the observation that the following diagram must commute up to homotopy:

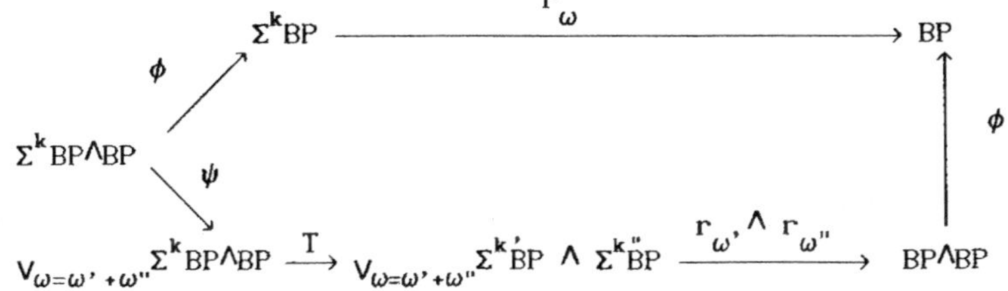

In this diagram ϕ is product map of BP and ψ is the pinching map. In each wedge summand $k = k'+k''$ and T is the switching map. ∎

3. The Inductive Procedure

In this section we will describe in detail the inductive procedure that we will use to compute the stable stems. However, before we apply this procedure in Chapters 5 to 7 we will digress to compute the first eight rows of the spectral sequence in Chapter 3 and to study two of the basic ingredients of our procedure: Toda brackets in Chapter 2 and the image of J in Chapter 4. This section concludes with an exposition of the notation that we will use to denote the elements of π_*^S.

Consider the Atiyah-Hirzebruch spectral sequence:

(1.3.1) $$E^2_{N,t} = H_N BP \otimes \pi_t^S \Longrightarrow \pi_{N+t} BP.$$

Since H_*BP is zero in odd degrees we see that in this spectral sequence:

(1.3.2)
$$E^r_{N,*} = 0 \text{ if } N \text{ is odd},$$
$$d^{2r+1} = 0 \text{ and}$$
$$E^{2r+1} = E^{2r+2} \text{ for all } r.$$

The Hurewicz homomorphism is given in terms of this spectral sequence by the following commutative square:

(1.3.3)
$$\begin{array}{ccc} \pi_N BP & \xrightarrow{\quad h \quad} & H_N BP \\ \downarrow & & \uparrow \cong \\ E^\infty_{N,0} & \rightarrowtail & E^2_{N,0} \end{array}$$

Since h is one-to-one, it follows that:

(1.3.4)
$$E^\infty_{N,t} = \begin{cases} 0 & \text{if } t \neq 0 \\ \pi_N BP & \text{if } t = 0 \end{cases} \quad \text{and}$$

(1.3.5)
$$E^\infty_{*,0} = Z_{(2)}[V_1,\ldots,V_N,\ldots].$$

Thus, there must be nonzero differentials originating on the 0 row so that each monomial $K(2^{-e}V_1^{e(1)}\cdots V_M^{e(M)})$ in E^2 survives to E^∞ if and only if K is divisible by 2^e where $e = e(1)+\cdots+e(M)$. We will prove in Chapter 4 that, in our range of computations, all nonzero differentials which originate on the 0 row land in $ImJ \otimes H_*BP$. We will assume that ImJ is known. The first step in our analysis of the spectral sequence (1.3.1) will be to compute all these differentials which originate on the 0 row in degrees 2 through 70. This computation is entirely algorithmic, is done by computer with no human assistance and is carried out in Section 4.4. The purpose of this computation is to record the cokernels of all of these differentials.

The behavior of the following elements in the spectral sequence is the key to the determination of differentials which originate above the 0 row.

DEFINITION 1.3.6 Let $\phi \in \pi_t^S$ have order q and let $V \in H_{2N}BP$. Assume that:

(a) $\phi\cdot V \in E^2_{2N,t}$ survives to an element of $E^{2r}_{2N,t}$ for some $2 \le r \le \infty$;

(b) if $r = \infty$ then $V = 0$;

(c) we know all differentials which originate or land on elements of $E^{2s}_{2k,t}$ which have a representative in $Z_q\phi \otimes H_*BP$ for all s and all $0 \le k < N'$ where $N' = N$ if $r < \infty$ or $N' = \infty$ if $r = \infty$.

We call such an element $\phi\cdot V$ a ϕ-leader.

Note: A ϕ-leader can be zero. In that case our assumption is that we know all differentials which originate or land in $Z_q\phi \otimes H_*BP$.

The following unfortunate phenomenon is the obstruction to using Theorem 1.2.6(e) to computing d^{2r}-differentials on $\phi\cdot V''$, degree V'' > degree V, from the d^{2r}-differential on a ϕ-leader $\phi\cdot V$.

DEFINITION 1.3.7 Let $\phi\cdot V$ be a ϕ-leader, and assume all the notation of Definition 1.3.6. A nonzero differential $d^{2u}(\phi\cdot V')$ is called a hidden differential if:

(a) $\phi\cdot V'$ is also a ϕ-leader;

(b) degree V' > degree V;

(c) $u < r$.

Thus, if there is a hidden differential, the d^{2u}-differentials determined by $d^{2u}(\phi\cdot V')$ must be computed before the d^{2r}-differentials determined by $d^{2r}(\phi\cdot V)$ even though degree $\phi\cdot V'$ > degree $\phi\cdot V$. The inductive computation of π_N^S now proceeds as follows. Assume that the information contained in the following induction hypothesis is known.

(1.3.8) INDUCTION HYPOTHESIS

(1_N) We know π_k^S for $0 \le k < N$.

(2_N) Write each nonzero differential on a ϕ-leader $\phi\cdot V \in E^{2r}_{2a,b}$, with $a+b \le N$, in the form $d^{2r}(\phi\cdot V) = \lambda V' \neq 0$ where $\phi \in \pi_b^S$, $\lambda \in \pi^S_{b+2r-1}$, $V \in H_{2a}BP$ and $V' \in H_{2a-2r}BP$. Assume that we have "computed" $d^{2r}(\phi\cdot V") = \sum \alpha_I \lambda V_I$ for all $V" \in H_{2a"}BP$.

(3_N) For each $\phi \in \pi_k^S$, $0 < k < N$, the ϕ-leader of largest known degree is $\phi\cdot V$ where either $V = 0$ or degree $\phi\cdot V \ge N+1$.

The information in (2_N) is called a "tentative differential table" and the information in (3_N) is called a "list of leaders". In condition (2_N), the word computed is in quotation marks because what we assume that we have done is that we have computed $r_\omega \circ d^{2r}(\phi\cdot V") = d^{2r} \circ r_\omega(\phi\cdot V")$ for all Quillen operations r_ω of degree 2a"-2a. This would give an accurate computation of $d^{2r}(\phi\cdot V")$ if there were no hidden differentials. Unfortunately, there are examples of hidden differentials.

To accomplish the inductive step we must go through the procedure below. We use the terminology "$A \in E^{2r}_{2N,t}$ transgresses" if A survives to E^{2N}. In that case $d^{2N}(A) \in E^{2N}_{0,2N+t-1}$, a subquotient of π^S_{2N+t-1}.

(1.3.9) INDUCTION STEP

(a) Construct the following list of leaders of degrees N+1 and N+2:

<u>Leaders in Degree N+1</u>	<u>Leaders in Degree N+2</u>
α_1	β_1
.	.
.	.
.	.
α_p	β_q

Each $\alpha_i \in E^{2a(i)}_{2a(i),N-2a(i)+1}$ will either be hit by some β_j or it will transgress to determine a nonzero element of π^S_N. In either case α_i transgresses to an element $d^{2a(i)}(\alpha_i) = \hat{\alpha}_i \in \pi^S_N$. In the former case $\hat{\alpha}_i = 0$, and in the latter case $\hat{\alpha}_i \neq 0$.

(b) Search for hidden differentials $d^{2u}(\beta) = \alpha_i$, where $d^{2r}(\beta) = \alpha'$ was one of the differentials in the tentative differential table of 1.3.8(2_N). If a hidden differential is found then α_i must be removed from the list in (a) and replaced with α'. Assume that any necessary adjustments of this sort have been made to the list in (a).

(c) Use Toda bracket methods from Chapter 2 and consequences of differentials which follow from Theorem 1.2.6(e) to make the following deductions:

(i) some of the $\hat{\alpha}_i$ are zero;

(ii) some of the β_j transgress.

This step is complete when

$$\text{card}\,\{\alpha_i \mid \hat{\alpha}_i = 0\} = \text{card}\,\{\beta_j \mid \beta_j \text{ is not known to transgress}\}.$$

(d) Construct the following list of all α_i, β_j such that $\hat{\alpha}_i = 0$ and β_j is not known to transgress:

$$\begin{array}{ll} \alpha_{i(1)} & \beta_{j(1)} \\ \cdot & \cdot \\ \cdot & \cdot \\ \cdot & \cdot \\ \alpha_{i(s)} & \beta_{j(s)} \end{array}$$

There is a nonzero differential on each $\beta_{j(k)}$ with image some $\alpha_{i(h)}$. Use Toda bracket methods from Chapter 2, consequences of differentials deduced from Theorem 1.2.6(e) and ad hoc monoid chain arguments to match which $\beta_{j(k)}$s hit which $\alpha_{i(h)}$s.

(e) Use Toda bracket methods from Chapter 2 to solve the additive extension problems to determine π^S_N from its composition series $\{E^{2r}_{0,N} \mid 1 \leq r \leq [(N+1)/2]\}$. This gives the information required in (1_{N+1}). This step is not absolutely essential and the computation can proceed even if all the additive extension problems can not be solved.

(f) Use the computer program of Section 9.3 to extend the tentative differential table for each of the nonzero differentials determined in (d). This gives the information required in (2_{N+1}).

(g) Update the list of leaders using the new information in the tentative differential table determined in (f). This gives the information required in (3_{N+1}).

In practice this inductive procedure is quite straightforward. There are usually no hidden differentials. Also there are usually very few matchings to be done in (d) and those matchings are obvious. In addition, there are never many possibilities for nontrivial additive extensions and many of these possibilities are quite easy to eliminate. As a final word of encouragement, the reader will see that the above procedure is merely the formalization of the straightforward common sense approach to the analysis of the spectral sequence. The following theorem is widely applicable. (See Appendix 2.)

THEOREM 1.3.10 Assume that $\xi \in \pi_N^S$ is defined as $\xi = d^r(X)$ where ξ is nonzero in $E^r_{N,0}$. If $r > N/2$ then ξ is indecomposable in the ring π_*^S.

PROOF. Assume that ξ is decomposable. Write $\xi = \Sigma\, \alpha_i\beta_i$, where $\alpha_i = d^{s(i)}(A_i)$, $\beta_i = d^{t(i)}(B_i)$ and $s(i) \le t(i)$ for all i. Then $s(i) < r$ for all i and $\xi = \Sigma\, d^s(\beta_i A_i)$ where s is the largest of all the s(i). Since $s < r$, $\xi = 0$ in E^r, a contradiction. Thus, ξ must be indecomposable. ∎

We conclude with the notation that we will use to describe elements of π_*^S. There are competing notations for the elements of the known stable stems. To add to the confusion, most methods of computing stable stems (including ours) only define elements of π_*^S modulo indeterminacy: the indeterminacy of a Toda bracket or of the filtration of a spectral sequence. We will use the usual notation for the elements of Hopf invariant one:

$$\eta \in \pi_1^S, \ \nu \in \pi_3^S \text{ and } \sigma \in \pi_7^S.$$

We will also use the following notation for elements in Im J: $\alpha_N \in \pi^S_{8N+1}$, $\beta_N \in \pi^S_{8N+3}$ and $\gamma_N \in \pi^S_{8N+7}$. If an element $X \in \pi_*^S$ is known to be decomposable then we will usually write it as a product. We will use the following notation for other elements of π_*^S.

DEFINITION 1.3.11 A[N,k] denotes the kth element of π_N^S of order two, B[N,k] denotes the kth element of π_N^S of order four, C[N,k] denotes the kth element of π_N^S of order eight, etc. If there is only one element of π_N^S of a given order then we drop the second entry.

The following examples will help to explain this notation.

1. The element usually denoted $\varepsilon \in \pi_8^S$ of order two will be denoted A[8].
2. The element usually denoted $\bar{\kappa} \in \pi_{20}^S$ of order eight will be denoted C[20].
3. If we write D[45] we are denoting an element of π_{45}^S which has order 16.

We will also use the following notation. Let R be a PID and $\mathcal{A}$ a commutative R-algebra which is a free R-module. If $B, X_1, \ldots, X_t \in \mathcal{A}$ then $RB\{X_1, \ldots, X_t\}$ denotes the free R-submodule of $\mathcal{A}$ with basis $\{BX_1, \ldots, BX_t\}$. For example, let $\xi \in \pi_*^S$ have order N. We may take $R = Z_N$, $\mathcal{A} = Z_N\xi \otimes H_*BP$ and $X_1, \ldots, X_t$ linearly independent elements of H_*BP.

If α, $\beta \in \pi_*^S$ and $\alpha \cdot \beta = 0$ then $B_{\alpha\beta}$ denotes a map H from a disc to a sphere such that H restricted to the boundary of its domain is $\alpha' \wedge \beta'$ where α', β' represents α, β respectively.

CHAPTER 2: TODA BRACKETS

1. Introduction

As we saw in the previous chapter, there is one very important step in our computation that is not algorithmic: the determination of the additive and multiplicative structure of π_*^S from the composition series which has been deduced from the Atiyah-Hirzebruch spectral sequence. One of the main tools we will use to determine these extensions is the relationship between Toda brackets in π_*^S and differentials in the spectral sequence. This idea was originated by J. P. May [40,Section 4]. May's three basic theorems regarding the behaviour of Massey products in spectral sequences defined from a filtered differential graded algebra were generalized to the Adams and Atiyah-Hirzebruch spectral sequences in [28]. In addition to these classical results, we will derive and use several new theorems of this type.

In Section 2 we give two definitions of Toda brackets in π_*^S: one using the composition product and one using the the smash product. By [29], these two Toda brackets are always equal. We will find that there are situations in which one point of view is advantageous over the other. In Section 3, we derive the basic properties of these Toda brackets. In Section 4, we prove several theorems which relate these Toda brackets to the differentials in the Atiyah-Hirzebruch spectral sequence. We will only be using three-fold and four-fold Toda brackets in our applications. Therefore, we do not hesitate to specialize to these cases.

2. Definitions

We will find it convenient to work with spectra in the coordinate-free setting of J. P. May [41]. After introducing coordinate-free notation, we give two defininitions of Toda brackets: one based on the smash product and one based on the composition product. These definitions were first given in [29]. Our composition Toda bracket generalizes Toda's orginal three-fold product [60] and Oguchi's four-fold product [51]. It agrees with Spanier's Toda bracket [58] but it is not clear whether it agree's with Gershenson's Toda bracket [21]. Our smash Toda bracket agrees with that of Porter [51] and corresponds under the Pontrjagin-Thom isomorphism to the Massey product of manifolds defined in [28]. In Theorem 2.2.3 we state the theorem from [29] that our two Toda brackets are equal. In addition, our Toda bracket is contained in Joel Cohen's Toda bracket [18]. We conclude this section with several practical criteria for concluding that a four-fold Toda bracket is defined.

The following notation will be used throughout. Let R^∞ be the real inner product space with orthonormal basis $B = \{b_1, b_2, \ldots\}$. We consider only finite dimensional subspaces W of R^∞ which have a subset of B as a basis. Internal direct sum is denoted by +, and if W' is a subspace of W then $W'^{\perp}$ denotes the orthogonal complement of W' in W. All spaces are based CW complexes, all maps are based and all homotopies, cones and suspensions are reduced. Let S denote one point compactification. The n-sphere is defined as $S^n \equiv S(R^n)$. The isomorphism from a subspace V to $R^{\dim V}$ which preserves the ordered standard bases induces a canonical homeomorphism from SV to $S^{\dim V}$. Thus a map from SV to SW determines an element of $\pi_{\dim V}(S^{\dim W})$. If $i_1 < \cdots < i_t$ then define the disc $D(Rb_{i_1} + \cdots + Rb_{i_t})$ as $CS(Rb_{i_1}) \wedge S(Rb_{i_2} + \cdots + Rb_{i_t})$ where $C(-) = (I, \{1\}) \wedge (-)$ is the cone functor. If $1 \le j_1 < \cdots < j_k \le t$ and $f: SU_1 \wedge \cdots \wedge SU_t \wedge X \longrightarrow SU_1 \wedge \cdots \wedge SU_t \wedge Y$ then define $C_{j_1 \ldots j_k}(f)$ as the canonical map from $C_{j_1, \ldots, j_k}(SU_1 \wedge \cdots \wedge SU_t \wedge X)$ $\equiv SU_1 \wedge \cdots \wedge DU_{j_1} \wedge \cdots \wedge DU_{j_k} \wedge \cdots \wedge SU_t \wedge X$ to $C_{j_1, \ldots, j_k}(SU_1 \wedge \cdots \wedge SU_t \wedge Y)$ $\equiv SU_1 \wedge \cdots \wedge DU_{j_1} \wedge \cdots \wedge DU_{j_k} \wedge \cdots \wedge SU_t \wedge Y$ induced by f. Define an equivalence relation on ∂I^{t-1} by $(a_1,\dots,a_{t-1}) \approx (b_1,\dots,b_{t-1})$ if $\max(a_1,\dots a_{t-1}) = 1$ and $\max(b_1,\dots,b_{t-1}) = 1$. For $t \geq 3$ choose homeomorphisms $h_t : S^{t-2} \longrightarrow (\partial I^{t-1})/\approx$. Let T denotes the canonical interchange map. Then the maps $T\circ(h_t \wedge 1_{SV_1\wedge\cdots\wedge SV_t})$ define homeomorphisms

$$h: S(R^{t-2} + V_1 + \cdots + V_t) \longrightarrow \partial[DV_1\wedge\cdots\wedge DV_{t-1}\wedge SV_t]$$

Our spectra will be functors E defined on all finite dimensional subspaces W of R^∞ with basis a subset of B. We will use the symbol ε to denote either the structure map $S \wedge E \longrightarrow E$ of a spectrum or the product $E \wedge E \longrightarrow E$ of a ring spectrum. Then $\pi_N E$ is defined as the direct limit over all W of the groups $[SW, EW']$ where W' is a subspace of W with $N = \dim(W/W')$. The structure maps of this direct limit are $\varepsilon\circ(SV \wedge -)$ where $V \perp W$. We now have the notation to give the two definitions of the Toda bracket $\langle X_1,\dots,X_t\rangle$ where $X_1,\dots,X_{t-1} \in \pi^S_*$, $X_t \in \pi_*(E)$ and E is any spectrum. We begin with the definition based on the composition of maps.

DEFINITION 2.2.1. Let E be a spectrum, let $X_1,\dots,X_{t-1} \in \pi^S_*$ and let $X_t \in \pi_* E$. Let $G_{i-1,i} : SV_i\wedge\cdots\wedge SV_t\wedge SU \longrightarrow SV_{i+1}\wedge\cdots\wedge SV_t\wedge E_iU$ represent X_i, $1 \leq i \leq t$, where $R^{t-2} \perp V_1 \perp \dots \perp V_t \perp U$, $E_i = S$ for $i \leq i \leq t-1$ and $E_t = E$. A defining system for $\langle G_{0,1},\dots,G_{t-1,t}\rangle'_\circ$ consists of maps

$$G_{ij} : DV_{i+1}\wedge\cdots\wedge DV_{j-1} \wedge SV_j\wedge\cdots\wedge SV_t\wedge SU \longrightarrow SV_{j+1}\wedge\cdots\wedge SV_t\wedge E_jU$$

for $0 \leq i < j-1 < t$, $(i,j) \neq (0,t)$, such that

$$G_{ij}|\partial(DV_{i+1}\wedge\cdots\wedge DV_{j-1}\wedge SV_j\wedge\cdots\wedge SV_t\wedge SU) = \cup_{k=i+1}^{j-1} G^k_{ij}$$

where G^k_{ij} is the composite map

$$DV_{i+1}\wedge\cdots\wedge DV_{k-1}\wedge SV_k\wedge DV_{k+1}\wedge\cdots\wedge DV_{j-1}\wedge SV_j\wedge\cdots\wedge SV_t\wedge SU \xrightarrow{C_{k+1,\dots,j-1}(G_{ik})}$$

$$DV_{k+1}\wedge\cdots\wedge DV_{j-1}\wedge SV_j\wedge\cdots\wedge SV_t\wedge SU \xrightarrow{G_{kj}} SV_{j+1}\wedge\cdots\wedge SV_t\wedge E_jU.$$

If $\langle G_{01},\dots,G_{t-1,t}\rangle'_\circ$ has a defining system then define $\langle G_{01},\dots,G_{t-1,t}\rangle'_\circ$ as the set of homotopy classes of the maps

$$\tilde{G}_{0t} \equiv \cup_{k=1}^{t-1} G_{0t}^{k} \circ (h \wedge 1_{SU}) : S(R^{t-2}+V_1+\cdots+V_t) \wedge SU \longrightarrow EU$$

for all defining systems $\{G_{ij}\}$ of $\langle G_{01},\ldots,G_{t-1,t}\rangle'_{\circ}$. Define

$\langle G_{01},\ldots,G_{t-1,t}\rangle_{\circ} = \varinjlim_{W} \langle G_{01} \wedge 1_{SW},\ldots,G_{t-2,t-1} \wedge 1_{SW}, \varepsilon\circ(G_{t-1,t} \wedge 1_{SW})\rangle'_{\circ}$.

This direct limit is taken over all W with $W \perp (R^{t-2}+V_1+\cdots+V_t+U)$. If W' is a subspace of W then the map $- \wedge 1_{S(W'^{\perp})}$ sends a defining system of $\langle G_{01}\wedge 1_{SW'},\ldots,\varepsilon\circ(G_{t-1,t}\wedge 1_{SW'})\rangle'_{\circ}$ to a defining system of $\langle G_{01}\wedge 1_{SW},\ldots,\varepsilon\circ(G_{t-1,t}\wedge 1_{SW})\rangle'_{\circ}$. Finally, define $\langle X_1,\ldots,X_t\rangle_{\circ}$ as the union of $\langle G_{01},\ldots,G_{t-1,t}\rangle_{\circ}$ for all choices of representatives $G_{i-1,i}$ of X_i, $1\le i\le t$.

The following definition of the Toda bracket based on the smash product is a direct analogue of the usual algebraic definition of the Massey product in the homology of a differential graded algebra.

DEFINITION 2.2.2 Let E be a spectrum, let $X_1,\ldots,X_{t-1} \in \pi_*^S$ and let $X_t \in \pi_*E$. Let $G_{i-1,i}: SV_i \wedge SU_i \longrightarrow E_iU_i$ represent X_i for $1 \le i \le t$ where $R^{t-2} \perp V_1 \perp U_1 \perp \cdots \perp V_t \perp U_t$, $E_i = S$ for $1 \le i \le t-1$ and $E_t = E$. A defining system for $\langle G_{01},\ldots,G_{t-1,t}\rangle'_{\wedge}$ consists of maps

$$G_{ij}: DV_{i+1} \wedge SU_{i+1} \wedge\cdots\wedge DV_{j-1} \wedge SU_{j-1} \wedge SV_j \wedge SU_j \longrightarrow E_j(U_{i+1}+\cdots+U_j)$$

for $0 \le i < j-1 < t$, $(i,j)\ne(0,t)$, such that

$$G_{ij}|\partial(DV_{i+1} \wedge SU_{i+1} \wedge\cdots\wedge DV_{j-1} \wedge SU_{j-1} \wedge SV_j \wedge SU_j) = \cup_{k=i+1}^{j-1} G_{ij}^{k}$$

where G_{ij}^{k} is the composite map $\varepsilon\circ T\circ(G_{ik} \wedge G_{kj})$. If $\langle G_{01},\ldots,G_{t-1,t}\rangle'_{\wedge}$ has a defining system, then define $\langle G_{01},\ldots,G_{t-1,t}\rangle'_{\wedge}$ as the set of homotopy classes of the maps

$$\tilde{G}_{0t} \equiv (\cup_{k=1}^{t-1} G_{0t}^{k})\circ T\circ(h \wedge 1_{SU_1\wedge\cdots\wedge SU_t}) : S(R^{t-2}+V_1+\cdots+V_t) \wedge SU_1\wedge\cdots\wedge SU_t \longrightarrow E(U_1+\cdots+U_t)$$

for all defining systems $\{G_{ij}\}$ of $\langle G_{01},\ldots,G_{t-1,t}\rangle'_{\wedge}$. Define

$$\langle G_{01},\dots,G_{t-1,t}\rangle_\wedge = \varinjlim_{W_1,\dots,W_t} \langle G_{01}\wedge 1_{SW_1},\dots,G_{t-2,t-1}\wedge 1_{SW_{t-1}},\varepsilon\circ(G_{t-1,t}\wedge 1_{SW_t})\rangle'_\wedge$$

where the direct limit is taken over all $W_1,\dots,W_t$ with $W_1 \perp \cdots \perp W_t \perp (R^{t-2}+V_1+U_1+\cdots+V_t+U_t)$. If W'_i is a subspace of W_i, $1 \le i \le t$, then the maps $\varepsilon\circ T\circ(-\wedge 1_{S(W'_{i+1}{}^\perp)}\wedge\cdots\wedge_{S(W'_j{}^\perp)})\circ T$ send a defining system of $\langle G_{01}\wedge 1_{SW'_1},\dots,\varepsilon\circ(G_{t-1,t}\wedge 1_{SW'_t})\rangle'_\wedge$ to a defining system of $\langle G_{01}\wedge 1_{SW_1},\dots,\varepsilon\circ(G_{t-1,t}\wedge 1_{SW_t})\rangle'_\wedge$. Finally, define $\langle X_1,\dots,X_t\rangle_\wedge$ as the union of $\langle G_{01},\dots,G_{t-1,t}\rangle_\wedge$ for all choices of representatives $G_{i-1,i}$ of X_i, $1\le i\le t$.

The reader can find the proof of the following theorem in [29,Theorem 3.2].

THEOREM 2.2.3 Let E be a spectrum, let $X_1,\dots,X_{t-1} \in \pi_*^S$ and let $X_t \in \pi_*E$. Then $\langle X_1,\dots,X_t\rangle_\circ$ is defined if and only if $\langle X_1,\dots,X_t\rangle_\wedge$ is defined. Moreover, if these Toda brackets are defined then they are equal.

NOTATION: In view of this theorem, we will use the symbol $\langle X_1,\dots,X_t\rangle$ to denote $\langle X_1,\dots,X_t\rangle_\circ = \langle X_1,\dots,X_t\rangle_\wedge$.

We will try to imitate proofs of results for algebraic Massey products to construct proofs of the corresponding results for Toda brackets with defining systems constructed with the smash product. An obvious ingredient which we will require is the ability to add maps defined on cones.

DEFINITION 2.2.4 Let f and g be two maps from $C_{j_1,\dots,j_k}(X\wedge SU_1\wedge\cdots\wedge SU_t)$ to Y, where $U_1 \perp \cdots \perp U_t$ and $0\le k\le t$. Let $\{b_{i_1},\dots,b_{i_N}\}$ be a basis for $U_1+\cdots+U_t$ with $i_1 < \cdots < i_N$ and let $\mu(f) = \mu(X \wedge SU_1\wedge\cdots\wedge SU_t) = i_1$. Define

$$f \oplus g: C_{j_1,\dots,j_k}(X \wedge SU_1\wedge\cdots\wedge SU_t) \longrightarrow Y$$

in the usual way by pinching in the $\mu(f) = i_1$ coordinate. Also define -f in the usual way reversing the $\mu(f) = i_1$ coordinate. Let $f \ominus g = f \oplus (-g)$.

Now we have a sum $\oplus$ and a product $\wedge$ defined for the maps that arise in defining systems of Toda brackets. Unfortunately most of the usual algebraic identities only hold up to homotopy for these operations. However, there are five identities which these operations do satisfy.

THEOREM 2.2.5 The following identities hold whenever the expressions appearing in them are defined.

(a) $f \wedge (g \wedge h) = (f \wedge g) \wedge h$

(b) $-(f \oplus g) = (-f) \oplus (-g)$

(c) If $\mu(f) < \mu(W)$ then $1_{SW} \wedge (f \oplus g) = (1_{SW} \wedge f) \oplus (1_{SW} \wedge g)$.

(d) If $\mu(f) > \mu(g)$ then $f \wedge (g \oplus h) = (f \wedge g) \oplus (f \wedge h)$.

(e) If $\mu(f) > \mu(g)$ then $-(f \wedge g) = f \wedge (-g)$.

PROOF: The proofs of these properties are straightforward and are left to the reader.∎

NOTATION: In view of property (e) above, $-f_1 \wedge \cdots \wedge f_t$ will mean $f_1 \wedge \cdots \wedge (-f_k) \wedge \cdots \wedge f_t$ where $\mu(f_k) = \min(\mu(f_1), \ldots, \mu(f_t))$.

We state next a useful technical result which says that $\langle X_1, \ldots, X_t \rangle_\wedge$ can be defined from any fixed set of representatives of $X_1, \ldots, X_t$.

THEOREM 2.2.6 Assume that $\langle X_1, \ldots, X_t \rangle$ is defined. Let $G_{i-1,i}$ represent X_i for $1 \le i \le t$. Then any element Z of $\langle X_1, \ldots, X_t \rangle$ has a representatives $\tilde{G}_{0t}$ where $\{G_{ij} | 0 \le i < j \le t,\ (i,j) \ne (0,t)\}$ is a defining system which contains the given $\{G_{i-1,i} | 1 \le i \le t\}$.

PROOF. Let $\{A_{ij} | 0 \le i < j \le t,\ (i,j) \ne (0,t)\}$ be a defining system such that $\tilde{A}_{ij}$ is a representative of Z. By induction on $k = j - i \ge 1$, we construct a defining system $\{G_{ij}\}$ and homotopies H_{ij} from A_{ij} to G_{ij} such that $H_{ij}|\text{Domain}(G_{ir} \wedge G_{rj}) = H_{ir} \wedge H_{rj}$ for $i < r < j$. When $k = 1$, the $G_{i-1,i}$ are given, and the $H_{i-1,i}$ can be found since $A_{i-1,i}$ and $G_{i-1,i}$ both represent X_i. Let $j-i = k$ and assume that the G_{st} and H_{st} have been constructed for $1 \le t-s < k$. Since (Domain G_{ij}, Domain $\tilde{G}_{ij}$) is homeomorphic to some (D^N, S^N), it has the homotopy extension property. By the induction hypothesis the homotopies $H_{ir} \wedge H_{rj}$, $i < r < j$, agree where their domains intersect and thus define a homotopy $H = \bigcup_{r=i+1}^{j-1} (H_{ir} \wedge H_{rj})$ from $\tilde{A}_{ij}$ to $\tilde{G}_{ij}$. By the homotopy extension property, there is a homotopy H_{ij} of A_{ij} which extends both H and A_{ij}. Define $G_{ij} = H_{ij}|\text{Domain}\ (G_{ij} \times \{1\})$. This completes the inductive step. Thus we have constructed a defining system $\{G_{ij}\}$ and a homotopy $\bigcup_{r=1}^{t-1} (H_{0r} \wedge H_{rt})$ from $\tilde{A}_{0t}$ to $\tilde{G}_{0t}$. ∎

Observe that the three-fold Toda bracket $\langle X_1, X_2, X_3\rangle$ is defined if and only if $X_1 \cdot X_2 = 0$ and $X_2 \cdot X_3 = 0$. The following theorem gives practical criteria for concluding that a four-fold Toda bracket is defined.

THEOREM 2.2.7 Assume that $0 \in \langle X_1, X_2, X_3\rangle$ and $0 \in \langle X_2, X_3, X_4\rangle$. Let N_i = Degree X_i, $1 \le i \le 4$. In addition assume that one of the following conditions is true.

(a) $\langle X_1, X_2, X_3\rangle = 0$.

(b) $\langle X_2, X_3, X_4\rangle = 0$.

(c) $X_1 \cdot \pi^S_{1+N_2+N_3} = 0$.

(d) $X_4 \cdot \pi^S_{1+N_2+N_3} = 0$.

(e) If $Y \in \pi^S_{1+N_2+N_3}$ then $Y = Y_1 + Y_2$ such that $X_1 \cdot Y_1 = 0$ and $X_4 \cdot Y_2 = 0$.

(f) $X_1 = X_3$.

(g) $X_2 = X_4$.

Then $\langle X_1, X_2, X_3, X_4\rangle$ is defined.

PROOF: We use the smash product and the smash product Toda bracket of Definition 2.2.2 throughout the proof.

(a) Let G_{12}, G_{23}, G_{34}, G_{13}, G_{24} be a defining system for $\langle X_2, X_3, X_4\rangle$ which defines 0 in $\langle X_2, X_3, X_4\rangle$. Extend this defining system by choosing any G_{01} and G_{02}. Then $\tilde{G}_{03} \in \langle X_1, X_2, X_3\rangle = 0$, and thus we can find G_{03} to complete the defining system.

(b) The proof of (b) is analogous to the proof of (a).

(c) As in the proof of (a) select G_{01}, G_{12}, G_{23}, G_{34}, G_{02}, G_{13}, G_{24} and G_{14}. By the previous theorem, there is a defining system G_{01}, G_{12}, G_{23}, G_{02}, G'_{13} of $\langle X_1, X_2, X_3\rangle$ which defines $0 \in \langle X_1, X_2, X_3\rangle$. Then $G_{01} \wedge (G_{13} \ominus G'_{13})$ represents an element of $X_1 \cdot \pi^S_{1+N_2+N_3} = 0$. Thus we can find G_{03} to complete the defining system.

(d) The proof of (d) is analogous to the proof of (c).

(e) As in the proof of (a) select G_{01}, G_{12}, G_{23}, G_{34}, G_{02}, G_{13}, G_{24} and G_{14}. By the previous theorem, there is a defining system G_{01}, G_{12}, G_{23}, G_{02}, G'_{13} of $\langle X_1, X_2, X_3\rangle$ which defines $0 \in \langle X_1, X_2, X_3\rangle$. Write $G_{13} \ominus G'_{13} = Y_2 \oplus Y_1$ where $X_1 \wedge Y_1$ and $X_4 \wedge Y_2$ are null homotopic. Then we can replace G_{13} by $(-Y_2 \oplus G_{13}) \oplus (-G'_{13} \oplus G'_{13})$ and find a new appropriate G_{14}. Since the new G_{13} equals $(-Y_2 \oplus Y_2) \oplus (Y_1 \oplus G'_{13})$ we can find a G_{03} to complete the defining system.

(f) Let G_{12}, G_{23}, G_{34}, G_{13}, G_{24} be a defining system for $\langle X_2, X_3, X_4\rangle$ which defines 0 in $\langle X_2, X_3, X_4\rangle$. Extend this defining system by choosing $G_{01} = G_{23}$ and any G_{02}. There are other choices $G'_{02} = G_{02} \oplus X$ and $G'_{13} = G_{13} \oplus Y$ such that the defining system G_{01}, G_{12}, G_{23}, G'_{02}, G'_{13} defines G which represents 0 in $\langle X_1, X_2, X_3\rangle$. Replace G_{02} by $(G_{02} \oplus X \oplus Y) \cup (Y \cup_1 G_{23})$. Now $\tilde{G}_{03} = G$, and we can find a G_{03} to complete the defining system.

(g) The proof of (g) is analogous to the proof of (f). ∎

3. Properties of the Toda Bracket

In this section, we derive the indeterminacy as well as the additive and associative properties of the three-fold and four-fold Toda brackets defined in the previous section. Most of these results are direct analogues of the algebraic results for Massey products given by May in [39]. As with algebraic Massey products we say that $<X_1,\ldots,X_t>$ is strictly defined if $<X_m,\ldots,X_n> = 0$ whenever $1 \le m < n \le t$ and $n-m < t-1$. Note that every triple product which is defined is automatically strictly defined. We define the indeterminacy of a Toda bracket by

$$\text{Indet} <X_1,\ldots,X_t> = <X_1,\ldots,X_t> - <X_1,\ldots,X_t>.$$

In all of the proofs of this section we use defining systems as in Definition 2.2.2 which are based upon the smash product.

Before embarking on manipulating our Toda brackets, we should remark that there is a hidden sign convention built into our definitions. The easiest way to deal with this problem is to consider a defining system $\{G_{ij}\}$ of $<X_1,\ldots,X_t>_\wedge$ in which the $G_{01},\ldots,G_{t-1,t}$ use subspace $V_1,\ldots,V_t$ of R^∞ such that V_i has basis $\{b_{N(i,j)} | 1 \le j \le \dim(V_i)\}$ and $\{b_{N(i,j)} | 1 \le i \le t,\ 1 \le j \le \dim(V_i)\}$ in the lexicographical order of the $N(i,j)$ is the same ordering as the given ordering of B. Now think of $\tilde{G}_{0t}$ as using t-2 additional basis vectors $b_{k_1},\ldots,b_{k_{t-2}}$ where $k_1 < N(1,j_1) < k_2 < N(2,j_2) < k_3 < \cdots < k_{t-2} < N(t-2,j_{t-2})$ for all $j_1,\ldots,j_{t-2}$.

THEOREM 2.3.1 Let $X_i \in \pi^S_{N_i}$ for $1 \le i \le t$.

(a) Indet $<X_1,X_2,X_3>$ is the ideal spanned by X_1 and X_3.

(b) If $X_3 \cdot \pi^S_{N_1+N_2+1} \cap X_1 \cdot \pi^S_{N_2+N_3+1} = 0$ and $X_2 \cdot \pi^S_{N_3+N_4+1} \cap X_4 \cdot \pi^S_{N_2+N_3+1} = 0$ then

$$\text{Indet} <X_1,X_2,X_3,X_4> = \bigcup_A <A,X_3,X_4> \cup \bigcup_B <X_1,B,X_4> \cup \bigcup_C <X_1,X_2,C>$$

where the first union is taken over all $A \in \pi^S_{N_1+N_2+1}$ such that $A \cdot X_3 = 0$, the second union is taken over all $B \in \pi^S_{N_2+N_3+1}$ such that $B \cdot X_1 = B \cdot X_4 = 0$ and the third union is taken over all $C \in \pi^S_{N_3+N_4+1}$ such that $C \cdot X_2 = 0$.

PROOF: The proof of is this theorem is a direct analogue of the proof of the corresponding algebraic result for Massey products [40, Prop. 2.4]. ∎

NOTE: The hypothesis in (b) above is satisfied if $\langle X, X, X, X\rangle$ is strictly defined.

THEOREM 2.3.2 Assume that $\langle X_1, \ldots, X'_k + X''_k, \ldots, X_t\rangle$ is defined and $\langle X_1, \ldots, X'_k, \ldots X_t\rangle$ is strictly defined. Then $\langle X_1, \ldots, X''_k, \ldots, X_t\rangle$ is defined and $\langle X_1, \ldots, X'_k + X''_k, \ldots, X_t\rangle \subset \langle X_1, \ldots, X'_k, \ldots, X_t\rangle + \langle X_1, \ldots, X''_k, \ldots, X_t\rangle$.

PROOF. The proof is a direct analoge of the algebraic proof of [40, Prop. 2.7]. ∎

The following associative properties of the three-fold Toda bracket are proved by Toda in [60].

THEOREM 2.3.3 Let degree $X_i = N(i)$ for $0 \le i \le 3$ and let degree $Y = M$.

(a) If $\langle X_1, X_2, X_3\rangle$ is defined then

$$Y \cdot \langle X_1, X_2, X_3\rangle \subset (-1)^M \langle Y \cdot X_1, X_2, X_3\rangle \text{ and } \langle X_1, X_2, X_3\rangle \cdot Y \subset \langle X_1, X_2, X_3 \cdot Y\rangle.$$

(b) If $X_0 \cdot X_1 = X_1 \cdot X_2 = X_2 \cdot X_3 = 0$ then

$$X_0 \cdot \langle X_1, X_2, X_3\rangle = (-1)^{N(0)+N(1)} \langle X_0, X_1, X_2\rangle \cdot X_3.$$

(c) If the second of the three Toda brackets below is defined then they are all defined and

$$0 \in (-1)^{N(0)} \langle\langle X_0, X_1, X_2\rangle, X_3, X_4\rangle + \langle X_0, \langle X_1, X_2, X_3\rangle, X_4\rangle + (-1)^{N(1)} \langle X_0, X_1, \langle X_2, X_3, X_4\rangle\rangle.$$

(d) If $X_1 \cdot Y \cdot X_2 = 0$ and $X_2 \cdot X_3 = 0$ then $\langle X_1 \cdot Y, X_2, X_3\rangle \subset (-1)^M \langle X_1, Y \cdot X_2, X_3\rangle$.

(e) If $X_1 \cdot X_2 = 0$ and $X_2 \cdot Y \cdot X_3 = 0$ then $\langle X_1, X_2, Y \cdot X_3\rangle \subset \langle X_1, X_2 \cdot Y, X_3\rangle$.

In the next three theorems we give the analogous results for four-fold Toda brackets. Most of these results were proved by Oguchi [51] for his composition four-fold products. However, his Toda brackets are only defined under more restrictive conditions than ours. As a result some of his conclusions are sharper than ours.

THEOREM 2.3.4 Let degree $X_i = N(i)$ for $1 \le i \le 4$ and let degree $Y = M$.

(a) If $\langle X_1,X_2,X_3,X_4\rangle$ is defined then $\langle X_1,X_2,X_3,X_4\rangle = (-1)^P \langle X_4,X_3,X_2,X_1\rangle$ where $P = N(4)[N(1)+N(2)+N(3)+1] + N(3)[N(1)+N(2)] + N(1)[N(2)+1]$.

(b) If $\langle X_1,X_2,X_3,X_4\rangle$ is defined then

$$Y\cdot\langle X_1,X_2,X_3,X_4\rangle \subset (-1)^M \langle Y\cdot X_1,X_2,X_3,X_4\rangle \text{ and}$$

$$\langle X_1,X_2,X_3,X_4\rangle\cdot Y \subset \langle X_1,X_2,X_3,X_4\cdot Y\rangle.$$

(c) If $\langle X_1\cdot Y,X_2,X_3,X_4\rangle$ is defined then $\langle X_1,Y\cdot X_2,X_3,X_4\rangle$ is defined and

$$\langle X_1\cdot Y,X_2,X_3,X_4\rangle \subset (-1)^M \langle X_1,Y\cdot X_2,X_3,X_4\rangle.$$

(d) If $\langle X_1,X_2,X_3,Y\cdot X_4\rangle$ is defined then $\langle X_1,X_2,X_3\cdot Y,X_4\rangle$ is defined and

$$\langle X_1,X_2,X_3,Y\cdot X_4\rangle \subset \langle X_1,X_2,X_3\cdot Y,X_4\rangle.$$

(e) Assume that $\langle X_1,X_2\cdot Y,X_3,X_4\rangle$ and $\langle X_1,X_2,Y\cdot X_3,X_4\rangle$ are defined, and that $\langle X_1,X_2,YX_3\rangle = 0$. Then $I \equiv \langle X_1,X_2\cdot Y,X_3,X_4\rangle \cap \langle X_1,X_2,Y\cdot X_3,X_4\rangle \ne \phi$. Moreover the indeterminacy is given by $\mathrm{Indet}(I) \equiv I-I = \bigcup_A \langle A,X_3,X_4\rangle \cup \bigcup_B \langle X_1,X_2,B\rangle$ where the first union is taken over all $A \in \pi^S_{N(1)+N(2)+M+1}/\, Y\cdot\pi^S_{N(1)+N(2)+1}$ with $AX_3 = 0$ and the second union is taken over all $B \in \pi^S_{N(3)+N(4)+M+1}/\, Y\cdot\pi^S_{N(3)+N(4)+1}$ with $X_2B = 0$.

PROOF. (a) If $\{G_{ij} | 0 \le i < j \le 4,\ (i,j) \ne (0,4)\}$ is a defining system for $\langle X_1,X_2,X_3,X_4\rangle$, let $A_{ij} = G_{4-j,4-i}$. Then $\{A_{ij} | 0 \le i < j \le 4,\ (i,j) \ne (0,4)\}$ is a defining system for $\langle X_4,X_3,X_2,X_1\rangle$. Since $\tilde{G}_{ij} = \tilde{A}_{ij}$, $\langle X_1,X_2,X_3,X_4\rangle \subset (-1)^P\langle X_4,X_3,X_2,X_1\rangle$, and by symmetry the two Toda brackets are equal.

(b) Let $\{G_{ij} | 0 \le i < j \le 4,\ (i,j) \ne (0,4)\}$ be a defining system for $\langle X_1,X_2,X_3,X_4\rangle$ and let J represent Y. Then the following display is a defining system for $\langle Y\cdot X_1, X_2, X_3, X_4\rangle$:

$$\begin{matrix} J \wedge G_{01} & & G_{12} & & G_{23} & & G_{34} \\ & J \wedge G_{02} & & G_{13} & & G_{24} & \\ & & J \wedge G_{03} & & G_{14} & & \end{matrix}$$

Thus, $\langle Y\cdot X_1, X_2, X_3, X_4\rangle$ is defined and contains $J \wedge \tilde{G}_{04}$. Therefore $Y\cdot\langle X_1, X_2, X_3, X_4\rangle \subset (-1)^M\langle Y\cdot X_1, X_2, X_3, X_4\rangle$. The second identity in (b) follows from the first one by (a).

(c) Let $\{G_{ij} | 0 \le i < j \le 4,\ (i,j) \ne (0,4)\}$ be a defining system for $\langle X_1\cdot Y, X_2, X_3, X_4\rangle$. Assume that $G_{01} = G'_{01} \wedge J$ where G'_{01}, J represents X_1, Y, resp. Then the following display is a defining system for $\langle X_1, Y\cdot X_2, X_3, X_4\rangle$:

$$\begin{matrix} G'_{01} & & J \wedge G_{12} & & G_{23} & & G_{34} \\ & G_{02} & & J \wedge G_{13} & & G_{24} & \\ & & G_{03} & & J \wedge G_{14} & & \end{matrix}$$

Thus $\langle X_1, Y\cdot X_2, X_3, X_4\rangle$ is defined and contains $\tilde{G}_{04}$ because $G'_{01} \wedge (J \wedge G_{14}) = G_{01} \wedge G_{14}$. Therefore $\langle X_1\cdot Y, X_2, X_3, X_4\rangle \subset (-1)^M \langle X_1, Y\cdot X_2, X_3, X_4\rangle$.

(d) This identity follows from the identity in (c) by applying the identity in (a).

(e) Let $G_{i-1,i}$ represent X_i for $1 \le i \le 4$, and let J represent Y. Extend G_{01}, $G_{12} \wedge J$, G_{23}, G_{34} to a defining system $\{G_{ij} | 0 \le i < j \le 4,\ (i,j) \ne (0,4)\}$ of $\langle X_1, X_2\cdot Y, X_3, X_4\rangle$. Extend G_{01}, G_{12}, $J \wedge G_{23}$, G_{13} by finding a G'_{02} to get a defining system of $\langle X_1, X_2, YX_3\rangle$. Since $\langle X_1, X_2, YX_3\rangle = 0$, we can find a G'_{03} such that $\partial G'_{03} = (G_{01} \wedge G_{13}) \cup (G'_{02} \wedge (J \wedge G_{23}))$. Then the following diagram exhibits two defining systems, one for $\langle X_1, X_2\cdot Y, X_3, X_4\rangle$ and the other for $\langle X_1, X_2, Y\cdot X_3, X_4\rangle$:

$$\begin{matrix} G_{01} & & G_{12} \wedge J & & G_{23} & & G_{34} \\ & G'_{02} \wedge J & & G_{13} & & G_{24} & \\ & & G'_{03} & & G_{14} & & \end{matrix} \qquad \begin{matrix} G_{01} & & G_{12} & & J \wedge G_{23} & & G_{34} \\ & G'_{02} & & G_{13} & & J \wedge G_{24} & \\ & & G'_{03} & & G_{14} & & \end{matrix}$$

Both of these defining systems define the same element, and thus the two Toda brackets have an element in common. The indeterminacy arises because not all defining systems of $\langle X_1\cdot Y, X_2, X_3, X_4\rangle$ have a (0,2) entry of the form ?∧J and not all defining systems of $\langle X_1, Y\cdot X_2, X_3, X_4\rangle$ have a (2,4) entry of the form J∧?.∎

THEOREM 2.3.5 Let degree $X_i = N(i)$ for $0 \le i \le 4$. Assume that $\langle X_1, X_2, X_3, X_4\rangle$ and $\langle X_0, X_1, X_2, X_3\rangle$ are strictly defined. Then

$$X_0\cdot\langle X_1, X_2, X_3, X_4\rangle = (-1)^{N(0)+N(1)}\ \langle X_0, X_1, X_2, X_3\rangle\cdot X_4.$$

PROOF. Let $\{G_{ij} | 0 \le i < j \le 4,\ (i,j) \ne (0,4)\}$ be a defining system for $\langle X_1, X_2, X_3, X_4\rangle$. Extend $\{G_{01}, G_{12}, G_{23}, G_{02}, G_{13}, G_{03}\}$ to a defining system $\{G_{ij} | -1 \le i < j \le 3,\ (i,j) \ne (-1,3)\}$ of $\langle X_0, X_1, X_2, X_3\rangle$. Then $(G_{-1,1} \wedge G_{14}) \cup (G_{-1,2} \wedge G_{24})$ restricted to the boundary of its domain is $(G_{-1,0} \wedge \tilde{G}_{04}) \cup (\tilde{G}_{-1,3} \wedge G_{34})$. Thus $X_0\cdot\langle X_1, X_2, X_3, X_4\rangle$ $\subset (-1)^{N(0)+N(1)}\langle X_0, X_1, X_2, X_3\rangle\cdot X_4$ and by symmetry the theorem follows.∎

THEOREM 2.3.6 Let degree $X_i = N(i)$ for $0 \le i \le 4$.

(a) Assume that $\langle X_1, X_2, X_3, X_4\rangle$ is defined and that $X_0\cdot X_1 = 0$. Then

$$X_0\cdot\langle X_1, X_2, X_3, X_4\rangle \subset (-1)^{N(1)+1}\ \langle\langle X_0, X_1, X_2\rangle, X_3, X_4\rangle.$$

(b) Assume that $\langle X_0, X_1, X_2, X_3\rangle$ is defined and that $X_3\cdot X_4 = 0$. Then

$$\langle X_0, X_1, X_2, X_3\rangle\cdot X_4 \subset (-1)^{N(1)+1}\ \langle X_0, X_1, \langle X_2, X_3, X_4\rangle\rangle.$$

(c) Assume that $X_0\cdot X_1 = 0$, $X_1\cdot X_2 = 0$, $X_3\cdot X_4 = 0$ and $0 \in \langle X_0, X_1, X_2\rangle\cdot X_3$. Then $\langle X_0, X_1, X_2\cdot X_3, X_4\rangle$ is defined and contains $(-1)^{N(0)+1}\ \langle\langle X_0, X_1, X_2\rangle, X_3, X_4\rangle$.

(d) Assume that $X_0\cdot X_1 = 0$, $X_2\cdot X_3 = 0$, $X_3\cdot X_4 = 0$ and $0 \in X_1\cdot\langle X_2, X_3, X_4\rangle$. Then $\langle X_0, X_1\cdot X_2, X_3, X_4\rangle$ is defined and contains $(-1)^{N(1)+1}\ \langle X_0, X_1, \langle X_2, X_3, X_4\rangle\rangle$.

PROOF. (a) Let $\{G_{ij} | 0 \le i < j \le 4,\ (i,j) \ne (0,4)\}$ be a defining system for $\langle X_1, X_2, X_3, X_4\rangle$ and let $G_{-1,0}$ represent X_0. Then the following display is a defining system for $\langle\langle X_0, X_1, X_2\rangle, X_3, X_4\rangle$:

$$\begin{matrix} \tilde{G}_{-1,2} & & G_{23} & & G_{34} \\ & (G_{-1,0}\wedge G_{03})\cup(G_{-1,1}\wedge G_{13}) & & G_{24} & \end{matrix}$$

Now $G_{-1,1}\wedge G_{14}$ restricted to the boundary of its domain is the element of $\langle\langle X_0,X_1,X_2\rangle,X_3,X_4\rangle$ determined by the above defining system union $G_{-1,0}\wedge \tilde{G}_{04}$. Thus $X_0\cdot\langle X_1,X_2,X_3,X_4\rangle \subset (-1)^{N(1)+1}\langle\langle X_0,X_1,X_2\rangle,X_3,X_4\rangle$.

(b) This identity follows from the one in (a) by Theorem 2.3.4(a).

(c) Let $\{G_{ij} \mid -1 \le i < j \le 2,\ (i,j) \ne (-1,2)\}$ be a defining system for $\langle X_0,X_1,X_2\rangle$. Let G_{23}, G_{34} represent X_3, X_4, respectively. Find G_{24} such that $\tilde{G}_{24} = G_{23}\wedge G_{34}$ and find $G_{-1,3}$ such that $\tilde{G}_{-1,3} = \tilde{G}_{-1,2}\wedge G_{23}$. Then the following display is a defining system for $\langle X_0,X_1,X_2\cdot X_3,X_4\rangle$:

$$\begin{array}{ccccccc}
G_{-1,0} & & G_{01} & & G_{12}\wedge G_{23} & & G_{34} \\
& G_{-1,1} & & G_{02}\wedge G_{23} & & G_{12}\wedge G_{24} & \\
& & G_{-1,3} & & G_{02}\wedge G_{24} & &
\end{array}$$

This defining system defines

$(G_{-1,0}\wedge G_{02}\wedge G_{24}) \cup (G_{-1,1}\wedge G_{12}\wedge G_{24}) \cup (G_{-1,3}\wedge G_{34})$

$= (\tilde{G}_{-1,2}\wedge G_{24}) \cup (G_{-1,3}\wedge G_{34})$, an arbitrary element of $\langle\langle X_0,X_1,X_2\rangle,X_3,X_4\rangle$. Thus $\langle\langle X_0,X_1,X_2\rangle,X_3,X_4\rangle \subset (-1)^{N(0)+1}\langle X_0,X_1,X_2\cdot X_3,X_4\rangle$.

(d) This identity follows from the identity in (c) by Theorem 2.3.4(a). ∎

We conclude this section by recording a useful theorem of Toda [60,3.10].

THEOREM 2.3.7 Let α and β be elements of π_*^S.

(a) If degree α is odd then $\langle\alpha,\beta,\alpha\rangle \cap (-1)^{\deg\beta}\langle\beta,\alpha,\ 2\alpha\rangle \ne \emptyset$.

(b) If degree α is even then $\langle\alpha,\beta,\alpha\rangle \cap \beta\cdot\pi_*^S \ne \emptyset$.

4. The Atiyah-Hirzebruch Spectral Sequence

Toda brackets in the limit of a spectral sequence are related to the differentials in the spectral sequence. In this section we prove several theorems which depict this relationship in the Atiyah-Hirzebruch spectral sequence for the homotopy of a spectrum B:

$$E^2_{pq} = H_p(B;\pi^S_q) \Longrightarrow \pi^S_{p+q}(B)$$

Of course, the case in which we are inerested is when B = BP, and we specialize to that case in the last three theorems of this section. The idea of the following theorem is to analyze a Toda bracket by passing to an appropriate mapping cone. This idea is due to Joel Cohen [18] where he used it to decompose elements of π^S_* as Toda brackets of Hopf classes.

THEOREM 2.4.1 Let $X_0 \in \pi^S_{N(0)}$, $X_2 \in \pi^S_{N(2)}$, $X_3 \in \pi^S_{N(3)}$, $Y \in H_*B$ and let $r \geq 2$. Let C be the mapping cone of X_2. Assume that:

(i) $X_2 \cdot X_3 = 0$ in π^S_*.

(ii) $d^r(X_3 \cdot Y) = X_0$.

(iii) Y transgresses to the projection of $X_{02} \in C_*$ into the Atiyah-Hirzebruch spectral sequence for C_*B.

Let $X_1 = \sigma_*(X_{02}) \in \pi^S_{N(1)}$ where $\sigma: C \longrightarrow S^{N(2)+1}$ is the canonical collapsing map. Then $\langle X_1, X_2, X_3\rangle$ is defined and contains X_0.

PROOF. We use the composition product Toda bracket of Definition 2.2.1 to prove this theorem. Let $G_{i-1,i}$ represent X_i for $0 \leq i \leq 3$, and let G_{02} represent X_{02}. Consider Figure 2.4.1. In that diagram, j is the canonical inclusion map and G_{13} exists by (i). Let G_{13*} be the map of spectral sequences induced by G_{13}. Then $X_0 = d^r(X_3 \cdot Y) = d^r \circ G_{13*}(Y) = G_{13*} \circ d^r(Y) = G_{13*}(X_{02})$. Thus X_0 is represented by $G_{13} \circ SG_{02}$ which is an element of $\langle X_1, X_2, X_3\rangle$. ∎

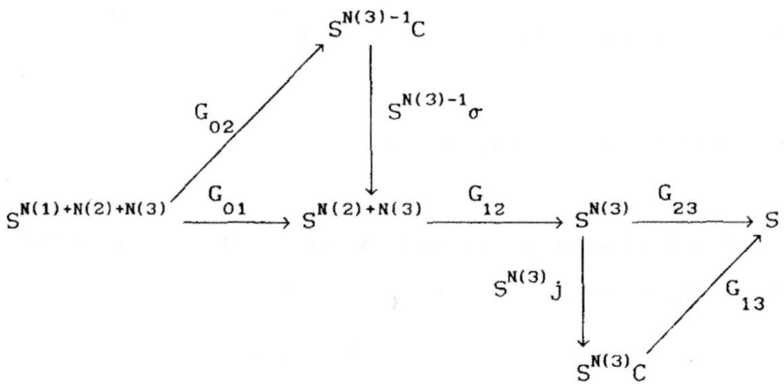

FIGURE 2.4.1

The next theorem is the most direct way of detecting a triple product in π_*^S from differentials in the Atiyah-Hirzebruch spectral sequence.

THEOREM 2.4.2 Assume that $<X_1, X_2, X_3>$ is defined in π_*^S. Assume that $d^{r(1)}(Y_1) = X_1$ and $d^{r(3)}(Y_3) = X_3$. Then $X_2 \cdot Y_1 \cdot Y_3$ survives to $E^{r(1)+r(3)}$ and there is an element of $<X_1, X_2, X_3>$ which projects to $d^{r(1)+r(3)}(X_2 \cdot Y_1 \cdot Y_3)$.

PROOF. We use the smash product Toda bracket of Definition 2.2.2 to prove this theorem. Let $N(i)$ = degree X_i. For $i=1,3$, represent $Y_i \in E^{r(i)}_{r(i),p(i)}$ by $G_i: (SV_i \wedge DU_i, SV_i \wedge SU_i) \longrightarrow (SU_i \wedge B^{[r(i)]}, SU_i)$ where $G_i | SV_i \wedge SU_i \equiv G_{i-1,i}$ represents X_i. Represent X_2 by $G_{12}: SV_2 \wedge SU_2 \longrightarrow SU_2$. Find maps G_{02} and G_{13} as in Definition 2.2.2 to complete the defining system $\{G_{ij}\}$ of $<X_1, X_2, X_3>$. Define

$$F: (SV_1 \wedge DU_1 \wedge SV_2 \wedge SU_2 \wedge SV_3 \wedge DU_3) \cup (DV_1 \wedge SU_1 \wedge SV_2 \wedge SU_2 \wedge SV_3 \wedge DU_3)$$
$$\cup (SV_1 \wedge DU_1 \wedge DV_2 \wedge SU_2 \wedge SV_3 \wedge SU_3) \longrightarrow SU_1 \wedge SU_2 \wedge SU_3 \wedge B^{[r(1)+r(3)]}$$

as $[\varepsilon \circ (G_1 \wedge G_{12} \wedge G_3)] \cup [\varepsilon \circ (G_{02} \wedge G_3)] \cup [\varepsilon \circ (G_1 \wedge G_{13})]$. Then Domain F is homeomorphic to a disc and

$$F: (\text{Domain } F, \partial \text{Domain } F) \longrightarrow (SU_1 \wedge SU_2 \wedge SU_3 \wedge B^{[r(1)+r(3)]}, SU_1 \wedge SU_2 \wedge SU_3)$$

represents $X_2 \cdot Y_1 \cdot Y_3$. Thus $X_1 \cdot Y_2 \cdot Y_3$ survives to $E^{r(1)+r(3)}_{r(1)+r(3),p(1)+p(3)+N(2)}$ and $d^{r(1)+r(3)}(X_2 \cdot Y_1 \cdot Y_3)$ is represented by $F | \partial\text{Domain}(F) = (G_{02} \wedge G_{23}) \cup (G_{01} \wedge G_{13}) = \tilde{G}_{03} \in <X_1, X_2, X_3>$. ∎

The previous theorem generalizes to longer Toda brackets. Unfortunately, technical hypotheses need to be added and the conclusion has indeterminacy. We give such a generalization for four-fold brackets.

THEOREM 2.4.3 Assume that $<X_1, X_2, X_3, X_4>$ is defined in π_*^S, and let $N(i)$ = degree X_i for $1 \le i \le 4$. Assume that $d^{r(i)}(Y_i) = X_i$ for $i=1,3,4$ where $Y_i \in E^{r(i)}_{r(i),p(i)}$. Assume that one of the following hypotheses hold:

(i) $E^{r(4)+h}_{r(1)-h,p(1)+N(2)+h} = 0$ for $0 \le h \le r(1)$.

(ii) $E^{r(4)+k}_{r(3)-k,p(3)+N(4)+k} = 0$ for $0 \leq k \leq r(3)$.

Then $X_2 \cdot Y_1 \cdot Y_3 \cdot Y_4$ survives to $E^{r(1)+r(3)+r(4)}_{N(1)+N(3)+N(4)+3,N(2)}$ and there is an element of $\langle X_1, X_2, X_3, X_4\rangle$ which projects to $d^{r(1)+r(3)+r(4)}(X_2 \cdot Y_1 \cdot Y_3 \cdot Y_4)$.

PROOF. We use the smash product Toda bracket of Definition 2.2.2 to prove this theorem. Let $\{G_{ij} | 0 \leq i < j \leq 4, (i,j) \neq (0,4)\}$ be a defining system for $\langle X_1, X_2, X_3, X_4\rangle$. For i=1,3,4 represent $Y_i \in E^{r(i)}_{r(i),p(i)}$ by $G_i: (SV_i \wedge DU_i, SV_i \wedge SU_i) \longrightarrow (SU_i \wedge B^{[r(i)]}, SU_i)$ where $G_i | SV_i \wedge SU_i \equiv G_{i-1,i}$ represents X_i. Let

$$F = (G_1 \wedge G_{12} \wedge G_3 \wedge G_4) \cup (G_{02} \wedge G_3 \wedge G_4) \cup (G_1 \wedge G_{13} \wedge G_4) \cup (G_{03} \wedge G_4) \cup (G_1 \wedge G_{14}):$$
$$\big[(SV_1 \wedge DU_1 \wedge SV_2 \wedge SU_2 \wedge SV_3 \wedge DU_3 \wedge SV_4 \wedge DU_4) \cup (DV_1 \wedge SU_1 \wedge SV_2 \wedge SU_2 \wedge SV_3 \wedge DU_3 \wedge SV_4 \wedge DU_4)$$
$$\cup (SV_1 \wedge DU_1 \wedge DV_2 \wedge SU_2 \wedge SV_3 \wedge SU_3 \wedge SV_4 \wedge DU_4) \cup (DV_1 \wedge SU_1 \wedge DV_2 \wedge SU_2 \wedge SV_3 \wedge SU_3 \wedge SV_4 \wedge DU_4)$$
$$\cup (SV_1 \wedge DU_1 \wedge DV_2 \wedge SU_2 \wedge DV_3 \wedge SU_3 \wedge SV_4 \wedge SU_4), (DV_1 \wedge SU_1 \wedge DV_2 \wedge SU_2 \wedge SV_3 \wedge SU_3 \wedge SV_4 \wedge SU_4)$$
$$\cup (SV_1 \wedge SU_1 \wedge DV_2 \wedge SU_2 \wedge DV_3 \wedge SU_3 \wedge SV_4 \wedge SU_4) \cup (SV_1 \wedge DU_1 \wedge SV_2 \wedge SU_2 \wedge SV_3 \wedge DU_3 \wedge SV_4 \wedge SU_4)$$
$$\cup (SV_1 \wedge DU_1 \wedge SV_2 \wedge SU_2 \wedge DV_3 \wedge SU_3 \wedge SV_4 \wedge SU_4) \cup (DV_1 \wedge SU_1 \wedge SV_2 \wedge SU_2 \wedge SV_3 \wedge DU_3 \wedge SV_4 \wedge SU_4)\big]$$
$$\longrightarrow (B^{[r(1)+r(3)+r(4)]}, B^{[r(1)+r(3)]}).$$

F has a disk as its domain and F restricted to the boundary of its domain is $[(G_{03} \wedge G_{34}) \cup (G_{01} \wedge G_{14})] \cup [(G_1 \wedge G_{12} \wedge G_3 \wedge G_{34}) \cup (G_1 \wedge G_{12} \wedge G_{24}) \cup (G_{02} \wedge G_3 \wedge G_{34})]$. Clearly F represents $X_2 \cdot Y_1 \cdot Y_3 \cdot Y_4$. Moreover, F restricted to the boundary of its domain is the sum of $(G_{01} \wedge G_{14}) \cup (G_{02} \wedge G_{24}) \cup (G_{03} \wedge G_{34})$ and the product $[(G_1 \wedge G_{12}) \cup G_{02}] \wedge [(G_3 \wedge G_{34}) \cup G_{24}]$. The first summand is an element of $\langle X_1, X_2, X_3, X_4\rangle$. Under hypothesis (i), the first factor of the product is the boundary of a map of filtration degree less than r(1)+r(4) while the second factor is in filtration degree r(3) so that the product is the boundary of a map of filtration degree less than [r(1)+r(4)]+r(3). Under hypothesis (ii), the second factor of the product is the boundary of a map of filtration degree less than r(3)+r(4) while the first factor is in filtarion degree r(1) so that the product is the boundary of a map of filtration degree less than r(1)+[r(3)+r(4)]. Thus, in either case we can represent $X_2 \cdot Y_1 \cdot Y_3 \cdot Y_4$ by a map whose boundary is an element of $\langle X_1, X_2, X_3, X_4\rangle$. Thus, $X_2 \cdot Y_1 \cdot Y_3 \cdot Y_4$ survives to $E^{r(1)+r(3)+r(4)}$ and $d^{r(1)+r(3)+r(4)}(X_2 \cdot Y_1 \cdot Y_3 \cdot Y_4)$ is the projection into $E^{r(1)+r(3)+r(4)}_{0,N(1)+N(2)+N(3)+N(4)+2}$ of an element of $\langle X_1, X_2, X_3, X_4\rangle$.∎

We conclude this section with three theorems that refer only to our Atiyah-Hirzebruch spectral sequence, i.e., we take $B = BP$. As we shall see, the Toda brackets constructed there are common and useful for detecting nontrivial extensions in our spectral sequence. In Chapter 3, we shall see that we have elements of H_*BP with the following differentials: $d^2(M_1) = \eta$, $d^2(M_2) = \eta M_1^2$, $d^4(M_1^2) = \nu$, $d^4(\bar{M}_2) = \nu M_1$, $d^4(M_2^2) = \nu M_1^4$, $d^8\langle M_1^4\rangle = \sigma$ and $d^8\langle M_2^2\rangle = \sigma M_1^2$. We will represent M_1, M_2, M_1^2, $\bar{M}_2$, M_2^2, $\langle M_1^4\rangle$, $\langle M_2^2\rangle$ by μ_1, μ_{01}, μ_2, $\bar{\mu}_{01}$, μ_{02}, μ_4, $\langle\mu_{02}\rangle$, respectively. The reader may prefer to read the remainder of this section after reading Chapter 3.

THEOREM 2.4.4 Let $X \in \pi_*^S$.

(a) $X \cdot M_1^3$ survive to E^6 if and only if $\eta \cdot X = 0$ and $\nu \cdot X = 0$. In that case $\langle \eta, X, \nu\rangle$ is defined and projects to $d^6(X \cdot M_1^3)$.

(b) $X \cdot M_2$ survives to E^6 if and only if $\eta \cdot X = 0$. In that case $\langle \nu, \eta, X\rangle$ is defined and projects to $d^6(X \cdot M_2)$.

(c) $X \cdot \bar{M}_2$ survives to E^6 if and only if $\nu \cdot X = 0$. In that case $\langle \eta, \nu, X\rangle$ is defined and projects to $d^6(X \cdot \bar{M}_2)$.

PROOF. Represent $M_1 \in E^2_{2,0}$ by $\mu_1 : (S^1 \wedge DA, S^1 \wedge SA) \longrightarrow (SA \wedge BP^{[2]}, SA)$ such that $\mu_1 | S^1 \wedge SA = \eta$. Represent $M_1^2 \in E^4_{4,0}$ by $\mu_2 : (S^3 \wedge DB, S^3 \wedge SB) \longrightarrow (SB \wedge BP^{[4]}, SB)$ such that $\mu_2 | S^3 \wedge SB = \nu$. Let $G : SV \wedge SU \longrightarrow SU$ represent X. We use the smash product Toda bracket of Definition 2.2.2 throughout the proof. Observe that all three Toda brackets in this theorem have indeterminacy contained in (η, ν) which projects to zero in E^6.

(a) $d^2(X \cdot M_1^3) = \eta \cdot X \cdot M_1^2$ and if $\eta \cdot X = 0$ then $d^4(X \cdot M_1^3) = \nu \cdot X \cdot M_1$. Thus, $X \cdot M_1^3$ survives to E^6 if and only if $\eta \cdot X = 0$ and $\nu \cdot X = 0$. The latter condition is equivalent to $\langle\eta, X, \nu\rangle$ being defined. In this case we can apply Theorem 2.4.2 to conclude that $d^6(X\cdot M_1^3)$ is the projection of $\langle\eta, X, \nu\rangle$ into E^6.

(b) Represent $M_2 \in E^2_{6,0}$ by

$$\mu_{01}: \left[D^4\wedge DB \wedge S^1\wedge SA \wedge SC, (S^3\wedge DB \wedge S^1\wedge SA \wedge SC) \cup (D^4\wedge SB \wedge S^1\wedge SA \wedge SC)\right] \longrightarrow (SB \wedge SA \wedge SC \wedge BP^{[6]}, SB \wedge SA \wedge SC \wedge BP^{[4]})$$

such that μ_{01} restricted to the boundary of its domain is $(\mu_2 \wedge \eta) \cup B_{\nu\eta}$ where $B_{\nu\eta'} \mid S^3 \wedge SB \wedge S^1\wedge SA \wedge SC = \nu \wedge \eta$. Let

$B_{\eta X}: D^2\wedge SA \wedge SC \wedge SV \wedge SU \longrightarrow SA \wedge SC \wedge SU$ such that

$B_{\eta X}|S^1\wedge SA \wedge SC \wedge SV \wedge SU = \eta \wedge G \wedge 1_{SC}$. Then $X\cdot M_2 \in E^6$ is represented by

$F = (\mu_{01}\wedge G \wedge 1_{SC}) \cup (\mu_2\wedge B_{\eta X})$:

$$\left[(D^4\wedge DB \wedge S^1\wedge SA \wedge SC \wedge SV \wedge SU) \cup (S^3\wedge DB \wedge D^2\wedge SA \wedge SC \wedge SV \wedge SU),\right.$$
$$\left.(D^4\wedge SB \wedge S^1\wedge SA \wedge SC \wedge SV \wedge SU) \cup (S^3\wedge SB \wedge D^2\wedge SA \wedge SC \wedge SV \wedge SU)\right]$$
$$\longrightarrow (SB \wedge SA \wedge SC \wedge SU \wedge BP^{[6]}, SB \wedge SA \wedge SC \wedge SU).$$

Thus, $d^6(X\cdot M_2)$ is represented by F restricted to the boundary of its domain which is $(B_{\nu\eta}\wedge G) \cup (\nu \wedge B_{\eta X}) \in \langle\nu, \eta, X\rangle$.

(c) Represent $\bar{M}_2 \in E^4_{6,0}$ by

$$\bar{\mu}_{01}: \left[D^2\wedge DA \wedge S^3\wedge SB \wedge SH, (S^1\wedge DA \wedge S^3\wedge SB \wedge SH) \cup (D^2\wedge SA \wedge S^3\wedge SB \wedge SH)\right] \longrightarrow (SA \wedge SB \wedge SH \wedge BP^{[6]}, SA \wedge SB \wedge SH \wedge BP^{[2]})$$

such that $\bar{\mu}_{01}$ restricted to the boundary of its domain is $[(\mu_1 \wedge \nu) \cup B_{\eta\nu}] \wedge 1_{SH}$. Let $B_{\nu X}: D^4 \wedge SB \wedge SH \wedge SV \wedge SU \longrightarrow SB \wedge SH \wedge SU$ such that $B_{\nu X} \mid S^3\wedge SB \wedge SH \wedge SV \wedge SU = \nu \wedge G \wedge 1_{SH}$. Then $X\cdot\bar{M}_2 \in E^6$ is represented by $F = (\bar{\mu}_{01}\wedge G) \cup (\mu_1\wedge B_{\nu X})$:

$$\left[(D^2\wedge DA \wedge S^3\wedge SB \wedge SH \wedge SV \wedge SU) \cup (S^1\wedge DA \wedge D^4\wedge SB \wedge SH \wedge SV \wedge SU),\right.$$
$$\left.(D^2\wedge SA \wedge S^3\wedge SB \wedge SH \wedge SV \wedge SU) \cup (S^1\wedge SA \wedge D^4\wedge SB \wedge SV \wedge SU)\right] \longrightarrow$$
$$(SA \wedge SB \wedge SH \wedge SU \wedge BP^{[6]}, SA \wedge SB \wedge SH \wedge SU).$$

Thus, $d^6(X\cdot\bar{M}_2)$ is represented by F restricted to the boundary of its domain which is $(B_{\eta\nu} \wedge G \wedge 1_{SH}) \cup (\eta \wedge B_{\nu X}) \in \langle\eta, \nu, X\rangle$. ∎

THEOREM 2.4.5 Let $X \in \pi_*^S$.

(a) $\langle\nu,\eta,X,\eta\rangle$ is defined if and only if $X\cdot M_1M_2$ survives to E^8. In this case $\langle\nu,\eta,X,\eta\rangle$ projects to $d^8(X\cdot M_1M_2)$. Moreover, νX is divisible by η.

(b) Assume that $\langle\eta,\nu,X,\nu\rangle$ is defined. Then $X\cdot M_1^2\overline{M}_2$ survives to E^{10}, and $\langle\eta,\nu,X,\nu\rangle$ projects to $d^{10}(X\cdot M_1^2\overline{M}_2)$. Moreover, σX is divisible by ν.

PROOF. Let $G{:}SV \wedge SU \longrightarrow SU$ represent $X \in \pi_*^S$. We use the smash product Toda bracket of Definition 2.2.2 throughout this proof and the notation of the proof of the preceding theorem.

(a) XM_1M_2 survives to E^4 if and only if $\eta X = 0$. In this case, XM_1M_2 survives to E^6 if and only if νX is divisible by η, i.e. $0 \in \langle\eta,X,\eta\rangle$. Then $d^6(XM_1M_2)$ $= d^6(XM_2)M_1$ and $d^6(XM_2) \in \langle\nu,\eta,X\rangle$. Thus XM_1M_2 survies to E^8 if and only if $d^6(XM_2) \in (v)$, i.e. $0 \in \langle\nu,\eta,X\rangle$. Therefore, XM_1M_2 survives to E^8 if and only if $0 \in \langle\eta,X,\eta\rangle$ and $0 \in \langle\nu,\eta,X\rangle$. Then by Theorem 2.2.7(f), XM_1M_2 survives to E^8 if and only if $\langle\nu,\eta,X,\eta\rangle$ is defined. In that case let the following diagram depict a defining system for $\langle\nu,\eta,X,\eta\rangle_\wedge$:

$$\begin{array}{ccccccc} \nu & & \eta & & G & & \eta \\ & B_{\nu\eta} & & B_{\eta X} & & B_{X\eta} & \\ & & B_{\langle\nu,\eta,X\rangle} & & B_{\langle\eta,X,\eta\rangle} & & \end{array}$$

Here $B_{X\eta}{:}DV \wedge SU \wedge S^1 \wedge SA' \longrightarrow SU \wedge SA'$ such that $B_{X\eta}|SV \wedge SU \wedge S^1 \wedge SA'$ $= G \wedge \eta$, $B_{\langle\nu,\eta,X\rangle}{:}D^4 \wedge SB \wedge D^2 \wedge SA \wedge SC \wedge SV \wedge SU \longrightarrow SB \wedge SA \wedge SC \wedge SU$ such that $B_{\langle\nu,\eta,X\rangle}|\partial\,[\text{Domain } B_{\langle\nu,\eta,X\rangle}] = (B_{\nu\eta} \wedge G) \cup (\nu \wedge B_{\eta X})$ and $B_{\langle\eta,X,\eta\rangle}{:}D^2 \wedge SA \wedge SC \wedge DV \wedge SU \wedge S^1 \wedge SA' \longrightarrow SA \wedge SC \wedge SU \wedge SA'$ such that $B_{\langle\eta,X,\eta\rangle}|\partial\,[\text{Domain } B_{\langle\eta,X,\eta\rangle}] = (\eta \wedge B_{X\eta}) \cup (B_{\eta X} \wedge \eta)$. Then the following map F represents $X\cdot M_1M_2$ in E^8. F =

$$(\mu_{02} \wedge G \wedge \mu_1) \cup (\mu_2 \wedge B_{\eta X} \wedge \mu_1) \cup (\mu_{02} \wedge B_{X\eta}) \cup (B_{\langle\nu,\eta,X\rangle} \wedge \mu_1) \cup (\mu_2 \wedge B_{\langle\eta,X,\eta\rangle}){:}$$
$$\big[(D^4 \wedge DB \wedge S^1 \wedge SA \wedge SC \wedge SV \wedge SU \wedge S^1 \wedge DA') \cup (S^3 \wedge DB \wedge D^2 \wedge SA \wedge SC \wedge SV \wedge SU \wedge S^1 \wedge DA')$$
$$\cup\ (D^4 \wedge DB \wedge S^1 \wedge SA \wedge SC \wedge DV \wedge SU \wedge S^1 \wedge SA') \cup (D^4 \wedge SB \wedge D^2 \wedge SA \wedge SC \wedge SV \wedge SU \wedge S^1 \wedge DA')$$
$$\cup\ (S^3 \wedge DB \wedge D^2 \wedge SA \wedge SC \wedge DV \wedge SU \wedge S^1 \wedge SA'), (S^3 \wedge SB \wedge D^2 \wedge SA \wedge SC \wedge DV \wedge SU \wedge S^1 \wedge SA')$$
$$\cup\ (D^4 \wedge SB \wedge S^1 \wedge SA \wedge SC \wedge DV \wedge SU \wedge S^1 \wedge SA') \cup (D^4 \wedge SB \wedge D^2 \wedge SA \wedge SC \wedge SV \wedge SU \wedge S^1 \wedge SA')\big]$$

$$\longrightarrow (SB \wedge SA \wedge SC \wedge SU \wedge SA' \wedge BP^{[8]}, SB \wedge SA \wedge SC \wedge SU \wedge SA').$$

Thus $d^8(X \cdot M_1 M_2)$ is represented by $F|\partial$ [Domain F]

$= (v \wedge B_{\langle\eta,X,\eta\rangle}) \cup (B_{\nu\eta} \wedge B_{X\eta}) \cup (B_{\langle\nu,\eta,X\rangle} \wedge \eta) \in \langle\nu,\eta,X,\eta\rangle$.

The indeterminacy of $\langle\nu,\eta,X,\eta\rangle$ is a sum of elements of the form ηA, νB, $\langle\nu,\eta,C\rangle$ and $\langle\nu,D,\eta\rangle$. All such elements project to zero in E^8. Thus, $\langle\nu,\eta,X,\eta\rangle$ projects to a singleton in E^8. By Theorem 2.4.2, $v \cdot X \in \langle\eta,X,\eta\rangle$. However, $0 \in \langle\eta,X,\eta\rangle$ since $\langle\nu,\eta,X,\eta\rangle$ is defined. Thus, $\nu \cdot X$ is in the indeterminacy of $\langle\eta,X,\eta\rangle$ which is the ideal geneerated by η.

(b) Let the following diagram depict a defining system for $\langle\eta,\nu,X,\nu\rangle$:

η $\quad\nu\quad$ G $\quad\nu$

$B_{\eta\nu}$ $\quad B_{\nu X}$ $\quad B_{X\nu}$

$B_{\langle\eta,\nu,X\rangle}$ $\quad B_{\langle\nu,X,\rangle\nu}$

Here $B_{X\nu}: DV \wedge SU \wedge S^3 \wedge SB' \longrightarrow SU \wedge SB'$ such that $B_{X\nu}|SV \wedge SU \wedge S^3 \wedge SB' = G \wedge \nu$, $B_{\langle\eta,\nu,X\rangle}: D^2 \wedge SA \wedge D^4 \wedge SB \wedge SH \wedge SV \wedge SU \longrightarrow SA \wedge SB \wedge SH \wedge SU$ such that $B_{\langle\eta,\nu,X\rangle}|[\partial \text{ Domain } B_{\langle\eta,\nu,X\rangle}] = (\eta \wedge B_{\nu X}) \cup (B_{\eta\nu} \wedge G)$ and $B_{\langle\nu,X,\nu\rangle}: D^4 \wedge SB \wedge SH \wedge DV \wedge SU \wedge S^3 \wedge SB' \longrightarrow SB \wedge SH \wedge SU \wedge SB'$ such that $B_{\langle\nu,X,\nu\rangle}|[\partial \text{ Domain } B_{\langle\nu,X,\nu\rangle}] = (\nu \wedge B_{X\nu}) \cup (B_{\nu X} \wedge \nu)$. Then the following map F represents $X \cdot M_1^2 \bar{M}_2$ in E^{10}: F =

$$(\bar{\mu}_{02} \wedge G \wedge \mu_2) \cup (\mu_1 \wedge B_{\nu X} \wedge \mu_2) \cup (B_{\langle\eta,\nu,X\rangle} \wedge \mu_2) \cup (\bar{\mu}_{02} \wedge B_{X\nu}) \cup (\mu_1 \wedge G_{\langle\nu,X,\nu\rangle}):$$
$$[(D^2 \wedge DA \wedge S^3 \wedge SB \wedge SH \wedge SV \wedge SU \wedge S^3 \wedge DB') \cup (S^1 \wedge DA \wedge D^4 \wedge SB \wedge SH \wedge SV \wedge SU \wedge S^3 \wedge DB')$$
$$\cup (D^2 \wedge SA \wedge D^4 \wedge SB \wedge SH \wedge SV \wedge SU \wedge S^3 \wedge DB') \cup (D^2 \wedge DA \wedge S^3 \wedge SB \wedge SH \wedge DV \wedge SU \wedge S^3 \wedge SB')$$
$$\cup (S^1 \wedge DA \wedge D^4 \wedge SB \wedge SH \wedge DV \wedge SU \wedge S^3 \wedge SB'), (S^1 \wedge SA \wedge D^4 \wedge SB \wedge SH \wedge DV \wedge SU \wedge S^3 \wedge SB')$$
$$\cup (D^2 \wedge SA \wedge S^3 \wedge SB \wedge SH \wedge DV \wedge SU \wedge S^3 \wedge SB') \cup (D^2 \wedge SA \wedge D^4 \wedge SB \wedge SH \wedge DV \wedge SU \wedge S^3 \wedge SB')]$$
$$\longrightarrow (SA \wedge SB \wedge SH \wedge SU \wedge SB' \wedge BP^{[10]}, SA \wedge SB \wedge SH \wedge SU \wedge SB').$$

Thus $X \cdot M_1^2 \bar{M}_2$ survives to E^{10} and $d^{10}(X \cdot M_1^2 \bar{M}_2)$ is represented by $F|\partial$ [Domain F] $= (1_{SH} \wedge \eta \wedge B_{\langle\nu,X,\nu\rangle}) \cup (1_{SH} \wedge B_{\eta\nu} \wedge B_{X\nu}) \cup (1_{SH} \wedge B_{\langle\eta,\nu,X\rangle} \wedge \nu) \in \langle\eta,\nu,X,\nu\rangle$. The indeterminacy of $\langle\eta,\nu,X,\nu\rangle$ is the sum of elements of the form ηA, νB, $\langle\eta,\nu,C\rangle$ and $\langle\eta,D,\nu\rangle$. All such elements project to 0 in E^8. Therefore, $\langle\eta,\nu,X,\nu\rangle$

projects to a singleton in E^{10}. By Theorem 2.4.2, $\sigma\cdot X \in \langle\nu, X, \nu\rangle$. However, $0 \in \langle\nu, X, \nu\rangle$ because $\langle\eta, \nu, X, \nu\rangle$ is defined. Therefore $\sigma\cdot X$ is in the indeterminacy of $\langle\nu, X, \nu\rangle$ which is the ideal generated by ν. ∎

The following theorem gives three special cases of Theorem 2.4.3 where no technical hypotheses are required.

THEOREM 2.4.6 (a) Let $X \in \pi_*^S$, and assume that $\langle\sigma, \nu, X, \eta\rangle$ is defined. Then $XM_1M_2^2$ survives to E^{14} and $d^{14}(XM_1M_2^2) \in \langle\sigma, \nu, X, \eta\rangle$.

(b) Let $X \in \pi_*^S$, and assume that $\langle\nu, \sigma, X, \eta\rangle$ is defined. Then $XM_1\langle M_2^2\rangle$ survives to E^{14} and $d^{14}(XM_1\langle M_2^2\rangle) \in \langle\nu, \sigma, X, \eta\rangle$.

(c) Let $d^{2r}(Y) = X \in \pi_*^S$ and let $\xi \in \pi_*^S$. Assume that $\langle X, \xi, \nu, \eta\rangle$ is defined. Then $\xi Y\bar{M}_2$ survives to E^{2r+6} and $d^{2r+6}(\xi Y\bar{M}_2) \in \langle X, \xi, \nu, \eta\rangle$.

(d) Let $d^{2r}(Y) = X \in \pi_*^S$ and let $\xi \in \pi_*^S$. Assume that $\langle X, \xi, \eta, \nu\rangle$ is defined. Then ξYM_2 survives to E^{2r+6} and $d^{2r+6}(\xi YM_2) \in \langle X, \xi, \eta, \nu\rangle$.

PROOF. (a) Let the following diagram depict a defining system for $\langle\sigma, \nu, X, \eta\rangle_\wedge$ using the same notation as in the previous theorems:

$$\begin{array}{ccccccc} \sigma & & \nu & & G & & \eta \\ & B_{\sigma\nu} & & B_{\nu X} & & B_{X\eta} & \\ & & B_{\langle\sigma,\nu,X\rangle} & & B_{\langle\nu,X,\eta\rangle} & & \end{array}$$

Let μ_4 represent $\langle M_1^4\rangle$ such that μ_4 restricted to the boundary of its domain is σ. Let μ_{02} represent M_2^2 such that μ_{02} restricted to the boundary of its domain is $(\mu_4 \wedge \nu) \cup B_{\sigma\nu}$. Then $XM_1M_2^2$ is represented by $\phi =$

$(\mu_{12} \wedge G \wedge \mu_1) \cup (\mu_{02} \wedge B_{X\eta}) \cup (\mu_4 \wedge B_{\langle\nu,X,\eta\rangle}) \cup (B_{\langle\sigma,\nu,X\rangle} \wedge \mu_1) \cup (\mu_4 \wedge B_{\nu X} \wedge \mu_1)$.

Note that ϕ restricted to the boundary of its domain is $(\sigma \wedge B_{\langle\nu,X,\eta\rangle}) \cup (B_{\sigma\nu} \wedge B_{X\eta}) \cup (B_{\langle\sigma,\nu,X\rangle} \wedge \eta)$ which is an element of $\langle\sigma, \nu, X, \eta\rangle$. Thus, $XM_1M_2^2$ survives to E^{14} and $d^{14}(XM_1M_2^2) \in \langle\sigma, \nu, X, \eta\rangle$.

(b) Let the following diagram depict a defining system for $\langle\nu, \sigma, X, \eta\rangle_\wedge$ using the above notation:

$\nu \quad \sigma \quad G \quad \eta$

$B_{\nu\sigma} \quad B_{\sigma X} \quad B_{X\eta}$

$B_{<\nu,\sigma,X>} \quad B_{<\sigma,X,\eta>}$

Let $<\mu_{02}>$ represent $<M_2^2>$ such that $<\mu_{02}>$ restricted to the boundary of its domain is $(\mu_2 \wedge \sigma) \cup B_{\nu\sigma}$. Then $XM_1<M_2^2>$ is represented by $\phi = (<\mu_{02}> \wedge G \wedge \mu_1) \cup (<\mu_{02}> \wedge B_{X\eta}) \cup (\mu_2 \wedge B_{\sigma X} \wedge \mu_1) \cup (\mu_2 \wedge B_{<\sigma,X,\eta>})$ $\cup (B_{<\nu,\sigma,X>} \wedge \mu_1)$. Note that ϕ restricted to the boundary of its domain is $(B_{\nu\sigma} \wedge B_{X\eta}) \cup (\nu \wedge B_{<\sigma,X,\eta>}) \cup (B_{<\nu,\sigma,X>} \wedge \eta)$ which is an element of $<\nu,\sigma,X,\eta>$. Thus, $XM_1<M_2^2>$ survives to E^{14} and $d^{14}(XM_1<M_2^2>) \in <\nu,\sigma,X,\eta>$.

(c) Let the following diagram depict a defining system for $<X,\xi,\nu,\eta>_\wedge$ using the above notation:

$G \quad \xi \quad \nu \quad \eta$

$B_{X\xi} \quad B_{\xi\nu} \quad B_{\nu\eta}$

$B_{<X,\xi,\nu>} \quad B_{<\xi,\nu,\eta>}$

Then $\xi Y\bar{M}_2$ is represented by $\phi = (Y \wedge \xi \wedge \bar{\mu}_{01}) \cup (Y \wedge B_{\xi\nu} \wedge \mu_1) \cup (B_{X\xi} \wedge \bar{\mu}_{01})$ $\cup (B_{<X,\xi,\nu>} \wedge \mu_1) \cup (Y \wedge B_{<\xi,\nu,\eta>})$. Note that ϕ restricted to the boundary of its domain is $(B_{X\xi} \wedge B_{\nu\eta}) \cup (B_{<X,\xi,\nu>} \wedge \eta) \cup (X \wedge B_{<\xi,\nu,\eta>})$ which is an element of $<X,\xi,\nu,\eta>$. Thus, $\xi Y\bar{M}_2$ survives to E^{2r+6} and $d^{2r+6}(\xi Y\bar{M}_2) \in <X,\xi,\nu,\eta>$.

(d) Let the following diagram depict a defining system for $<X,\xi,\eta,\nu>_\wedge$ using the above notation:

$G \quad \xi \quad \eta \quad \nu$

$B_{X\xi} \quad B_{\xi\eta} \quad B_{\eta\nu}$

$B_{<X,\xi,\eta>} \quad B_{<\xi,\eta,\nu>}$

Then ξYM_2 is represented by $\phi = (Y \wedge \xi \wedge \mu_{01}) \cup (Y \wedge B_{\xi\eta} \wedge \mu_2) \cup (B_{X\xi} \wedge \mu_{01})$ $\cup (B_{<X,\xi,\eta>} \wedge \mu_2) \cup (Y \wedge B_{<\xi,\eta,\nu>})$. Note that ϕ restricted to the boundary of its domain is $(B_{X\xi} \wedge B_{\eta\nu}) \cup (B_{<X,\xi,\eta>} \wedge \nu) \cup (X \wedge B_{<\xi,\eta,\nu>})$ which is an element of $<X,\xi,\eta,\nu>$. Thus, ξYM_2 survives to E^{2r+6} and $d^{2r+6}(\xi YM_2) \in <X,\xi,\eta,\nu>$. ∎

CHAPTER 3: LOW DIMENSIONAL COMPUTATIONS

1. Introduction

In this chapter, we illustrate how our algorithm works by computing π_N^S for $0 \le N \le 8$. In addition we introduce notation and derive results that will be useful in our higher dimensional computations. Moreover, we are able to make gloabal computations of $E^4_{n,t}$ and $E^6_{n,t}$ for all n, t as well as global computations of $E^r_{n,t}$ for all r, n and $0 < t \le 8$.

In Section 2, we compute π_1^S and π_2^S and determine the behavior of the spectral sequence on the entire 1 and 2 rows. We also determine all d^2-differentials in the spectral sequence and give the global computation of E^4. In Section 3, we compute π_N^S and determine the behavior of the spectral sequence on the entire N row for $3 \le N \le 6$. We also determine all d^4 differentials in the spectral sequence and give the global computation of E^6. Then in Section 4, we compute π_7^S and π_8^S and determine the behavior of the spectral sequence on the entire 7 and 8 rows. We conclude Section 4 with a summary of some important notation which is introduced in this chapter. The results about π_*^S derived in this chapter are summarized in the initial parts of the tables in Appendices 1 to 4.

As the reader may have noticed, the computations of this chapter are in the range where the elements η, ν and σ of Hopf invariant one exist. That may explain why the spectral sequence has such a simple description in this range.

2. d^2 Differentials and the Determination of E^4

Consider the following diagram of E^2:

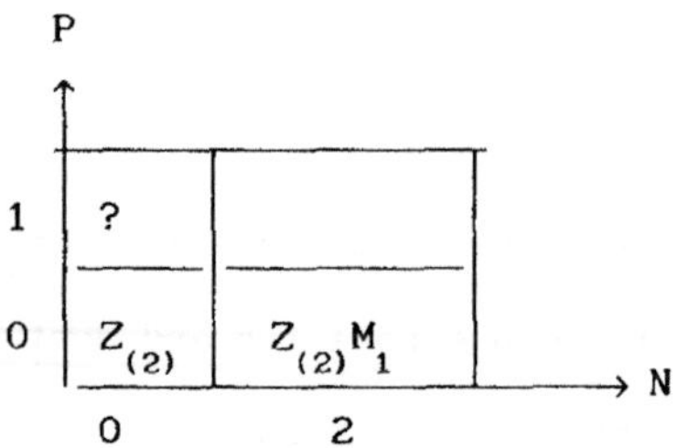

Figure 3.2.1: $E^2_{N,P}$

Now $E^4_{2,0} = E^\infty_{2,0} = \pi_2 BP = Z_{(2)}V_1 = Z_{(2)}(2M_1)$. Therefore $d^2(M_1)$ must be a nonzero element η of π^S_1 of order two. There are no other possibilities for nonzero differentials to land in $E^r_{0,1}$. Therefore, $\pi^S_1 = Z_2\eta$. Consider the following diagram of E^2.

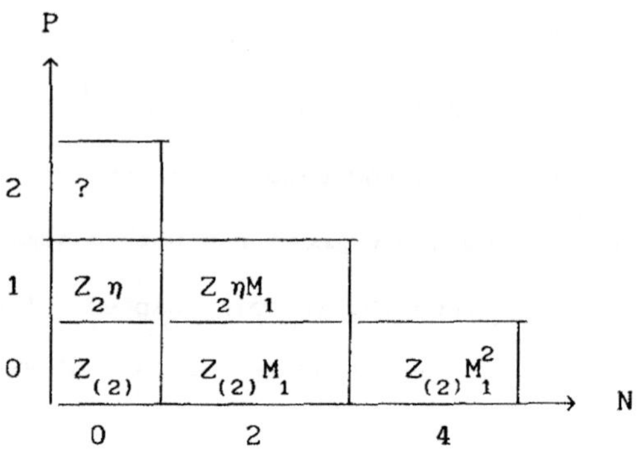

Figure 3.2.2: $E^2_{N,P}$

Now $E^4_{2,1} = E^\infty_{2,1} = 0$, and $d^2(M_1^2) = 0$. Thus, $d^2(\eta\cdot M_1) = \eta^2$ must be a nonzero element of π^S_2 of order two. There are no other possibilities for nonzero differentials to land in $E^r_{0,2}$. Thus, $\pi^S_2 = Z_2\eta^2$. We have thus proved the first part of the following theorem.

THEOREM 3.2.1 (a) $\pi^S_1 = Z_2\eta$ and $\pi^S_2 = Z_2\eta^2$.

(b) $\eta^2 \in \langle 2,\eta,2\rangle$.

PROOF. (b) Consider the Atiyah-Hirzebruch spectral sequence:

$$'E^2_{N,p} = H_N(K(Z_2);\pi^S_p) \Longrightarrow \pi^S_{N+p}(K(Z_2))$$

Here $K(Z_2)$ is the Eilenberg-MacLane spectrum with $\pi^S_*(K(Z_2)) = \pi^S_0(K(Z_2)) = Z_2$. Let $F: BP \longrightarrow K(Z_2)$ such that $F^*(\iota) = 1$. Then F induces a map of spectral sequences $F^r: E^r_{N,p} \longrightarrow {}'E^r_{N,p}$. From [30, Theorem 3.2] we see that the cellular chains of $K(Z_2)$ can be written as $C_*K(Z_2) = Z[\xi_1, \bar{\xi}_k \mid k \geq 2]$ with $\partial(\xi_1) = 2$. Recall from [20] that $E^1_{N,p} = C_N K(Z_2) \otimes \pi^S_p$. Thus $'E^1_{1,1} = Z\xi_1$ and $d^1(\xi_1) = 2$. Now $F^2(M_1) = \xi_1^2$ and hence $d^2(\eta\xi_1^2) = \eta^2$. By Theorem 2.4.2, $\eta^2 \in \langle 2, \eta, 2\rangle$. ∎

We repeat our argument one more time. Consider the following diagram of E^2.

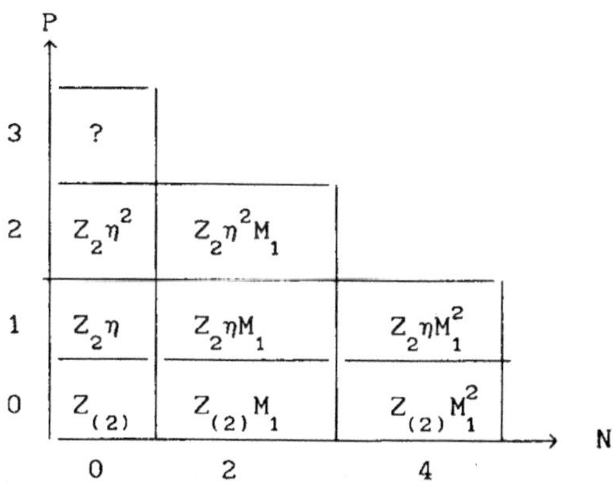

Figure 3.2.3: $E^2_{N,P}$

Now $0 = E^\infty_{2,2} = E^4_{2,2}$ and $d^2(\eta M_1^2) = 0$. Thus, $\eta^3 = d^2(\eta^2 M_1)$ is a nonzero element of order two in π^S_3. We now use Quillen operations to extend our computation to compute the d^2 differentials on the entire 1 and 2 rows.

THEOREM 3.2.2 (a) $d^2(M_N) = \eta M^2_{N-1}$ for $N \geq 1$.

(b) Let $\bar{M}_1 = M_1$, $\bar{M}_2 = 3M_2 - M_1^3$ and $\bar{M}_N = M_N - M_1 M^2_{N-1}$ for $N \geq 3$. Then $\bar{M}_N$ is a d^2 cycle for $N \geq 2$.

PROOF. (a) The only nonzero Quillen operation of degree $2^{N+1}-2$ on M_N is

$r_{2\Delta_{N-1}}$ and $r_{2\Delta_{n-1}} \circ d^2(M_N) = d^2 \circ r_{2\Delta_{N-1}}(M_N) = d^2(M_1) = \eta$. Thus, $d^2(M_N) = \eta M^2_{N-1}$.

(b) $d^2(\bar{M}_2) = 3d^2(M_2) - d^2(M_1^3) = 0$. For $N \geq 3$, $d^2(\bar{M}_N) = d^2(M_N) - d^2(M_1 M^2_{N-1})$ $= \eta M^2_{N-1} - d^2(M_1)M^2_{N-1} = 0$. ∎

Observe that $H_*BP = Z_{(2)}[\bar{M}_N \mid N \geq 1]$. Thus, we can use these new polynomial generators to describe the behavior of the d^2-differentials in the entire spectral sequence.

THEOREM 3.2.3 Let $\mu_q(\eta): \pi^S_q \longrightarrow \pi^S_{q+1}$ denote multiplication by η. For every $p \geq 1$, π^S_p can be written as a direct sum of cyclic groups $Z_{2^{N(k)}}X_k$, $1 \leq k \leq t$, such that:

(i) $Z_{2^{N(k)}}X_k \otimes H_*BP$ is a direct summand of the p row, $E^2_{*,p}$;

(ii) Kernel $\mu_p(\eta)$ is generated by $\{X_k \mid \eta X = 0\} \cup \{2X_k \mid \eta X \neq 0\}$;

(iii) Image $\mu_{p-1}(\eta)$ has a Z_2 basis $\{2^{N(k)-1}X_k \mid k \in A\}$ for some subset A of $\{1,\ldots,t\}$.

Define B<2> as the subalgebra $Z_{(2)}[M_1^2, \bar{M}_N \mid N \geq 2]$ of $E^4_{*,0}$. Then E^4 is the direct sum of the following summands:

(a) if $\eta X_k \neq 0$ and $k \notin A$ then

$$\left(Z_{2^{N(k)}}X_k \oplus Z_{2^{N(k)-1}}(2X_k M_1)\right) \otimes B\langle 2\rangle$$

is a direct summand of the p row of E^4;

(b) if $\eta X_k \neq 0$ and $k \in A$ then

$$\left(Z_{2^{N(k)-1}}X_k \oplus Z_{2^{N(k)-1}}(2X_k M_1)\right) \otimes B\langle 2\rangle$$

is a direct summand of the p row of E^4;

(c) if $\eta X_k = 0$ and $k \in A$ then

$$\left(Z_{2^{N(k)-1}}X_k \oplus Z_{2^{N(k)}}X_k M_1\right) \otimes B\langle 2\rangle$$

is a direct summand of the p row of E^4;

(d) if $\eta X_k = 0$ and $k \notin A$ then $Z_{2^{N(k)}}X_k \otimes H_*BP$ is a direct summand of the p row of E^4.

PROOF. The decomposition of π_P^S as a direct sum of cyclic groups with the required properties follows routinely from the fundamental theorem of abelian groups. Note that Image $d^2 = \eta \cdot \pi_*^S \otimes B\langle 2\rangle$ and Kernel $d^2 =$

$(\oplus_{\{k|\eta X_k=0\}} Z_{2^{N(k)}} X_k \otimes H_*BP) \oplus (\oplus_{\{k|\eta X_k \neq 0\}} (Z_{2^{N(k)-1}} 2X_k M_1 \oplus Z_{2^{N(k)}} X_k) \otimes B\langle 2\rangle$.

The description of E^4 in (a)-(d) is a direct consequence of this observation. ∎

COROLLARY 3.2.4 $E^4_{N,1} = E^4_{N,2} = 0$ for all $N \geq 0$.

PROOF. In the remarks preceeding Theorem 3.2.2 we saw that $\eta^3 \neq 0$. Thus, this corollary follows from Theorem 3.2.3(b). ∎

COROLLARY 3.2.5 $E^4_{*,0} = (Z_{(2)} \oplus Z_{(2)} 2M_1) \otimes B\langle 2\rangle$.

PROOF. This corollary follows from Theorem 3.2.3(a). ∎

3. d^4 Differentials and the Determination of E^6

We continue our analysis of the spectral sequence by considering the following diagram of E^4.

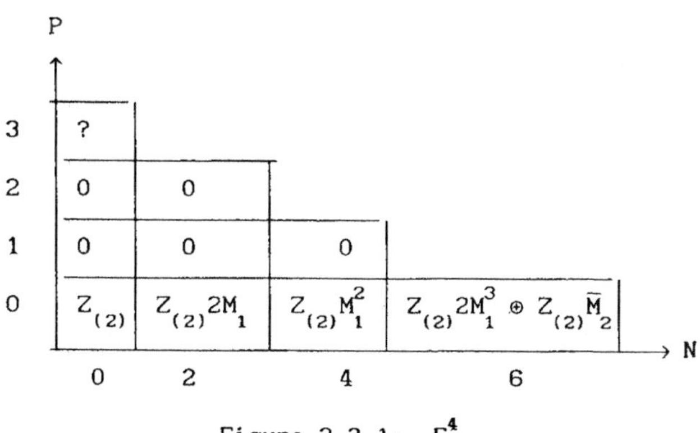

Figure 3.3.1: $E^4_{N,P}$

Note that $Z_{(2)}(4M_1^2) = E^\infty_{4,0} = E^6_{4,0}$. Thus, $d^4(M_1^2)$ must be a nonzero element

$\bar{\nu} \in E^4_{0,3}$ of order 4. Thus we have the following composition series for π^S_3:

$$0 \longrightarrow Z_2\eta^3 \longrightarrow \pi^S_3 \longrightarrow Z_4\bar{\nu} \longrightarrow 0$$

Let $\nu \in \pi^S_3$ be a lifting of $\bar{\nu}$. Either $4\nu = \eta^3$ and $\pi^S_3 = Z_8$ or $4\nu = 0$ and $\pi^S_3 = Z_4\nu \oplus Z_2\eta^3$. By Theorem 2.4.2, we see that

$$2\nu \in \langle\eta, 2, \eta\rangle. \qquad [3.1]$$

By Theorem 3.2.1(b), $\eta^2 \in \langle 2, \eta, 2\rangle$. By Theorem 2.3.3(b), $4\nu \in 2\langle\eta, 2, \eta\rangle$ $= \langle 2, \eta, 2\rangle\eta = \eta^3$. We have thus proved the following theorem.

THEOREM 3.3.1 $\pi^S_3 = Z_8\nu$ and $\eta^3 = 4\nu$.

We use Quillen operations to extend our computation of d^4-differentials to the entire 0 row.

THEOREM 3.3.2 (a) $d^4(\bar{M}_2) = \nu M_1$ and $d^4(2M_1^3) = 2\nu M_1$.
(b) $d^4(\bar{M}_N) = 5\nu M_1 M^4_{N-2} + 2\nu\bar{M}_{N-1} M^2_{N-2}$.

PROOF. (a) $r_1 \circ d^4(\bar{M}_2) = d^4 \circ r_1(\bar{M}_2) = -3\nu$ and $r_1 \circ d^4(2M_1^3) = d^4 \circ r_1(2M_1^3) = 6\nu$. Thus $d^4(\bar{M}_2) \equiv 5\nu M_1$ and $d^4(2M_1^3) \equiv 6\nu$ mod (4) in $E^4_{2,3} = Z_8\nu M_1$. The Hazewinkel generator $V_2 = 2M_2 - 4M_1^3 = \frac{2}{3}\bar{M}_2 - \frac{5}{3}(2M_1^3)$ is a d^4-cycle. Thus $2\nu M_1 = 2d^4(\bar{M}_2)$ $= 5d^4(2M_1^3)$ and $d^4(2M_1^3) = 2\nu M_1$. Note that our definition of ν in Theorem 3.3.1 was only made modulo (4). Thus, define ν so that $d^4(\bar{M}_2)$ is ν and not 5ν.
(b) The only nonzero Quillen operations of degree $2^{N+1} - 6$ on $\bar{M}_N$ are $r_{\Delta_1 + 4\Delta_{N-2}}(\bar{M}_N) = -M_1^2$ and $r_{\Delta_{N-1} + 2\Delta_{N-2}}(\bar{M}_N) = -2M_1^2$. Thus, $d^4(\bar{M}_N) \equiv$ $3\nu M_1 M^4_{N-2} + 2\nu M_{N-1} M^2_{N-2} \equiv 2\nu\bar{M}_{N-1} M^2_{N-2} + \nu M_1 M^4_{N-2}$ mod (4). Then $r_{4\Delta_{N-2}} \circ d^4(\bar{M}_N) =$ $d^4 \circ r_{4\Delta_{N-2}}(\bar{M}_N) = d^4(\frac{1}{3}\bar{M}_2 - \frac{2}{3}M_1^3) = 5\nu M_1$. Note that $r_{4\Delta_{N-2}}(\bar{M}_{N-1} M^2_{N-2}) = 0$ and $r_{4\Delta_{N-2}}(M_1 M^4_{N-2}) = M_1$. Therefore, $d^4(\bar{M}_N) = 2\nu\bar{M}_{N-1} M^2_{N-2} + 5\nu M_1 M^4_{N-2}$. ∎

We use the preceding theorem to define several very important d^4-cycles.

COROLLARY 3.3.3 The following elements are all d^4-cylces.

(a) $\langle M_1^4\rangle = M_1^4 + 2M_1M_2$.

(b) $\langle M_2^2\rangle = M_2^2 + M_1^6$.

(c) $\langle M_3\rangle = \bar{M}_3 + 5\bar{M}_2M_1^4 + 14M_1^7$.

PROOF. It is straightforward to use the previous theorem to compute $d^4\langle M_1^4\rangle$, $d^4\langle M_2^2\rangle$, $d^4\langle M_3\rangle$ and to find that each of these differentials is zero.▮

The following two lemmas will be used to compute E^6 and to prove the existence of polynomial generators $\langle M_N\rangle$, $N \geq 3$, which are d^4-cycles.

LEMMA 3.3.4 In the notation of Theorem 3.2.3, assume that νX_k has order 2^B where B = 1, 2 or 3. Let $\varepsilon = 1$ if $\eta X_k \neq 0$, $\varepsilon = 0$ if $\eta X_k = 0$ and $\beta = B - \delta_3^B$. Then the kernel of d^4 restricted to $Z_{2^{N(k)}}X_k \otimes Z[\langle M_1^4\rangle,\langle M_2^2\rangle]$ is the $Z[\langle M_1^4\rangle,\langle M_2^2\rangle]$- module spanned by

$$\{X_k,\ 2^{\varepsilon}M_1X_k,\ 2^{\beta}M_1^2X_k,\ 2^BM_1^3X_k,\ V_2X_k,\ V_1V_2X_k,\ 2^BM_1^2\bar{M}_2X_k,\ 2^BM_1^3\bar{M}_2X_k\}.$$

PROOF. Observe that $d^4(X_k) = 0$, $d^4(M_1^2X_k) = \nu X_k$, $d^4(\bar{M}_2X_k) = \nu M_1X_k$ and $d^4(M_1^2\bar{M}_2X_k) = \nu(\bar{M}_2+M_1^3)X_k$. If $\eta X_k = 0$ then $d^4(M_1X_k) = 0$, $d^4(M_1^3X_k) = \nu M_1X_k$, $d^4(M_1\bar{M}_2X_k) = \nu M_1^2X_k$ and $d^4(M_1^3\bar{M}_2X_k) = \nu(M_1\bar{M}_2+M_1^4)X_k$. If $\eta X_k \neq 0$ then $d^4(M_1X_k) = 0$, $d^4(2M_1^3X_k) = 2\nu M_1X_k$, $d^4(2M_1\bar{M}_2X_k) = 2\nu M_1^2X_k$ and $d^4(2M_1^3\bar{M}_2X_k) = 2\nu(M_1\bar{M}_2+M_1^4)X_k$. Since $\eta^3 = 4\nu$, $\nu M_1^A\bar{M}_2^CX_k$ has order 2^B if A is odd and has order 2^{β} if A is even. The conclusion of the lemma follows from these observations.▮

LEMMA 3.3.5 In the notation of Theorem 3.2.3 and Lemma 3.3.4, assume that νX_k has order 2^B where B is 1, 2 or 3. Then the image of d^4 in $Z_{2^B}(\nu X_k) \otimes Z[\langle M_1^4\rangle,\langle M_2^2\rangle]$ is the $Z[\langle M_1^4\rangle,\langle M_2^2\rangle]$-module spanned by

$$\{\nu X_k,\ \nu M_1X_k,\ 2^{\varepsilon}\nu M_1^2X_k,\ \nu(\bar{M}_2+M_1^3)X_k,\ 2^{\varepsilon}\nu M_1\bar{M}_2X_k\}.$$

PROOF. This result follows from the computations of d^4-differentials in the proof of Lemma 3.3.4. ∎

Our first applications of these lemmas is to compute π_4^S and π_5^S.

THEOREM 3.3.6 $\pi_4^S = 0$ and $\pi_5^S = 0$.

PROOF. It follows from Lemma 3.3.5 that $E^6_{2,3} = 0$. It follows from Lemma 3.3.4 that $E^6_{6,0} = Z_{(2)}V_2 \oplus Z_{(2)}(8M_1^3) = E^\infty_{6,0}$. We saw in Corollary 3.2.4 that $E^4_{*,1} = E^4_{*,2} = 0$. Therefore, there are no possibilities for nonzero differentials to land in $E^r_{0,4}$ or $E^r_{0,5}$. Thus, $\pi_4^S = \pi_5^S = 0$. ∎

It follows from Lemma 3.3.5 that νM_1^2, $2\nu M_1^3$ and $4\nu M_1^3\overline{M}_2$ are not d^4-boundaries. This leads to the following theorem.

THEOREM 3.3.7 (a) Let $\nu^2 = d^4(\nu M_1^2)$. Then $\pi_6^S = Z_2\nu^2$.

(b) $A[8] = d^6(2\nu M_1^3)$ is a nonzero element of order 2 in π_8^S.

(c) $A[14] = d^{12}(4\nu M_1^3\overline{M}_2)$ is a nonzero element of π_{14}^S.

PROOF. (a) Since νM_1^2 is not a d^4-boundary and $E^\infty_{4,3} = 0$, $\nu^2 \neq 0$. Since ν has odd degree, $2\nu^2 = 0$. By Theorem 2.4.2,

$$\nu^2 \in \langle \eta, \nu, \eta \rangle \qquad [3.2]$$

(b) Since $2\nu M_1^3$ is not a d^4-boundary and $E^\infty_{6,3} = 0$, $A[8] \neq 0$. From Theorem 2.4.4(a), we see that

$$A[8] \in \langle \eta, \nu, 2\nu \rangle. \qquad [3.3]$$

Therefore $2A[8] \in 2\langle \eta, \nu, 2\nu \rangle = \langle 2, \eta, \nu \rangle 2\nu = 0$ because $\langle 2, \eta, \nu \rangle \in \pi_5^S = 0$.

(c) Since $4\nu M_1^3\overline{M}_2$ is not a d^4-boundary and $E^\infty_{12,3} = 0$, $A[14] \neq 0$. ∎

We next apply our lemmas is to produce the d^4-cylces $\langle M_N \rangle$, $N \geq 3$, which will be used in our computation of E^6.

THEOREM 3.3.8 For $N \geq 3$, there are polynomial generators $\langle M_N \rangle$ of H_*BP which survive to E^6.

PROOF. We construct the $\langle M_N \rangle$ by induction on $N \geq 3$. We already found $\langle M_3 \rangle$ in Theorem 3.3.4(c). Assume that $N \geq 4$ and that we have found $\langle M_k \rangle$ for $3 \leq k < N$. Then $d^4(\bar{M}_N) \in \{Z_8 \nu M_1 \bar{M}_2^{\delta} \oplus Z_4 \nu \bar{M}_2^{\delta} | \delta=0,1\} \otimes Z[M_1^2, \langle M_2^2 \rangle, \langle M_3 \rangle, \ldots, \langle \bar{M}_{N-1} \rangle]$ and $d^4(\bar{M}_N)$ is a d^4-cycle, a d^6-cycle and a d^{12}-cycle. The tables in Figure 3.3.2 shows that all such cycles are in the image of d^4 on $S = \{Z_{(2)} \oplus Z_{(2)} M_1\} \otimes Z_{(2)}[M_1^2, \bar{M}_2, \ldots, \bar{M}_{N-1}]$. In those tables an entry $\longleftarrow X$ in row $k\nu$, column M means that X hits $k\nu M$ and an entry $\longrightarrow Y$ in row $k\nu$, column M means that $k\nu M$ hits Y. Thus there is an element $U \in S$ such that $d^4(U) = d^4(\bar{M}_N)$. Let $\langle M_N \rangle = \bar{M}_N - U$. ∎

4ν	$\leftarrow \eta^2 M_1$	$\leftarrow 4\bar{M}_2$	$\leftarrow \eta^2 M_1^3$	$\leftarrow \eta^2 M_1 \bar{M}_2$	$\leftarrow 4M_1^2 \bar{M}_2$	$\leftarrow 4(M_1^3 \bar{M}_2 - M_1^6)$
2ν	$\leftarrow 2M_1^2$	$\leftarrow 2\bar{M}_2$	$\leftarrow 2M_1 \bar{M}_2$	$\rightarrow A[8]$	$\leftarrow 2M_1^2 \bar{M}_2$	$\leftarrow 2(M_1^3 \bar{M}_2 - M_1^6)$
ν	$\leftarrow M_1^2$	$\leftarrow \bar{M}_2$	$\rightarrow \nu^2$	$\rightarrow \nu^2 M_1$	$\leftarrow M_1^2 \bar{M}_2$	$\rightarrow \nu^2 M_1^2$
	1	M_1	M_1^2	$\bar{M}_2$	$M_1^3 + \bar{M}_2$	$M_1 \bar{M}_2$

4ν	$\leftarrow \eta^2 M_1^3 \bar{M}_2$	$\rightarrow A[14]$
2ν	$\rightarrow A[8] M_1^2$	$\rightarrow A[8] \bar{M}_2$
ν	$\rightarrow \nu^2 M_2$	$\rightarrow \nu^2 (M_1 \bar{M}_2 + M_1^4)$
	$M_1^2 \bar{M}_2$	$M_1^3 \bar{M}_2$

Figure 3.3.2: d^4 on $E^4_{*,3}$

The final application of our two lemmas is to compute E^6. We begin by strengthening the two lemmas to obtain a global calculation of all d^4-cycles and all d^4-boundaries. First we introduce an important algebra.

DEFINITION 3.3.9 Let B<4) be the subalgebra $Z_{(2)}[\langle M_1^4\rangle, \langle M_2^2\rangle, \langle M_3\rangle, \ldots \langle M_N\rangle, \ldots]$ of $E^8_{*,0}$.

LEMMA 3.3.10 In the notation of Theorem 3.2.3, assume that νX_k has order 2^B where B = 1, 2 or 3. Let $\varepsilon = 1$ if $\eta X_k \neq 0$, let $\varepsilon = 0$ if $\eta X_k = 0$ and let $\beta = B - \delta_3^B$. Then the kernel of d^4 on $Z_{2^{N(k)}} X_k \otimes H_*BP$ is the B<4)- module spanned by $\{X_k,\ 2^{\varepsilon} M_1 X_k,\ 2^{\beta} M_1^2 X_k,\ 2^B M_1^3 X_k,\ V_2 X_k,\ V_1 V_2 X_k,\ 2^B M_1^2 \bar{M}_2 X_k,\ 2^B M_1^3 \bar{M}_2 X_k\}$.
The image of d^4 in $Z_{2^B}\, \nu\, X_k \otimes H_*BP$ is the $B\langle 4\rangle$-module spanned by
$\{\nu X_k,\ \nu M_1 X_k,\ 2^{\varepsilon}\nu M_1^2 X_k,\ \nu(\bar{M}_2 + M_1^3) X_k,\ 2^{\varepsilon}\nu M_1 \bar{M}_2 X_k\}$.
PROOF. Recall that π_*^S is the direct sum of the X_k and therefore E^2 is the direct sum of the $Z_{2^k} X_k \otimes Z_{(2)}\{M_1^e, M_2^f \mid 0 \le e \le 3,\ 0 \le f \le 1\} \otimes B\langle 4\rangle$. Since all the elements of $B\langle 4\rangle$ survive to E^8,

$$E^4 = H_*(\oplus Z_{2^k} X_k \otimes Z_{(2)}\{M_1^e M_2^f \mid 0 \le e \le 3,\ 0 \le f \le 1\}, d^2) \otimes B\langle 4\rangle.$$

Now $H\langle 2\rangle = H_*(\oplus Z_{2^k} X_k \otimes Z_{(2)}\{M_1^e M_2^f \mid 0 \le e \le 3,\ 0 \le f \le 1\}, d^2)$ is given by Theorem 3.2.3. Then $E^6 = H_*(H\langle 2\rangle, d^4) \otimes B\langle 4\rangle$. Thus, Kernel d^4 = (Kernel $d^4 | H\langle 2\rangle) \otimes B\langle 4\rangle$ and Image d^4 = (Image $d^4 | H\langle 2\rangle) \otimes B\langle 4\rangle$. Therefore, this lemma follows from Lemmas 3.3.5 and 3.3.6. ∎

THEOREM 3.3.11 Let $\mu_q(\nu): \pi_q^S \longrightarrow \pi_{q+3}^S$ denote multiplication by ν. In the notation of Theorem 3.2.3, assume that the X_k have the following two addditional properties:

(i) Kernel $\mu_p(\nu)$ is generated by $\{X_k \mid \nu X_k = 0\} \cup \{2^{B(k)} X_k \mid \nu X_k \neq 0\}$ where $\nu \cdot X_k$ has order $2^{B(k)}$;

(ii) Image $\mu_{p-3}(\nu)$ is the direct sum of $\{Z_{2^{N(k)-\gamma(k)}}\, 2^{N(k)-\gamma(k)}X_k \mid k \in S\}$ for some subset S of $\{1,\ldots,t\}$.

Abuse notation by denoting B(k) by B and $\gamma(k)$ by γ. Let $\varepsilon = 1$ if $\eta\cdot X_k \neq 0$ and let $\varepsilon = 0$ if $\eta\cdot X_k = 0$. Let $\delta = 1$ if $2^{N(k)-1}X_k$ is divisible by η and let $\delta = 0$ if $2^{N(k)-1}X_k$ is not divisible by η. Then E^4 is the direct sum of the following B<4>-modules $X_k\langle 4\rangle$.

(a) If $2^{N(k)-1}X_k$ is not divisible by ν and $\nu X_k = 0$ then

$$X_k\langle 4\rangle = [Z_{2^{N(k)-\delta}}\, X_k \oplus Z_{2^{N(k)-\varepsilon}}\, (2^{\varepsilon}XM_1)] \otimes B\langle 2\rangle.$$

(b) If $2^{N(k)-1}X_k$ is not divisible by ν and $\nu X_k \neq 0$ then

$$\begin{aligned} X_k\langle 4\rangle = [&Z_{2^{N(k)-\delta}}\, X_k \oplus Z_{2^{N(k)-\varepsilon}}\, 2^{\varepsilon}M_1X_k \oplus Z_{2^{N(k)-\beta-\delta}}\, 2^{\beta}M_1^2X_k \oplus Z_{2^{N(k)-B}}\, 2^{B}M_1^3X_k \\ &\oplus Z_{2^{N(k)-\delta}}\, (\bar{M}_2+M_1^3)X_k \oplus Z_{2^{N(k)-\beta}}\, 2^{\beta}M_1\bar{M}_2X_k \oplus Z_{2^{N(k)-\delta-B}}\, 2^{B}M_1^2\bar{M}_2X_k \\ &\oplus Z_{2^{N(k)-B}}\, 2^{B}M_1^3\bar{M}_2X_k] \otimes B\langle 4\rangle. \end{aligned}$$

(c) Assume that X_k is divisible by ν, $\nu Y = 2^{N(k)-\gamma}X_k$ and $\nu X_k = 0$.

Let $\lambda = 0$ if $\eta Y = 0$ and let $\lambda = 1$ if $\eta Y \neq 0$. Then

$$\begin{aligned} X_k\langle 4\rangle = [&Z_{2^{N(k)-\gamma}}\, X_k \oplus Z_{2^{N(k)-\gamma-\varepsilon}}\, 2^{\varepsilon}M_1X_k \oplus Z_{2^{N(k)-\gamma+\lambda}}\, M_1^2X_k \oplus Z_{2^{N(k)-\varepsilon}}\, 2^{\varepsilon}M_1^3X_k \\ &\oplus Z_{2^{N(k)-\gamma}}\, (\bar{M}_2+M_1^3)X_k \oplus Z_{2^{N(k)-\gamma-\varepsilon+\lambda}}\, 2^{\varepsilon}M_1\bar{M}_2X_k \oplus Z_{2^{N(k)-\delta}}\, 2^{\delta}M_1^2\bar{M}_2X_k \\ &\oplus Z_{2^{N(k)-\varepsilon}}\, 2^{\varepsilon}M_1^3\bar{M}_2X_k] \otimes B\langle 4\rangle. \end{aligned}$$

(d) Assume that X_k is divisible by ν, $\nu Y = 2^{N(k)-\gamma}X_k$, and $\nu X_k \neq 0$.

Define λ as in (c). Then

$$\begin{aligned} X_k\langle 4\rangle = [&Z_{2^{N(k)-\gamma}}X_k \oplus Z_{2^{N(k)-\gamma-\varepsilon}}2^{\varepsilon}M_1X_k \oplus Z_{2^{N(k)-\beta-\gamma+\lambda}}2^{\beta}M_1^2X_k \oplus Z_{2^{N(k)-B-\delta}}2^{B}\bar{M}_2X_k \\ &\oplus Z_{2^{N(k)-\gamma}}(\bar{M}_2+M_1^3)X_k \oplus Z_{2^{N(k)-\beta-\gamma+\lambda}}2^{\beta}M_1\bar{M}_2X_k \oplus Z_{2^{N(k)-B-\delta}}2^{B}M_1^2\bar{M}_2X_k \\ &\oplus Z_{2^{N(k)-B}}2^{B}M_1^3\bar{M}_2X_k] \otimes B\langle 4\rangle. \end{aligned}$$

PROOF. This theorem follows from considering the various cases of the preceding lemma and Theorem 3.2.3. ∎

COROLLARY 3.3.12 $E^6_{*,0} = E^8_{*,0} = Z_{(2)}\{V_1, V_1^2, V_1^3, V_2, V_1V_2, V_1^2V_2, 8M_1^3M_2\} \otimes B\langle 4\rangle$.

PROOF. $E^6_{*,0}$ follows from Theorem 3.3.11(b). $E^8_{*,0} = E^6_{*,0}$ because $\pi_5^S = 0$. ∎

In the next theorem, we analyze the behavior of the spectral sequence on the 3 row .

THEOREM 3.3.13 (a) $E^6_{*,3} = \{Z_2(2\nu\bar{M}_2) \oplus Z_2(2\nu M_1^2\bar{M}_2) \oplus Z_4(2\nu M_1^3\bar{M}_2)\} \otimes B\langle 4\rangle$.

(b) $d^6(E^6_{*,3}) = \{Z_2A[8] \oplus Z_2A[8]M_1^2 \oplus Z_2A[8]\bar{M}_2\} \otimes B\langle 4\rangle$.

(c) $E^8_{*,3} = Z_2(4\nu M_1^3\bar{M}_2) \otimes B\langle 4\rangle$.

(d) $d^{12}(E^{12}_{*,3}) = Z_2A[14] \otimes B\langle 4\rangle$.

(e) $E^{14}_{*,3} = 0$.

PROOF. (a) This follows from Theorem 3.3.11(d).

(b) This follows from Theorem 3.3.7(b) and Figure 3.3.2.

(c) This is an immediate consequence of (a) and (b).

(d) This follows from Theorem 3.3.7(c).

(e) This is an immediate consequence of (c) and (d). ∎

We conclude this section with the determination of the behavior of the spectral sequence on the 6 row. We will require a relation in π_9^S which is a nontrivial extension. The relation will be deduced from the following lemma.

LEMMA 3.3.14 If $X \in \pi_*^S$ such that $2X = 0$ and $\eta^2X \neq 0$ then there is an element $Y \in \langle\eta, 2, X\rangle$ such that $2Y = \eta^2X$.

PROOF. By Theorem 3.2.1(b), $\eta^2 \in \langle 2, \eta, 2\rangle$, and $\eta^2\cdot X \in \langle 2, \eta, 2\rangle X = 2\langle\eta, 2, X\rangle$. ∎

Although the proof of the following two theorems require several technical results which will be proved later, the results seem more appropriate to this chapter than to Chapter 5. Therefore, we record them here.

THEOREM 3.3.15 (a) $\nu^3 = \eta A[8]$ is a nonzero element of π_9^S of order two.

(b) $E^6_{*,6} = 0$.

PROOF. (a) $A[8]M_1$ is not a d^6-boundary and therfore can not bound because $E^4_{*,1} = 0$. Therefore $\eta\cdot A[8] = d^2(A[8]M_1)$ is nonzero. If $\eta^2 A[8] \neq 0$ then by Lemma 3.3.14 there is $Y \in \langle\eta,2,A[8]\rangle$ such that $2Y = \eta^2 A[8]$. Thus,

modulo $\nu\cdot\pi_7^S \oplus \eta\cdot\pi_9^S$, $Y \in \langle\eta,2,A[8]\rangle$

$\subset \langle\eta,2,\langle\eta,\nu,2\nu\rangle\rangle$ by the proof of Theorem 3.3.7(b)

$= \langle\langle\eta,2,\eta\rangle,\nu,2\nu\rangle + \langle\eta,\langle 2,\eta,\nu\rangle,2\nu\rangle$ by Theorem 2.3.3

$= \langle 2\nu,\nu,2\nu\rangle$ by Theorem 2.4.1, noting $\langle 2,\eta,\nu\rangle \in \pi_5^S = 0$

$= \langle\nu,4\nu,\nu\rangle$ by Theorem 2.3.3(d),(e)

$= \langle\nu,\eta^3,\nu\rangle = \eta^2\langle\nu,\eta,\nu\rangle$ by Theorem 2.3.3(a),(d).

Therefore $2Y \in 2\eta^2\langle\nu,\eta,\nu\rangle = 0 \bmod 2\nu\cdot\pi_7^S$. As we shall see in the next section, $\pi_7^S = Z_{16}\sigma$ and $\nu\cdot\sigma = 0$. Thus $\eta^2 A[8] = 2Y = 0$. Therefore $\eta A[8]M_1$ must be a boundary. Since $\pi_4^S = 0$ and $E^4_{*,2} = 0$, $\eta A[8]M_1$ can only bound from the 0 row or the 6 row. In the next section we shall see that $E^{10}_{12,0}$ $= Z_{(2)}V_2^2 \oplus Z_{(2)}16V_1^3V_2 \oplus Z_{(2)}16M_1^6$ and in Chapter 4 we will see that $16M_1^6$ survives to E^{12} and $d^{12}(16M_1^6) = \beta_1$. Thus $\eta A[8]M_1$ must bound by a d^4-differential from the 6 row. Since $\pi_6^S = Z_2\nu^2$, it follows that $\nu^3 = \eta A[8]$. By Theorem 2.4.2,

$$\eta A[8] \in \langle\eta,\nu^2,\eta\rangle \qquad [3.4]$$

(b) This result is now an immediate consequence of Theorem 3.3.11(d).∎

THEOREM 3.3.16 $2A[14] = 0$

PROOF. $0 \in \langle\eta,2,A[8]\rangle$ because $\langle\eta,2,A[8]\rangle \in \pi_{10}^S$ which, as we shall see, equals $\eta\cdot\pi_9^S$. Also $\langle 2,A[8],\nu\rangle \in \pi_{12}^S$ which, as we shall see, is zero. Note that

$$\nu A[8] \in \nu\langle\eta,2,\nu^2\rangle = \langle\nu,\eta,2\rangle\nu^2 = 0. \qquad [3.5]$$

Thus by Theorem 2.2.7, $\langle\eta,2,A[8],\nu\rangle$ is defined. Note that hypothesis (i) of Theorem 2.4.3 is satisfied. Therefore,

$$A[14] \in \langle\eta,2,A[8],\nu\rangle \qquad [3.6]$$

Note that $\nu A[8] \in \nu\langle\eta,2,\nu^2\rangle = \langle\nu,\eta,2\rangle\nu^2 = 0$. By Theorem 2.3.6(a), $2A[14] \in 2\langle\eta,2,A[8],\nu\rangle \subset \langle\langle 2,\eta,2\rangle,A[8],\nu\rangle = \langle\eta^2,A[8],\nu\rangle \subset \langle\eta,\eta A[8],\nu\rangle$ $= \langle\eta,\nu^3,\nu\rangle \supset \langle\eta,\nu,\nu^3\rangle = \langle\eta,\nu,\eta A[8]\rangle \supset \langle\eta,\nu,\eta\rangle A[8] = \nu^2 A[8] = 0$. Thus, $2A[14] = 0$ modulo $\eta\cdot\pi^S_{13} + \nu\cdot\pi^S_{11}$ which as we shall see is zero.∎

4. d^8 Differentials and the Seven Row

We continue the study of our spectral sequence with an analysis of $d^8: E^8_{*,0} \longrightarrow E^8_{*,7}$ and an analysis of all the differentials which originate on the 7 row: a d^8-differential which defines $\sigma^2 \in \pi^S_{14}$, a d^{10}-differential which defines $A[16] \in \pi^S_{16}$, a d^{12}-differential which defines $C[18] \in \pi^S_{18}$ and a d^{24}-differential which defines $A[30] \in \pi^S_{30}$. In the process of this analysis, we construct polynomial generators $\{M_N\}$ of H_*BP for $N \geq 5$ which survive to E^{10}. We conclude the section with a complete analysis of the 8 row.

Observe that $E^8_{8,0} = Z_{(2)}\langle M_1^4\rangle$ and $E^{10}_{8,0} = E^{\infty}_{8,0} = Z_{(2)}(16M_1^4)$. Thus, $\sigma = d^8\langle M_1^4\rangle$ is a nonzero element of π^S_7 of order 16. Since $E^4_{*,2} = 0$, $\pi^S_4 = 0$ and $\eta\cdot\pi^S_6 = \eta\cdot\nu^2 = 0$, there are no other nonzero differentials which land in $E^r_{0,7}$. Thus, we have proven the first part of the following theorem.

THEOREM 3.4.1 $\pi^S_7 = Z_{16}\,\sigma$ and $\eta\cdot\sigma \neq 0$.

PROOF. To show that $\eta\cdot\sigma \neq 0$, it suffices to show that σM_1 is not a d^8-boundary. Observe that $E^8_{10,0} = Z_{(2)}(2M_1\langle M_1^4\rangle) \oplus Z_{(2)}(V_1^2V_2)$. Note that $d^8(2M_1\langle M_1^4\rangle) = 10\sigma M_1$ and $V_1^2V_2$ is an infinite cycle. Thus σM_1 does not bound.∎

We next compute d^8 on several of the key elements of $E^8_{*,0}$.

LEMMA 3.4.2 (a) $d^8\langle M_1^4\rangle = \sigma$.

(b) $d^8\langle M_2^2\rangle = 15\sigma M_1^2$.

(c) $d^8\langle M_3\rangle = 5\sigma\bar{M}_2$.

(d) We can choose $\langle M_4\rangle$ such that $d^8\langle M_4\rangle \equiv 2\sigma M_1M_2M_3$ modulo (4σ).

PROOF. (a) This differential defines σ.

(b) $d^8\langle M_2^2\rangle = d^8(M_2^2+M_1^6) = 15\sigma M_1^2$, using the Quillen operation $r_{2\Delta_1}$.

(c) $d^8\langle M_3\rangle = d^8(\bar{M}_3+5M_1^4\bar{M}_2+14M_1^7) = 15\sigma M_2 + 11\sigma M_1^3 = 5\sigma\bar{M}_2$ using the Quillen operations r_{Δ_2} and $r_{3\Delta_1}$.

(d) Observe that we can choose $\langle M_4\rangle \equiv M_4 - M_1M_3^2 + \bar{M}_2M_2^4 + 2M_1^2M_2^2M_3 + 2M_1^5\bar{M}_2M_3 + 5M_1^8\langle M_3\rangle$ modulo (4). Then d^8 of such an $\langle M_4\rangle$ is $2\sigma\, M_1M_2M_3$ modulo (4σ). ∎

Next we analyze the d^8-differentials which originate on the 7 row.

THEOREM 3.4.3 (a) $\sigma^2 = d^8(\sigma\, M_1^4)$ is a nonzero element of order 2 in π_{14}^S.

(b) $[Z_8(2\sigma) \otimes H_*BP] \oplus [d^8(E_{*,0}^8) \otimes Z_2] = \text{Kernel}\ [d^8: E_{*,7}^8 \longrightarrow E_{*,14}^8]$.

(c) $[Z_2\sigma^2 \otimes H_*BP]\ /\ \text{Image}\ [d^8: E_{*,7}^8 \longrightarrow E_{*,14}^8]$

$$= \left[Z_2\sigma^2\{M_1^3,\ M_1^2\bar{M}_2,\ M_1^3\bar{M}_2\} \otimes B\langle 4\rangle\right]$$

$$\oplus \left[Z_2\sigma^2\{\langle M_1^4\rangle\langle M_2^2\rangle,\ M_1\langle M_1^4\rangle\langle M_2^2\rangle,\ M_1^2\langle M_1^4\rangle\langle M_2^2\rangle,\ \bar{M}_2\langle M_1^4\rangle\langle M_2^2\rangle,\ M_1^2\bar{M}_2\langle M_1^4\rangle\langle M_2^2\rangle\} \otimes Z[\langle M_1^4\rangle^2, \langle M_2^2\rangle^2, \langle M_3\rangle, \dots, \langle M_N\rangle, \dots]\right]$$

$$\oplus \left[Z_2\sigma^2\{\bar{M}_2\langle M_1^4\rangle,\ M_1^2\bar{M}_2\langle M_1^4\rangle\} \otimes Z[\langle M_1^4\rangle^2, \langle M_2^2\rangle, \langle M_3\rangle, \dots, \langle M_N\rangle, \dots]\right]$$

$$\oplus \left[Z_2\sigma^2(M_1^6\langle M_3\rangle)[\langle M_1^4\rangle^2, \langle M_2^2\rangle^2, \langle M_3\rangle^2, \langle M_4\rangle, \dots, \langle M_N\rangle, \dots]\right].$$

(d) For $N \ge 4$, there are polynomial generators $\{M_N\}'$ of H_*BP which survive to E^8 and $d^8\{M_N\}' \in (2\sigma)$. Let $B'\langle 8\rangle = Z[\langle M_1^4\rangle^2, \langle M_2^2\rangle^2\langle M_3\rangle^2, \{M_4\}', \dots, \{M_N\}', \dots]$.

(e) $E_{*,7}^{10} = [Z_8(2\sigma) \otimes H_*BP]\ /\ [d^8(E_{*,0}^8) \cap (Z_8(2\sigma) \otimes H_*BP)]$.

PROOF. (a) $E_{16,0}^8 = Z_{(2)}\langle M_1^4\rangle^2 \oplus Z_{(2)}V_1V_2\langle M_1^4\rangle \oplus Z_{(2)}V_1^2\langle M_2^2\rangle \oplus Z_{(2)}V_1\langle M_3\rangle$. Clearly $d^8(E_{16,0}^8) \subset (2\sigma)$. Therefore, σM_1^4 is not a boundary and $\sigma^2 = d^8(\sigma M_1^4) \neq 0$. By Theorem 2.4.2,

$$\sigma^2 \in \langle \nu, \sigma, \nu\rangle \qquad [3.7]$$

(b), (c), (d) These parts of the theorem will follow from the table in Figure 3.4.1. The symbol $\equiv$ in the right hand column of that table means "congruent modulo $d^8(E^8_{*,7})$".

$E^8_{*,0}$		$E^8_{*,7}$		$E^8_{*,14}$
$\langle M_1^4\rangle$	$\longrightarrow$	σ		
$\langle M_2^2\rangle$	$\longrightarrow$	σM_1^2		
$\langle M_3\rangle$	$\longrightarrow$	$\sigma\bar{M}_2$		
		$\sigma M_1^2\bar{M}_2$	$\longrightarrow$	$\sigma^2 M_1$
		$\sigma\langle M_1^4\rangle$	$\longrightarrow$	σ^2
		$\sigma M_1^2\langle M_1^4\rangle$	$\longrightarrow$	$\sigma^2 M_1^2$
		$\sigma\bar{M}_2\langle M_1^4\rangle$	$\longrightarrow$	$\sigma^2\bar{M}_2$
		$\sigma M_1^2\bar{M}_2\langle M_1^4\rangle$	$\longrightarrow$	$\sigma^2 M_1^2 M_2$
$\langle M_1^4\rangle\langle M_2^2\rangle$	$\longrightarrow$	$\sigma(\langle M_2^2\rangle+M_1^2\langle M_1^4\rangle)$		
		$\sigma M_1^2\langle M_2^2\rangle$	$\longrightarrow$	$\sigma^2 M_1^4$
		$\sigma\bar{M}_2\langle M_2^2\rangle$	$\longrightarrow$	$\sigma^2(M_1^2M_2+M_1^5) \equiv \sigma^2 M_1^5$
		$\sigma M_1^2\bar{M}_2\langle M_2^2\rangle$	$\longrightarrow$	$\sigma^2(M_1M_2^2+M_2M_1^4)$
$\langle M_1^4\rangle\langle M_3\rangle$	$\longrightarrow$	$\sigma(\langle M_3\rangle+\bar{M}_2\langle M_1^4\rangle)$		
$\langle M_2^2\rangle\langle M_3\rangle$	$\longrightarrow$	$\sigma(M_1^2\langle M_3\rangle+\bar{M}_2\langle M_2^2\rangle)$		
		$\sigma\bar{M}_2\langle M_3\rangle$	$\longrightarrow$	$\sigma^2(M_2^2+M_1^6) \equiv \sigma^2 M_1^6$
		$\sigma M_1^2\bar{M}_2\langle M_3\rangle$	$\longrightarrow$	$\sigma^2(M_1\langle M_3\rangle+M_1^2\langle M_2^2\rangle) \equiv \sigma^2 M_1\langle M_3\rangle$
		$\sigma\langle M_1^4\rangle\langle M_2^2\rangle$	$\longrightarrow$	$\sigma^2 M_2^2$
		$\sigma M_1^2\langle M_1^4\rangle\langle M_2^2\rangle$	$\longrightarrow$	$\sigma^2 M_1^2 M_2^2$
		$\sigma\bar{M}_2\langle M_1^4\rangle\langle M_2^2\rangle$	$\longrightarrow$	$\sigma^2\bar{M}_2 M_2^2$
		$\sigma M_1^2\bar{M}_2\langle M_1^4\rangle\langle M_2^2\rangle$	$\longrightarrow$	$\sigma^2(M_1^2\bar{M}_2M_2^2+M_1^5\langle M_2^2\rangle)$
		$\sigma\langle M_1^4\rangle\langle M_3\rangle$	$\longrightarrow$	$\sigma^2(\langle M_3\rangle+\bar{M}_2M_1^4)$
		$\sigma M_1^2\langle M_1^4\rangle\langle M_3\rangle$	$\longrightarrow$	$\sigma^2(M_1^2\langle M_3\rangle+M_1^6\bar{M}_2)$
		$\sigma\bar{M}_2\langle M_1^4\rangle\langle M_3\rangle$	$\longrightarrow$	$\sigma^2(\bar{M}_2\langle M_3\rangle+M_1^4\langle M_2^2\rangle)$

$$
\begin{array}{lll}
 & \sigma M_1^2\bar{M}_2\langle M_1^4\rangle\langle M_3\rangle & \longrightarrow \ \sigma^2(M_1^2\bar{M}_2\langle M_3\rangle+M_1^6\langle M_2^2\rangle+M_1^5\langle M_3\rangle) \\
\langle M_1^4\rangle\langle M_2^2\rangle\langle M_3\rangle \longrightarrow & \sigma(\langle M_2^2\rangle\langle M_3\rangle+M_1^2\langle M_1^4\rangle\langle M_3\rangle+\bar{M}_2\langle M_1^4\rangle\langle M_2^2\rangle) & \\
 & \sigma M_1^2\langle M_2^2\rangle\langle M_3\rangle & \longrightarrow \ \sigma^2(M_1^4\langle M_3\rangle+M_1^2\bar{M}_2\langle M_2^2\rangle) \\
 & \sigma\bar{M}_2\langle M_2^2\rangle\langle M_3\rangle & \longrightarrow \ \sigma^2(\langle M_2^2\rangle^2+M_1^2\bar{M}_2\langle M_3\rangle) \equiv \sigma^2M_1^2\bar{M}_2\langle M_3\rangle \\
 & \sigma M_1^2\bar{M}_2\langle M_2^2\rangle\langle M_3\rangle & \longrightarrow \ \sigma^2(M_1\langle M_2^2\rangle\langle M_3\rangle+\bar{M}_2M_1^4\langle M_3\rangle+M_1^2\langle M_2^2\rangle^2) \\
 & & \qquad \equiv \sigma^2(M_1\langle M_2^2\rangle\langle M_3\rangle+\bar{M}_2M_1^4\langle M_3\rangle) \\
 & \sigma\langle M_1^4\rangle\langle M_2^2\rangle\langle M_3\rangle & \longrightarrow \ \sigma^2(\langle M_2^2\rangle\langle M_3\rangle+M_1^6\langle M_3\rangle+\bar{M}_2M_1^4\langle M_2^2\rangle) \\
 & \sigma M_1^2\langle M_1^4\rangle\langle M_2^2\rangle\langle M_3\rangle & \rightarrow \ \sigma^2(M_1^2\langle M_2^2\rangle\langle M_3\rangle+M_1^8\langle M_3\rangle+M_1^6\bar{M}_2\langle M_2^2\rangle) \\
 & \sigma\bar{M}_2\langle M_1^4\rangle\langle M_2^2\rangle\langle M_3\rangle & \rightarrow \ \sigma^2(\bar{M}_2\langle M_2^2\rangle\langle M_3\rangle+M_1^6\bar{M}_2\langle M_3\rangle+M_1^4\langle M_2^2\rangle^2) \\
 & & \qquad \equiv \sigma^2(\bar{M}_2\langle M_2^2\rangle\langle M_3\rangle+M_1^6\bar{M}_2\langle M_3\rangle) \\
 & \sigma M_1^2M_2\langle M_1^4\rangle\langle M_2^2\rangle\langle M_3\rangle & \longrightarrow \\
 & & \sigma^2(M_1^5\langle M_2^2\rangle\langle M_3\rangle+M_1^2M_2\langle M_2^2\rangle\langle M_3\rangle+M_2M_1^8\langle M_3\rangle+M_1^6\langle M_2^2\rangle^2) \\
 & & \qquad \equiv \sigma^2(M_1^5\langle M_2^2\rangle\langle M_3\rangle+M_1^2\bar{M}_2\langle M_2^2\rangle\langle M_3\rangle+\bar{M}_2M_1^8\langle M_3\rangle)
\end{array}
$$

Figure 3.4.1: d^8-Boundaries

We prove that (b), (c) and (d) are valid for $E^8_{N,7}$ simultaneously by induction on N. Of course we only need to worry about (d) when N is of the form $2^p - 10$, and in that case we have to show that $\{M_p\}'$ exists. Clearly (b), (c) and (d) are valid for small values of N. Assume that $2^q-10 < N \le 2^{q+1}-10$ and that (b), (c), (d) are valid for $E^8_{K,7}$ if $k < N$. Let $A = Z[M_1^8, \langle M_2^2\rangle^2, \langle M_3\rangle^2, \{M_4\}', \ldots, \{M_q\}']$, and let B be the set of all monomials in the given polynomial generators of A. Let R be the set of elements in the right hand column of the above table, and let L be the set of elements in the left hand column of the above table. The elements of R are summarized in the table of Figure 3.4.2 below. In that table an entry * indicates that the given element bounds modulo other entries of the table. Note that this table is valid when tensored with $Z_2[\langle M_1^4\rangle^2, \langle M_2^2\rangle^2, \langle M_3\rangle^2, \langle M_4\rangle, \ldots, \langle M_N\rangle, \ldots]$. In particular, it is evident from the table that $\{r\cdot b | r \in R \text{ and } b \in B\}$ is linearly independent. Therefore $\{ l\cdot b \mid l \in L \text{ and } b \in B \}$ are showing how all the elements in the kernel of $d^8: E^8_{t,7} \longrightarrow E^8_{t,14}$ are boundaries for $t \le N$. This proves (b). The image of $d^8: E^8_{*,7} \longrightarrow E^8_{*,14}$ is the $Z_2[\langle M_1^4\rangle^2, \langle M_2^2\rangle^2, \langle M_3\rangle^2, \langle M_4\rangle, \ldots \langle M_N\rangle, \ldots]$-module spanned by the elements in the right hand column of Figure 3.4.1. Examination of that column, as depicted in Figure 3.4.2, verifies that the assertion made in (c) is true. If $N = 2^{q+1}-10$ then $d^8\langle M_{q+1}\rangle$ is a d^8-cycle. By our proof of (b), this d^8-cycle is in $[Z_8(2\sigma) \otimes H_*BP] \oplus d^8(Z_{(2)}[\langle M_1^4\rangle, \langle M_2^2\rangle, \langle M_3\rangle, \{M_4\}', \ldots, \{M_q\}'])$. So write $d^8\langle M_{q+1}\rangle = 2\sigma X + d^8(\gamma)$ for some $\gamma \in Z_{(2)}[\langle M_1^4\rangle, \langle M_2^2\rangle, \langle M_3\rangle, \{M_4\}', \ldots, \{M_q\}']$. Let $\{M_{q+1}\}' = \langle M_{q+1}\rangle - \gamma$. This completes the proof of the induction step.

(e) This statement is a consequence of (b). ∎

σ^2	1	$\langle M_1^4\rangle$	$\langle M_2^2\rangle$	$\langle M_3\rangle$	$\langle M_1^4\rangle\langle M_2^2\rangle$	$\langle M_1^4\rangle\langle M_3\rangle$	$\langle M_2^2\rangle\langle M_3\rangle$	$\langle M_1^4\rangle\langle M_2^2\rangle\langle M_3\rangle$
1	*	*	*	*		*	*	
M_1	*	*	*	*		*	*	
M_1^2	*	*	*	*			*	
$\bar{M}_2$	*		*	*			*	
$M_1^2\bar{M}_2$	*		*	*			*	

Figure 3.4.2: Summary of d^8- Boundaries

We digress to prove a result from which it will follow that $\nu\cdot\sigma = 0$.

THEOREM 3.4.4 Let $\phi \in \pi_*^S$ such that ϕM_1^2 and $\phi\bar{M}_2$ are both boundaries. Then $\nu\cdot\phi = 0$.

PROOF. Since ϕM_1^2 bounds, $d^4(\phi M_1^2) = 0$. Therefore $\nu\cdot\phi$ is a d^2-boundary which means $\nu\cdot\phi$ is divisible by η. Then $d^4(\phi\bar{M}_2) = \nu\phi M_1$ is zero if and only if $\nu\cdot\phi = 0$ in π_*^S. Since $\phi\bar{M}_2$ bounds, it follows that $\nu\cdot\phi = 0$. ∎

COROLLARY 3.4.5 $\nu\cdot\sigma = 0$ and $\eta\cdot\sigma^2 = 0$.

PROOF. Note that $\sigma M_1^2 = d^8\langle M_2^2\rangle$ and $\sigma \bar{M}_2 = d^8\langle M_3\rangle$. By Theorem 3.4.4, $\nu\cdot\sigma = 0$. Now $\sigma^2 M_1 = d^8(\sigma M_1^2\bar{M}_2)$. Thus, $\eta\sigma^2 = d^2(\sigma^2 M_1) = 0$. ▮

We next prove that the analogue Theorem 3.4.3 which analyzes the d^{10}-differentials which originate on the 7 row.

THEOREM 3.4.6 (a) $A[16] = d^{10}(\sigma M_1^5)$ is a nonzero element of order 2 in π^S_{16}.

(b) $E^{12}_{*,7}$ consists of the $[d^8(E^8_{*,0}) \cap (Z_8(2\sigma)\otimes H_*BP)]$-cosets of

$$E(7,12) \equiv [Z_8(2\sigma)\{A\} \otimes B'\langle 8\rangle]$$
$$\oplus\ [Z_8\{2\sigma M_1^{2\alpha}\bar{M}_2^{\beta}\langle M_1^4\rangle^{\gamma}\langle M_2^2\rangle^{\delta}\langle M_3\rangle^{\varepsilon} \mid 0 \le \alpha,\beta,\gamma,\delta,\varepsilon \le 1,\ (\alpha,\beta) \ne (1,1) \text{ and } (\delta,\varepsilon) \ne (1,1)\} \otimes B'\langle 8\rangle].$$

Here A is the following set:

$\{M_1^7\bar{M}_2 + M_1\bar{M}_2\langle M_2^2\rangle$, $M_1^3\bar{M}_2\langle M_3\rangle + M_1^7 M_2^2$, $M_1^6\bar{M}_2 + M_1^3\langle M_2^2\rangle$,

$M_1^2\bar{M}_2\langle M_2^2\rangle + M_1\bar{M}_2\langle M_3\rangle$, $M_2^2\bar{M}_3 + M_1^{13}$, $M_2^2\langle M_3\rangle M_1^2 + M_1^{12}\bar{M}_2$,

$\bar{M}_2 M_2^2\bar{M}_3 + M_1^6\bar{M}_2\langle M_3\rangle$, $\langle M_1^4\rangle M_2^2\bar{M}_3 + M_1^{11}\langle M_2^2\rangle$, $M_1^6 M_2^2\bar{M}_3 + M_1^9\bar{M}_2\langle M_3\rangle$,

$M_1^6\bar{M}_2\langle M_3\rangle + M_1^3\langle M_2^2\rangle\langle M_3\rangle + M_1^7\bar{M}_2\langle M_2^2\rangle$, $\bar{M}_2\langle M_1^4\rangle M_2^2\bar{M}_3 + M_1^{10}\bar{M}_2\langle M_3\rangle\}$.

(c) $d^{10}(E^{10}_{*,7}) \subset Z_2A[16] \otimes B\langle 2\rangle$ and

$$[Z_2A[16] \otimes B\langle 2\rangle] \,/\, \text{Image } [d^{10}: E^{10}_{*,7} \longrightarrow E^{10}_{*,16}]$$
$$= Z_2A[16]\{M_1^2\bar{M}_2\langle M_1^4\rangle,\ M_1^2\bar{M}_2\langle M_1^4\rangle\langle M_2^2\rangle,\ M_1^2\langle M_1^4\rangle\langle M_3\rangle,\ M_1^2\bar{M}_2\langle M_1^4\rangle\langle M_3\rangle,$$
$$M_1^2\langle M_1^4\rangle\langle M_2^2\rangle\langle M_3\rangle,\ \bar{M}_2\langle M_1^4\rangle\langle M_2^2\rangle\langle M_3\rangle,\ M_1^2\bar{M}_2\langle M_1^4\rangle\langle M_2^2\rangle\langle M_3\rangle\} \otimes B'\langle 8\rangle.$$

(d) For $N \ge 4$, there are polynomial generators $\{M_N\}''$ of H_*BP which survive to E^8 and $d^8\{M_N\}'' \in E(7,12)$.

Let $B''\langle 8\rangle = Z[\langle M_1^4\rangle^2, \langle M_2^2\rangle^2\langle M_3\rangle^2, \{M_4\}'', \ldots, \{M_N\}'', \ldots]$.

PROOF. (a) $E^8_{18,0} = Z_{(2)}\{V_1^2\langle M_3\rangle, V_2\langle M_2^2\rangle, V_1^3\langle M_2^2\rangle, V_1^2V_2\langle M_1^4\rangle, V_1\langle M_1^4\rangle^2\}$. Therefore, $d^8(E^8_{18,0}) = Z_8(2\sigma M_1^2 M_2) \oplus Z_4(4\sigma M_1^5)$ and $E^{10}_{10,7} = Z_2(2\sigma M_1^5)$. It follows that there must be a nonzero differential originating on $2\sigma M_1^5 = \sigma V_1\langle M_1^4\rangle$. Now $2\sigma M_1^5$ survives to E^{10} and $d^{10}(2\sigma M_1^5)$ defines a nonzero element $A[16]$ of π^S_{16}. By Theorem 2.4.2,

$$A[16] \in \langle \eta, 2, \sigma^2 \rangle. \qquad [3.8]$$

Then $2A[16] \in 2\langle \eta, 2, \sigma^2 \rangle = \langle 2, \eta, 2 \rangle \sigma^2 = \eta^2\sigma^2 = 0$. Hence $2A[16] = 0$.

(b), (c), (d) The proof of the remainder of this theorem is analogous to the proof of the last four parts of Theorem 3.4.3. We create a new table in Figure 3.4.3 from the table in Figure 3.4.1 by changing each σ^2 to A[16] and multiplying each monomial in the first and second columns by V_1. In addition there are monomials μ in $Z[M_1^2, \bar{M}_2, \bar{M}_3, \{M_4\}', \ldots, \{M_N\}', \ldots] \otimes Z_8(2\sigma)$ which map to $2\sigma M_1^2 \bar{M}_2$ under a Quillen operation and hence $d^{10}(2\sigma\mu) \neq 0$. Such monomials μ must be divisible by $M_1^2\bar{M}_2$ or by $M_2^2\bar{M}_3$. We include the $2\sigma\mu$ of this form in the table in Figure 3.4.3. Note that all $2\sigma\mu$ which are not listed in Figure 3.4.3 are d^{10}-cycles.

$E^8_{*,0}$		$E^r_{*,7}$		$E^{10}_{*,16}$
$V_1\langle M_1^4\rangle$	$\longrightarrow$	$2\sigma M_1$		
$V_1\langle M_2^2\rangle$	$\longrightarrow$	$2\sigma M_1^3$		
$V_1\langle M_3\rangle$	$\longrightarrow$	$2\sigma M_1\bar{M}_2$		
$V_2\langle M_3\rangle+2\langle M_1^4\rangle\langle M_2^2\rangle$	$\longrightarrow$	$2\sigma(M_1^3\bar{M}_2+M_1^6)$		
		$2\sigma M_1\langle M_1^4\rangle$	$\longrightarrow$	$A[16]$
		$2\sigma M_1^3\langle M_1^4\rangle$	$\longrightarrow$	$A[16]M_1^2$
		$2\sigma M_1\bar{M}_2\langle M_1^4\rangle$	$\longrightarrow$	$A[16]\bar{M}_2$
		$2\sigma M_1^3\bar{M}_2\langle M_1^4\rangle$	$\longrightarrow$	$A[16]M_1^2\bar{M}_2$
$V_1\langle M_1^4\rangle\langle M_2^2\rangle$	$\longrightarrow$	$2\sigma M_1M_2^2$		
		$2\sigma M_1^3\langle M_2^2\rangle$	$\longrightarrow$	$A[16]M_1^4$
		$2\sigma(M_1\bar{M}_2\langle M_2^2\rangle+M_1^3\bar{M}_2\langle M_1^4\rangle)$	$\longrightarrow$	0
		$2\sigma M_1^3\bar{M}_2\langle M_2^2\rangle$	$\longrightarrow$	$A[16]\bar{M}_2M_1^4$
$V_1\langle M_1^4\rangle\langle M_3\rangle$	$\longrightarrow$	$2\sigma M_1(\langle M_3\rangle+\bar{M}_2\langle M_1^4\rangle)$		
$V_1\langle M_2^2\rangle\langle M_3\rangle$	$\longrightarrow$	$2\sigma M_1(M_1^2\langle M_3\rangle+\bar{M}_2\langle M_2^2\rangle)$		
		$2\sigma(M_1\bar{M}_2\langle M_3\rangle+M_1^5\langle M_2^2\rangle)$	$\longrightarrow$	$A[16]M_1^6$

$$2\sigma(M_1^3\bar{M}_2\langle M_3\rangle+M_1^7M_2^2) \longrightarrow 0$$

$$2\sigma M_1\langle M_1^4\rangle\langle M_2^2\rangle \longrightarrow A[16]M_2^2$$

$$2\sigma M_1^3\langle M_1^4\rangle\langle M_2^2\rangle \longrightarrow A[16]M_1^2M_2^2$$

$$2\sigma M_1\bar{M}_2\langle M_1^4\rangle\langle M_2^2\rangle \longrightarrow A[16]\bar{M}_2M_2^2$$

$$2\sigma M_1^3\bar{M}_2\langle M_1^4\rangle\langle M_2^2\rangle \longrightarrow A[16]M_1^2\bar{M}_2M_2^2$$

$$2\sigma M_1\langle M_1^4\rangle\langle M_3\rangle \longrightarrow A[16]\bar{M}_3$$

$$2\sigma M_1^3\langle M_1^4\rangle\langle M_3\rangle \longrightarrow A[16]M_1^2\bar{M}_3$$

$$2\sigma M_1\bar{M}_2\langle M_1^4\rangle\langle M_3\rangle \longrightarrow A[16](\bar{M}_2\langle M_3\rangle+M_1^{10})$$

$$2\sigma(M_1^3\bar{M}_2\langle M_1^4\rangle\langle M_3\rangle+M_1\bar{M}_2\langle M_2^2\rangle\langle M_3\rangle+M_1^5\langle M_2^2\rangle^2) \longrightarrow A[16]M_1^6M_2^2$$

$$V_1\langle M_1^4\rangle\langle M_2^2\rangle\langle M_3\rangle \longrightarrow 2\sigma M_1(M_2^2\langle M_3\rangle+\bar{M}_2\langle M_1^4\rangle\langle M_2^2\rangle)$$

$$2\sigma(M_1^3\langle M_2^2\rangle\langle M_3\rangle+M_1^3\bar{M}_2\langle M_1^4\rangle\langle M_2^2\rangle) \longrightarrow A[16]M_1^4\langle M_3\rangle$$

$$2\sigma(M_1\bar{M}_2\langle M_2^2\rangle\langle M_3\rangle+M_1^5\langle M_2^2\rangle^2+M_1^{11}\langle M_2^2\rangle) \longrightarrow A[16]M_1^2\bar{M}_2\langle M_3\rangle$$

$$2\sigma(M_1^3\bar{M}_2\langle M_2^2\rangle\langle M_3\rangle+M_1^7\langle M_2^2\rangle^2+M_1^9\bar{M}_2\langle M_3\rangle) \longrightarrow A[16]\bar{M}_2M_1^4\langle M_3\rangle$$

$$2\sigma M_1\langle M_1^4\rangle\langle M_2^2\rangle\langle M_3\rangle \longrightarrow A[16](M_2^2\langle M_3\rangle+\bar{M}_2M_1^4\langle M_2^2\rangle)$$

$$2\sigma M_1^3\langle M_1^4\rangle\langle M_2^2\rangle\langle M_3\rangle \longrightarrow A[16](M_1^2M_2^2\langle M_3\rangle+M_1^6\bar{M}_2\langle M_2^2\rangle)$$

$$2\sigma(M_1\bar{M}_2\langle M_1^4\rangle\langle M_2^2\rangle\langle M_3\rangle+M_1^3\langle M_2^2\rangle^3) \longrightarrow A[16]\bar{M}_2M_2^2\langle M_3\rangle$$

$$2\sigma(M_1^3\bar{M}_2\langle M_1^4\rangle\langle M_2^2\rangle\langle M_3\rangle+M_1\bar{M}_2\langle M_3\rangle\langle M_2^2\rangle^2+M_1^{23}) \longrightarrow A[16]M_1^2\bar{M}_2M_2^2\langle M_3\rangle$$

$$V_2\langle M_2^2\rangle \longrightarrow 2\sigma(M_1^2\bar{M}_2+M_1^5)$$

$$2\sigma(M_1^2\bar{M}_2\langle M_1^4\rangle+M_1^3\langle M_2^2\rangle) \longrightarrow 0$$

$$2\sigma(M_1^2\bar{M}_2\langle M_2^2\rangle+M_1\bar{M}_2\langle M_3\rangle \longrightarrow 0$$

$$V_2\langle M_2^2\rangle\langle M_3\rangle \longrightarrow 2\sigma(M_1^2\bar{M}_2\langle M_3\rangle+M_1^5\langle M_3\rangle+\bar{M}_2\langle M_2^2\rangle)$$

$$2\sigma M_1^2\bar{M}_2\langle M_1^4\rangle\langle M_2^2\rangle \longrightarrow A[16]M_1^4\langle M_2^2\rangle$$

$$2\sigma(M_1^2\bar{M}_2\langle M_1^4\rangle\langle M_3\rangle+M_1^3\langle M_2^2\rangle\langle M_3\rangle+M_1^7\bar{M}_2\langle M_2^2\rangle) \longrightarrow 0$$

$$2\sigma(M_1^2\bar{M}_2\langle M_2^2\rangle\langle M_3\rangle+M_1^5\langle M_2^2\rangle\langle M_3\rangle) \longrightarrow A[16](M_1^6\langle M_3\rangle+M_1^4\bar{M}_2^3)$$

$$2\sigma M_1^2\bar{M}_2\langle M_1^4\rangle\langle M_2^2\rangle\langle M_3\rangle \longrightarrow A[16]\langle M_1^4\rangle\langle M_2^2\rangle\langle M_3\rangle$$

$$2\sigma(M_2^2\bar{M}_3+M_1^{13}) \longrightarrow 0$$

$$2\sigma(M_1^2M_2^2\bar{M}_3+M_1^6\bar{M}_2\langle M_2^2\rangle) \longrightarrow 0$$

$$2\sigma(\bar{M}_2M_2^2\bar{M}_3+M_1^6\bar{M}_2\langle M_3\rangle) \longrightarrow 0$$

$$2\sigma(\langle M_1^4\rangle M_2^2\bar{M}_3+M_1^{11}\langle M_2^2\rangle) \longrightarrow 0$$

$$2\sigma(M_1^2\langle M_1^4\rangle M_2^2\bar{M}_3+M_1^9\bar{M}_2\langle M_3\rangle) \longrightarrow 0$$

$$2\sigma(\bar{M}_2\langle M_1^4\rangle M_2^2\bar{M}_3+M_1^{10}\bar{M}_2\langle M_3\rangle) \longrightarrow 0$$

Figure 3.4.3: d^{10}-Boundaries

The analysis of the table in Figure 3.4.1 which we used to prove Theorem 3.4.3(b)-(e) applies to the table in Figure 3.4.3 to prove (b), (c) and (d) of our Theorem. The right hand column of the table in Figure 3.4.3 is summarized in the table of Figure 3.4.4 below to assist in verifying (c).∎

A[16]	1	$\langle M_1^4\rangle$	$\langle M_2^2\rangle$	$\langle M_3\rangle$	$\langle M_1^4\rangle\langle M_2^2\rangle$	$\langle M_1^4\rangle\langle M_3\rangle$	$\langle M_2^2\rangle\langle M_3\rangle$	$\langle M_1^4\rangle\langle M_2^2\rangle\langle M_3\rangle$
1	*	*	*	*	*	*	*	*
M_1^2	*	*	*	*	*		*	
$\bar{M}_2$	*	*	*	*	*	*	*	
$M_1^2\bar{M}_2$	*		*	*			*	

Figure 3.4.4: Summary of d^{10}- Boundaries

We continue our analysis of differentials which originate on the 7 row by considering the d^{12}-differentials which originate there. We will eventually see in Chapter 4 that they land in a direct sum of Z_8s and Z_4s. In the next theorem we analyze these d^{12}-differentials tensored with Z_2. We use the notation $\tilde{E}(S)$ to denote the reduced exterior algebra generated by the set S.

THEOREM 3.4.7 (a) There is an element $C[18] \in \pi_{18}^S$ such that the projection of C[18] into $E_{0,18}^{12}$ has order four and equals $d^{12}(2\sigma M_1^8)$.

(b) Let $K(12) = \{X \in E^8_{*,7} | X$ survives to E^{12} and

$X \in \text{Kernel } [d^{12}: E^{12}_{*,7} \longrightarrow E^{12}_{*,18}]\}$.

Then $K(12) / \{[d^8(E^8_{*,0}) \oplus Z_4(4\sigma) \otimes H_*BP] \cap K(12)]\} =$

$K_2 \otimes Z[\langle M_1^4\rangle^4, \langle M_2^2\rangle^4, \langle M_3\rangle^4, \langle M_4\rangle^2, \{M_5\}', \ldots, \{M_N\}', \ldots\}]$

where $K_2 = Z_2(2\sigma)\{\langle M_1^4\rangle, M_1^2\langle M_2^2\rangle, \bar{M}_2\langle M_3\rangle, M_1M_2M_3\} \otimes \tilde{E}(\langle M_1^4\rangle^2, \langle M_2^2\rangle^2, \langle M_3\rangle^2, \langle M_4\rangle)$

modulo the following relations:

(i) $2\sigma\langle M_2^2\rangle + 2\sigma M_1^2\langle M_1^4\rangle = 0$;

(ii) $2\sigma\langle M_3\rangle + 2\sigma\bar{M}_2\langle M_1^4\rangle = 0$;

(iii) $2\sigma\langle M_4\rangle + 2\sigma M_1M_2M_3\langle M_1^4\rangle = 0$;

(iv) $2\sigma M_1^2\langle M_3\rangle + 2\sigma\bar{M}_2\langle M_2^2\rangle = 0$;

(v) $2\sigma M_1^2\langle M_4\rangle + 2\sigma M_1M_2M_3\langle M_2^2\rangle = 0$;

(vi) $2\sigma\bar{M}_2\langle M_4\rangle + 2\sigma M_1M_2M_3\langle M_3\rangle = 0$;

(vii) $2\sigma\langle M_2^2\rangle\langle M_3\rangle + 2\sigma M_1^2\langle M_1^4\rangle\langle M_3\rangle + 2\sigma\bar{M}_2\langle M_1^4\rangle\langle M_2^2\rangle = 0$;

(viii) $2\sigma\langle M_2^2\rangle\langle M_4\rangle + 2\sigma M_1^2\langle M_1^4\rangle\langle M_4\rangle + 2\sigma M_1M_2M_3\langle M_1^4\rangle\langle M_2^2\rangle = 0$;

(ix) $2\sigma\langle M_3\rangle\langle M_4\rangle + 2\sigma\bar{M}_2\langle M_1^4\rangle\langle M_4\rangle + 2\sigma M_1M_2M_3\langle M_1^4\rangle\langle M_3\rangle = 0$;

(x) $2\sigma M_1^2\langle M_3\rangle\langle M_4\rangle + 2\sigma\bar{M}_2\langle M_2^2\rangle\langle M_4\rangle + 2\sigma M_1M_2M_3\langle M_2^2\rangle\langle M_3\rangle = 0$;

(xi) $2\sigma\langle M_2^2\rangle\langle M_3\rangle\langle M_4\rangle + 2\sigma M_1^2\langle M_1^4\rangle\langle M_3\rangle\langle M_4\rangle + 2\sigma\bar{M}_2\langle M_1^4\rangle\langle M_2^2\rangle\langle M_4\rangle$

$+ 2\sigma M_1M_2M_3\langle M_1^4\rangle\langle M_2^2\rangle\langle M_3\rangle = 0$;

(c) Let $T = Z_2 \otimes [Z_8C[18] \otimes H_*BP]$. Then

$T / \{T \cap \text{Image } [d^{12}: E^{12}_{*,7} \longrightarrow E^{12}_{*,18}]\}$

$= Z_2\{M_1^\alpha M_2^\beta \langle M_1^4\rangle^{e(1)}\langle M_2^2\rangle^{e(2)}\langle M_3\rangle^{e(3)} \cdots \langle M_N\rangle^{e(N)} \cdots | 0 \le \alpha \le 3,\ 0 \le \beta \le 1, e(N) \ge 0$ for $N \ge 1$

and either (i) $\alpha+\beta \ge 2$ or (ii) $e(1)e(2)$ is odd while $e(3)$ is even$\}$.

(d) For $N \ge 5$, there are polynomial generators $\{M_N\}'''$ of H_*BP which survive to E^8 such that $d^8\{M_N\}''' \in (4\sigma)$.

PROOF. (a) $E^8_{20,0} = Z_{(2)}V_1^2\langle M_1^4\rangle^2 \oplus Z_{(2)}(8M_1^3M_2\langle M_1^4\rangle) \oplus Z_{(2)}\langle M_1^4\rangle\langle M_2^2\rangle \oplus Z_{(2)}V_1V_2\langle M_2^2\rangle \oplus Z_{(2)}V_1^3\langle M_3\rangle \oplus Z_{(2)}V_2\langle M_3\rangle$. Thus, $d^8(E^8_{20,0}) = Z_2(8\sigma M_1^6) \oplus Z_{16}(\sigma M_2^2) \oplus Z_8(2\sigma M_1^3M_2)$. Hence $E^8_{12,7} = Z_4(2\sigma M_1^6)$. Now $2\sigma M_1^2\langle M_1^4\rangle$ survives to E^{10}. Since $2\sigma M_1^2\langle M_1^4\rangle \in \text{Image } r_{\Delta_1}$, $2\sigma M_1^2\langle M_1^4\rangle$ must be a d^{10}-cycle.

Therefore $d^{12}(2\sigma M_1^6)$ is a nonzero element of $E^{12}_{0,18}$ of order four. This element is the projection into $E^{12}_{0,18}$ of an element $C[18] \in \pi^S_{18}$. By Theorem 2.4.2,

$$C[18] \in \langle \nu, \sigma, 2\sigma \rangle. \qquad [3.9]$$

(b)-(d) The proof will be analogous to the proof of Theorem 3.4.3(b)-(e) and will use the table in Figure 3.4.5 below. The entries in the middle column are all the B"<8>-module generators of $E^{12}_{*,7}$ according to Theorem 3.4.4(b). The only change that we have made is to replace $2\sigma(M_1^2\bar{M}_2\langle M_2^2\rangle + M_1\bar{M}_2\langle M_3\rangle)$ by $2M_1M_2M_3$ since they are equal modulo other entries in the table.

$E^8_{*,0}$		$E^r_{*,7}$		$E^{12}_{*,18}$
$2\langle M_1^4\rangle$	$\longrightarrow$	2σ		
$2\langle M_2^2\rangle$	$\longrightarrow$	$2\sigma M_1^2$		
$2\langle M_3\rangle$	$\longrightarrow$	$2\sigma\bar{M}_2$		
$\langle M_1^4\rangle^2$	$\longrightarrow$	$2\sigma\langle M_1^4\rangle$		
		$2\sigma M_1^2\langle M_1^4\rangle$	$\longrightarrow$	$C[18]$
		$2\sigma\bar{M}_2\langle M_1^4\rangle$	$\longrightarrow$	$C[18]M_1$
$V_2\langle M_1^4\rangle\langle M_2^2\rangle + V_1\langle M_1^4\rangle^3$	$\longrightarrow$	$2\sigma(M_1^2\bar{M}_2\langle M_1^4\rangle + M_1^3\langle M_2^2\rangle + \bar{M}_2\langle M_2^2\rangle)$		
$2\langle M_1^4\rangle\langle M_2^2\rangle$	$\longrightarrow$	$2\sigma\langle M_2^2\rangle + M_1^2\langle M_1^4\rangle$		
$\langle M_2^2\rangle^2$	$\longrightarrow$	$2\sigma M_1^2\langle M_2^2\rangle$		
		$2\sigma\bar{M}_2\langle M_2^2\rangle$	$\longrightarrow$	$C[18]M_2$
		$2\sigma(M_1^2\bar{M}_2\langle M_2^2\rangle + M_1\bar{M}_2\langle M_3\rangle)$	$\longrightarrow$	$C[18]M_1^5$
$2\langle M_1^4\rangle\langle M_3\rangle$	$\longrightarrow$	$2\sigma(\langle M_3\rangle + \bar{M}_2\langle M_1^4\rangle)$		
$2\langle M_2^2\rangle\langle M_3\rangle$	$\longrightarrow$	$2\sigma(M_1^2\langle M_3\rangle + \bar{M}_2\langle M_2^2\rangle)$		
$\langle M_3\rangle^2$	$\longrightarrow$	$2\sigma\bar{M}_2\langle M_3\rangle$		
		$2\sigma\langle M_1^4\rangle\langle M_2^2\rangle$	$\longrightarrow$	$C[18]M_1^4$
		$2\sigma M_1^2\langle M_1^4\rangle\langle M_2^2\rangle$	$\longrightarrow$	$C[18]\langle M_2^2\rangle$
		$2\sigma\bar{M}_2\langle M_1^4\rangle\langle M_2^2\rangle$	$\longrightarrow$	$C[18](M_1^4M_2 + M_1\langle M_2^2\rangle)$
$\{M_4\}'$	$\longrightarrow$	$2\sigma M_1M_2M_3$		

$$2\sigma M_1^2\langle M_1^4\rangle\langle M_3\rangle \longrightarrow C[18](\bar{M}_3+M_1^7)$$

$$2\sigma\bar{M}_2\langle M_1^4\rangle\langle M_3\rangle \longrightarrow C[18]M_1\langle M_3\rangle$$

$$2\sigma(M_1^2\bar{M}_2\langle M_1^4\rangle\langle M_3\rangle+M_1^3\langle M_2^2\rangle\langle M_3\rangle+M_1^7\bar{M}_2\langle M_2^2\rangle) \longrightarrow C[18]M_2\langle M_3\rangle$$

$$2\langle M_1^4\rangle\langle M_2^2\rangle\langle M_3\rangle \rightarrow 2\sigma(\langle M_2^2\rangle\langle M_3\rangle+M_1^2\langle M_1^4\rangle\langle M_3\rangle+\bar{M}_2\langle M_1^4\rangle\langle M_2^2\rangle)$$

$$2\sigma M_1^2\langle M_2^2\rangle\langle M_3\rangle \longrightarrow C[18]M_2\langle M_2^2\rangle$$

$$V_2V_3\langle M_1^4\rangle\langle M_2^2\rangle \longrightarrow 2\sigma(M_2M_3\langle M_2^2\rangle+M_1^2M_2\langle M_1^4\rangle M_3)$$

$$2\sigma(\langle M_1^4\rangle M_2^2\bar{M}_3+M_1^{11}\langle M_2^2\rangle) \longrightarrow C[18]M_1^4M_3$$

$$2\sigma(M_1^2\langle M_1^4\rangle M_2^2\bar{M}_3+M_1^9\bar{M}_2\langle M_3\rangle) \longrightarrow C[18](\langle M_2^2\rangle\langle M_3\rangle+M_2\langle M_1^4\rangle\langle M_2^2\rangle)$$

$$2\sigma(\bar{M}_2\langle M_1^4\rangle M_2^2\bar{M}_3+M_1^{10}\bar{M}_2\langle M_3\rangle) \longrightarrow C[18](M_1^4\bar{M}_2M_3+M_1M_2^2M_3+M_1^{14})$$

$$\langle M_3\rangle^2+V_1\langle M_2^2\rangle\langle M_3\rangle+V_2\langle M_1^4\rangle\langle M_3\rangle \rightarrow 2\sigma(M_1^7\bar{M}_2+M_1\bar{M}_2\langle M_2^2\rangle+\langle M_1^4\rangle\langle M_2^2\rangle)$$

$$2\sigma(M_1^3\bar{M}_2\langle M_3\rangle+M_1^7M_2^2+M_1^2\langle M_1^4\rangle\langle M_3\rangle \longrightarrow C[18]M_1^4M_2$$

Figure 3.4.5: d^{12}-Boundaries Mod Two

There is a new phenomenon in the table of Figure 3.4.5. The nonzero d^8-differentials on $\langle M_1^4\rangle^2$, $\langle M_2^2\rangle^2$, $\langle M_3\rangle^2$ and $\langle M_4\rangle$ extend to nonzero d^8-differentials on

$$\langle M_1^4\rangle^{2\alpha}\langle M_2^2\rangle^{2\beta}\langle M_3\rangle^{2\gamma}\langle M_4\rangle^{\delta}\langle M_1^4\rangle^{4e(1)}\langle M_2^2\rangle^{4e(2)}\langle M_3\rangle^{4e(3)}\langle M_4\rangle^{2e(4)}\{M_5\}^{e(5)}\cdots\{M_N\}^{e(N)}$$

for all α, β, γ, $\delta \in \{0,1\}$, $\alpha+\beta+\gamma+\delta > 0$ and $e(k) \geq 0$, $1 \leq k \leq N$. That is, the images of these d^8-differentials are the only d^8-boundaries in

$$Z_2\otimes \{Z_8(2\sigma)\{\langle M_1^4\rangle, M_1^2\langle M_2^2\rangle, \bar{M}_2\langle M_3\rangle, M_1M_2M_3\} \otimes E(\langle M_1^4\rangle^2, \langle M_2^2\rangle^2, \langle M_3\rangle^2, \langle M_4\rangle)$$
$$\otimes Z[\langle M_1^4\rangle^4, \langle M_2^2\rangle^4, \langle M_3\rangle^4, \langle M_4\rangle^2, \{M_5\}', \ldots, \{M_N\}', \ldots\}]\}.$$

This determines the 11 relations in $K(12)/\{[d^8(E^8_{*,0})\otimes Z_4(4\sigma)\otimes H_*BP]\cap K(12)]\}$. Now the proof of the remainder of this theorem is a direct analogue of the proof of Theorem 3.4.3(b)-(e). The right hand column of the table in Figure 3.4.5 is summarized in the table of Figure 3.4.6 to assist in the proof of (c). The only additional observation required to prove (d) is that all of the $d^8\{M_n\}''$, $N \geq 5$, must be elements of $[Z_4(4\sigma) \otimes H_*BP] \cap E^{14}_{*,7}$ because if that were not so then we could apply a Quillen operation to see that $2\sigma M_1^{12}$ or $2\sigma(M_2^3\bar{M}_3+M_1^9\langle M_3\rangle)$ is a d^8-boundary which we know is not the case. Thus, the $\{M_N\}'''$ exist. ∎

C[18]	1	$\langle M_1^4\rangle$	$\langle M_2^2\rangle$	$\langle M_3\rangle$	$\langle M_1^4\rangle\langle M_2^2\rangle$	$\langle M_1^4\rangle\langle M_3\rangle$	$\langle M_2^2\rangle\langle M_3\rangle$	$\langle M_1^4\rangle\langle M_2^2\rangle\langle M_3\rangle$
1	*	*	*	*		*	*	
M_1	*	*	*	*			*	
M_2	*	*	*	*				

Figure 3.4.6: Summary of d^{12}- Boundaries Modulo Two

In the previous theorem we analyzed the d^{12}-differentials which originate on the 7 row modulo two. In the next theorem we analyze these d^{12}-differentials modulo four. We continue with the notation of the previous theorem.

THEOREM 3.4.8 (a) $K(12) / \{d^8(E^8_{*,0}) \oplus Z_2(8\sigma)] \cap K(12)\}$ =

$$\left[K_2 \otimes Z[\langle M_1^4\rangle^4, \langle M_2^2\rangle^4, \langle M_3\rangle^4, \langle M_4\rangle^2, \{M_5\}', \ldots, \{M_N\}', \ldots\}\right]$$

$$\oplus \left[Z_2(4\sigma(M_1\langle M_1^4\rangle\langle M_3\rangle + M_1^3M_2\langle M_2^2\rangle)) \otimes B'\langle 8\rangle\right].$$

(b) Let $U = Z_2 \otimes [Z_4(2\nu^*) \otimes H_*BP]$. Then $U / \{U \cap \text{Image } [d^{12}: E^{12}_{*,7} \longrightarrow E^{12}_{*,18}]\}$

$= Z_2\{2\nu^* M_1^{\alpha} M_2^{\beta} \langle M_1^4\rangle^{e(1)} \langle M_2^2\rangle^{e(2)} \langle M_3\rangle^{e(3)} \cdots \langle M_N^{e(N)}\rangle \cdots \mid 0 \le \alpha \le 3,\ 0 \le \beta \le 1,$

$e(N) \ge 0$ for $N \ge 1$, $\alpha + \beta \ge 3$ and if $e(3)$ is odd then either

(i) $e(1)$ is even or (ii) $e(1)$ is odd, $e(2)$ is even, $\alpha = 1$, $\beta = 0\}$.

(c) For $N \ge 5$, there are polynomial generators $\{M_N\}^{iv}$ of H_*BP which survive to E^8 such that $d^8\{M_N\}^{iv} \in (8\sigma)$.

PROOF. Again the proof is analogous to the proof of Theorem 3.4.3(b)-(e). It makes use of the table in Figure 3.4.7 below. In addition, observe that $d^{12}(K_2) \subset Z_2(4C[18]) \otimes H_*BP$.

$E^8_{*,0}$		$E^r_{*,7}$	$E^{12}_{*,18}$
$4\langle M_1^4\rangle$	$\longrightarrow$	4σ	
$2V_1\langle M_1^4\rangle$	$\longrightarrow$	$4\sigma M_1$	

$$4\langle M_2^2\rangle \longrightarrow 4\sigma M_1^2$$

$$2V_1\langle M_2^2\rangle \longrightarrow 4\sigma M_1^3$$

$$4\langle M_3\rangle \longrightarrow 4\sigma \bar{M}_2$$

$$2V_1\langle M_3\rangle \longrightarrow 4\sigma M_1\bar{M}_2$$

$$2V_2\langle M_2^2\rangle \longrightarrow 4\sigma M_1^2M_2$$

$$V_1V_2\langle M_2^2\rangle \longrightarrow 4\sigma M_1^3M_2$$

$$V_1^2\langle M_2^2\rangle \longrightarrow 4\sigma\langle M_1^4\rangle$$

$$V_1^2\langle M_3\rangle+2V_2\langle M_2^2\rangle \longrightarrow 4\sigma M_1\langle M_1^4\rangle$$

$$4\sigma M_1^2\langle M_1^4\rangle \longrightarrow 2C[18]$$

$$4\sigma M_1^3\langle M_1^4\rangle \longrightarrow 2C[18]M_1$$

$$V_1V_2\langle M_3\rangle+2V_1\langle M_1^4\rangle\langle M_2^2\rangle \longrightarrow 4\sigma M_2\langle M_1^4\rangle$$

$$4\sigma M_1M_2\langle M_1^4\rangle \longrightarrow 2C[18]M_1^2$$

$$4\sigma M_1^2M_2\langle M_1^4\rangle \longrightarrow 2C[18](\bar{M}_2+M_1^3)$$

$$4\sigma M_1^3M_2\langle M_1^4\rangle \longrightarrow 2C[18]M_1\bar{M}_2$$

$$4\langle M_1^4\rangle\langle M_2^2\rangle \longrightarrow 4\sigma(\langle M_2^2\rangle+M_1^2\langle M_1^4\rangle)$$

$$2V_1\langle M_1^4\rangle\langle M_2^2\rangle \longrightarrow 4\sigma(M_1\langle M_2^2\rangle+M_1^3\langle M_1^4\rangle)$$

$$V_1^2\langle M_1^4\rangle\langle M_2^2\rangle+\langle M_1^4\rangle^3 \longrightarrow 4\sigma M_1^2\langle M_2^2\rangle$$

$$4\langle M_2^2\rangle\langle M_3\rangle+2V_2\langle M_1^4\rangle\langle M_2^2\rangle+V_1^2\langle M_1^4\rangle\langle M_3\rangle+2V_1\langle M_1^4\rangle^3 \longrightarrow 4\sigma M_1^3\langle M_2^2\rangle$$

$$2V_2\langle M_1^4\rangle\langle M_2^2\rangle \longrightarrow 4\sigma(M_2\langle M_2^2\rangle+M_1^2M_2\langle M_1^4\rangle)$$

$$V_1V_2\langle M_1^4\rangle\langle M_2^2\rangle \longrightarrow 4\sigma(M_1M_2\langle M_2^2\rangle+M_1^3M_2\langle M_1^4\rangle)$$

$$4V_1^2\langle M_2^2\rangle\langle M_3\rangle \longrightarrow 4\sigma M_1^2\bar{M}_2\langle M_2^2\rangle+4\bar{M}_2\langle M_1^4\rangle\langle M_3\rangle$$

$$4\sigma M_1^3M_2\langle M_2^2\rangle \longrightarrow 2C[18]M_1^6$$

$$4\langle M_1^4\rangle\langle M_3\rangle \longrightarrow 4\sigma(\langle M_3\rangle+\bar{M}_2\langle M_1^4\rangle)$$

$$2V_1\langle M_1^4\rangle\langle M_3\rangle \longrightarrow 4\sigma(M_1\langle M_3\rangle+M_1\bar{M}_2\langle M_1^4\rangle)$$

$$4\langle M_2^2\rangle\langle M_3\rangle \longrightarrow 4\sigma(M_1^2\langle M_3\rangle+\bar{M}_2\langle M_2^2\rangle)$$

$$2V_1\langle M_2^2\rangle\langle M_3\rangle \longrightarrow 4\sigma(M_1^3\langle M_3\rangle+M_1\bar{M}_2\langle M_2^2\rangle)$$

$$2V_2\langle M_1^4\rangle\langle M_3\rangle \longrightarrow 4\sigma(M_2\langle M_3\rangle+M_2\bar{M}_2\langle M_1^4\rangle)$$

$$V_1V_3\langle M_3\rangle \longrightarrow 4\sigma M_1\bar{M}_2M_3$$

$$V_2V_3\langle M_2^2\rangle \longrightarrow 4\sigma M_1^2M_2M_3$$

$$V_1V_2\langle M_2^2\rangle\langle M_3\rangle+2V_1\langle M_1^4\rangle\langle M_2^2\rangle^2 \longrightarrow 4\sigma(M_1^3M_2\langle M_3\rangle+\bar{M}_2\langle M_1^4\rangle\langle M_2^2\rangle$$

$$4\sigma\langle M_1^4\rangle\langle M_2^2\rangle \longrightarrow 2C[18]M_1^4$$

$$4\sigma M_1\langle M_1^4\rangle\langle M_2^2\rangle \longrightarrow 2C[18]M_1^5$$

$$4\sigma M_1^2\langle M_1^4\rangle\langle M_2^2\rangle \longrightarrow 2C[18]\langle M_2^2\rangle$$

$$4\sigma M_1^3\langle M_1^4\rangle\langle M_2^2\rangle \longrightarrow 2C[18]M_1\langle M_2^2\rangle$$

$$4\sigma M_2\langle M_1^4\rangle\langle M_2^2\rangle \longrightarrow 2C[18](M_1^4\bar{M}_2+M_1^7)$$

$$4\sigma M_1M_2\langle M_1^4\rangle\langle M_2^2\rangle \longrightarrow 2C[18](M_1^5\bar{M}_2+M_1^2M_2^2+M_1^8)$$

$$4\sigma M_1^2M_2\langle M_1^4\rangle\langle M_2^2\rangle \longrightarrow 2C[18](\bar{M}_2\langle M_2^2\rangle+M_1^3\langle M_2^2\rangle)$$

$$4\sigma M_1^3M_2\langle M_1^4\rangle\langle M_2^2\rangle \longrightarrow 2C[18](M_1\bar{M}_2\langle M_2^2\rangle+M_1^{10})$$

$$V_1^2\langle M_2^2\rangle\langle M_3\rangle \longrightarrow 4\sigma(\langle M_1^4\rangle\langle M_3\rangle+M_1^2\bar{M}_2\langle M_2^2\rangle)$$

$$4\sigma(M_1\langle M_1^4\rangle\langle M_3\rangle+M_1^3M_2\langle M_2^2\rangle) \rightarrow 0$$

$$4\sigma M_1^2\langle M_1^4\rangle\langle M_3\rangle \longrightarrow 2C[18](\langle M_3\rangle+M_1^4\bar{M}_2+M_1^7)$$

$$4\sigma M_1^3\langle M_1^4\rangle\langle M_3\rangle \longrightarrow 2C[18](M_1\langle M_3\rangle+M_1^5\bar{M}_2+M_1^8)$$

$$4\sigma M_2\langle M_1^4\rangle M_3 \longrightarrow 2C[18]M_1^2M_2^2$$

$$4\sigma M_1M_2\langle M_1^4\rangle\langle M_3\rangle \longrightarrow 2C[18](M_1^2\langle M_3\rangle+M_1^9)$$

$$4\sigma M_1^2M_2\langle M_1^4\rangle\langle M_3\rangle \longrightarrow 2C[18](\bar{M}_2\langle M_3\rangle+M_1^3\langle M_3\rangle+M_1^4M_2^2)$$

$$4\sigma M_1^3M_2\langle M_1^4\rangle\langle M_3\rangle \longrightarrow 2C[18](M_1\bar{M}_2\langle M_3\rangle+M_1^5M_2^2+M_1^8\bar{M}_2)$$

$$4\langle M_1^4\rangle\langle M_2^2\rangle\langle M_3\rangle \longrightarrow 4\sigma(\langle M_2^2\rangle\langle M_3\rangle+M_1^2\langle M_1^4\rangle\langle M_3\rangle+\bar{M}_2\langle M_1^4\rangle\langle M_2^2\rangle)$$

$$2V_1\langle M_1^4\rangle\langle M_2^2\rangle\langle M_3\rangle \longrightarrow 4\sigma(M_1\langle M_2^2\rangle\langle M_3\rangle+M_1^3\langle M_1^4\rangle\langle M_3\rangle+M_1\bar{M}_2\langle M_1^4\rangle\langle M_2^2\rangle)$$

$$V_1^2\langle M_1^4\rangle\langle M_2^2\rangle\langle M_3\rangle \longrightarrow 4\sigma(M_1^2\langle M_2^2\rangle\langle M_3\rangle+\langle M_1^4\rangle^2\langle M_3\rangle+M_1^2\bar{M}_2\langle M_1^4\rangle\langle M_2^2\rangle)$$

$$4\sigma M_1^3\langle M_2^2\rangle\langle M_3\rangle \longrightarrow 2C[18](M_1\bar{M}_2\langle M_2^2\rangle+M_1^4\langle M_2^2\rangle)$$

$$2V_2\langle M_1^4\rangle\langle M_2^2\rangle\langle M_3\rangle \longrightarrow 4\sigma(M_2\langle M_2^2\rangle\langle M_3\rangle+M_1^2M_2\langle M_1^4\rangle\langle M_3\rangle+M_2\bar{M}_2\langle M_1^4\rangle\langle M_2^2\rangle)$$

$$V_1V_2\langle M_1^4\rangle\langle M_2^2\rangle\langle M_3\rangle \longrightarrow 4\sigma(M_1M_2\langle M_2^2\rangle\langle M_3\rangle+M_1^3M_2\langle M_1^4\rangle\langle M_3\rangle+M_1M_2\bar{M}_2\langle M_1^4\rangle\langle M_2^2\rangle)$$

$$4\sigma(M_1^2M_2\langle M_2^2\rangle\langle M_3\rangle+M_1^6M_2^4) \longrightarrow 2C[18]M_1^6\langle M_2^2\rangle$$

$$4\sigma M_1^3M_2\langle M_2^2\rangle\langle M_3\rangle \longrightarrow 2C[18](M_1^6\bar{M}_3+M_1M_2^4+\bar{M}_2\langle M_1^4\rangle M_2^2+M_1^7M_2^2)$$

$$4\sigma\langle M_1^4\rangle\langle M_2^2\rangle\langle M_3\rangle \longrightarrow 2C[18](M_1^4\langle M_3\rangle+M_1\langle M_1^4\rangle\langle M_2^2\rangle)$$

$$4\sigma M_1\langle M_1^4\rangle\langle M_2^2\rangle\langle M_3\rangle \longrightarrow 2C[18](M_1^5\langle M_3\rangle+M_1^6\langle M_2^2\rangle)$$

$$4\sigma M_1^2\langle M_1^4\rangle\langle M_2^2\rangle\langle M_3\rangle \longrightarrow 2C[18](\langle M_2^2\rangle\bar{M}_3+M_1^7\langle M_2^2\rangle)$$

$$4\sigma M_1^3\langle M_1^4\rangle\langle M_2^2\rangle\langle M_3\rangle \longrightarrow 2C[18](M_1\langle M_2^2\rangle\bar{M}_3+M_1^8\langle M_2^2\rangle)$$

$$4\sigma M_2\langle M_1^4\rangle\langle M_2^2\rangle\langle M_3\rangle \longrightarrow 2C[18](\bar{M}_2\langle M_1^4\rangle\langle M_3\rangle+M_1^3\langle M_1^4\rangle\langle M_3\rangle+M_1\bar{M}_2\langle M_1^4\rangle\langle M_2^2\rangle)$$

$$4\sigma M_1M_2\langle M_1^4\rangle\langle M_2^2\rangle\langle M_3\rangle \longrightarrow 2C[18](M_1^2\langle M_2^2\rangle\langle M_3\rangle+M_1^5\bar{M}_2\bar{M}_3)$$

$$4\sigma M_1^2\bar{M}_2\langle M_1^4\rangle\langle M_2^2\rangle\langle M_3\rangle \longrightarrow 2C[18](\bar{M}_2\langle M_2^2\rangle\bar{M}_3+M_1^3\langle M_2^2\rangle\langle M_3\rangle+M_1^{10}\langle M_2^2\rangle)$$

$$4\sigma M_1^3\bar{M}_2\langle M_1^4\rangle\langle M_2^2\rangle\langle M_3\rangle \longrightarrow 2C[18](M_1M_2M_2^2M_3+M_1^7M_2\langle M_3\rangle+M_1^{10}M_3+M_1^8M_2M_2^2)$$

$$2\sigma(M_2^3\bar{M}_3+M_1^9\langle M_3\rangle+2M_1M_2\langle M_1^4\rangle^3) \longrightarrow 0$$

Figure 3.4.7: d^{12} Boundaries Modulo Four

To prove (b) we summarize the boundaries of the right hand column of the table in Figure 3.4.7 in the table of Figure 3.4.8 below. The proofs of (b) and (c) are analogous to the proofs in the previous theorems. ∎

2C[18]	1	$\langle M_1^4\rangle$	$\langle M_2^2\rangle$	$\langle M_3\rangle$	$\langle M_1^4\rangle\langle M_2^2\rangle$	$\langle M_1^4\rangle\langle M_3\rangle$	$\langle M_2^2\rangle\langle M_3\rangle$	$\langle M_1^4\rangle\langle M_2^2\rangle\langle M_3\rangle$
1	*	*	*	*	*		*	
M_1	*	*	*	*	*	*	*	
M_1^2	*	*	*	*	*		*	
$\bar{M}_2$	*	*	*	*	*		*	
$M_1\bar{M}_2$	*	*	*	*	*		*	

Figure 3.4.8: Summary of d^{12}- Boundaries Modulo Four

We will see in Chapter 5 that C[18] has order eight. At this point, however, we can not yet eliminate the possibility that C[18] has order four. This accounts for the indeterminate symbol ξ in the following theorem. With the insight from Chapter 5, the reader can replace ξ by $4C[18]M_1$.

THEOREM 3.4.9 If $4C[18] \neq 0$ then let $r = 12$ and $\xi = 4C[18]M_1$. If $4C[18] = 0$ then $r = 14$ and there is $\xi \in \pi^S_{20}$ which projects to a nonzero element of $E^{14}_{0,20}$. In both cases, the following is true.

(a) $E^{2r+2}_{*,7} = Z_2(2\sigma)\{\langle M_1^4\rangle^3, M_1^2\langle M_2^2\rangle^3, \bar{M}_2\langle M_3\rangle^3, M_1M_2M_3\langle M_4\rangle, \bar{M}_2\langle M_3\rangle\langle M_1^4\rangle^2\langle M_3\rangle^2,$

$M_1M_2M_3\langle M_1^4\rangle\langle M_4\rangle, M_1M_2M_3\langle M_2^2\rangle\langle M_4\rangle, M_1M_2M_3\langle M_1^4\rangle\langle M_2^2\rangle\langle M_4\rangle\}$

$\otimes Z[\langle M_1^4\rangle^4, \langle M_2^2\rangle^4, \langle M_3\rangle^4, \langle M_4\rangle^2, \{M_5\}', \ldots, \{M_N\}', \ldots].$

(b) $[Z_2(\xi) \otimes H_*BP] \;/\; d^r[Z_2(8\sigma) \otimes H_*BP]$

$= [Z_2(\xi M_1) \otimes B\langle 2\rangle]$

$\oplus [Z_2(\xi M_1^2\bar{M}_2\langle M_4\rangle) \otimes B\langle 4\rangle \;/\; (\langle M_2^2\rangle^3\langle M_3\rangle\langle M_4\rangle, \langle M_1^4\rangle^2\langle M_2^2\rangle^3\langle M_3\rangle\langle M_4\rangle)]$

$\oplus [Z_2(\xi\langle M_1^4\rangle\langle M_2^2\rangle\langle M_3\rangle\{1, M_1^2, \bar{M}_2\}) \otimes B''\langle 8\rangle] \oplus [Z_2(\xi M_1^2\langle M_1^4\rangle\langle M_3\rangle) \otimes B''\langle 8\rangle].$

(c) For $N \geq 5$, there are polynomial generators $\{M_N\}$ of H_*BP which survive to E^{10}.

PROOF. We begin by computing $d^8\colon E^8_{22,0} \longrightarrow E^8_{14,7}$ to see that $E^{12}_{14,7} = Z_8(2\sigma)M_1^7$. Observe that $E^8_{22,0} = Z_{(2)}\langle M_1^4\rangle\langle M_3\rangle \oplus Z_{(2)}V_1V_2\langle M_3\rangle \oplus Z_{(2)}V_1^2V_2\langle M_2^2\rangle$ $\oplus Z_{(2)}V_1\langle M_1^4\rangle\langle M_2^2\rangle \oplus Z_{(2)}V_2\langle M_1^4\rangle^2 \oplus Z_{(2)}V_1^3\langle M_1^4\rangle^2$. Thus, $d^8(E^8_{22,0})$ $= Z_8(\sigma\bar{M}_3) \oplus Z_4(2\sigma M_1M_2^2) \oplus Z_2(4\sigma M_1^4M_2)$. Note that $d^8(\sigma M_1^4\bar{M}_2) = \sigma^2\bar{M}_2$ and $d^{10}(2\sigma M_1^4M_2) = \eta^*M_1^2$. Thus $E^{12}_{14,7} = Z_8(2\sigma)M_1^7$ as asserted. Therefore if $4\nu^* \neq 0$ then $d^{12}(8\sigma M_1^7) = 4\nu^*M_1 = \xi$. On the other hand, if $4\nu^* = 0$ then $E^{14}_{14,7} = Z_2(4\sigma M_1^7)$ and $d^{14}(4\sigma M_1^7)$ is a nonzero element ξ of $E^{14}_{0,20}$.

(a), (b), (c) Once again the proof is analogous to the proof of Theorem 3.4.3(b)-(e). It makes use of the table in Figure 3.4.9 below.

$E^8_{*,0}$		$E^r_{*,7}$		$E^r_{*,r+6}$
$V_1^3\langle M_3\rangle+2V_1V_2\langle M_2^2\rangle$	$\longrightarrow$	$8\sigma M_1^2\langle M_1^4\rangle$		
		$8\sigma M_1^3\langle M_1^4\rangle$	$\longrightarrow$	ξ
$V_1^2V_2\langle M_3\rangle+2V_1^2\langle M_1^4\rangle\langle M_2^2\rangle$	$\longrightarrow$	$8\sigma M_1M_2\langle M_1^4\rangle$		
		$8\sigma M_1^2M_2\langle M_1^4\rangle$	$\longrightarrow$	ξM_1^2
		$8\sigma M_1^3M_2\langle M_1^4\rangle$	$\longrightarrow$	$\xi\bar{M}_2$
$V_1^3\langle M_2^2\rangle\langle M_3\rangle+V_1V_3\langle M_1^4\rangle^2+2V_1V_2\langle M_1^4\rangle^3$	$\longrightarrow$	$8\sigma M_1^3M_2\langle M_2^2\rangle$		

$$4V_1\langle M_2^2\rangle\langle M_3\rangle+V_1^3\langle M_1^4\rangle\langle M_3\rangle+2V_1V_2\langle M_1^4\rangle\langle M_2^2\rangle+\langle M_2^2\rangle\langle M_1^4\rangle^2 \longrightarrow 8\sigma\langle M_1^4\rangle\langle M_2^2\rangle$$

$$8\sigma M_1\langle M_1^4\rangle\langle M_2^2\rangle \longrightarrow \xi M_1^4$$

$$8(M_2^3M_3+M_1^2\overline{M}_3^2) \longrightarrow 8\sigma M_1^2\langle M_1^4\rangle\langle M_2^2\rangle$$

$$8\sigma M_1^3\langle M_1^4\rangle\langle M_2^2\rangle \longrightarrow \xi\langle M_2^2\rangle$$

$$8\sigma M_2\langle M_1^4\rangle\langle M_2^2\rangle \longrightarrow \xi M_1^6$$

$$8\sigma M_1M_2\langle M_1^4\rangle\langle M_2^2\rangle \longrightarrow \xi M_1^4\overline{M}_2$$

$$8\sigma M_1^2M_2\langle M_1^4\rangle\langle M_2^2\rangle \longrightarrow \xi M_1^2\langle M_2^2\rangle$$

$$8\sigma M_1^3M_2\langle M_1^4\rangle\langle M_2^2\rangle \longrightarrow \xi\overline{M}_2\langle M_2^2\rangle$$

$$4\sigma(M_1\langle M_1^4\rangle\langle M_3\rangle+M_1^3M_2\langle M_2^2\rangle) \longrightarrow \xi M_1^2\overline{M}_2$$

$$V_1^2V_3\langle M_3\rangle+2V_2V_3\langle M_2^2\rangle \longrightarrow 8\sigma M_1M_3\langle M_1^4\rangle$$

$$8(M_1M_2^3M_3+M_1^{11}M_2^2) \longrightarrow 8\sigma(M_1^2\langle M_1^4\rangle\langle M_3\rangle+M_2\langle M_1^4\rangle\langle M_2^2\rangle)$$

$$8\sigma M_1^3\langle M_1^4\rangle\langle M_3\rangle \longrightarrow \xi(\langle M_3\rangle+M_1^4\overline{M}_2)$$

$$8(M_1^2M_2^3M_3+M_1M_2M_3^2) \longrightarrow 8\sigma(M_2\langle M_1^4\rangle\langle M_3\rangle+M_1M_2\langle M_1^4\rangle\langle M_2^2\rangle)$$

$$8(M_1^3M_2^3M_3+M_1^6M_2^2M_3+M_1^4M_2^5+M_1^7M_2^4) \longrightarrow 8\sigma(M_1M_2\langle M_1^4\rangle\langle M_3\rangle+M_1^{15})$$

$$8\sigma M_1^2M_2\langle M_1^4\rangle\langle M_3\rangle \longrightarrow \xi M_1^2\langle M_3\rangle$$

$$8\sigma M_1^3M_2\langle M_1^4\rangle\langle M_3\rangle \longrightarrow \xi(\overline{M}_2\langle M_3\rangle+M_1^4M_2^2)$$

$$V_1^3\langle M_1^4\rangle\langle M_2^2\rangle\langle M_3\rangle+2V_1V_3\langle M_1^4\rangle^3+8M_1^3M_2\langle M_2^2\rangle\langle M_1^4\rangle^2+2V_2^2\langle M_2^2\rangle\langle M_1^4\rangle^2+8\langle M_1^4\rangle^5 \longrightarrow 8\sigma M_1^3\langle M_2^2\rangle\langle M_3\rangle$$

$$V_1^2V_2\langle M_1^4\rangle\langle M_2^2\rangle\langle M_3\rangle+V_2V_3\langle M_1^4\rangle^3+V_1^3\langle M_3\rangle\langle M_2^2\rangle^2+V_1V_2\langle M_2^2\rangle^3+8\langle M_2^2\rangle\langle M_1^4\rangle^4 \longrightarrow 8\sigma M_1^2M_2\langle M_2^2\rangle\langle M_3\rangle$$

$$8\sigma M_1^3M_2\langle M_2^2\rangle\langle M_3\rangle \longrightarrow \xi(M_2^4+M_1^6M_2^2)$$

$$8\sigma\langle M_1^4\rangle\langle M_2^2\rangle\langle M_3\rangle \longrightarrow \xi\langle M_1^4\rangle\langle M_2^2\rangle$$

$$8\sigma M_1\langle M_1^4\rangle\langle M_2^2\rangle\langle M_3\rangle \longrightarrow \xi M_1^4\langle M_3\rangle$$

$$8M_1^7M_2^3M_3 \longrightarrow 8\sigma(M_1^2\langle M_1^4\rangle\langle M_2^2\rangle\langle M_3\rangle+M_1^3M_2\langle M_2^2\rangle\langle M_3\rangle+M_1^3\langle M_1^4\rangle\langle M_2^2\rangle^2)$$

$$8\sigma M_1^3\langle M_1^4\rangle\langle M_2^2\rangle\langle M_3\rangle \longrightarrow \xi\langle M_2^2\rangle\overline{M}_3$$

$$8\sigma M_2\langle M_1^4\rangle\langle M_2^2\rangle\langle M_3\rangle \longrightarrow \xi(M_1^2\langle M_1^4\rangle\langle M_3\rangle+\overline{M}_2\langle M_1^4\rangle\langle M_2^2\rangle)$$

$$8\sigma M_1M_2\langle M_1^4\rangle\langle M_2^2\rangle\langle M_3\rangle \longrightarrow \xi M_1^4\overline{M}_2\overline{M}_3$$

$$8\sigma M_1^2\overline{M}_2\langle M_1^4\rangle\langle M_2^2\rangle\langle M_3\rangle \longrightarrow \xi M_1^2\langle M_2^2\rangle\langle M_3\rangle$$

$$8\sigma M_1^3\overline{M}_2\langle M_1^4\rangle\langle M_2^2\rangle\langle M_3\rangle \longrightarrow \xi(\overline{M}_2M_2^2\overline{M}_3+M_1^6\overline{M}_2\langle M_3\rangle)$$

$$2\sigma\langle M_1^4\rangle^3 \longrightarrow 0$$

$$2\sigma\langle M_1^4\rangle\langle M_2^2\rangle^2 \longrightarrow \xi M_1^2\bar{M}_2\langle M_1^4\rangle$$

$$2\sigma M_1^2\langle M_2^2\rangle^3 \longrightarrow 0$$

$$2\sigma\langle M_1^4\rangle\langle M_3\rangle^2 \longrightarrow \xi M_1^2\bar{M}_2\langle M_2^2\rangle$$

$$2\sigma M_1^2\langle M_2^2\rangle\langle M_3\rangle^2 \longrightarrow \xi M_1^2\bar{M}_2\langle M_1^4\rangle\langle M_2^2\rangle$$

$$2\sigma\bar{M}_2\langle M_3\rangle^3 \longrightarrow 0$$

$$2\sigma\langle M_1^4\rangle\langle M_4\rangle \longrightarrow \xi M_1^2\bar{M}_2\langle M_3\rangle$$

$$2\sigma M_1^2\langle M_2^2\rangle\langle M_4\rangle \longrightarrow \xi M_1^2\bar{M}_2\langle M_1^4\rangle\langle M_3\rangle$$

$$2\sigma\bar{M}_2\langle M_3\rangle\langle M_4\rangle \longrightarrow \xi M_1^2\bar{M}_2\langle M_2^2\rangle\langle M_3\rangle$$

$$2\sigma M_1M_2M_3\langle M_4\rangle \longrightarrow 0$$

$$2\sigma\langle M_1^4\rangle^3\langle M_2^2\rangle^2 \longrightarrow \xi M_1^2\bar{M}_2\langle M_1^4\rangle^3$$

$$2\sigma M_1^2\langle M_1^4\rangle^2\langle M_2^2\rangle^3 \longrightarrow \xi M_1^2\bar{M}_2\langle M_1^4\rangle\langle M_2^2\rangle^2$$

$$2\sigma\langle M_1^4\rangle^3\langle M_3\rangle^2 \longrightarrow \xi M_1^2\bar{M}_2\langle M_2^2\rangle\langle M_1^4\rangle^2$$

$$2\sigma M_1^2\langle M_1^4\rangle^2\langle M_2^2\rangle\langle M_3\rangle^2 \longrightarrow \xi M_1^2\bar{M}_2\langle M_1^4\rangle^3\langle M_2^2\rangle$$

$$2\sigma\bar{M}_2\langle M_1^4\rangle^2\langle M_3\rangle^3 \longrightarrow 0$$

$$2\sigma\langle M_1^4\rangle^3\langle M_4\rangle \longrightarrow \xi M_1^2\bar{M}_2\langle M_3\rangle\langle M_1^4\rangle^2$$

$$2\sigma M_1^2\langle M_1^4\rangle^2\langle M_2^2\rangle\langle M_4\rangle \longrightarrow \xi M_1^2\bar{M}_2\langle M_3\rangle\langle M_1^4\rangle^3$$

$$2\sigma\bar{M}_2\langle M_1^4\rangle^2\langle M_3\rangle\langle M_4\rangle \longrightarrow \xi M_1^2\bar{M}_2\langle M_2^2\rangle\langle M_3\rangle\langle M_1^4\rangle^2$$

$$2\sigma M_1M_2M_3\langle M_1^4\rangle^2\langle M_4\rangle \longrightarrow 0$$

$$2\sigma\langle M_1^4\rangle\langle M_2^2\rangle^2\langle M_3\rangle^2 \longrightarrow \xi M_1^2\bar{M}_2[\langle M_1^4\rangle\langle M_3\rangle^2+\langle M_2^2\rangle^3]$$

$$2\sigma M_1^2\langle M_2^2\rangle^3\langle M_3\rangle^2 \longrightarrow \xi M_1^2\bar{M}_2\langle M_1^4\rangle\langle M_2^2\rangle^3$$

$$2\sigma\bar{M}_2\langle M_2^2\rangle^2\langle M_3\rangle^3 \longrightarrow \xi M_1^2\bar{M}_2\langle M_1^4\rangle\langle M_2^2\rangle\langle M_3\rangle^2$$

$$2\sigma\langle M_1^4\rangle\langle M_2^2\rangle^2\langle M_4\rangle \longrightarrow \xi M_1^2\bar{M}_2[\langle M_3\rangle\langle M_2^2\rangle^2+\langle M_1^4\rangle\langle M_4\rangle]$$

$$2\sigma M_1^2\langle M_2^2\rangle^3\langle M_4\rangle \longrightarrow \xi M_1^2\bar{M}_2\langle M_1^4\rangle\langle M_3\rangle\langle M_2^2\rangle^2$$

$$2\sigma\bar{M}_2\langle M_2^2\rangle^2\langle M_3\rangle\langle M_4\rangle \longrightarrow \xi M_1^2\bar{M}_2[\langle M_2^2\rangle^3\langle M_3\rangle+\langle M_1^4\rangle\langle M_3\rangle^3]$$

$$2\sigma M_1M_2M_3\langle M_2^2\rangle^2\langle M_4\rangle \longrightarrow 0$$

$$2\sigma\langle M_1^4\rangle\langle M_3\rangle^2\langle M_4\rangle \longrightarrow \xi M_1^2\bar{M}_2[\langle M_3\rangle^3+\langle M_2^2\rangle\langle M_4\rangle]$$

$$2\sigma M_1^2\langle M_2^2\rangle\langle M_3\rangle^2\langle M_4\rangle \longrightarrow \xi M_1^2\bar{M}_2[\langle M_1^4\rangle\langle M_3\rangle^3+\langle M_1^4\rangle\langle M_2^2\rangle\langle M_4\rangle]$$

$$2\sigma\bar{M}_2\langle M_3\rangle^3\langle M_4\rangle \longrightarrow \xi M_1^2\bar{M}_2\langle M_2^2\rangle\langle M_3\rangle^3$$

$$2\sigma M_1M_2M_3\langle M_3\rangle^2\langle M_4\rangle \longrightarrow \xi M_1^2\bar{M}_2\langle M_1^4\rangle^2\langle M_2^2\rangle\langle M_3\rangle^2$$

$$2\sigma\langle M_1^4\rangle^3\langle M_2^2\rangle^2\langle M_3\rangle^2 \longrightarrow \xi M_1^2\bar{M}_2[\langle M_1^4\rangle^3\langle M_3\rangle^2+\langle M_1^4\rangle^2\langle M_2^2\rangle^3]$$

$$2\sigma M_1^2\langle M_1^4\rangle^2\langle M_2^2\rangle^3\langle M_3\rangle^2 \longrightarrow \xi M_1^2\bar{M}_2[\langle M_1^4\rangle^3\langle M_2^2\rangle^3+\langle M_1^4\rangle\langle M_2^2\rangle^2\langle M_3\rangle^2]$$

$$2\sigma\bar{M}_2\langle M_1^4\rangle^2\langle M_2^2\rangle^2\langle M_3\rangle^3 \longrightarrow \xi M_1^2\bar{M}_2[\langle M_1^4\rangle^3\langle M_2^2\rangle\langle M_3\rangle^2+\langle M_2^2\rangle^3\langle M_3\rangle^2]$$

$$2\sigma\langle M_1^4\rangle^3\langle M_2^2\rangle^2\langle M_4\rangle \longrightarrow \xi M_1^2\bar{M}_2[\langle M_1^4\rangle^2\langle M_2^2\rangle^2\langle M_3\rangle+\langle M_1^4\rangle^3\langle M_4\rangle]$$

$$2\sigma M_1^2\langle M_1^4\rangle^2\langle M_2^2\rangle^3\langle M_4\rangle \longrightarrow \xi M_1^2\bar{M}_2[\langle M_1^4\rangle^3\langle M_2^2\rangle^2\langle M_3\rangle+\langle M_1^4\rangle\langle M_2^2\rangle^2\langle M_4\rangle]$$

$$2\sigma\bar{M}_2\langle M_1^4\rangle^2\langle M_2^2\rangle^2\langle M_3\rangle\langle M_4\rangle \longrightarrow \xi M_1^2\bar{M}_2[\langle M_1^4\rangle^2\langle M_2^2\rangle^3\langle M_3\rangle+\langle M_1^4\rangle^3\langle M_3\rangle^3$$
$$+\langle M_2^2\rangle^3\langle M_4\rangle+\langle M_1^4\rangle\langle M_3\rangle^2\langle M_4\rangle]$$

$$2\sigma M_1M_2M_3\langle M_1^4\rangle^2\langle M_2^2\rangle^2\langle M_4\rangle \longrightarrow 0$$

$$2\sigma\langle M_1^4\rangle^3\langle M_3\rangle^2\langle M_4\rangle \longrightarrow \xi M_1^2\bar{M}_2[\langle M_3\rangle^3\langle M_1^4\rangle^2+\langle M_1^4\rangle^2\langle M_2^2\rangle\langle M_4\rangle]$$

$$2\sigma M_1^2\langle M_1^4\rangle^2\langle M_2^2\rangle\langle M_3\rangle^2\langle M_4\rangle \longrightarrow \xi M_1^2\bar{M}_2[\langle M_1^4\rangle^3\langle M_3\rangle^3+\langle M_1^4\rangle^3\langle M_2^2\rangle\langle M_4\rangle+\langle M_1^4\rangle\langle M_3\rangle^2\langle M_4\rangle]$$

$$2\sigma\bar{M}_2\langle M_1^4\rangle^2\langle M_3\rangle^3\langle M_4\rangle \longrightarrow \xi M_1^2\bar{M}_2[\langle M_2^2\rangle\langle M_3\rangle^3\langle M_1^4\rangle^2+\langle M_2^2\rangle\langle M_3\rangle^2\langle M_4\rangle]$$

$$2\sigma M_1M_2M_3\langle M_1^4\rangle^2\langle M_3\rangle^2\langle M_4\rangle \longrightarrow \xi M_1^2\bar{M}_2\langle M_1^4\rangle\langle M_2^2\rangle^3\langle M_3\rangle^2$$

$$2\sigma\langle M_1^4\rangle\langle M_2^2\rangle^2\langle M_3\rangle^2\langle M_4\rangle \longrightarrow \xi M_1^2\bar{M}_2[\langle M_2^2\rangle^2\langle M_3\rangle^3+\langle M_2^2\rangle^3\langle M_4\rangle+\langle M_1^4\rangle\langle M_3\rangle^2\langle M_4\rangle]$$

$$2\sigma M_1^2\langle M_2^2\rangle^3\langle M_3\rangle^2\langle M_4\rangle \longrightarrow \xi M_1^2\bar{M}_2[\langle M_1^4\rangle\langle M_2^2\rangle^2\langle M_3\rangle^3+\langle M_1^4\rangle\langle M_2^2\rangle^3\langle M_4\rangle]$$

$$2\sigma\bar{M}_2\langle M_2^2\rangle^2\langle M_3\rangle^3\langle M_4\rangle \longrightarrow \xi M_1^2\bar{M}_2[\langle M_2^2\rangle^3\langle M_3\rangle^3+\langle M_1^4\rangle\langle M_2^2\rangle\langle M_3\rangle^2\langle M_4\rangle]$$

$$2\sigma M_1M_2M_3\langle M_2^2\rangle^2\langle M_3\rangle^2\langle M_4\rangle \longrightarrow \xi M_1^2\bar{M}_2[\langle M_2^2\rangle^3\langle M_3\rangle\langle M_4\rangle+\langle M_1^4\rangle\langle M_3\rangle^3\langle M_4\rangle]$$

$$2\sigma\langle M_1^4\rangle^3\langle M_2^2\rangle^2\langle M_3\rangle^2\langle M_4\rangle \longrightarrow \xi M_1^2\bar{M}_2[\langle M_1^4\rangle^2\langle M_2^2\rangle^2\langle M_3\rangle^3+\langle M_1^4\rangle^2\langle M_2^2\rangle^3\langle M_4\rangle$$
$$+\langle M_1^4\rangle^3\langle M_3\rangle^2\langle M_4\rangle]$$

$$2\sigma M_1^2\langle M_1^4\rangle^2\langle M_2^2\rangle^3\langle M_3\rangle^2\langle M_4\rangle \longrightarrow \xi M_1^2\bar{M}_2[\langle M_1^4\rangle^3\langle M_2^2\rangle^2\langle M_3\rangle^3+\langle M_1^4\rangle^3\langle M_2^2\rangle^3\langle M_4\rangle$$
$$+\langle M_1^4\rangle\langle M_2^2\rangle^2\langle M_3\rangle^2\langle M_4\rangle]$$

$$2\sigma\bar{M}_2\langle M_1^4\rangle^2\langle M_2^2\rangle^2\langle M_3\rangle^3\langle M_4\rangle \longrightarrow \xi M_1^2\bar{M}_2[\langle M_1^4\rangle^2\langle M_2^2\rangle^3\langle M_3\rangle^3+\langle M_1^4\rangle^3\langle M_2^2\rangle\langle M_3\rangle^2\langle M_4\rangle$$
$$+\langle M_2^2\rangle^3\langle M_3\rangle^2\langle M_4\rangle]$$

$$2\sigma M_1M_2M_3\langle M_1^4\rangle^2\langle M_2^2\rangle^2\langle M_3\rangle^2\langle M_4\rangle \longrightarrow \xi M_1^2\bar{M}_2[\langle M_1^4\rangle^2\langle M_2^2\rangle^3\langle M_3\rangle\langle M_4\rangle+\langle M_1^4\rangle^3\langle M_3\rangle^3\langle M_4\rangle$$
$$+\langle M_2^2\rangle^2\langle M_3\rangle^3\langle M_4\rangle]$$

Figure 3.4.9: d^r Boundaries

To prove (b) we summarize the boundaries of the right hand column above in the table below. The proof is analogous to the proofs of the previous theorems.█

ξ	1	$\langle M_1^4\rangle$	$\langle M_2^2\rangle$	$\langle M_3\rangle$	$\langle M_1^4\rangle\langle M_2^2\rangle$	$\langle M_1^4\rangle\langle M_3\rangle$	$\langle M_2^2\rangle\langle M_3\rangle$	$\langle M_1^4\rangle\langle M_2^2\rangle\langle M_3\rangle$
1	*	*	*	*	*	*	*	
M_1^2	*	*	*	*	*		*	
$\overline{M}_2$	*	*	*	*	*	*	*	
$M_1^2\overline{M}_2$	*	‡	‡	‡	‡	‡	‡	

‡ Some of these elements bound and some do not bound. The bounding elements are marked by in the table below. An entry * indicates that the box bounds. In addition, the sum of all the boxes with the same letter as entry bound.

$\xi M_1^2\overline{M}_2$	$\langle M_1^4\rangle$	$\langle M_2^2\rangle$	$\langle M_3\rangle$	$\langle M_1^4\rangle\langle M_2^2\rangle$	$\langle M_1^4\rangle\langle M_3\rangle$	$\langle M_2^2\rangle\langle M_3\rangle$
1	*	*	*	*	*	*
$\langle M_1^4\rangle^2$	*	*	*	*	*	*
$\langle M_2^2\rangle^2$	*	A	B	*	*	E
$\langle M_3\rangle^2$	A		C	*	D, E	*
$\langle M_4\rangle$	B	C		D		
$\langle M_1^4\rangle^2\langle M_2^2\rangle^2$		F	G	I	J	N
$\langle M_1^4\rangle^2\langle M_3\rangle^2$	F	*	H	K	L, N	O
$\langle M_1^4\rangle^2\langle M_4\rangle$	G	H		L		
$\langle M_2^2\rangle^2\langle M_3\rangle^2$	I	K	M	*	P	Q
$\langle M_2^2\rangle^2\langle M_4\rangle$	J	M, N		P		R
$\langle M_3\rangle^2\langle M_4\rangle$	N, M, L	O		Q	R	
$\langle M_1^4\rangle^2\langle M_2^2\rangle^2\langle M_3\rangle^2$			S		T	U
$\langle M_1^4\rangle^2\langle M_2^2\rangle^2\langle M_4\rangle$		S		T		V
$\langle M_1^4\rangle^2\langle M_3\rangle^2\langle M_4\rangle$	S			U	V	
$\langle M_2^2\rangle^2\langle M_3\rangle^2\langle M_4\rangle$	T	U	V			
$\langle M_1^4\rangle^2\langle M_2^2\rangle^2\langle M_3\rangle^2\langle M_4\rangle$						

Figure 3.4.10: Summary of d^r- Boundaries

The previous theorem has left a remnant of a few elements on the "2σ row". We see in the next theorem that these elements map monomorphicly under d^s-differentials to elements of $Z_2\theta \otimes H_*BP$ in the s+7 row. We shall show in Chapter 6 that s = 24, and in Chapter 8 we will show that this element $\theta \in \pi^S_{30}$ is the (in)famous element θ_4 of Arf invariant one.

THEOREM 3.4.10 (a) There is an s > 6 and a nonzero element $\theta \in E^s_{24-s,s+6}$ of order two such that $d^s(2\sigma M_1^{12}) = \theta$.

(b) Image $[d^{24}: E^{24}_{*,7} \longrightarrow E^{24}_{*,7}] = Z_2\theta[1, \langle M_1^4\rangle^2, \langle M_2^2\rangle^2, \langle M_3\rangle^2]$

$$\otimes Z[\langle M_1^4\rangle^4, \langle M_2^2\rangle^4, \langle M_3\rangle^4, \langle M_4\rangle^2, \{M_5\}', \ldots, \{M_N\}', \ldots].$$

(c) $E^{s+2}_{*,7} = 0$.

PROOF. By Theorem 3.4.9(b), $E^{r+2}_{24,7} = Z_2(2\sigma M_1^{12})$. Thus, there must be a nonzero d^s-differential originating on $2\sigma M_1^{12}$ hitting an element θ of order 2. Part (b) is a consequence of Theorem 3.4.9(a) and the computations in Figure 3.4.11. By Theorem 3.4.9(a), d^s is an isomorphism from $E^s_{*,7}$ to the elements listed in (b), and $E^{s+2}_{*,7} = 0$.∎

$E^s_{*,7}$		$E^s_{*,s+6}$
$2\sigma\langle M_1^4\rangle^3$	$\longrightarrow$	θ
$2\sigma M_1^2\langle M_2^2\rangle^3$	$\longrightarrow$	$\theta\langle M_1^4\rangle^2$
$2\sigma\bar{M}_2\langle M_3\rangle^3$	$\longrightarrow$	$\theta\langle M_2^2\rangle^2$
$2\sigma M_1M_2M_3\langle M_4\rangle$	$\longrightarrow$	$\theta\langle M_3\rangle^2$
$2\sigma\bar{M}_2\langle M_3\rangle\langle M_1^4\rangle^2\langle M_3\rangle^2$	$\longrightarrow$	$\theta\langle M_1^4\rangle^2\langle M_2^2\rangle^2$
$2\sigma M_1M_2M_3\langle M_1^4\rangle\langle M_4\rangle$	$\longrightarrow$	$\theta\langle M_1^4\rangle^2\langle M_3\rangle^2$
$2\sigma M_1M_2M_3\langle M_2^2\rangle\langle M_4\rangle$	$\longrightarrow$	$\theta\langle M_2^2\rangle^2\langle M_3\rangle^2$
$2\sigma M_1M_2M_3\langle M_1^4\rangle\langle M_2^2\rangle\langle M_4\rangle$	$\longrightarrow$	$\theta\langle M_1^4\rangle^2\langle M_2^2\rangle^2\langle M_3\rangle^2$

Figure 3.4.11: Summary of d^s-Boundaries

We conclude with a complete analysis of the 8 row. The element λ mentioned in the theorem will be shown to be $\eta A[14]M_1$ in Chapter 5.

THEOREM 3.4.11 (a) $\pi_8^S = Z_2A[8] \oplus Z_2\eta\sigma$.

(b) $\eta A[8]$ and $\eta^2\sigma$ are both nonzero.

(c) The only differentials which land on the 8 row are determined by the leading differentials $d^2(\sigma M_1) = \eta\sigma$ and $d^{12}(2\nu M_1^3) = A[8]$.

(d) The only differentials which originate from the 8 row are determined by the leading differentials $d^2(\eta\sigma M_1) = \eta^2\sigma$, $d^2(A[8]M_1) = \eta A[8]$ and $d^t(A[8]M_1^2\bar{M}_2) = \lambda$.

PROOF. (a), (b), (c) $\pi_5^S = 0$ and $E^4_{*,1} = 0$. Thus the only nonzero differentials which can land in $E^r_{0,8}$ originate from the 7 row and the 3 row. These differentials were analyzed in Theorems 3.3.13 and 3.4.1. It was shown in Theorems 3.3.7 and 3.3.15 that $2A[8] = 0$ and $\eta A[8] \neq 0$. Note that $\eta\sigma M_1$ is not a d^2-boundary, $\eta\sigma M_1$ is not a d^4-boundary because $\pi_5^S = 0$, $\eta\sigma M_1$ is not a d^6-boundary because $E^6_{8,3} = 0$ and $\eta\sigma M_1$ is not a d^8-boundary because $E^4_{10,1} = 0$. Thus, $\eta^2\sigma \neq 0$.

(d) $E^4_{*,8} = Z_2A[8][M_1^2, \bar{M}_2, \ldots, \bar{M}_N, \ldots]$ and by Theorem 3.3.13(b) it follows that $E^8_{*,8} = Z_2(A[8]M_1^2\bar{M}_2) \otimes B\langle 4\rangle$. Thus there must be a nonzero differential $d^t(A[8]M_1^2\bar{M}_2) = \lambda$. Clearly d^t is a monomorphism with image $Z_2\lambda \otimes B\langle 4\rangle$. Thus, $E^{t+2}_{*,8} = 0$. ∎

The following Toda bracket is not easily seen from the Atiyah-Hirzebruch spectral sequence. We therefore record it as a corollary.

COROLLARY 3.4.12 $A[8] = \langle\nu, \eta, \nu\rangle$

PROOF. Multiplication by η is a monomorphism on π_8^S. Now $\eta\langle\nu, \eta, \nu\rangle = \langle\eta, \nu, \eta\rangle\nu = \nu^2\cdot\nu = \eta A[8]$. Thus $\langle\nu, \eta, \nu\rangle = A[8]$. ∎

We conclude by summarizing the notation we introduced in this chapter. This notation will be used throughout the remainder of this work.

DEFINITION 3.4.13 (a) We have the following d^2-cycles:

$$\bar{M}_2 = 3M_2 - M_1^3$$

$$\bar{M}_N = M_N - M_1 M_{N-1}^2 \quad \text{for } N \geq 3.$$

(b) We have the following d^4-cycles:

$$\langle M_1^4 \rangle = M_1^4 + 2M_1 M_2$$

$$\langle M_2^2 \rangle = M_2^2 + M_1^6$$

$$\langle M_3 \rangle = \bar{M}_3 + 5M_1^4 \bar{M}_2 + 14M_1^7$$

$\langle M_4 \rangle$ = a polynomial generator in $H_{30}BP$.

(c) We have the following d^8-cycles:

$\{M_N\}$, $N \geq 5$, which are polynomial generators of H_*BP.

(d) We have the following subalgebras of H_*BP:

$B\langle 2\rangle = Z_{(2)}[M_1^2, \bar{M}_2, \ldots, \bar{M}_N, \ldots]$ is a subalgebra of $E^4_{*,0}$;

$B\langle 4\rangle = Z_{(2)}[\langle M_1^4\rangle, \langle M_2^2\rangle, \langle M_3\rangle, \langle M_4\rangle, \ldots, \langle M_N\rangle, \ldots]$ is a subalgebra of $E^8_{*,0}$;

$B\langle 8\rangle = Z_{(2)}[\langle M_1^4\rangle^2, \langle M_2^2\rangle^2, \langle M_3\rangle^2, \langle M_4\rangle, \{M_5\}, \ldots, \{M_N\}, \ldots]$ is a subalgebra of $E^8_{*,0}$ such that $d^8(B\langle 8\rangle) \subset Z_8(2\sigma) \otimes H_*BP$.

CHAPTER 4: THE IMAGE OF J

1. Introduction

The study of the canonical map $J: O \longrightarrow F$ inducing $J_*: \pi_* O \longrightarrow \pi_*^S$, called the J-homomorphism, was initiated by Adams in a series of four papers [3], [4], [5], [6]. He determined the image of J_* modulo the "Adams conjecture", and the determination was completed with the solution of this conjecture by Quillen [53]. Mahowald [34] has constructed a spectrum J and a map of spectra $j: S \longrightarrow J$ such that the induced map $j_*: \pi_*^S \longrightarrow \pi_* J$ is a split epimorphism. Moreover, j_* is an isomorphsim from ImJ to $\pi_* J$ where ImJ equals Image J_* direct sum with an additional family of Z_2s.

In Section 2, we collect the known facts about $\pi_* J$ which we will use to simplify our computation. In addition, we use the Adams spectral sequence to derive several relations and Toda brackets involving elements in ImJ. The relevance of ImJ to our computation is that all differentials in our spectral sequence which originate on the 0 row land in ImJ $\otimes$ $H_* BP$. In Section 3, we prove that this theorem is true through degree 66 which suffices for the computation of the first 64 stable stems. In Section 4, we give the computer printout of the computation of the cokernels of these differentials through degree 70. This computation is one of the essential ingredients in our inductive determination of the first 64 stable stems. The computer program itself is discussed in Appendix 5.

2. ImJ and the Adams Spectral Sequence

Mahowald [34] defined the spectrum J as the fiber of a map $bo \longrightarrow \Sigma^4 bsp$. The homotopy of J is periodic with period eight and is given by the table in Figure 4.2.1.

DEGREE	$\pi_N J$
1	$Z_2\alpha_0$
8N+1 (N ≥ 1)	$Z_2\alpha_N \oplus Z_2\eta^2\gamma_{N-1}$
8N+2	$Z_2\eta\cdot\alpha_N$
8N+3	$Z_8\beta_N,\ 4\beta_N = \eta^2\alpha_N$
8N+4	0
8N+5	0
8N+6	0
8N+7	$Z_{C(N)}\gamma_N$
8N+8	$Z_2\eta\cdot\gamma_N$

FIGURE 4.2.1: The Homotopy of J

In the above table $C(N) = 2^{\mathscr{C}(N)}$ denotes the largest power of two which divides 16N+16. In each row of this table we take N ≥ 0 except in the second row where N ≥ 1. These elements of π_*J include the elements of Hopf invariant one: $\eta = \alpha_0$, $\nu = \beta_0$ and $\sigma = \gamma_0$. We apologize for the new notation, but this notation is very convenient. The following theorem describes the how π_*J is related to Image J_*, and how $j:S \longrightarrow J$ induces a split epimorphism j_*.

THEOREM 4.2.1 (a) The map $j:S \longrightarrow J$ induces an isomorphsim j_* from Image $[J_*:\pi_*O \longrightarrow \pi_*^S]$ to the subgroup of π_*J generated by $\{\alpha_N, \eta\alpha_N, \beta_N, \gamma_N, \eta\gamma_N \mid N \geq 0\}$.

(b) Let $\bar{\gamma}_N = J_*[(J_*\circ j_*)^{-1}(\gamma_N)]$. Then the map $j:S \longrightarrow J$ induces an isomorphism between π_*J and

$$\mathrm{ImJ} = \mathrm{Image}\ [J_*:\pi_*O \longrightarrow \pi_*^S] \oplus Z_2\{\eta^2\bar{\gamma}_N \mid N \geq 0\}.$$

(c) ImJ is a direct summand of π_*^S.

Let $\bar{\alpha}_N = J_*[(J_*\circ j_*)^{-1}(\alpha_N)]$ and let $\bar{\beta}_N = J_*[(J_*\circ j_*)^{-1}(\beta_N)]$. From now on we will abuse our notation by denoting $\bar{\alpha}_N$ by α_N, $\bar{\beta}_N$ by β_N and $\bar{\gamma}_N$ by γ_N. Thus, the table in Figure 4.2.1 can be thought of as giving a description of ImJ. Define CokJ as Kernel j_*. Then we have the direct sum decomposition:

$$\pi_*^S = \mathrm{ImJ}_* \oplus \mathrm{CokJ}_* \qquad [4.2.1]$$

Consider the classical mod two Admas spectral sequence [1]:

$$E_2^{s,t} = \mathrm{Ext}_{\mathfrak{A}}^{t}(Z_2, Z_2)_s \Longrightarrow {}_2\pi_s^S \qquad [4.2.2]$$

Here $\mathfrak{A}$ denotes the mod two Steendrod algebra and ${}_2\pi_*^S$ denotes π_*^S modulo the subgroup of all elements of odd degree. This spectral sequence has a vanishing line [8], and directly below the vanishing line are the elements depicted in Figure 4.2.2 which survive to ImJ as well as several "self-destructing families", i.e. elements which are nonzero in E_2 but only zero survives from them to E_S. Below these families is the self-destructing wedge [38]. In this fiugre, vertical lines denote multiplication by h_0, lines of positive slope denote multiplication by h_1 and lines of negative slope -r denote d^r-differentials.

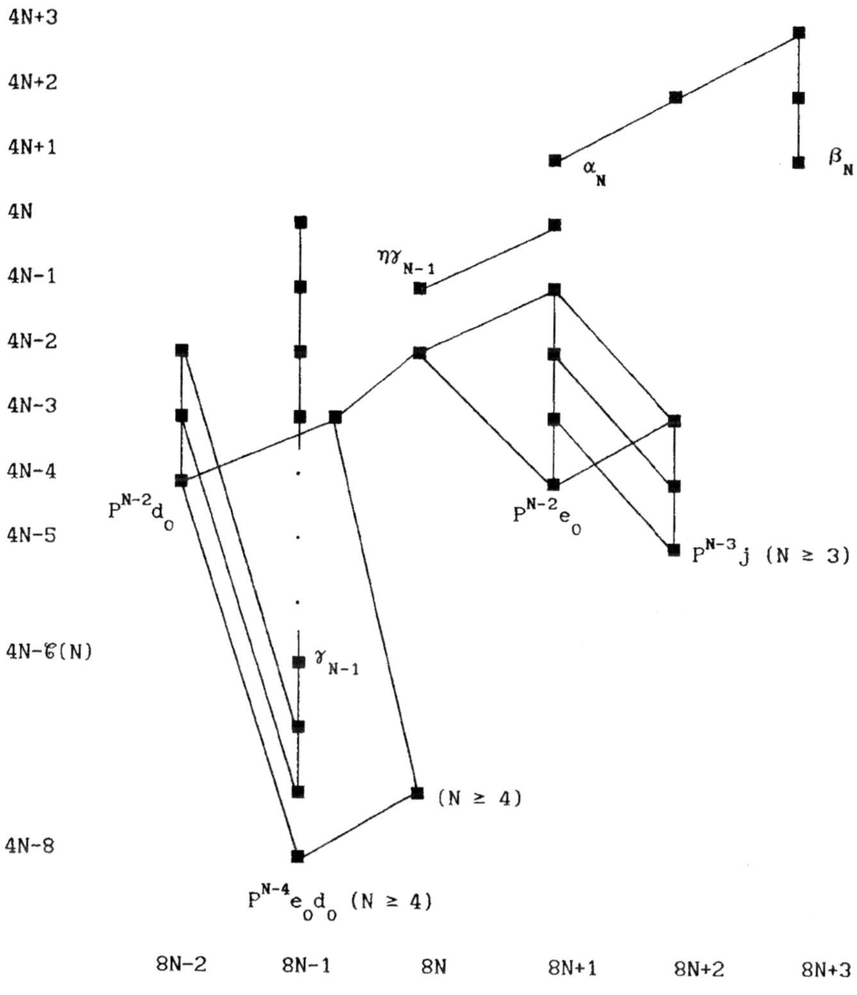

In addition, $h_2\alpha_N = 0$, $h_2\beta_{2N+1} = 0$, $h_2\beta_{2N} = h_0^2P^{N-2}d_0$ and $h_2\gamma_N = 0$.

FIGURE 4.2.2: ImJ in the Adams Spectral Sequence (N ≥ 2)

The following theorem gives the structure of ImJ as a module over the subring of π_*^S generated by η, ν and σ.

THEOREM 4.2.2 (a) $\nu\alpha_N = 0$;

(b) $\sigma\alpha_N = k\eta\gamma_N$;

(c) $\eta\beta_N = 0$;

(d) $\nu\beta_N = 0$ for $N \geq 1$;

(e) $\sigma\beta_N = 0$;

(f) $\nu\gamma_N = 0$;

(g) $\sigma\gamma_N = 0$ for $N \geq 1$.

PROOF. (a), (c), (d), (g) There are no possible nonzero elements of E_2 of the Adams spectral sequence to represent any of these products with the exception of $\nu\beta_{2N}$ which can only be the boundary $h_0^2P^{2N-1}d_0$.

(b) From Figure 4.2.2, we see that the only possibility for $\sigma\alpha_N$ to be nonzero is if it equals $\eta\gamma_N$. Thus, $\sigma\alpha_N = k\eta\gamma_N$.

(e) From Figure 4.2.2, we see that the only possibility for $\sigma\beta_N$ to be nonzero is if it equals $\eta\alpha_{N+1}$. However, $\eta\sigma\beta_N = 0$ while $\eta^2\alpha_{N+1} \neq 0$.

(f) From Figure 4.2.2, we see that the only possibility for $\nu\gamma_N$ to be nonzero is if it equals $\eta\alpha_{N+1}$. However, $\eta\nu\gamma_N = 0$ while $\eta^2\alpha_{N+1} \neq 0$.∎

We will use the following theorem to identify certain Toda brackets in the Adams spectral sequence (*).

THEOREM 4.2.3 (a) Let A', B', C' be an element of E_2 of the Admas spectral sequence (*) which converges to A, B, C, respectively. Assume that $\langle A,B,C\rangle$ is defined in π_*^S and $\langle A',B',C'\rangle$ is defined in E_r where $r \geq 2$. Let (p,q), (s,t) be the bidegree of the product A'B', B'C', respectively. Assume that the only d_u-boundary in $E_{N+r}^{p,q+N}$ and $E_{N+r-1}^{s,t+N}$, $N \geq 1$, $u \geq N + r$, is zero. Then there is an element X of $\langle A',B',C'\rangle$ in E_r of (*) which is an infinite cycle and represents an element of $\langle A,B,C\rangle$.

(b) Let A', B', C', D' be an element of E_2 of the Adams spectral sequence (*) which converges to A, B, C, D, respectively. Assume that $\langle A,B,C,D\rangle$ is defined in π_*^S, and $\langle A',B',C',D'\rangle$ is defined in E_r where $r \geq 2$. Let (p_1,q_1),

(p_2,q_2), (p_3,q_3), (p_4,q_4), (p_5,q_5) be the bidegree of A'B', B'C', C'D', <A',B',C'>, <B',C',D'>, respectively. Asume that

(i) $E_{N+r}^{p_i,q_i+N} = 0$, $N \geq 1$, if $4 \leq i \leq 5$ and

(ii) the only d_u-boundary in $E_{N+r}^{p_i,q_i+N}$, $N \geq 1$, $u \geq N + r$, is zero if $1 \leq i \leq 3$.

Then there is an element X of <A',B',C',D'> in E_r of (*) which is an infinite cycle and represents an element of <A,B,C,D>.

The first part of this theorem is due to Moss [47]. (Our technical hypothesis is equivalent to his, i.e. there are no "crossing differentials".) The second part of this theorem is a variation of Lawrence's generalization [33] of Moss's theorem. However, we do not assume that <A,B,C,D> is strictly defined, and we replace two of the five cases of the technical hypothesis of no crossing differentials by the more stringent hypothesis that the target groups of such differentials are zero. Part (b) of this theorem can be proved either by the methods of Moss and Lawrence or by using bordism chains as in [28].

Observe that the conclusion in (a) or (b) does not preclude the possibility that X is zero in $E_\infty^{a,b}$ and is the projection into E_∞ of a nonzero element ξ of the Toda bracket which lies in $F^{b+1}\pi_a^S$. However, if $E_\infty^{a,c} = 0$ for all $c > b$ then X = 0 implies that 0 is an elements of the Toda bracket. This explains why we require the stronger hypothesis in (b) when i equals 4 or 5.

THEOREM 4.2.4 (a) $\langle\eta,\nu,\alpha_N\rangle$ contains 0 for $N \geq 1$;

(b) $\langle\eta,\nu,\beta_N\rangle$ contains 0 for $N \geq 1$;

(c) $\langle\nu,\eta,\beta_N\rangle$ contains 0 for $N \geq 1$;

(d) $\langle\eta,\nu,\gamma_N\rangle$ contains 0;

(e) $\langle\nu,\eta,\eta^2\gamma_N\rangle$ contains 0;

(f) $\langle\nu,\eta,\beta_N,\eta\rangle$ contains 0 when N is odd, 2, 4, or 6;

(g) $\langle\nu,\eta,\eta^2\gamma_N,\eta\rangle$ contains 0;

(h) $\langle\eta,\nu,\alpha_N,\nu\rangle$ contains 0 for $N \geq 1$;

(i) $\langle\eta,\nu,\beta_N,\nu\rangle$ contains 0 for $N \geq 1$;

(j) $\langle\eta,\nu,\gamma_N,\nu\rangle$ contains 0 for $N \geq 1$;

(k) $\langle\eta,\beta_N,\sigma\rangle$ contains 0;

(l) $\langle\eta,\eta^2\gamma_N,\sigma\rangle$ contains 0 for $N \geq 1$.

PROOF. (a), (c), (d), (e), (k), (l) From Figure 4.2.2, we see that Theorem 4.2.3(a) with $r = 2$ applies to show that each of these Toda brackets has an element which projects to zero in $E_\infty^{8N+6,4N+2}$, $E_\infty^{8N+8,4N+2}$, $E_\infty^{8N+12,4N-\mathscr{E}(N)+5}$, $E_\infty^{8N+14,4N+5}$, $E_\infty^{8N+12,4N+2}$, $E_\infty^{8N+18,4N+5}$ in case (a), (c), (d), (e), (k), (l), respectively. With two exceptions, E_∞ is zero in each of these degrees in higher filtration degrees. The first exception is that possibly $\eta\gamma_{N+1}$ might be an element of $\langle\nu,\eta,\beta_N\rangle$. However, $\eta^2\gamma_{N+1} \neq 0$ and $\eta\langle\nu,\eta,\beta_{N+1}\rangle = \langle\eta,\nu,\eta\rangle\beta_{N+1} = \nu^2\beta_{N+1} = 0$. It follows that each of these Toda brackets contains 0. The second exception is that $\eta\alpha_{N+2}$ might be an element of $\langle\eta,\eta^2\gamma_N,\sigma\rangle$. However, $\eta\alpha_{N+2}$ is in the indeterminacy of $\langle\eta,\eta^2\gamma_N,\sigma\rangle$. Thus, $0 \in \langle\eta,\eta^2\gamma_N,\sigma\rangle$ in all cases.

(b) Since $h_2\beta_{2N+1} = 0$ in E_2, the preceeding argument applies to $\langle\eta,\nu,\beta_{2N+1}\rangle$ which is defined in E_2. It shows that $0 \in \langle\eta,\nu,\beta_{2N+1}\rangle$. However, $h_2\beta_{2N} = h_0^2P^{N-2}d_0$ which is a d^3-boundary. Now the analogue of the preceeding argument with $r = 3$ applies to $\langle\eta,\nu,\beta_{2N}\rangle$ to show that it contains zero.

(f), (g), (h), (i), (j) Note that by Theorem 2.2.7(g) all of the four-fold Toda brackets in this theorem are defined: In (h), $\eta\langle\nu,\alpha_N,\nu\rangle = \langle\eta,\nu,\alpha_N\rangle\nu$ which contains 0. Since $\eta\langle\nu,\alpha_N,\nu\rangle$ is a singleton, it must equal $\{0\}$. Thus, $\eta\gamma_N$ can not be an element of $\langle\nu,\alpha_N,\nu\rangle$ because $\eta^2\gamma_N \neq 0$. From Figure 4.2.2, we see that there is now no possibility for $\langle\nu,\alpha_N,\nu\rangle$ to contain a nonzero element. In (i), $\eta\langle\nu,\beta_N,\nu\rangle = \langle\eta,\nu,\beta_N\rangle\nu = \{0\}$ while $\eta(\eta\alpha_{N+1}) \neq 0$. Hence $\eta\alpha_{N+1}$ can not be an element of $\langle\nu,\beta_N,\nu\rangle$. Thus, we see from Figure 4.2.2 that there is now no possibility for $<\nu,\beta_N,\nu>$ to contain a nonzero element. All the other triple products in (f) - (j) contain zero by (a) - (e) or must equal zero by Theorem 4.2.3(a) and Figure 4.2.2.

Assume that N is odd in case (f). Let r = 2 in cases (f), (g), (j) and in case (i) when N is odd. Let r = 3 in case (h) and in case (i) when N is even. From Figure 4.2.2, we see that Theorem 4.2.3(b) applies to show that each of these Toda brackets has an element which projects to zero in $E_\infty^{8N+10,4N+2}$, $E_\infty^{8N+16,4N+1}$, $E_\infty^{8N+10,4N}$, $E_\infty^{8N+12,4N+2}$, $E_\infty^{8N+12,4N}$, $E_\infty^{8N+16,4N-\mathcal{C}(N)+1}$ in case (f), (g), (h), (i) with N odd, (i) with N even, (j), respectively. With four exceptions, E_∞ is zero in each of these degrees in higher filtration degrees: $\eta\alpha_{N+1}$ could be in $<\nu,\eta,\beta_N,\eta>$, $\eta\gamma_{N+1}$ could be in $<\nu,\eta,\eta^2\gamma_N,\eta>$, $\eta\alpha_{N+1}$ could be in $<\eta,\nu,\alpha_N,\nu>$ and $\eta\gamma_{N+1}$ could be in $<\eta,\nu,\gamma_N,\nu>$. However, $\eta\alpha_{N+1}$, $\eta\gamma_{N+1}$, $\eta\alpha_{N+1}$, $\eta\gamma_{N+1}$ is in the indeterminacy of $<\nu,\eta,\beta_N,\eta>$, $<\nu,\eta,\eta^2\gamma_N,\eta>$, $<\eta,\nu,\alpha_N,\nu>$, $<\eta,\nu,\gamma_N,\nu>$, respectively. Thus, each of these Toda brackets contains 0.

Now consider (f) when N is 2, 4 or 6. We shall see that there are only two elements of π_{26}^S, π_{42}^S and π_{58}^S that are not contained in (η,ν), and (η,ν) is contained in the indetermincacy of $<\nu,\eta,\beta_{2N},\eta>$. The two exceptions are C[42] and 2C[42]. Thus, $<\nu,\eta,\beta_2,\eta>$ and $<\nu,\eta,\beta_6,\eta>$ contain 0. Now $2<\nu,\eta,\beta_4,\eta> \subset <\nu,\eta,<\beta_4,\eta,2>> = <\nu,\eta,0>$ because $<\beta_4,\eta,2> \in \pi_{37}^S$, $\eta<\beta_4,\eta,2> = \beta_4<\eta,2,\eta> = 2\nu\beta_4 = 0$, $\nu<\beta_4,\eta,2> = 2<\nu,\beta_4,\eta>$ and only 0 in π_{37}^S has these properties. Thus, $2<\nu,\eta,\beta_4,\eta> = (\nu)$ which does not contain 2C[42] or 4C[42]. Therefore, $<\nu,\eta,\beta_4,\eta>$ contains 0.∎

The reader should not worry that we may get involved in circular reasoning when we use the facts that $<\eta,\nu,\beta_2,\nu>$, $<\eta,\nu,\beta_4,\nu>$ and $<\eta,\nu,\beta_6,\nu>$ contain 0. We will only use these facts to show that no leader XV can be of one of the following forms:

(i) $X \in \mathrm{CokJ}_{17}$, $V \in H_{12}BP$ and XV is in the image of d^{18};

(ii) $X \in \mathrm{CokJ}_{33}$, $V \in H_{12}BP$ and XV is in the image of d^{34};

(iii) $X \in \mathrm{CokJ}_{49}$, $V \in H_{12}BP$ and XV is in the image of d^{50}.

The only leader with such a bidegree is $\nu A[30]M_1^6$. Since twice $d^{12}(\nu A[30]M_1^6)$ equals $\sigma^2 A[30] = 4C[44] \neq 0$, $\nu A[30]M_1^6$ can not bound. Therefore, we have an alternate proof that there is no leader XV as in (i), (ii) or (iii).

3. Differentials Originating on the 0 Row - Theory

The material in this section is divided into two parts. First, we study the map of Atiyah-Hirzebruch spectral sequences induced by j. We deduce that j induces an isomorphism j_∞ on $E^\infty_{*,0}$. Second, we prove that j_r induces an isomorphism on $E^r_{N,0}$ for $r \geq 1$ and $N \leq 66$. That is, all differentials on the 0 row of our spectral sequence land in $\mathrm{ImJ} \otimes H_*BP$. Consider the Atiyah-Hirzebruch spectral sequence:

$$'E^2_{N,t} = H_NBP \otimes \pi_tJ \Longrightarrow J_{N+t}BP \qquad [4.3.1]$$

The map of spectra $j:S \longrightarrow J$ induces a map of spectral sequences

$$j_r:E^r_{N,t} \longrightarrow {}'E^r_{N,t},\quad 2 \leq r \leq \infty. \qquad [4.3.2]$$

Moreover, $j_2:H_*BP \otimes \pi^S_* \longrightarrow H_*BP \otimes \pi_*J$ equals $1 \otimes j_*$. Since H_NBP is zero when N is odd, we have the following result:

$$E^r_{N,t} = 0 \text{ if } N \text{ is odd,}$$
$$d^r = 0 \text{ if } r \text{ is odd and}$$
$$E^{2r+1} = E^{2r+2} \text{ for } r \geq 1.$$

The following simple theorem is the basis for many of the results of this section.

THEOREM 4.3.3 (a) If $j_2 \circ d^2(X) \neq 0$ modulo CokJ then X is an element of $Z_{(2)}\{M_1, \alpha_NM_1, \eta\alpha_NM_1, \gamma_NM_1, \eta\gamma_NM_1\} \otimes B\langle 2\rangle$ which reduces to a nonzero element modulo (2).

(b) If $j_{2r} \circ d^{2r}(X) \neq 0$ and $r \geq 2$ then $X \in E^{2r}_{*,0}$.

PROOF. (a) By Theorem 3.2.3, $E^2 = [Z_2 \oplus Z_2M_1] \otimes \pi_*^S \otimes B\langle 2\rangle$ and d^2 is the homomorphsim of $\pi_*^S \otimes B\langle 2\rangle$ - modules given by $d^2(1) = 0$ and $d^2(M_1) = \eta$. Hence if $d^2(X) \neq 0$ then $X = \xi M_1 B$ where $\xi \in \pi_*^S$, $B \in B\langle 2\rangle$ and $\eta\xi \neq 0$. Thus, if $j^2 \circ d^2(X) \neq 0$ then $\eta\xi$ is a nonzero element of ImJ. This part of the theorem now follows from the table in Figure 4.2.1.

(b) By Theorem 3.2.3:

$$'E^4 = [Z_2\alpha_K \oplus Z_4\beta_N \oplus Z_8\beta_N M_1 \oplus Z_{C(N)}\gamma_N \oplus Z_{C(N)/2}\gamma_N M_1 \oplus Z_2\eta^2\gamma_N M_1 \mid K \geq 1,\ N \geq 0] \otimes B\langle 2\rangle.$$

Thus, $'E^{2r}_{2N,2t} = 0$ for all N, $r \geq 2$ and $t > 0$. Therefore for all $r \geq 4$, $d^{2r}: {'E}^{2r}_{2N,2t} \longrightarrow {'E}^{2r}_{2N-2r,2t+2r-1}$ is zero if $t \neq 0$. That is, for $r \geq 4$ the only nonzero differentials in $'E^{2r}$ originate on the 0 row. Thus, if $0 \neq j^{2r} \circ d^{2r}(X) = d^{2r} \circ j^{2r}(X)$ then $j^{2r}(X)$ and hence X must be on the 0 row.∎

We deduce that all the elements of $\mathrm{ImJ}/(\eta \cdot \mathrm{ImJ})$ are hit by nonzero transgressions originating on the 0 row.

COROLLARY 4.3.4 All of the α_N, β_N, γ_N, $\beta_N M_1$ and $\eta^2\gamma_N M_1$ are hit by differentials which originate on the 0 row.

PROOF. All of the elements listed in this corollary are nonzero in E^4 and map to a nonzero element of $'E^4$. Of course, they are zero in E^∞ and thus must be hit by differentials. By Theorem 4.3.3(b), they can only be hit by differentials which originate on the 0 row.∎

The next theorem specifies bounds on when the nonzero differentials must occur on the 0 row to turn $E^2_{*,0} = H_*BP$ into $E^\infty_{*,0} = \pi_*BP$. Let $U_N = V_N/2$, $N \geq 1$, be polynomial generators of H_*BP.

THEOREM 4.3.5 Let F_t denote the $Z_{(2)}$-submodule of H_*BP generated by all $U_{N_1}\cdots U_{N_q}$ with $q < t$.

(a) No element of $2^{4t}U_{N_1}\cdots U_{N_{4t+1}} + F_{4t+1}$ survives to $E^{8t+4}_{*,0}$.

(b) No element of $2^{4t+1}U_{N_1}\cdots U_{N_{4t+2}} + F_{4t+2}$ or of $2^{4t+2}U_{N_1}\cdots U_{N_{4t+3}} + F_{4t+3}$ survives to $E^{8t+6}_{*,0}$.

(c) No element of $2^{4t-1}U_{N_1}\cdots U_{N_{4t}} + F_{4t}$ survives to $E^{8t+2}_{*,0}$.

PROOF. We use induction on t to prove that there is some $X_t \in F_t$ such that:

(i) no element of $2^{t-1}U_{N_1}\cdots U_{N_t} + F_t$ survives to the $E^r_{*,0}$ specified above;

(ii) $d^{2t}(2^{t-1}U_1^t + X_t) = \alpha_k$ if $t = 4k+1$;

(iii) $d^{2t}(2^{t-1}U_1^t + X_t) = 2\beta_k$ if $t = 4k+2$;

(iv) $d^{2t-2}(2^{t-1}U_1^t + X_t) = 4\beta_k M_1$ if $t = 4k+3$;

(v) $d^{2t}(2^{t-1}U_1^t + X_t) = (C(k)/2)\gamma_k$ if $t=4k+4$.

When $t = 1$, we know that $d^2(U_1) = \alpha_1$ and that no U_N survives to $E^4_{*,0}$. Assume that the above five conditions are true for $t < T$. Let $\varepsilon = 2$ if $T \equiv 3 \bmod 4$ and let $\varepsilon = 0$ otherwise. Note that $2^qU_{N_1}\cdots U_{N_q} = V_{N_1}\cdots V_{N_q} \in \pi_*BP$ and is thus an infinite cycle. By the induction hypothesis, $E^{2T-\varepsilon}_{2T,0} \equiv Z_{(2)}(2^kU_1^T + X)$ modulo $\pi_{2T}BP$ for some $X \in F_T$ and some $k \le T$. It follows from Corollary 4.3.4 that $k \le T-1$ and $d^{2T-\varepsilon}(2^{T-1}U_1^T + 2^{T-k-1}X)$ is α_k, $2\beta_k$, $4\beta_kM_1$, $(C(k)/2)\gamma_k$ if T is 4k+1, 4k+2, 4k+3, 4k+4, respectively. This proves whichever of conditions (ii) - (v) is relevant. Now assume that $Y = 2^{T-1}U_{N_1}\cdots U_{N_T} + f$ for some $f \in F_T$ survives to $E_{*,0}$. Let $I = 2\Delta_{N_1-1} + \cdots + 2\Delta_{N_T-1}$. Then $N_1+\cdots+N_T > T$ and degree $r_I > 0$. Observe that $U_N \equiv M_N$ modulo 2. Thus in $E^{2T-\varepsilon}_{*,0}$, $r_I(Y) = 2^{T-1}U_1^T + f'$ for some $f' \in F_T + (2^T)$. Thus $r_I \circ d^{2T-\varepsilon}(Y) \neq 0$ and Y does not survive to $E^{2T-\varepsilon+2}_{*,0}$, a contradiction. This verifies condition (i) and completes the proof of the induction step.∎

We can now specify the elements on the 0 row which transgress to hit the elements of $\mathrm{ImJ}/(\eta\cdot\mathrm{ImJ})$.

COROLLARY 4.3.6 (a) For all $t \geq 0$, there is $A_t \in F_{4t+1}$ such that $2^{4t}M_1^{4t+1} + A_t$ survives to $E^{8t+2}_{4t+2,0}$ and $d^{8t+2}(2^{4t}M_1^{4t+1} + A_t) = \alpha_t$.

(b) For all $t \geq 0$, there are B_t and C_t in F_{4t+2} such that $2^{4t}M_1^{4t+2} + B_t$ and $2^{4t}M_1^{4t+3} + C_t$ survive to $E^{8t+4}_{*,0}$, $d^{8t+4}(2^{4t}M_1^{4t+2} + B_t) = \beta_t$ and $d^{8t+4}(2^{4t}M_1^{4t+3} + C_t) = 3\beta_t M_1$.

(c) For all $t \geq 1$, there is $D_t \in F_{4t}$ such that $(2^{4t}/C(t))M_1^{4t} + D_t$ survives to $E^{8t}_{8t,0}$ and $d^{8t}((2^{4t}/C(t))M_1^{4t} + D_t) = \gamma_{t-1}$.

(d) For all $t \geq 1$, there is $G_t \in F_{4t+2}$ such that $2^{4t-1}M_1^{4t+2} + G_t$ survives to $E^{8t+2}_{8t+4,0}$ and $d^{8t+4}(2^{4t-1}M_1^{4t+2} + G_t) = \eta^2\gamma_{t-1}M_1$.

PROOF. Parts (a), (b) and (c) follow from the proof of Theorem 4.3.5. Note that we can choose C_t in F_{4t+2} because $F_{4t+2} \cap E^{8t+4}_{8t+6,0} = F_{4t+3} \cap E^{8t+4}_{8t+6,0}$. Also observe that $r_{\Delta_1}(M_1^{4t+3}) = (4t+3)M_1^{4t+2}$ and therefore $d^{8t+4}(2^{4t}M_1^{4t+3} + C_t)$ is either $3\beta_t M_1$ or $7\beta_t M_1$. Define β_t so that $d^{8t+4}(2^{4t}M_1^{4t+3} + C_t) = 3\beta_t$. By Corollary 4.3.4, $\eta^2\gamma_{t-1}M_1$ must bound from the 0 row. Thus the reasoning used to prove Theorem 4.3.5 also applies to prove (d).∎

As a consequence of our theorem we have another proof of the famous theorem of Adams [2] of the nonexistence of elements of Hopf invariant one in degrees 2^k-1 for $k \geq 4$.

COROLLARY 4.3.7 If $\xi \in \pi^S_{2^N-1}$ has Hopf invariant one then either $N = 0$, $\xi = 2$ or $N = 1$, $\xi = \eta$ or $N = 2$, $\xi = \nu$ or $N = 3$, $\xi = \sigma$.

PROOF. Recall that ξ has Hopf invariant one if and only if Sq^{2^N} is nonzero in the mapping cone C_ξ of ξ. In that case there must be an element

$X \in Z_{(2)}[M_1,\ldots,M_{N-1}]_{2^N}$ such that the coefficient of $M_1^{2^N}$ in X is odd and X transgresses to ξ. By Corollary 4.3.6(c), N must be 0, 1 or 2.∎

COROLLARY 4.3.8 The following Hurewicz homomorphisms have the same image:

$$\text{Image }[h:\pi_*BP \longrightarrow H_*BP] = \text{Image }[h:J_*BP \longrightarrow H_*BP].$$

PROOF. Recall that for a generalized homology theory F and a spectrum X, the image of the Hurewicz homomorphism h is given by $E^{\infty}_{*,0}$ of the Atiyah-Hirzebruch spectral sequence for F_*X. Moreover, the Hurewicz homomorphism for BP is a monomorphism, and thus $E^{\infty}_{*,0}$ in the spectral sequence for π_*BP equals π_*BP. The proof of Theorem 4.3.5 is valid in the Atiyah-Hirzebruch spectral sequence for J_*BP. In that context it says that in the spectral sequence for J_*BP, the intersection of the kernels of all differentials from the 0 row that land in $\pi_*J \otimes H_*BP$, equals $Z_{(2)}\{2^qU_{N_1}\cdots U_{N_q} \mid 0 \le q \text{ and } 1 \le N_1 \le \cdots \le N_q\}$ $= Z_{(2)}\{V_{N_1}\cdots V_{N_q} \mid 0 \le q \text{ and } 1 \le N_1 \le \cdots \le N_q\} = h(\pi_*BP)$.∎

By Theorem 4.3.5, if none of the differentials in our spectral sequence have image in CokJ $\otimes$ H_*BP then there would be enough image to these differentials so that $E^{\infty}_{*,0}$, the intersection of their kernels, would be π_*BP. It remains to show that this is indeed the case A differential which originates on the 0 row and lands in CokJ $\otimes$ H_*BP is what we called in Definition 1.3.7 a hidden differential originating on the 0 row. The proof of the following theorem can only be understood after reading Chapters 5, 6 and 7. There is no circular reasoning: we compute π_N^S for $N \le 64$ assuming Theorem 4.3.9 is true. The proof of Theorem 4.3.9 demonstrates that considering the leaders that occurred in the computation, no error could have been made through the assumption that Theorem 4.3.9 is true.

THEOREM 4.3.9 All differentials on $E^r_{N,0}$ have image in ImJ ⊗ H_*BP when $N \le 66$.

PROOF. Assume that $d^r(V) = XU$ is a hidden differential originating on the 0 row where $U \in H_{2M}BP$, $V \in H_{2N}BP$ and $X \in$ CokJ. Applying Landweber-Novikov operations shows that XU must be a leader. As a consequence of this hidden differential there is an element ζW, $\zeta \in$ ImJ and $W \in H_*BP$ which we incorrectly thought was hit by V, but in fact can not bound. Note that:

$W \ne 1$ lest ζ would nonzero in E^∞;

$W \ne M_1$ lest $\eta\zeta \ne 0$ in π_*^S while $\eta\zeta = 0$ in π_*ImJ, contradicting Theorem 4.2.2;

$W \ne M_1^2$ lest $\nu\zeta \ne 0$ in π_*^S while $\nu\zeta = 0$ in π_*ImJ, contradicting Theorem 4.2.2;

$W \ne M_2$ lest $0 \notin \langle\nu,\eta,\zeta\rangle$ in π_*^S while $0 \in \langle\nu,\eta,\zeta\rangle$ in π_*ImJ,
contradicting Theorem 4.2.4(c),(e);

$W \ne \bar{M}_2$ lest $0 \notin \langle\eta,\nu,\zeta\rangle$ in π_*^S while $0 \in \langle\eta,\nu,\zeta\rangle$ in π_*ImJ,
contradicting Theorem 4.2.4(a),(b),(d);

$W \ne M_1^4$ lest $\sigma\zeta \ne 0$ in π_*^S while $\sigma\zeta = 0$ in π_*ImJ, contradicting Theorem 4.2.2 (if $\zeta = \alpha_N$ and $\sigma\alpha_N = \eta\gamma_N$ then $\alpha_N M_1^4$ must bound because it transgresses to $\sigma\alpha_N = \eta\gamma_N$ which is zero in $E^8_{0,8N+8}$);

$W \ne M_1\bar{M}_2$ lest $0 \notin \langle\nu,\eta,\zeta,\eta\rangle$ in π_*^S while $0 \in \langle\nu,\eta,\zeta,\eta\rangle$ in π_*ImJ,
contradicting Theorem 4.2.4(f),(g);

$W \ne M_1^2\bar{M}_2$ lest $0 \notin \langle\eta,\nu,\zeta,\nu\rangle$ in π_*^S while $0 \in \langle\eta,\nu,\zeta,\nu\rangle$ in π_*ImJ,
contradicting Theorem 4.2.4(h),(i),(j);

$W \ne M_1^5$ lest $0 \notin \langle\eta,\zeta,\sigma\rangle$ in π_*^S while $0 \in \langle\eta,\zeta,\sigma\rangle$ in π_*ImJ,
contradicting Theorem 4.2.4(k),(l).

If we incorrectly assumed that this hidden differential $d^r(V) = XU$ did not occur then we would have drawn one of the following types of false conclusions:

(a) $d^{2M}(XU) = \xi$ is a nonzero element of π^S_{2M} ; or

(b) $d^{2Q}(Y) = XU$ where we had correctly proved that XU must be a bounding leader but we had incorrectly thought that Y was the only possible leader that could hit XU.

From the above observations it follows that $2M$ = degree $U \geq 14$. Checking the leaders of odd degree we see that the only possibilities for such an XU are $\eta A[30]M_1M_2^2$ and $\sigma C[44]M_1^4\bar{M}_2$. Since $d^{14}(\eta A[30]M_1M_2^2) = C[44]$ and $4C[44] = \sigma^2 A[30] \neq 0$, $\eta A[30]M_1M_2^2$ can not bound. If $\sigma C[44]M_1^4\bar{M}_2$ transgresses then $d^{14}(\sigma C[44]M_1^4\bar{M}_2) = B[64]$ and $2B[64] = \eta^2 B[62,1] \neq 0$. Thus, $\sigma C[44]M_1^4\bar{M}_2$ can not bound. Therefore in degrees less than 67, there is no leader XU as in (a) or (b), and there is no possibility for a hidden differential originating on the 0 row. ▮

4. Differentials which Originate on the 0 Row - Computation

In this section we reproduce the computer printout of the "cokernels" of the differentials which originate on the zero row. We can not actually compute the cokernels at this point. We do know E^4 of the spectral sequence but for each element $\xi \in \mathrm{ImJ}_{2r-1}$ of order k we do not know which differentials d^{2s}, $r \geq s \geq 2$, originate from elements with representatives in $Z_k\xi \otimes H_*BP$. By Theorem 4.3.9, the only differentials d^{2s}, $s \geq 2$, which can hit an element with a representative in $Z_k\xi \otimes H_*BP$ must originate on the 0 row. Therefore, $\{X \in E^{2r}_{*,2r-1} \mid X \text{ has a representative in } Z_k\xi \otimes H_*BP\}$ is a subgroup of $(Z_k\xi \otimes H_*BP)/(\text{Image } d^2)$. Let $\pi_r: E^2 \longrightarrow E^r$ denote the canonical projection. Then Cokernel $[d^{2r}: E^{2r}_{*,0} \longrightarrow \pi_{2r}(\mathrm{ImJ}_{2r-1} \otimes H_*BP)]$ is a subgroup of $\pi_4(\mathrm{ImJ}_{2r-1} \otimes H_*BP)$ / Image $[d^{2r}: E^{2r}_{*,0} \longrightarrow E^{2r}_{*,2r-1}]$. We thus make the following definition.

DEFINITION 4.4.1 Let "Cokernel $[d^{2r}: E^{2r}_{*,0} \longrightarrow E^{2r}_{*,2r-1}]$"

$$= \pi_4(\mathrm{ImJ}_{2r-1} \otimes H_*BP) \,/\, \text{Image } [d^{2r}: E^{2r}_{*,0} \longrightarrow E^{2r}_{*,2r-1}].$$

We order the data in this section by rows for convenience. The computer program which produced this data is discussed in Appendix 5. We will use the following notation.

1. When the range of the differentials is a Z_2-subspace of $Z_2\xi \otimes H_*BP$ then an entry "a b c d" on the list means that $\xi\ M_1^a\ \overline{M}_2^b\ \overline{M}_3^c\ \overline{M}_4^d$ is an element of the Z_2-basis of the "cokernel" of the differential.

2. The following will be used to denote a direct summand $Z_k(M_1^{e(1,1)}\overline{M}_2^{e(1,2)}\overline{M}_3^{e(1,3)}\overline{M}_4^{e(1,4)} + \cdots + M_1^{e(N,1)}\overline{M}_2^{e(N,2)}\overline{M}_3^{e(N,3)}\overline{M}_4^{e(N,4)})$:

Z_k e(1,1) e(1,2) e(1,3) e(1,4)
.
.
.
e(N,1) e(N,2) e(N,3) e(N,4).

3. The following will be used to denote a direct summand $Z_k(A_1 \cdot M_1^{e(1,1)}\overline{M}_2^{e(1,2)}\overline{M}_3^{e(1,3)}\overline{M}_4^{e(1,4)} + \cdots + A_N \cdot M_1^{e(N,1)}\overline{M}_2^{e(N,2)}\overline{M}_3^{e(N,3)}\overline{M}_4^{e(N,4)})$:

Z_k A_1 / e(1,1) e(1,2) e(1,3) e(1,4)
.
.
.
A_N / e(N,1) e(N,2) e(N,3) e(N,4).

We begin by listing the "cokernels" of the d^{10}-differentials from the 0 row to the 9 row. Note that $E^4_{*,9} = [Z_2\alpha_1 \oplus Z_2\eta^2\sigma\ M_1] \otimes B\langle 2\rangle$. Thus, the monomials below with an odd power of M_1 have coefficient $\eta^2\sigma$ and the monomials with an even power of M_1 have coefficient α_1

(4.4.2) "COKERNEL $[d^{10}: E^{10}_{*,0} \longrightarrow E^{10}_{*,9}]$":

DEGREE	BASIS	DEGREE	BASIS	DEGREE	BASIS
(18,9)	6 1 0 0	(20,9)	7 1 0 0	(22,9)	11 0 0 0
(24,9)	5 0 1 0		9 1 0 0	(26,9)	7 2 0 0

	6 0 1 0	(28,9)	11 1 0 0		5 3 0 0
	7 0 1 0	(30,9)	15 0 0 0		5 1 1 0
	6 3 0 0	(32,9)	13 1 0 0		7 3 0 0
	6 1 1 0	(34,9)	11 2 0 0		7 1 1 0
	14 1 0 0	(36,9)	11 0 1 0		15 1 0 0
	3 5 0 0		5 2 1 0		9 3 0 0
(38,9)	13 2 0 0		3 3 1 0		9 1 1 0
	6 2 1 0	(40,9)	7 2 1 0		11 3 0 0
	13 0 1 0		5 5 0 0		4 3 1 0
(42,9)	11 1 1 0		15 2 0 0		7 0 2 0
	5 3 1 0		14 0 1 0		6 5 0 0
(44,9)	15 0 1 0		5 1 2 0		7 0 0 1
	7 5 0 0		9 2 1 0		13 3 0 0
	6 3 1 0	(46,9)	5 1 0 1		11 4 0 0
	13 1 1 0		5 6 0 0		7 3 1 0
	6 1 2 0		14 3 0 0	(48,9)	3 7 0 0
	5 4 1 0		7 1 2 0		3 2 0 1
	9 5 0 0		11 2 1 0		15 3 0 0
	6 1 0 1		14 1 1 0	(50,9)	5 2 2 0
	7 1 0 1		7 6 0 0		9 3 1 0
	11 0 2 0		3 5 1 0		15 1 1 0
	6 4 1 0		22 1 0 0	(52,9)	9 1 2 0
	11 0 0 1		11 5 0 0		13 2 1 0
	1 6 1 0		3 3 2 0		23 1 0 0
	5 0 3 0		5 2 0 1		5 7 0 0
	7 4 1 0	(54,9)	15 4 0 0		5 5 1 0
	21 2 0 0		27 0 0 0		7 2 2 0
	9 1 0 1		3 3 0 1		11 3 1 0

	5 0 1 1		6 0 3 0	6 7 0 0
	14 2 1 0	(56,9)	11 1 2 0	3 6 1 0
	13 5 0 0		15 2 1 0	5 3 2 0
	21 0 1 0		25 1 0 0	7 0 3 0
	7 2 0 1		7 7 0 0	3 1 1 1
	12 3 1 0		6 5 1 0	6 0 1 1
(58,9)	5 3 0 1		1 7 1 0	23 2 0 0
	7 0 1 1		7 5 1 0	5 1 3 0
	11 1 0 1		11 6 0 0	13 3 1 0
	15 0 2 0		14 5 0 0	22 0 1 0
	6 3 2 0	(60,9)	13 1 2 0	15 0 0 1
	15 5 0 0		5 1 1 1	21 3 0 0
	23 0 1 0		27 1 0 0	5 6 1 0
	7 3 2 0		3 4 0 1	3 2 3 0
	9 7 0 0		11 4 1 0	14 3 1 0
	6 1 3 0		6 3 0 1	

Next we list the elements in the "cokernels" of the d^{12}-differentials from the 0 row to the 11 row. Recall that $E^4_{*,11} = [Z_4\beta_1 \oplus Z_8\beta_1 M_1] \otimes B\langle 2\rangle$. Thus, all the monomials below have coefficient β_1.

(4.4.3) "COKERNEL $[d^{12}: E^{12}_{*,0} \longrightarrow E^{12}_{*,11}]$":

DEGREE	GROUP	GENERATOR	DEGREE	GROUP	GENERATOR	DEGREE	GROUP	GENERATOR
(20,11)	Z_2	10 0 0 0	(22,11)	Z_4	8 1 0 0	(24,11)	Z_2	9 1 0 0
	Z_2	3/ 6 2 0 0	(26,11)	Z_2	1/ 6 0 1 0 2/ 10 1 0 0		Z_4	10 1 0 0
(28,11)	Z_2	2/ 11 1 0 0 1/ 4 1 1 0 3/ 14 0 0 0		Z_2	2/ 11 1 0 0 3/ 14 0 0 0		Z_8	11 1 0 0

(30,11) Z_4 6 3 0 0	Z_4 12 1 0 0	(32,11) Z_2 7 3 0 0 6 1 1 0
Z_2 3/ 13 1 0 0 2/ 6 1 1 0	Z_2 2/ 6 1 1 0 1/ 10 2 0 0	Z_4 6 1 1 0
(34,11) Z_2 3/ 7 1 1 0 2/ 14 1 0 0	Z_2 2/ 8 3 0 0 3/ 10 0 1 0 1/ 14 1 0 0	Z_4 8 3 0 0 Z_4 14 1 0 0
(36,11) Z_2 3/ 9 3 0 0 1/ 2 3 1 0 3/ 8 1 1 0 2/ 12 2 0 0	Z_2 2/ 15 1 0 0 3/ 2 3 1 0 3/ 8 1 1 0 1/ 12 2 0 0	Z_2 2/ 15 1 0 0 3/ 12 2 0 0 Z_4 2 3 1 0
Z_8 15 1 0 0	(38,11) Z_2 3/ 9 1 1 0 2/ 6 2 1 0 2/ 12 0 1 0	Z_4 6 2 1 0 10 3 0 0
Z_4 6 2 1 0 12 0 1 0	Z_4 6 2 1 0	(40,11) Z_2 1/ 7 2 1 0 2/ 11 3 0 0 2/ 13 0 1 0 3/ 4 3 1 0 1/ 10 1 1 0 2/ 14 2 0 0
Z_2 2/ 11 3 0 0 3/ 13 0 1 0 2/ 4 3 1 0 2/ 6 0 2 0 3/ 14 2 0 0	Z_2 2/ 11 3 0 0 2/ 4 3 1 0 3/ 6 0 2 0 1/ 14 2 0 0	Z_2 6 0 2 0
Z_4 4 3 1 0	Z_4 10 1 1 0	Z_8 11 3 0 0
(42,11) Z_2 2/ 5 3 1 0 2/ 11 1 1 0 1/ 6 0 0 1 1/ 8 2 1 0 1/ 12 3 0 0 1/ 14 0 1 0	Z_2 2/ 5 3 1 0 2/ 11 1 1 0 3/ 6 0 0 1 2/ 14 0 1 0	Z_4 5 3 1 0 11 1 1 0
Z_4 12 3 0 0 14 0 1 0	Z_4 12 3 0 0	Z_8 5 3 1 0
(44,11) Z_2 1/ 13 3 0 0 3/ 15 0 1 0 1/ 4 6 0 0 1/ 10 4 0 0 3/ 12 1 1 0	Z_2 3/ 4 1 0 1 2/ 6 3 1 0 2/ 12 1 1 0	Z_2 3/ 4 6 0 0 1/ 10 4 0 0
Z_2 2/ 13 3 0 0 1/ 4 6 0 0 2/ 6 3 1 0 2/ 12 1 1 0	Z_4 6 3 1 0 12 1 1 0	Z_4 6 3 1 0 Z_8 13 3 0 0

(46,11)	Z_2 3/ 13 1 1 0 1/ 4 4 1 0 1/ 6 1 2 0 2/ 8 5 0 0 2/ 14 3 0 0	Z_2 3/ 4 4 1 0 3/ 6 1 2 0 1/ 10 2 1 0 3/ 14 3 0 0	Z_4 6 1 2 0 8 5 0 0 14 3 0 0 Z_4 4 4 1 0
	Z_4 4 4 1 0 14 3 0 0	Z_4 6 1 2 0	Z_8 7 3 1 0
(48,11)	Z_2 3/ 7 1 2 0 2/ 9 5 0 0 1/ 11 2 1 0 3/ 15 3 0 0 3/ 4 2 2 0 1/ 6 1 0 1 1/ 6 6 0 0 3/ 8 3 1 0 2/ 14 1 1 0	Z_2 3/ 7 1 2 0 2/ 9 5 0 0 2/ 15 3 0 0 1/ 4 2 2 0 1/ 6 1 0 1 2/ 8 3 1 0 3/ 10 0 2 0	Z_2 3/ 9 5 0 0 1/ 4 2 2 0 2/ 6 1 0 1 3/ 6 6 0 0 2/ 14 1 1 0
	Z_2 3/ 4 2 2 0 2/ 6 1 0 1 1/ 6 6 0 0 2/ 8 3 1 0 3/ 10 0 2 0	Z_2 3/ 4 2 2 0 2/ 6 1 0 1 1/ 6 6 0 0 2/ 14 1 1 0	Z_2 2/ 6 1 0 1 1/ 6 6 0 0 2/ 8 3 1 0 2/ 14 1 1 0 Z_4 8 3 1 0
	Z_4 8 3 1 0 14 1 1 0	Z_4 6 1 0 1 8 3 1 0	Z_8 15 3 0 0
(50,11)	Z_2 1/ 7 1 0 1 2/ 15 1 1 0 3/ 4 2 0 1 2/ 4 7 0 0 2/ 6 4 1 0 1/ 10 0 0 1 2/ 10 5 0 0	Z_2 3/ 9 3 1 0 1/ 4 2 0 1 2/ 4 7 0 0 2/ 6 4 1 0 2/ 8 1 2 0 3/ 10 0 0 1 2/ 10 5 0 0 2/ 12 2 1 0	Z_2 3/ 4 2 0 1 1/ 10 0 0 1 2/ 10 5 0 0 Z_2 1/ 4 2 0 1 1/ 4 7 0 0 2/ 6 4 1 0 3/ 12 2 1 0
	Z_2 2/ 15 1 1 0 3/ 4 7 0 0 2/ 8 1 2 0 1/ 10 5 0 0 1/ 12 2 1 0	Z_4 4 2 0 1 4 7 0 0 6 4 1 0	Z_4 8 1 2 0
	Z_4 4 7 0 0	Z_4 6 4 1 0	Z_8 15 1 1 0
(52,11)	Z_2 3/ 5 7 0 0 1/ 7 4 1 0 3/ 11 5 0 0 2/ 4 0 1 1 1/ 4 5 1 0 1/ 6 2 2 0	Z_2 1/ 9 1 2 0 2/ 11 5 0 0 2/ 13 2 1 0 1/ 2 3 0 1 3/ 4 0 1 1 2/ 4 5 1 0 2/ 6 2 2 0 3/ 8 1 0 1 1/ 14 4 0 0 3/ 26 0 0 0	Z_2 1/ 5 7 0 0 3/ 11 5 0 0 1/ 13 2 1 0 3/ 2 3 0 1 3/ 4 5 1 0 3/ 8 1 0 1 1/ 26 0 0 0

Z_2 1/ 2 3 0 1
2/ 4 0 1 1
2/ 4 5 1 0
1/ 6 2 2 0
1/ 8 1 0 1
2/ 26 0 0 0

Z_2 3/ 4 0 1 1
2/ 4 5 1 0
3/ 6 2 2 0
3/ 26 0 0 0

Z_2 2/ 11 5 0 0
2/ 13 2 1 0
2/ 2 3 0 1
1/ 6 2 2 0
1/ 14 4 0 0
1/ 26 0 0 0

Z_2 26 0 0 0

Z_2 2/ 4 5 1 0
3/ 6 2 2 0

Z_4 4 5 1 0

Z_4 2 3 0 1

Z_4 10 3 1 0

Z_8 11 5 0 0

Z_8 13 2 1 0

(54,11) Z_2 3/ 5 5 1 0
1/ 9 1 0 1
1/ 4 3 2 0
2/ 6 0 3 0
2/ 6 2 0 1
3/ 6 7 0 0
1/ 10 1 2 0
3/ 12 5 0 0
1/ 20 0 1 0
1/ 24 1 0 0

Z_2 1/ 5 5 1 0
1/ 2 6 1 0
1/ 6 7 0 0
2/ 12 5 0 0
2/ 14 2 1 0
2/ 24 1 0 0

Z_2 1/ 2 6 1 0
3/ 4 3 2 0
1/ 6 2 0 1
3/ 6 7 0 0
1/ 10 1 2 0
3/ 12 5 0 0
1/ 14 2 1 0
3/ 20 0 1 0
3/ 24 1 0 0

Z_2 2/ 11 3 1 0
1/ 2 6 1 0
3/ 4 3 2 0
2/ 6 0 3 0
2/ 6 7 0 0
1/ 10 1 2 0
1/ 12 5 0 0
2/ 14 2 1 0
1/ 20 0 1 0
3/ 24 1 0 0

Z_2 2/ 4 3 2 0
2/ 6 7 0 0
2/ 14 2 1 0
3/ 20 0 1 0

Z_4 10 1 2 0
12 5 0 0
24 1 0 0

Z_4 4 3 2 0
6 0 3 0
14 2 1 0

Z_4 6 0 3 0
12 5 0 0
24 1 0 0

Z_4 6 0 3 0
6 7 0 0
14 2 1 0
24 1 0 0

Z_4 6 0 3 0
6 7 0 0

Z_4 6 7 0 0
24 1 0 0

Z_4 24 1 0 0

Z_8 11 3 1 0

(56,11) Z_2 3/ 5 3 2 0
1/ 7 0 3 0
1/ 11 1 2 0
2/ 13 5 0 0
3/ 15 2 1 0
1/ 25 1 0 0
1/ 4 1 3 0
1/ 12 3 1 0
2/ 22 2 0 0

Z_2 3/ 13 5 0 0
2/ 15 2 1 0
2/ 25 1 0 0
2/ 0 7 1 0
2/ 4 3 0 1
2/ 6 5 1 0
2/ 10 1 0 1
2/ 10 6 0 0
2/ 14 0 2 0

Z_2 3/ 7 0 3 0
3/ 7 7 0 0
2/ 11 1 2 0
3/ 15 2 1 0
1/ 25 1 0 0
3/ 0 7 1 0
2/ 4 1 3 0
2/ 6 0 1 1
2/ 6 5 1 0
3/ 22 2 0 0

Z_2 1/ 7 0 3 0
2/ 7 7 0 0
2/ 11 1 2 0
2/ 25 1 0 0
3/ 0 7 1 0
1/ 6 5 1 0
3/ 10 6 0 0
3/ 12 3 1 0
2/ 14 0 2 0
2/ 22 2 0 0

Z_2 1/ 25 1 0 0
2/ 0 7 1 0
2/ 4 3 0 1
2/ 6 0 1 1
2/ 6 5 1 0
2/ 10 6 0 0
2/ 12 3 1 0
2/ 14 0 2 0
3/ 22 2 0 0

Z_2 2/ 7 7 0 0
1/ 0 7 1 0
2/ 4 1 3 0
1/ 4 3 0 1
1/ 6 5 1 0
1/ 10 1 0 1
2/ 10 6 0 0
3/ 12 3 1 0
2/ 14 0 2 0
2/ 22 2 0 0

Z_2 2/ 7 7 0 0
2/ 11 1 2 0
3/ 0 7 1 0
1/ 4 1 3 0
1/ 6 0 1 1
2/ 10 1 0 1
2/ 10 6 0 0
1/ 12 3 1 0
1/ 14 0 2 0
3/ 22 2 0 0

Z_2 2/ 7 7 0 0
2/ 0 7 1 0
2/ 4 3 0 1
1/ 10 6 0 0
2/ 12 3 1 0
2/ 14 0 2 0

Z_2 2/ 11 1 2 0
2/ 6 0 1 1
2/ 12 3 1 0
1/ 14 0 2 0

Z_2 2/ 0 7 1 0
2/ 4 3 0 1
2/ 10 1 0 1
2/ 12 3 1 0
3/ 22 2 0 0

Z_4 6 0 1 1
10 1 0 1

Z_4 6 0 1 1

Z_4 6 0 1 1
12 3 1 0

Z_4 0 7 1 0
4 3 0 1
6 0 1 1
12 3 1 0

Z_8 7 7 0 0

Z_4 0 7 1 0
6 0 1 1
12 3 1 0

Z_8 11 1 2 0

(58, 11) Z_2 1/ 7 0 1 1
2/ 13 3 1 0
2/ 4 1 1 1
2/ 12 1 2 0
1/ 14 0 0 1
1/ 14 5 0 0

Z_2 2/ 7 5 1 0
2/ 8 7 0 0
3/ 10 4 1 0
1/ 14 5 0 0
1/ 20 3 0 0
1/ 26 1 0 0

Z_2 2/ 4 6 1 0
2/ 6 3 2 0
2/ 8 7 0 0
1/ 20 3 0 0
2/ 22 0 1 0
3/ 26 1 0 0

Z_2 1/ 5 3 0 1
3/ 7 5 1 0
1/ 11 1 0 1
1/ 13 3 1 0
1/ 4 1 1 1
1/ 4 6 1 0
1/ 8 7 0 0
2/ 14 0 0 1
2/ 14 5 0 0
3/ 22 0 1 0
1/ 26 1 0 0

Z_2 2/ 7 5 1 0
2/ 11 1 0 1
2/ 4 1 1 1
2/ 8 7 0 0
2/ 12 1 2 0
1/ 14 0 0 1
3/ 14 5 0 0
3/ 20 3 0 0
2/ 22 0 1 0
3/ 26 1 0 0

Z_2 2/ 5 3 0 1
2/ 11 1 0 1
2/ 6 3 2 0
2/ 8 7 0 0
2/ 12 1 2 0
1/ 22 0 1 0

Z_4 5 3 0 1
11 1 0 1
4 1 1 1
4 6 1 0
8 7 0 0
14 5 0 0

Z_4 4 6 1 0
6 3 2 0
14 5 0 0
26 1 0 0

Z_4 8 7 0 0
12 1 2 0

Z_4 6 3 2 0
8 7 0 0
26 1 0 0

Z_4	14 5 0 0	Z_4	8 7 0 0 26 1 0 0	Z_4	4 1 1 1 26 1 0 0
Z_4	26 1 0 0	Z_8	7 5 1 0	Z_8	11 1 0 1

Next we list the elements of the "cokernels" of the d^{16}-differentials from the 0 row to the 15 row. Note that $E^4_{*,15} = [Z_{32}\gamma_1 \oplus Z_{16}(2\gamma_1 M_1)] \otimes B<2>$. Thus, all the monomials below have coefficient γ_1.

(4.4.4) "COKERNEL $[d^{16}: E^{16}_{*,0} \longrightarrow E^{16}_{*,15}]$":

DEGREE	GROUP	GENERATOR	DEGREE	GROUP	GENERATOR	DEGREE	GROUP	GENERATOR
(22,15)	Z_2	8 1 0 0	(24,15)	Z_4	12 0 0 0	(26,15)	Z_8	10 1 0 0
(28,15)	Z_2	10/ 11 1 0 0 14/ 14 0 0 0		Z_{16}	14 0 0 0	(30,15)	Z_2	24/ 15 0 0 0 15/ 8 0 1 0 28/ 12 1 0 0
	Z_4	26/ 15 0 0 0 16/ 8 0 1 0 30/ 12 1 0 0		Z_{32}	12 1 0 0	(32,15)	Z_4	18/ 13 1 0 0 3/ 10 2 0 0
	Z_8	18/ 13 1 0 0 2/ 10 2 0 0	(34,15)	Z_2	9/ 8 3 0 0 1/ 10 0 1 0 9/ 14 1 0 0		Z_8 Z_{32}	10 0 0 1 14 1 0 0
(36,15)	Z_2	22/ 11 0 1 0 20/ 15 1 0 0 18/ 12 2 0 0		Z_4	10/ 15 1 0 0 1/ 8 1 1 0 19/ 12 2 0 0		Z_{16}	8 1 1 0
	Z_{16}	14/ 15 1 0 0 10/ 8 1 1 0 4/ 12 2 0 0	(38,15)	Z_2	2/ 13 2 0 0 25/ 6 2 1 0 10/ 10 3 0 0 3/ 12 0 1 0		Z_4	26/ 13 2 0 0 26/ 6 2 1 0 8/ 10 3 0 0 4/ 12 0 1 0
	Z_{32}	6 2 1 0	(40,15)	Z_2	30/ 11 3 0 0 22/ 13 0 1 0 12/ 8 4 0 0 26/ 10 1 1 0 30/ 14 2 0 0		Z_4	12/ 13 0 1 0 29/ 8 4 0 0 20/ 10 1 1 0 30/ 14 2 0 0
	Z_8	22/ 13 0 1 0 12/ 8 4 0 0 26/ 10 1 1 0 30/ 14 2 0 0		Z_{16}	10 1 1 0 14 2 0 0		Z_{16}	10 1 1 0

(42,15)	Z_2	14/ 11 1 1 0 9/ 6 5 0 0 13/ 8 2 1 0 6/ 12 3 0 0 13/ 14 0 1 0	Z_4	6/ 11 1 1 0 2/ 15 2 0 0 30/ 6 5 0 0 22/ 8 2 1 0 30/ 12 3 0 0 8/ 14 0 1 0	Z_8	18/ 15 2 0 0 24/ 6 5 0 0 6/ 8 2 1 0 14/ 12 3 0 0 20/ 14 0 1 0
	Z_8	1/ 6 5 0 0 2/ 12 3 0 0	Z_{32}	14 0 1 0	Z_{32}	12 3 0 0
(44,15)	Z_2	22/ 9 2 1 0 8/ 13 3 0 0 18/ 15 0 1 0 14/ 6 3 1 0 8/ 10 4 0 0 26/ 12 1 1 0	Z_4	20/ 13 3 0 0 30/ 15 0 1 0 7/ 6 3 1 0 13/ 10 4 0 0 27/ 12 1 1 0	Z_8	6/ 13 3 0 0 10/ 15 0 1 0 24/ 6 3 1 0 20/ 10 4 0 0 2/ 12 1 1 0
	Z_{16}	14/ 15 0 1 0 8/ 6 3 1 0 6/ 10 4 0 0 6/ 12 1 1 0	Z_{16}	10 4 0 0	Z_{32}	6 3 1 0
(46,15)	Z_2	12/ 11 4 0 0 24/ 13 1 1 0 24/ 4 4 1 0 3/ 8 0 0 1 4/ 8 5 0 0 4/ 10 2 1 0 8/ 14 3 0 0	Z_2	8/ 11 4 0 0 7/ 4 4 1 0 7/ 8 0 0 1 18/ 10 2 1 0	Z_4	10/ 11 4 0 0 28/ 13 1 1 0 12/ 4 4 1 0 12/ 8 0 0 1 14/ 8 5 0 0 12/ 10 2 1 0 4/ 14 3 0 0
	Z_8	30/ 13 1 1 0 14/ 4 4 1 0 18/ 8 0 0 1 16/ 8 5 0 0 14/ 10 2 1 0 16/ 14 3 0 0	Z_8	3/ 8 5 0 0 1/ 10 2 1 0 2/ 14 3 0 0	Z_{32} Z_{32}	14 3 0 0 10 2 1 0
(48,15)	Z_2	22/ 9 5 0 0 14/ 11 2 1 0 12/ 15 3 0 0 22/ 6 6 0 0 2/ 8 3 1 0 10/ 10 0 2 0 10/ 12 4 0 0 14/ 14 1 1 0	Z_4	16/ 11 2 1 0 28/ 15 3 0 0 25/ 6 6 0 0 26/ 8 3 1 0 31/ 10 0 2 0 20/ 12 4 0 0 18/ 14 1 1 0	Z_4 Z_8	6/ 11 2 1 0 1/ 6 6 0 0 2/ 8 3 1 0 7/ 12 4 0 0 6/ 14 1 1 0 2/ 11 2 1 0 3/ 6 6 0 0
	Z_8	2/ 11 2 1 0	Z_{16}	30/ 15 3 0 0 28/ 6 6 0 0 26/ 8 3 1 0 14/ 10 0 2 0 8/ 12 4 0 0 22/ 14 1 1 0	Z_{16} Z_{32}	8 3 1 0 14 1 1 0 8 3 1 0

(50,15) Z_2

10/ 9 3 1 0
24/ 13 4 0 0
18/ 15 1 1 0
21/ 4 7 0 0
29/ 6 4 1 0
11/ 8 1 2 0
1/ 10 0 0 1
8/ 10 5 0 0
14/ 12 2 1 0

Z_2

24/ 9 3 1 0
6/ 13 4 0 0
8/ 15 1 1 0
31/ 4 7 0 0
19/ 6 4 1 0
4/ 8 1 2 0
1/ 10 5 0 0
2/ 12 2 1 0

Z_4

2/ 9 3 1 0
10/ 13 4 0 0
14/ 15 1 1 0
12/ 4 7 0 0
4/ 6 4 1 0
12/ 8 1 2 0
22/ 10 0 0 1
2/ 10 5 0 0
2/ 12 2 1 0

Z_4

6/ 13 4 0 0
4/ 4 7 0 0
14/ 6 4 1 0
4/ 8 1 2 0
24/ 10 0 0 1
12/ 10 5 0 0
24/ 12 2 1 0

Z_8

1/ 4 7 0 0
1/ 8 1 2 0
2/ 12 2 1 0

Z_8

3/ 6 4 1 0
2/ 12 2 1 0

Z_{16}

26/ 15 1 1 0
4/ 4 7 0 0
22/ 6 4 1 0
26/ 8 1 2 0
26/ 10 0 0 1
26/ 10 5 0 0
28/ 12 2 1 0

Z_{32} 12 2 1 0

Z_{32} 8 1 2 0

(52,15) Z_2

6/ 7 4 1 0
12/ 11 0 0 1
24/ 13 2 1 0
2/ 4 5 1 0
22/ 8 1 0 1
12/ 8 6 0 0
4/ 10 3 1 0
14/ 12 0 2 0
26/ 14 4 0 0
27/ 26 0 0 0

Z_2

30/ 7 4 1 0
30/ 11 0 0 1
8/ 11 5 0 0
24/ 13 2 1 0
30/ 4 5 1 0
20/ 8 1 0 1
8/ 8 6 0 0
20/ 10 3 1 0
26/ 12 0 2 0
14/ 14 4 0 0
16/ 26 0 0 0

Z_2

18/ 11 0 0 1
28/ 11 5 0 0
28/ 13 2 1 0
22/ 4 5 1 0
16/ 8 1 0 1
22/ 8 6 0 0
28/ 10 3 1 0
18/ 12 0 2 0
24/ 14 4 0 0

Z_4

26/ 11 5 0 0
18/ 13 2 1 0
11/ 4 5 1 0
7/ 8 1 0 1
31/ 8 6 0 0
30/ 10 3 1 0
11/ 12 0 2 0
6/ 14 4 0 0
24/ 26 0 0 0

Z_4

14/ 11 5 0 0
16/ 13 2 1 0
19/ 4 5 1 0
5/ 8 6 0 0
18/ 10 3 1 0
28/ 12 0 2 0
19/ 14 4 0 0
18/ 26 0 0 0

Z_8

4/ 11 5 0 0
22/ 13 2 1 0
8/ 4 5 1 0
7/ 8 1 0 1
1/ 8 6 0 0
26/ 10 3 1 0
24/ 12 0 2 0
6/ 14 4 0 0
2/ 26 0 0 0

Z_8

18/ 11 5 0 0
6/ 13 2 1 0
22/ 4 5 1 0
30/ 8 1 0 1
28/ 8 6 0 0
10/ 10 3 1 0
20/ 12 0 2 0
20/ 14 4 0 0
20/ 26 0 0 0

Z_{16}

22/ 13 2 1 0
4/ 4 5 1 0
24/ 8 1 0 1
26/ 8 6 0 0
26/ 10 3 1 0
16/ 12 0 2 0
2/ 14 4 0 0
6/ 26 0 0 0

Z_{16} 4 5 1 0

Z_{16} 8 6 0 0

Z_{32} 10 3 1 0

(54,15)

Z_2	4/ 9 6 0 0 4/ 11 3 1 0 22/ 13 0 2 0 14/ 15 4 0 0 9/ 6 2 0 1 16/ 6 7 0 0 28/ 8 4 1 0 2/ 10 1 2 0 7/ 12 0 0 1 22/ 12 5 0 0 20/ 14 2 1 0 9/ 24 1 0 0	Z_4	8/ 9 6 0 0 8/ 11 3 1 0 22/ 15 4 0 0 14/ 6 2 0 1 18/ 6 7 0 0 8/ 8 4 1 0 6/ 10 1 2 0 2/ 12 0 0 1 4/ 12 5 0 0 12/ 14 2 1 0 19/ 24 1 0 0	Z_4	20/ 9 6 0 0 4/ 11 3 1 0 22/ 13 0 2 0 24/ 15 4 0 0 26/ 6 2 0 1 4/ 6 7 0 0 28/ 8 4 1 0 6/ 10 1 2 0 22/ 12 0 0 1 8/ 12 5 0 0 8/ 14 2 1 0 22/ 24 1 0 0
Z_4	16/ 9 6 0 0 16/ 11 3 1 0 22/ 15 4 0 0 18/ 6 2 0 1 20/ 6 7 0 0 20/ 8 4 1 0 12/ 10 1 2 0 2/ 12 0 0 1 2/ 12 5 0 0 12/ 14 2 1 0 18/ 24 1 0 0	Z_4	14/ 9 6 0 0 20/ 11 3 1 0 20/ 8 4 1 0 26/ 10 1 2 0 6/ 12 5 0 0 4/ 14 2 1 0	Z_8	3/ 6 7 0 0 3/ 10 1 2 0 3/ 12 5 0 0
		Z_8	2/ 11 3 1 0 3/ 8 4 1 0 3/ 10 1 2 0 2/ 14 2 1 0	Z_8	2/ 11 3 1 0 1/ 6 2 0 1 1/ 6 7 0 0
				Z_8	2/ 11 3 1 0 2/ 10 1 2 0
Z_{32}	14 2 1 0	Z_{32}	6 7 0 0	Z_{32}	10 1 2 0

Next we list the "cokernels" of the d^{18}-differentials from the 0 row to the 17 row. Note that $E^4_{*,17} = [Z_2\alpha_2 \oplus Z_2\eta^2\gamma_1 M_1] \otimes B<2>$. Thus, the monomials below with an odd power of M_1 have coefficient $\eta^2\gamma_1$ and the monomials with an even power of M_1 have coefficient α_2.

(4.4.5) "COKERNEL $[d^{18}: E^{18}_{*,0} \longrightarrow E^{18}_{*,17}]$":

DEGREE	BASIS	DEGREE	BASIS	DEGREE	BASIS
(34,17)	14 1 0 0	(36,17)	15 1 0 0	(38,17)	19 0 0 0
(40,17)	17 1 0 0	(42,17)	15 2 0 0		14 0 1 0
(44,17)	19 1 0 0		15 0 1 0	(46,17)	23 0 0 0
	13 1 1 0		14 3 0 0	(48,17)	21 1 0 0
	15 3 0 0		14 1 1 0	(50,17)	15 1 1 0
	19 2 0 0		12 2 1 0	(52,17)	23 1 0 0
	13 2 1 0		17 3 0 0		19 0 1 0

Next we list the elements of the "cokernels" of the d^{20}-differentials from the 0 row to the 19 row. Recall that $E^4_{*,19} = [Z_4\beta_2 \oplus Z_8\beta_2M_1] \otimes B\langle 2\rangle$. Thus, all the monomials below have coefficient β_2.

(4.4.6) "COKERNEL $[d^{20}: E^{20}_{*,0} \longrightarrow E^{20}_{*,19}]$":

DEGREE	GROUP	GENERATOR	DEGREE	GROUP	GENERATOR	DEGREE	GROUP	GENERATOR
(36,19)	Z_2	18 0 0 0	(38,19)	Z_4	16 1 0 0	(40,19)	Z_2	17 1 0 0
(42,19)	Z_2	14 0 1 0		Z_4	18 1 0 0	(44,19)	Z_2	2/ 19 1 0 0 3/ 22 0 0 0
	Z_8	19 1 0 0	(46,19)	Z_2	3/ 14 3 0 0 2/ 20 1 0 0		Z_4	20 1 0 0
(48,19)	Z_2	3/ 21 1 0 0 2/ 14 1 1 0 2/ 18 2 0 0		Z_2	2/ 14 1 1 0 1/ 18 2 0 0		Z_4	14 1 1 0
(50,19)	Z_2	1/ 15 1 1 0 2/ 22 1 0 0		Z_2	18 0 1 0 22 1 0 0		Z_4	16 3 0 0
							Z_4	22 1 0 0

Next we list the elements of the "cokernels" of the d^{24}-differentials from the 0 row to the 23 row. Note that $E^4_{*,23}$ contains $[Z_{16}\gamma_2 \oplus Z_8(2\gamma_2M_1)] \otimes B\langle 2\rangle$ as a direct summand. Thus, all the monomials below have coefficient γ_2.

(4.4.7) "COKERNEL $[d^{24}: E^{24}_{*,0} \longrightarrow E^{24}_{*,23}]$":

DEGREE	GROUP	GENERATOR	DEGREE	GROUP	GENERATOR	DEGREE	GROUP	GENERATOR
(40,23)	Z_2	20 0 0 0	(42,23)	Z_4	18 1 0 0	(44,23)	Z_8	22 0 0 0
(46,23)	Z_2	2/ 23 0 0 0 6/ 20 1 0 0		Z_{16}	20 1 0 0			

The d^r-differentials, $r > 24$, which originate on the 0 row have zero "cokernels" in degrees less than 70.

CHAPTER 5: The Japanese Stems (π_N, $9 \le N \le 31$)

1. Introduction

In 1962 Toda [60] introduced the idea of a triple bracket. Using this innovation and the EHP sequence he calcualted the first nineteen stable stems. We will compute these stable stems in Section 2 using our spectral sequence. However, we will not use Toda's notation. Instead we will use the new notation that was introduced at the end of Chapter 1 as well as the notation from Chapter 4 for elements in the image of J. The stable stems in degrees 20 through 31 were computed by Mahowald, May, Mimura , Mori, Oda, Tangora and Toda [10], [37], [39], [44], [45], [46], [50]. The Japanese authors on this list gave very careful expositions, along the lines of Toda, of the computations including unstable calculations while the American authors used the Adams spectral sequence. We compute these stems from our spectral sequence in Section 3. We name these stems after Oda who computed the last seven of them. In Section 4, we collect the tables of tentative differentials which are calculated by computer from the leading differentials of this chapter. The results of this chapter are summarized in Appendices 1 - 4.

2. The Toda Stems (π^S_N, $9 \le N \le 19$)

We begin by recalling, from Chapter 3, the leaders of degree at least 10 in rows one through eight which are not determining elements of ImJ. (These leaders are of the form $\alpha_N M_1$, $\eta\alpha_N M_1$, $\eta^2\alpha_N M_1$, $\gamma_N M_1$, $\eta\gamma_N M_1$, $\eta^2\gamma_N M_1$. In Chapter 4 we showed that the third and sixth of these leaders always bounds from the 0 row, and that the others never bound.) Moreover, if a leader arises of degree greater than 21 we will list it only once with an asterisk in our tables of leaders. However, in the last table at the end of this section the leaders of all degrees will be listed.

Row	Degree	Leader	Row	Degree	Leader
3	15	$4\nu M_1^3\bar{M}_2$	7	15	σM_1^4
6	12	$\nu^2 M_1^3$	8	10	$A[8]M_1$

FIGURE 5.2.1: Leaders in Rows 1 to 8 of Degree at Least 10

There is no possibility for $A[8]M_1$ to bound, and thus $\eta A[8] = d^2(A[8]M_1)$ is a nonero element of order two in π_9^S. In Theorem 3.3.15(a) we showed that $\eta A[8] = \nu^3$. This is the only possible nonzero element of $\mathrm{Cok}J_9$. Thus, we have proved the following theorem.

THEOREM 5.2.1 $\pi_9^S = Z_2\eta^2\sigma \oplus Z_2\alpha_1 \oplus Z_2\eta A[8]$

The computations from Section 4 show that we have the following leaders.

Row	Degree	Leader	Row	Degree	Leader
3	15	$4\nu M_1^3\bar{M}_2$	*9	27	$\eta^2\sigma M_1^9$
7	15	σM_1^4	9	21	$\nu^3 M_1^3\bar{M}_2$
8	18	$A[8]M_1^2\bar{M}_2$			

FIGURE 5.2.2: Leaders in Rows 1 to 9 of Degree at Least 11

It follows from the table in Figure 5.2.2 that $\pi_N^S = \mathrm{Im}J_N$ for $10 \le N \le 13$. Moreover, $d^8(\sigma M_1^4) = \sigma^2$ determines a nonzero element of order two in π_{14}^S. By Theorem 3.3.7(c), $d^{12}(4\nu M_1^3\bar{M}_2)$ has a nonzero representative $A[14]$ in π_{14}^S of order two. We have thus proved the following theorem.

THEOREM 5.2.2 (a) $\pi_{10}^S = Z_2\eta\alpha_1$

(b) $\pi_{11}^S = Z_8\beta_1$

(c) $\pi_{12}^S = \pi_{13}^S = 0$

(d) $\pi_{14}^S = Z_2\sigma^2 \oplus Z_2A[14]$

The computations of Section 4 show that we have the following leaders.

Row	Degree	Leader	Row	Degree	Leader
7	17	$2\sigma M_1^5$	*11	31	$\beta_1 M_1^{10}$
8	18	$A[8]M_1^2\overline{M}_2$	14	20	$\sigma^2 M_1^3$
9	21	$\nu^3 M_1^3 \overline{M}_2$	14	16	$A[14]M_1$

FIGURE 5.2.3: Leaders in Rows 1 to 14 of Degree at Least 16

By Theorem 3.4.6(a), $2\sigma M_1^5$ transgresses. Thus $\eta A[14] = d^2(A[14]M_1)$ is the only nonzero element of CokJ_{15}. We have thus proved the following theorem.

THEOREM 5.2.3 $\pi_{15}^S = Z_2 \eta A[14] \oplus Z_{32}\gamma_1$

The computations of Section 4 show that we have the following leaders.

Row	Degree	Leader	Row	Degree	Leader
7	17	$2\sigma M_1^5$	14	18	$A[14]M_1^2$
8	18	$A[8]M_1^2\overline{M}_2$	15	17	$\eta A[14]M_1$
9	21	$\nu^3 M_1^3 \overline{M}_2$	*15	37	$\gamma_1 M_1^8 \overline{M}_2$
14	20	$\sigma^2 M_1^3$			

FIGURE 5.2.4: Leaders in Rows 1 to 15 of Degree at Least 17

There are two leaders of degree 17 and two leaders of degree 18. Clearly $A[14]M_1^2$ transgresses. Now $A[14]\eta^2 = A[14]\langle 2,\eta,2\rangle = \langle A[14],2,\eta\rangle 2$ by Theorem 2.3.3(b). However, $\mathrm{ImJ}_{16} = Z_2(\eta\gamma_1)$, and $A[16]$ has order two. Thus $\eta^2 A[14] = 0$ and $\eta A[14]M_1$ must be a boundary. The only possiblity is $d^{10}(A[8]M_1^2\overline{M}_2) = \eta A[14]M_1$. By Theorem 3.4.6(a), $d^{10}(2\sigma M_1^5) = A[16]$ is a nonzero element of π_{16}^S of order two. We have thus proved the following theorem.

THEOREM 5.2.4 $\pi^S_{16} = Z_2A[16] \oplus Z_2\eta\gamma_1$ and $\eta^2A[14] = 0$.

The computations of Section 4 show that we have the following leaders.

Row	Degree	Leader	Row	Degree	Leader
7	19	$2\sigma M_1^6$	14	18	$A[14]M_1^2$
9	21	$\nu^3M_1^3\bar{M}_2$	15	21	$\eta A[14]M_1^3$
14	20	$\sigma^2M_1^3$	16	18	$A[16]M_1$

FIGURE 5.2.5: Leaders in Rows 1 to 16 of Degree at Least 18

By Theorem 3.4.7(a), the only leader of degree 19, $2\sigma M_1^6$, transgresses . Thus, both leaders of degree 18 transgress to nonzero elements, $\eta A[16]$ and $\nu A[14]$, both clearly of order two. We have thus proved the following theorem.

THEOREM 5.2.5 $\pi^S_{17} = Z_2\eta A[16] \oplus Z_2\nu A[14] \oplus Z_2\alpha_2 \oplus Z_2\eta^2\gamma_1$

Observe that $\sigma A[8] \in \sigma\langle\nu,2\nu,\eta\rangle = \langle\sigma,\nu,2\nu\rangle\eta$ and $\nu\langle\sigma,\nu,2\nu\rangle = \langle\nu,\sigma,\nu\rangle 2\nu = \sigma^2(2\nu) = 0$. Thus, $\langle\sigma,\nu,2\nu\rangle$ can not be $A[14]$. Since $\pi^S_{14} = Z_2A[14] \oplus Z_2\sigma^2$ and $\eta\sigma^2 = 0$, we have $\langle\sigma,\nu,2\nu\rangle\eta = 0$ and

$$\sigma A[8] = 0. \qquad [5.1]$$

The computations of Section 4 show that we have the following leaders.

Row	Degree	Leader	Row	Degree	Leader
7	19	$2\sigma M_1^6$	16	34	$A[16]M_1^6\bar{M}_2$
9	21	$\nu^3M_1^3\bar{M}_2$	17	19	$\eta A[16]M_1$
14	20	$\sigma^2M_1^3$	17	21	$\nu A[14]M_1^2$
15	21	$\eta A[14]M_1^3$	*17	51	$\alpha_2M_1^{14}\bar{M}_2$

FIGURE 5.2.6: Leaders in Rows 1 to 17 of Degree at Least 19

Note that $\sigma^2M_1^3$ clearly survives to E^6 since $\eta\sigma^2 = 0$ and $\nu\sigma = 0$. Thus, both leaders of degree 19 transgress to nonzero elements of π_{18}^S. Thus we have the following composition series for $CokJ_{18}$:

$$0 \longrightarrow Z_2\eta^2A[16] \longrightarrow \pi_{18}^S \longrightarrow Z_4d^{12}(2\sigma M_1^6) \longrightarrow 0$$

By Lemma 3.3.14, $\eta^2A[16]$ must be divisible by two. The only possiblity is that $d^{12}(2\sigma M_1^6)$ is represented by an element C[18] of order eight such that $4C[18] = \eta^2A[16]$. Thus, we have proved the following theorem.

THEOREM 5.2.6 $\pi_{18}^S = Z_8C[18] \oplus Z_2\eta\alpha_2$ and $\eta^2A[16] = 4C[18]$.

The computations of Section 4 show that we have the following leaders.

Row	Degree	Leader	Row	Degree	Leader
*7	31	$2\sigma M_1^{12}$	15	21	$\eta A[14]M_1^3$
9	21	$\nu^3M_1^3\overline{M}_2$	17	21	$\nu A[14]M_1^2$
14	20	$\sigma^2M_1^3$	18	22	$C[18]M_1^2$

FIGURE 5.2.7: Leaders in Rows 1 to 18 of Degree at Least 20

There is one leader of degree 20 and three leaders of degree 21. Clearly $\eta A[14]M_1^3$ is a d^4-cycle and thus transgresses. Also, $\nu A[14]M_1^2$ clearly transgresses. Thus, the only possibility for the leader $\sigma^2M_1^3$ to bound is from $\nu^3M_1^3\overline{M}_2$, which is in the 9 row. If that were the case then σ^2 would be $d^6(\nu^3\overline{M}_2)$. However, σ^2 bounds from the 8 row. Thus, $\sigma^2M_1^3$ does not bound and transgresses to a nonzero element A[19]. By Theorem 2.4.4(a),(b),

$$A[19] \in \langle\eta,\sigma^2,\nu\rangle = \langle\nu,\eta,\sigma^2\rangle \qquad [5.2]$$

We have thus proved the following theorem.

THEOREM 5.2.7 $\pi_{19}^S = Z_2A[19] \oplus Z_8\beta_2$ and $\eta C[18] = \nu A[16] = 0$.

The computations of Section 4 show that we have the following table of leaders. Those leaders of degree greater than 21 which were omitted from the previous tables are included in the table below.

Row	Degree	Leader	Row	Degree	Leader
7	31	$2\sigma M_1^{12}$	16	34	$A[16]M_1^6\overline{M}_2$
9	27	$\eta^2\sigma M_1^9$	17	21	$\nu A[14]M_1^2$
9	21	$\nu^3 M_1^3\overline{M}_2$	17	51	$\alpha_2 M_1^{14}\overline{M}_2$
11	31	$\beta_1 M_1^{10}$	18	22	$C[18]M_1^2$
14	28	$\sigma^2 M_1^4\overline{M}_2$	19	23	$A[19]M_1^2$
15	21	$\eta A[14]M_1^3$	19	55	$\beta_2 M_1^{18}$
15	37	$\gamma_1 M_1^8\overline{M}_2$			

FIGURE 5.2.8: Leaders in Rows 1 to 19 of Degree at Least 21

3. The Oda Stems (π_N^S, $20 \le N \le 31$)

We continue the computations of the preceeding section. In the tables of leaders of this section we only list leaders of degree greater than 33 once with an asterisk. The final table of leaders at the end of this section will list the leaders of all degrees.

Note that the only leader of degree 22 in the table of Figure 5.2.8 is $C[18]M_1^2$ which transgresses to $\nu C[18]$. Thus, the three leaders of degree 21 all transgress to nonzero elements. Therefore, π_{20}^S has a composition series consisting of three Z_2s. Let $C[20] = d^{12}(\nu^3 M_1^3\overline{M}_2)$. Note that there is only one leader of degree 23, 24 or 25, $A[19]M_1^2$, which transgresses to $\nu A[19]$. Therefore $0 \ne \eta^3 C[20] = 4\nu C[20]$ and $4C[20] \ne 0$. Thus, $C[20]$ has order eight. Observe that $\nu C[18]$ has order two because $2\nu C[18]$ can not be $\eta C[20]$ as $\eta^2 C[20] \ne 0$. We have thus proved the following theorem.

THEOREM 5.3.1 (a) $\pi^S_{20} = Z_8 C[20]$, $4C[20] = \nu^2 A[14]$ and $\eta A[19] = 0$.

(b) $\pi^S_{21} = Z_2 \eta C[20] \oplus Z_2 \nu C[18]$

(c) $\pi^S_{22} = Z_2 \eta^2 C[20] \oplus Z_2 \nu A[19]$

The computations of Section 4 show that we have the following leaders.

Row	Degree	Leader	Row	Degree	Leader
7	31	$2\sigma M_1^{12}$	18	24	$2C[18]\bar{M}_2$
9	27	$\eta^2\sigma M_1^9$	*19	55	$\beta_2 M_1^{18}$
11	31	$\beta_1 M_1^{10}$	20	24	$C[20]M_1^2$
14	28	$\sigma^2 M_1^4 \bar{M}_2$	21	26	$\nu C[18]M_1^3$
*15	37	$\gamma_1 M_1^8 \bar{M}_2$	22	24	$\eta^2 C[20]M_1$
*16	34	$A[16]M_1^6\bar{M}_2$	22	28	$\nu A[19]M_1^3$
*17	51	$\alpha_2 M_1^{14}\bar{M}_2$			

FIGURE 5.3.1: Leaders in Rows 1 to 22 of Degree at Least 24

Note that there are no leaders of degree 25. Therefore, the leaders of degree 24 must transgress to nonzero elements. We already observed that $\nu C[20]$ has order eight and $4\nu C[20] = \eta^3 C[20]$. It remains to check that $A[23] = d^6(2C[18]\bar{M}_2)$ has order two. By Theorem 2.4.4(c),

$$A[23] \in \langle \eta, \nu, 2C[18] \rangle. \qquad [5.3]$$

By Theorem 2.3.3(b), $2A[23] \in 2\langle \eta, \nu, 2C[18]\rangle = \langle 2, \eta, \nu\rangle 2C[18] = 0$ because $\langle 2, \eta, \nu\rangle \in \pi^S_5 = 0$. We have thus proved the following theorem.

THEOREM 5.3.2 $\pi^S_{23} = Z_8 \nu C[20] \oplus Z_2 A[23] \oplus Z_{16}\gamma_2$ and $4\nu C[20] = \eta^3 C[20]$.

The computations of Section 4 show that we have the following leaders.

Row	Degree	Leader	Row	Degree	Leader
7	31	$2\sigma M_1^{12}$	21	27	$\nu C[18]M_1^3$
9	27	$\eta^2\sigma M_1^9$	22	28	$\nu A[19]M_1^3$
11	31	$\beta_1 M_1^{10}$	23	27	$\nu C[20]M_1^2$
14	28	$\sigma^2 M_1^4\bar{M}_2$	23	25	$A[23]M_1$
*18	38	$C[18](M_1^4\bar{M}_2^2+2M_1^7\bar{M}_2)$	*23	63	$\gamma_1 M_1^{20}$

FIGURE 5.3.2: Leaders in Rows 1 to 23 of Degree at Least 25

There are no leaders of degree 26. Therefore, $d^2(A[23]M_1) = \eta A[23]$ is nonzero. We have thus proved the following theorem.

THEOREM 5.3.3 $\pi_{24}^S = \eta A[23] \oplus \eta\gamma_2$

The computations of Section 4 show that we have the following leaders.

Row	Degree	Leader	Row	Degree	Leader
7	31	$2\sigma M_1^{12}$	22	28	$\nu A[19]M_1^3$
9	27	$\eta^2\sigma M_1^9$	23	27	$\nu C[20]M_1^2$
11	31	$\beta_1 M_1^{10}$	23	33	$A[23]M_1^2\bar{M}_2$
14	28	$\sigma^2 M_1^4\bar{M}_2$	24	26	$\eta A[23]M_1$
21	27	$\nu C[18]M_1^3$			

FIGURE 5.3.3: Leaders in Rows 1 to 24 of Degree at Least 26

There is one leader of degree 26 and three leaders of degree 27. By Lemma 3.3.14, if $\eta^2 A[23]$ is nonzero then it must be divisible by two. However, there are no other elements of CokJ_{25}. Thus $\eta^2 A[23] = 0$ and $\eta A[23]M_1$ must be a boundary. Note that if $d^4(\nu C[18]M_1^3) = \eta A[23]M_1$ then $\nu^2 C[18] = \eta A[23]$. Since $\nu^2 C[18] \in \langle\eta,\nu,\eta\rangle C[18] = \eta\langle\nu,\eta,C[18]\rangle$ by Theorem 2.3.3(b), it would follow that $A[23] \in \langle\nu,\eta,C[18]\rangle$ and $d^6(C[18]M_2) = A[23]$ by

Theorem 2.4.4(b). However, $C[18]M_2 = d^{12}(2\sigma(M_1^6M_2+M_1^3M_2^2))$. Thus, $\nu C[18]M_1^3$ transgresses and

$$\nu^2C[18] = 0 \qquad [5.4]$$

Clearly, $\nu C[20]M_1^2$ transgresses. The only remaining possibility for $\eta A[23]M_1$ to bound is $d^{16}(\eta^2\sigma M_1^9) = \eta A[23]M_1$. Thus we have proved the following theorem.

THEORM 5.3.4 $\pi_{25}^S = Z_2\alpha_3 \oplus Z_2\eta^2\gamma_2$.

The computations of Section 4 show that we have the following leaders.

Row	Degree	Leader	Row	Degree	Leader
7	31	$2\sigma M_1^{12}$	22	28	$\nu A[19]M_1^3$
9	29	$\eta^2\sigma M_1^7\bar{M}_2$	23	27	$\nu C[20]M_1^2$
11	31	$\beta_1 M_1^{10}$	23	33	$A[23]M_1^2\bar{M}_2$
14	28	$\sigma^2M_1^4\bar{M}_2$	*24	60	$\eta A[23]M_1^{15}\bar{M}_2$
21	27	$\nu C[18]M_1^3$			

FIGURE 5.3.4: Leaders in Rows 1 to 25 of Degree at Least 27

LEMMA 5.3.5 (a) $\nu C[18] = \sigma^3$.

(b) $\sigma A[14] = 0$.

PROOF. (a) We showed in [3.9] that $C[18] \in \langle\nu,\sigma,2\sigma\rangle$. By Theorem 2.3.7(a),

$$C[18] \in \langle\sigma,\nu,\sigma\rangle \qquad [5.5]$$

Therefore, $\nu C[18] \in \nu\langle\sigma,\nu,\sigma\rangle = \langle\nu,\sigma,\nu\rangle\sigma = \sigma^3$ by Theorems 2.3.3(b) and 2.4.2.

(b) We showed in [3.6] that $A[14] \in \langle\nu,A[8],2,\eta\rangle$. Thus,

$\sigma A[14] \in \sigma\langle\nu,A[8],2,\eta\rangle \subset \langle\langle\sigma,\nu,A[8]\rangle,2,\eta\rangle = \langle 0,2,\eta\rangle$ because $\nu\langle\sigma,\nu,A[8]\rangle = \langle\nu,\sigma,\nu\rangle A[8] = \sigma^2A[8] = 0$ and $\mathrm{Cok}_{19} = Z_2A[19]$, $\nu A[19] \neq 0$. Therefore, $\sigma A[14] \in \eta\cdot\pi_{20}^S = Z_2\eta C[20]$. Now $\eta^2\sigma A[14] = 0$ while $\eta^2(\eta C[20]) = 4\nu C[20] \neq 0$. Thus, $\sigma A[14] = 0$. ∎

It follows from the lemma that $d^8(\sigma^2M_1^4\bar{M}_2) = \nu C[18]\bar{M}_2 = \nu C[18]M_1^3$. Since there are no other leaders of degree 28, the other leader, $\nu C[20]M_1^2$, of degree 27 must transgress to a nonzero element. Observe that $\sigma A[19] \in \sigma\langle\nu,\eta,\sigma^2\rangle = \langle\sigma,\nu,\eta\rangle\sigma^2 = 0$. We have thus proved the following theorem.

THEOREM 5.3.6 $\pi^S_{26} = Z_2\nu^2C[20] \oplus Z_2\eta\alpha_3$ and $\sigma A[19] = 0$.

The computations of Section 4 show that we have the following leaders.

Row	Degree	Leader	Row	Degree	Leader
7	31	$2\sigma M_1^{12}$	22	28	$\nu A[19]M_1^3$
9	29	$\eta^2\sigma M_1^7\bar{M}_2$	23	29	$2\nu C[20]M_1^3$
11	31	$\beta_1M_1^{10}$	23	33	$A[23]M_1^2\bar{M}_2$
*14	34	$\sigma^2M_1^4\bar{M}_2^2$	26	32	$\nu^2C[20]M_1^3$
21	33	$\nu C[18]M_1^3\bar{M}_2$			

FIGURE 5.3.5: Leaders in Rows 1 to 26 of Degree at Least 28

There is one leader of degree 28 and two leaders of degree 29. Let $X = d^6(\nu A[19]M_1^3) = d^6(\nu A[19]\bar{M}_2)$. Assume that X is not zero. By Theorems 2.4.2 and 2.4.4(c), $X \in \langle\eta,\nu,A[19]\rangle = \langle\eta,\nu,\langle\eta\sigma,\sigma,\nu\rangle\rangle$. Note that $\langle\eta,\nu,\eta\sigma\rangle = 0$ because $\pi^S_{13} = 0$. Also by Theorem 2.4.2, $0 = d^{12}(\eta\sigma M_1^6) \in \langle\nu,\eta\sigma,\sigma\rangle$ because $\eta\sigma M_1^6 = d^2(\sigma M_1^7)$. By Theorem 2.2.7(a), $\langle\eta,\nu,\eta\sigma,\sigma\rangle$ is defined. By Theorem 2.3.6(a), $X \in \langle\eta,\nu,\langle\eta\sigma,\sigma,\nu\rangle\rangle \supset \langle\eta,\nu,\eta\sigma,\sigma\rangle\nu$. Since X is not a d^4-boundary, $X \in \text{Indet}\,\langle\eta,\nu,\langle\eta\sigma,\sigma,\nu\rangle\rangle = (\eta)$ as $\pi^S_5 = 0$. Thus, X is a d^2-boundary, a contradiction. Therefore $X = 0$ and $\nu A[19]M_1^3$ must be a boundary. Since $2\nu C[20]M_1^3$ transgresses, the only other leader of degree 29, $\eta^2\sigma M_1^7\bar{M}_2$, must hit $\nu A[19]M_1^3$. Also observe that there are no leaders of degree 30. Thus, $A[8]C[20] = d^6(2\nu C[20]M_1^3)$ must be nonzero. We have thus proved the following theorem.

THEOREM 5.3.7 (a) $\pi^S_{27} = Z_8\beta_3$

(b) $\pi^S_{28} = Z_2A[8]C[20]$

The computations of Section 4 show that we have the following leaders.

Row	Degree	Leader	Row	Degree	Leader
7	31	$2\sigma M_1^{12}$	*23	35	$2\nu C[20]M_1^3\bar{M}_2$
9	33	$\eta^2\sigma M_1^5\bar{M}_3$	23	33	$A[23]M_1^2\bar{M}_2$
11	31	$\beta_1 M_1^{10}$	26	32	$\nu^2C[20]M_1^3$
21	33	$\nu C[18]M_1^3\bar{M}_2$	28	30	$A[8]C[20]M_1$
22	32	$\nu A[19]M_1^2\bar{M}_2$			

FIGURE 5.3.6: Leaders in Rows 1 to 27 of Degree at Least 29

To continue our analysis we will need the following relations and Toda brackets in π^S_*.

LEMMA 5.3.8 (a) $\eta A[8] \in \langle 2, A[8], 2\rangle$

(b) $A[8]A[14] = \eta^2C[20]$

(c) $\eta C[20] \in \langle \nu^2, 2, A[14]\rangle$

(d) $\eta A[14] \in \langle 2, A[14], 2\rangle$

(e) $C[20] \in \langle \nu, \eta, 2, A[14]\rangle$ and $2C[20] \in \langle \eta A[14], \eta, \nu\rangle$

(f) $\eta A[8]C[20] = 0$

PROOF. (a) By Theorem 2.4.2, $\eta A[8] \in \eta\langle \eta, \nu, 2\nu\rangle \subset \eta\langle \eta, \nu^2, 2\rangle \subset \langle \eta^2, \nu^2, 2\rangle$ $= \langle\langle 2, \eta, 2\rangle, \nu^2, 2\rangle = \langle 2, \langle \eta, 2, \nu^2\rangle, 2\rangle + \langle 2, \eta, \langle 2, \nu^2, 2\rangle\rangle$. By Theorem 2.4.2, $A[8] \in \langle \eta, 2, \nu^2\rangle$. Since $\pi^S_8 = Z_2\eta\sigma \oplus Z_2A[8]$, $\eta\langle 2, \nu^2, 2\rangle = \langle \eta, 2, \nu^2\rangle 2 = 0$ and $\langle 2, \nu^2, 2\rangle \in (2\sigma)$. Thus, for some integer k, $\eta A[8] \in \langle 2, A[8], 2\rangle + \langle 2, \eta, 2k\sigma\rangle$ $= \langle 2, A[8], 2\rangle + k\sigma\langle 2, \eta, 2\rangle = \langle 2, A[8], 2\rangle + k\eta^2\sigma$. By Theorem 2.3.7(b), $\eta A[8] \in \langle 2, A[8], 2\rangle$.

(b), (c) $\eta A[8]A[14] = \nu^3 A[14] = 4\nu C[20] = \eta^3 C[20]$. Since multiplication by η on π_{22}^S has kernel $Z_2\nu A[19]$, $A[8]A[14] = \eta^2 C[20]$ modulo $(\nu A[19])$. By Theorem 2.4.2, $A[8] \in \langle\nu^2, 2, \eta\rangle$ so $\eta A[8] \in \eta\langle\nu^2, 2, \eta\rangle = \langle\eta, \nu^2, 2\rangle\eta$, and hence $A[8] \in \langle\eta, \nu^2, 2\rangle$. Thus, $A[8]A[14] \in A[14]\langle 2, \nu^2, \eta\rangle = \langle A[14], 2, \nu^2\rangle\eta$. Therefore, $A[8]A[14] = \eta^2 C[20]$ and $\eta C[20] \in \langle\nu^2, 2, A[14]\rangle$.

(d), (e) Note that by Theorem 2.2.7(a), $\langle\nu, \eta, 2, A[14]\rangle$ is defined because $\langle\nu, \eta, 2\rangle \in \pi_5^S = 0$ and $\langle\eta, 2, A[14]\rangle$ contains 0. (It can not contain A[16] because as we shall see in Chapter 6, $\sigma A[16] = A[23]$ and $\sigma\langle\eta, 2, A[14]\rangle = \eta\langle 2, A[14], \sigma\rangle$ which can not be A[23].) Now $\eta\langle\nu, \eta, 2, A[14]\rangle \subset \langle\langle\eta, \nu, \eta\rangle, 2, A[14]\rangle = \langle\nu^2, 2, A[14]\rangle$ which contains $\eta C[20]$. This proves (e). Now $2C[20] \in 2\langle A[14], 2, \eta, \nu\rangle \subset \langle\langle 2, A[14], 2\rangle, \eta, \nu\rangle$, and by Theorem 2.4.4(b), $2C[20] \in \langle\eta A[14], \eta, \nu\rangle$. By Theorem 2.3.7(b), $\langle 2, A[14], 2\rangle$ contains $\eta A[14]$.

(f) $\eta A[8]C[20] \in C[20]\langle 2, A[8], 2\rangle \subset \langle 2C[20], A[8], 2\rangle \subset \langle\langle\eta A[14], \eta, \nu\rangle, A[8], 2\rangle \supset \eta A[14]\langle\eta, \nu, A[8], 2\rangle$. This four fold Toda bracket is defined by Theorem 2.2.7(a) because $\langle\eta, \nu, A[8]\rangle \subset \pi_{13}^S = 0$ and $\langle\nu, A[8], 2\rangle \subset \pi_{12}^S = 0$. Thus, $\eta A[8]C[20] \in \eta A[14]\cdot\pi_{14}^S$ modulo Indet $\langle\langle\eta A[14], \eta, \nu\rangle, A[8], 2\rangle = 2\pi_{29}^S$ and $2\pi_{29}^S$ will be zero whether (f) holds or not. (If (f) holds then $\pi_{29}^S = 0$ and if (f) is false then $\pi_{29}^S = Z_2\eta A[8]C[20]$.) Note that $\eta A[14]\cdot\pi_{14}^S = (\eta A[14]^2)$. Thus, $\eta A[8]C[20] \in (\eta A[14]^2)$. By Barratt's argument [25], $\eta A[14]^2 \in A[14]\langle 2, A[14], 2\rangle = \langle A[14], 2, A[14]\rangle 2 \subset 2\cdot\pi_{29}^S = 0$, as we remarked above. ∎

THEOREM 5.3.9 (a) $\pi_{29}^S = 0$

(b) $d^{18}(\beta_2 M_1^{10}) = A[8]C[20]M_1$

PROOF. By Lemma 5.3.8(f), $A[8]C[20]M_1$ must bound and $\pi_{29}^S = 0$. From the table in Figure 5.3.6, we see that there are two possibilities: either $\beta_2 M_1^{10}$ or $2\sigma M_1^{12}$ hits $A[8]C[20]M_1$. We will show that $2\sigma M_1^{12}$ transgresses. D. Kahn [25] defined a cup one product $\cup_1$ such that if B is a ring spectrum, $F: DU \longrightarrow B$ and $G: DV \longrightarrow B$ then $F\cup_1 G: I \ltimes DU\wedge DV \longrightarrow B$ with the following properties:

(a) $(F\cup_1 G)\,|\,\{0\}\times DU\wedge DV = F\wedge G$;

(b) $(F\cup_1 G)\,|\,\{1\}\times DU\wedge DV = G\wedge F$;

(c) $(F\cup_1 G)\,|\,I \ltimes SU\wedge DV = (\partial F)\cup_1 G$;

(d) $(F\cup_1 G)\,|\,I \ltimes DU\wedge SV = F\cup_1(\partial G)$;

(e) if Image $F\subset B^{(h)}$ and Image $G\subset B^{(k)}$ then Image $(F\cup_1 G)\subset B^{(h+k)}$.

It follows that there is an induced map $\cup_1 : E^1_{M,s}\otimes E^1_{N,t}\longrightarrow E^1_{M+N,s+t}$ on our spectral sequence such that

$$d^1(X\cup_1 Y) = X\cdot Y - (-1)^{(\deg X)(\deg Y)}Y\cdot X - \partial(X)\cup_1 Y - (-1)^{\deg X}X\cup_1 Y.$$

Using this cup-one product we can construct a representative F of $2\sigma M_1^{12}$. For $1\le i\le 5$, let $G_i:(DU_i,SU_i)\longrightarrow (BP^{(8)},S)$ be a representative of $\langle M_1^4\rangle$, i.e. $G_i' = G_i|SU_i$ represents σ. For $1\le i\le 4$, let $H_i(2\sigma^2):DV\longrightarrow SV'$ such that $H_i(2\sigma^2)|SV = 2G_i\wedge G_{i+1}$. Since $\pi^S_{29} = 0$, $0 = \langle\sigma^2,2,\sigma^2\rangle$. Thus, let $K_1:DW_1\longrightarrow SW_1'$ such that $K_1|SW_1 = [H_1(2\sigma^2)\wedge G_3'\wedge G_4']\cup[G_1'\wedge G_2'\wedge H_3(2\sigma^2)]$ and let $K_2:DW_2\longrightarrow SW_2'$ such that $K_2|SW_2 = [G_2'\wedge G_3'\wedge H_4(2\sigma^2)]\cup[H_2(2\sigma^2)\wedge G_4'\wedge G_5']$. Note that $(2\sigma^2)\cup_1\sigma\in 2\pi^S_{22} = 0$. Thus, let $L:DW\longrightarrow SW'$ such that $L|\partial W = (2G_2'\wedge G_3')\cup_1 G_4'$. To save on long and hideous notation we supress the domains and ranges of the maps in the following definition of F. Define

$$F = 2G_1\wedge G_2'\wedge G_3\wedge G_4'\wedge G_5 \cup H_1(2\sigma^2)\wedge G_3\wedge G_4'\wedge G_5 \cup G_1\wedge H_2(2\sigma^2)\wedge G_4'\wedge G_5$$
$$\cup\, G_1\wedge G_2'\wedge G_3\wedge H_4(2\sigma^2) \cup G_1\wedge K_2 \cup H_1(2\sigma^2)\wedge[(G_3'\wedge G_4')\cup_1 G_5]$$
$$\cup\, G_1'\wedge[H_2(2\sigma^2)\cup_1 G_4']\wedge G_5 \cup G_1'\wedge G_2'\wedge[H_4(2\sigma^2)\cup_1 G_3] \cup K_1\wedge G_5 \cup G_1'\wedge L\wedge G_5.$$

Clearly all the unionands except the first one have range in $BP^{(8)}$. Therefore, F represents $2\sigma M_1^{12}$. A straightforward calculation shows that ∂F maps to $BP^{(0)}$. Thus $2\sigma M_1^{12}$ transgresses and $d^{18}(\beta_2 M_1^{10}) = A[8]C[20]M_1$.∎

We are about to stumble over the first hidden differential! Therefore instead of updating our table of leaders incorrectly, we compute π^S_{30} first and note this hidden differential. There is only one leader of degree 31, $2\sigma M_1^{12}$ which clearly can not bound and therefore transgresses to define the unique nonzero element $A[30] = d^{24}(2\sigma M_1^{12})$ of π_{30}^S. Observe that $\beta_1 M_1^{11}$ can not survive to E^{18} because if it did then $r_{\Delta_1} \circ d^{18}(\beta_1 M_1^{11}) = d^{18}(\beta_1 M_1^{10}) = A[8]C[20]M_1$ and the latter element is not in Image r_{Δ_1}. There are two leaders, $\nu A[19]M_1^2\overline{M}_2$ and $\nu^2 C[20]M_1^3$, of degree 32 below the 28 row. However, $2^e\beta_1 M_1^{11}$, e = 0,1, are in Image $r_{2\Delta_1}$ and $\nu A[19]M_1^2\overline{M}_2$ is not in Image $r_{2\Delta_1}$. Therefore $d^{12}(2^e\beta_1 M_1^{11})$ can not be $\nu A[19]M_1^2\overline{M}_2$ for e = 0,1. Thus, $d^{16}(\beta_1 M_1^{11}) = \nu^2 C[20]M_1^3$ and $2\beta_1 M_1^{11}$ transgresses. Let $A[31] = d^{22}(2\beta_1 M_1^{11})$. Since A[30] is not a d^8-boundary, $\sigma A[23] = 0$. We have proved the first part of the following theorem.

THEOREM 5.3.10 (a) $\pi_{30}^S = Z_2 A[30]$ and $\sigma A[23] = 0$.
(b) $\pi_{31}^S = Z_2\eta A[30] \oplus Z_2 A[31] \oplus Z_{64}\gamma_3$.

PROOF. (b) We observed above that $\nu A[19]M_1^2\overline{M}_2$ transgresses to a nonzero element A[31] of π_{31}^S. The only other nonzero leader of degree 32 is $A[30]M_1$ which could only bound from an element below the 7 row, and there are no leaders of degree 33 there. Thus, $\eta A[30] = d^2(A[30]M_1) \neq 0$. Therefore π_{31}^S has a composition series of two Z_2s. By Theorem 2.2.7(b), $\langle\eta,\nu,\nu A[19],\nu\rangle$ is defined because $\langle\nu,\nu A[19],\nu\rangle \in \pi_{29}^S = 0$ and $\langle\eta,\nu,\nu A[19]\rangle \in \pi_{27}^S = Z_8\beta_3$. Thus, $\langle\eta,\nu,\nu A[19]\rangle$ contains 0 because $4\beta_3 = \eta^2\alpha_3 \in \text{Indet}\langle\eta,\nu,\nu A[19]\rangle$ and $2\langle\eta,\nu,\nu A[19]\rangle = \langle 2,\eta,\nu\rangle\nu A[19] = 0$. By Theorem 2.4.5(b),

$$A[31] \in \langle\eta,\nu,\nu A[19],\nu\rangle. \qquad [5.6]$$

Thus, $2A[31] \in 2\langle\eta,\nu,\nu A[19],\nu\rangle \subset \langle\langle 2,\eta,\nu\rangle,\nu A[19],\nu\rangle = \langle 0,\nu A[19],\nu\rangle = (\nu)$. However, $\nu\cdot\pi_{28}^S = \nu A[8]C[20] = 0$. Hence $2A[31] = 0$. ∎

The computations of Section 4 show that we have the following leaders. Since this is the last table of leaders of this chapter, we include the leaders of all degrees.

Row	Degree	Leader	Row	Degree	Leader
9	33	$\eta^2\sigma M_1^5\bar{M}_3$	22	62	$\nu A[19]M_1^7M_2^2\langle M_3\rangle$
11	33	$2\beta_1M_1^{11}$	23	33	$A[23]M_1^2\bar{M}_2$
14	34	$\sigma^2M_1^4\bar{M}_2^2$	23	35	$4\nu C[20]M_1^3\bar{M}_2$
15	37	$\gamma_1M_1^8\bar{M}_2$	23	63	$\gamma_2M_1^{20}$
16	34	$A[16]M_1^6\bar{M}_2$	24	60	$\eta A[23]M_1^{15}\bar{M}_2$
17	51	$\alpha_2M_1^{14}\bar{M}_2$	28	36	$A[8]C[20]M_1\bar{M}_2$
18	38	$C[18](M_1^4\bar{M}_2^2+2M_1^7\bar{M}_2)$	30	34	$A[30]M_1^2$
19	55	$\beta_2M_1^{18}$	31	33	$\eta A[30]M_1$
21	33	$\nu C[18]M_1^3\bar{M}_2$	31	35	$A[31]M_1^2$

FIGURE 5.3.7: Leaders in Rows 1 to 31 of Degree at Least 33

4. Tentative Differentials

In this section we give the tentative differentials determined by the differentials on leaders of degrees less than or equal to 32 which were determined in this chapter. We omit the differentials originating on the 7 row since they were determined in Chapter 3. Recall that these differentials are tentative in the sense that they are only valid under the assumption that there are no hidden differentials interfering with the computation.

We order the differentials by row for easy reference. We use the same notation as in Section 4.4 to display the bases in the various bidegrees. In a Z_2-vector space we omit the group in front of each basis element, and monomials which are to be added are bunched together. For example, see the first basis element in degree (28,9) below.

<u>DEGREE 9:</u> α_1 and $\eta\ \sigma$

The leading differential $d^{14}(\eta^2\sigma M_1^7\overline{M}_2) = \nu A[19]M_1^3$ determines tentative differentials by assigning the following values to monomials of degree 29 of $[Z_2(\eta^2\sigma M_1) \oplus Z_2\alpha_1] \otimes B<2>$: $\eta^2\sigma M_1\overline{M}_2^3$ and $\eta^2\sigma M_1^7\overline{M}_2$ are assigned 1 and all other monomials are assigned 0. The kernel of these tentative differentials is given by the table below. In this table as well as in the following one, the monomials with an even factor of M_1 have coefficient α_1 while the monomials with an odd factor of M_1 have coefficient $\eta^2\sigma$. The new leader is $\alpha_1 M_1^6\overline{M}_2$.

DEGREE	BASIS	DEGREE	BASIS	DEGREE	BASIS
(18,9)	6 1 0 0	(22,9)	11 0 0 0	(24,9)	5 0 1 0
	9 1 0 0	(26,9)	6 0 1 0		7 2 0 0
(28,9)	5 3 0 0 7 0 1 0		11 1 0 0	(30,9)	5 1 1 0
	6 3 0 0		15 0 0 0	(32,9)	6 1 1 0
	13 1 0 0	(34,9)	11 2 0 0		14 1 0 0
(36,9)	3 5 0 0		5 2 1 0 15 1 0 0		9 3 0 0 15 1 0 0
	11 0 1 0	(38,9)	3 3 1 0		6 2 1 0
	9 1 1 0		13 2 0 0	(40,9)	4 3 1 0
	5 5 0 0		11 3 0 0		13 0 1 0
(42,9)	7 0 2 0		6 5 0 0		11 1 1 0
	14 0 1 0		15 2 0 0	(44,9)	5 1 2 0 15 0 1 0
	7 0 0 1 7 5 0 0 15 0 1 0		6 3 1 0		9 2 1 0 15 0 1 0
	13 3 0 0 15 0 1 0	(46,9)	5 1 0 1		6 1 2 0
	5 6 0 0		11 4 0 0		13 1 1 0
	14 3 0 0	(48,9)	3 2 0 1		6 1 0 1
	3 7 0 0		5 4 1 0		9 5 0 0

11 2 1 0

5 2 2 0

9 3 1 0
15 1 1 0

(52,9) 1 6 1 0
5 2 0 1
13 2 1 0

5 2 0 1
5 7 0 0
13 2 1 0

11 0 0 1

(54,9) 3 3 0 1

5 5 1 0

6 7 0 0

15 4 0 0

(56,9) 3 1 1 1

5 3 2 0
7 0 3 0
15 2 1 0

11 1 2 0

21 0 1 0

5 1 3 0
5 3 0 1
7 0 1 1
7 5 1 0

11 1 0 1

14 5 0 0

(60,9) 3 2 3 0

6 1 3 0

9 7 0 0
15 5 0 0

15 0 0 1
15 5 0 0
23 0 1 0

14 1 1 0

6 4 1 0

11 0 2 0

3 3 2 0

5 7 0 0
7 4 1 0

11 5 0 0

5 0 1 1

7 2 2 0

11 3 1 0

21 2 0 0

6 0 1 1

7 2 0 1
7 7 0 0
15 2 1 0

12 3 1 0

25 1 0 0

5 3 0 1
7 0 1 1
7 5 1 0
13 3 1 0

11 6 0 0

22 0 1 0

3 4 0 1

6 3 0 1

11 4 1 0

14 3 1 0

(50,9) 3 5 1 0

7 6 0 0

22 1 0 0

5 0 3 0
13 2 1 0

9 1 2 0
13 2 1 0

13 2 1 0
23 1 0 0

6 0 3 0

9 1 0 1

14 2 1 0

27 0 0 0

3 6 1 0

6 5 1 0

13 5 0 0

(58,9) 1 7 1 0
7 5 1 0

6 3 2 0

15 0 2 0

23 2 0 0

5 1 1 1
15 0 0 1

5 6 1 0
15 5 0 0

13 1 2 0
15 0 0 1
15 5 0 0

21 3 0 0
23 0 1 0

27 1 0 0

The leading differential $d^{16}(\eta^2\sigma M_1^9) = \eta A[23]M_1$ determines tentative differentials by assigning the following values to monomials of degree 27 of $[Z_2(\eta^2\sigma M_1) \oplus Z_2\alpha_1] \otimes B\langle 2\rangle$: $\alpha_1 M_1^2\overline{M}_3$ is assigned 0 and all other monomials are assigned 1. The kernel of these tentative differentials is given by the table below. The new leader is $\eta^2\sigma M_1^5\overline{M}_3$.

DEGREE	BASIS	DEGREE	BASIS	DEGREE	BASIS
(24,9)	5 0 1 0	(26,9)	7 2 0 0	(28,9)	5 3 0 0
					7 0 1 0
					11 1 0 0
(30,9)	5 1 1 0	(36,9)	3 5 0 0	(38,9)	3 3 1 0
	15 0 0 0		5 2 1 0		9 1 1 0
			11 0 1 0		13 2 0 0
			15 1 0 0		
(40,9)	5 5 0 0	(42,9)	7 0 2 0	(44,9)	5 1 2 0
	13 0 1 0				7 0 0 1
	4 3 1 0				7 5 0 0
(46,9)	5 1 0 1	(48,9)	3 2 0 1	(50,9)	5 2 2 0
	11 4 0 0		9 5 0 0		7 6 0 0
			6 1 0 1		11 0 2 0
(52,9)	1 6 1 0		3 3 2 0		5 0 3 0
	7 4 1 0		5 0 3 0		5 2 0 1
	13 2 1 0		5 7 0 0		7 4 1 0
	23 1 0 0		7 4 1 0		11 0 0 1
			9 1 2 0		
			11 5 0 0		
(54,9)	3 3 0 1		5 0 1 1		7 2 2 0
	5 0 1 1		15 4 0 0		21 2 0 0
	9 1 0 1				
	6 7 0 0				
	14 2 1 0				
	21 2 0 0	(56,9)	3 1 1 1		5 3 2 0
	27 0 0 0		13 5 0 0		7 0 3 0
			6 0 1 1		7 2 0 1
					7 7 0 0
					11 1 2 0

	21 1 0 1	(58,9)	5 1 3 0		23 2 0 0
			5 3 0 1		
			7 0 1 1		
			7 5 1 0		
			11 1 0 1		
			11 6 0 0		
			15 0 2 0		
			14 5 0 0		
			22 0 1 0		
(60,9)	3 2 3 0		3 4 0 1		21 3 0 0
	3 4 0 1		5 1 1 1		23 0 1 0
	5 6 1 0		11 4 1 0		27 1 0 0
	11 4 1 0		15 5 0 0		
	13 1 2 0		23 0 1 0		
	15 0 0 1				
	21 3 0 0				
	23 0 1 0				
	6 1 3 0				

<u>DEGREE 9:</u> $\eta A[8] = \nu^3$

The leading differential $d^2(A[8]M_1) = \eta A[8]$ determines tentative differentials with cokernel $Z_2\eta A[8]M_1 \otimes B<2>$, and the $\eta A[8]$-leader is $\eta A[8]M_1$.

The leading differential $d^4(\nu^2 M_1^3) = \eta A[8]M_1$ determines tentative differentials with image $Z_2\eta A[8]\{M_1, M_1^3, M_1\bar{M}_2\} \otimes B<4>$. The remaining elements are $Z_2(\eta A[8]M_1^3\bar{M}_2) \otimes B<4>$, and the new $\eta A[8]$-leader is $\eta A[8]M_1^3\bar{M}_2$.

The leading differential $d^{12}(\eta A[8]M_1^3\bar{M}_2) = C[20]$ determines tentative differentials which are a monomorphism on $Z_2(\eta A[8]M_1^3\bar{M}_2) \otimes B<4>$. Thus, there are no remaining elements.

<u>DEGREE 11:</u> β_1

The leading differential $d^{16}(\beta_1 M_1^{11}) = \nu^2 C[20]M_1^3$ determines tentative differentials by assigning the following values to monomials of degree 33 of $Z_8\beta_1 \otimes H_*BP$: $\beta_1 M_1^{11}$ is assigned 1 and all other monomials are assigned 0. The kernel of these tentative differentials is given by the table below. The new β_1-leader is $\beta_1 M_1^{10}$.

DEGREE	GROUP	GENERATOR
(20,11)	Z_2	10 0 0 0
	Z_2	9 1 0 0
(28,11)	Z_2	2/ 11 1 0 0 1/ 4 1 1 0 3/ 14 0 0 0
(30,11)	Z_2	2/ 12 1 0 0
	Z_2	2/ 7 3 0 0 1/ 13 1 0 0
(34,11)	Z_2	1/ 7 1 1 0 6/ 14 1 0 0
	Z_2	2/ 14 1 0 0
	Z_2	1/ 9 3 0 0 2/ 15 1 0 0 3/ 12 2 0 0
(38,11)	Z_2	2/ 6 2 1 0 2/ 10 3 0 0
(40,11)	Z_2	3/ 7 2 1 0 6/ 11 3 0 0 6/ 13 0 1 0 1/ 4 3 1 0 3/ 10 1 1 0 6/ 14 2 0 0
	Z_2	2/ 11 3 0 0 1/ 14 2 0 0
	Z_4	1/ 7 2 1 0 2/ 10 1 1 0 1/ 14 2 0 0
	Z_2	6/ 5 3 1 0 6/ 11 1 1 0 1/ 6 0 0 1 6/ 14 0 1 0
(44,11)	Z_2	1/ 4 1 0 1 6/ 6 3 1 0 6/ 12 1 1 0

DEGREE	GROUP	GENERATOR
(22,11)	Z_2	2/ 8 1 0 0
(26,11)	Z_2	1/ 6 0 1 0 2/ 10 1 0 0
	Z_2	2/ 11 1 0 0 3/ 14 0 0 0
	Z_4	6 3 0 0
	Z_2	5/ 13 1 0 0 1/ 10 2 0 0
	Z_2	2/ 8 3 0 0
(36,11)	Z_2	3/ 9 3 0 0 1/ 2 3 1 0 3/ 8 1 1 0 2/ 12 2 0 0
	Z_2	6/ 15 1 0 0 1/ 12 2 0 0
	Z_2	1/ 9 1 1 0 2/ 10 3 0 0 6/ 12 0 1 0
	Z_2	6/ 7 2 1 0 2/ 11 3 0 0 4/ 13 0 1 0 1/ 6 0 2 0 6/ 10 1 1 0 7/ 14 2 0 0
	Z_2	1/ 13 0 1 0 2/ 14 2 0 0
(42,11)	Z_2	2/ 5 3 1 0 2/ 11 1 1 0 6/ 12 3 0 0
	Z_2	2/ 12 3 0 0
	Z_4	2/ 11 1 1 0
	Z_2	1/ 4 6 0 0 3/ 10 4 0 0

DEGREE	GROUP	GENERATOR
(24,11)	Z_2	6 2 0 0
	Z_2	2/ 10 1 0 0
	Z_4	14 0 0 0
(32,11)	Z_2	7 3 0 0 6 1 1 0
	Z_4	1/ 7 3 0 0 5/ 13 1 0 0
	Z_2	1/ 10 0 1 0 3/ 14 1 0 0
	Z_2	2/ 9 3 0 0 2/ 8 1 1 0
	Z_4	2/ 15 1 0 0
	Z_2	1/ 10 3 0 0 6/ 12 0 1 0
	Z_4	6 2 1 0
	Z_2	2/ 7 2 1 0 6/ 11 3 0 0 4/ 13 0 1 0 2/ 10 1 1 0 1/ 14 2 0 0
	Z_4	14 2 0 0
	Z_2	1/ 8 2 1 0 5/ 12 3 0 0 3/ 14 0 1 0
	Z_4	5 3 1 0 11 1 1 0 12 3 0 0 14 0 1 0
	Z_2	2/ 13 3 0 0 2/ 6 3 1 0 5/ 10 4 0 0 2/ 12 1 1 0

	Z_2 2/ 13 3 0 0 1/ 10 4 0 0	Z_2 7/ 13 3 0 0 1/ 15 0 1 0 1/ 12 1 1 0	Z_4 6 3 1 0 Z_4 2/ 13 3 0 0
(46,11)	Z_2 1/ 4 4 1 0 1/ 6 1 2 0 3/ 10 2 1 0 1/ 14 3 0 0	Z_2 2/ 6 1 2 0 6/ 10 2 1 0	Z_2 2/ 8 5 0 0 2/ 10 2 1 0 2/ 14 3 0 0
	Z_2 1/ 13 1 1 0 1/ 10 2 1 0 5/ 14 3 0 0	Z_2 2/ 14 3 0 0 Z_4 6 1 2 0	Z_8 1/ 7 3 1 0 5/ 13 1 1 0
(48,11)	Z_2 1/ 4 2 2 0 6/ 6 1 0 1 3/ 6 6 0 0 6/ 8 3 1 0 1/ 10 0 2 0	Z_2 5/ 7 1 2 0 6/ 9 5 0 0 7/ 11 2 1 0 5/ 15 3 0 0 1/ 6 1 0 1 7/ 8 3 1 0 3/ 10 0 2 0 6/ 14 1 1 0	Z_2 2/ 7 1 2 0 4/ 9 5 0 0 6/ 11 2 1 0 2/ 15 3 0 0 3/ 6 6 0 0 6/ 10 0 2 0 2/ 14 1 1 0
	Z_2 1/ 6 6 0 0 2/ 8 3 1 0	Z_2 7/ 11 2 1 0 7/ 15 3 0 0 1/ 8 3 1 0 5/ 10 0 2 0 6/ 14 1 1 0	Z_2 2/ 11 2 1 0 2/ 15 3 0 0 1/ 10 0 2 0 2/ 14 1 1 0
	Z_2 2/ 11 2 1 0 2/ 15 3 0 0 6/ 14 1 1 0	Z_2 1/ 9 5 0 0 Z_4 2/ 15 3 0 0	Z_4 1/ 7 1 2 0 6/ 15 3 0 0 6/ 14 1 1 0
(50,11)	Z_2 3/ 7 1 0 1 6/ 15 1 1 0 1/ 4 2 0 1 6/ 4 7 0 0 6/ 6 4 1 0 3/ 10 0 0 1 6/ 10 5 0 0	Z_2 1/ 7 1 0 1 2/ 15 1 1 0 2/ 4 7 0 0 2/ 6 4 1 0 Z_2 2/ 6 4 1 0	Z_2 6/ 15 1 1 0 1/ 4 7 0 0 6/ 8 1 2 0 3/ 10 5 0 0 3/ 12 2 1 0
	Z_2 2/ 15 1 1 0 1/ 10 0 0 1 3/ 10 5 0 0	Z_2 1/ 9 3 1 0 4/ 15 1 1 0 2/ 10 5 0 0	Z_2 4/ 15 1 1 0 2/ 10 5 0 0 2/ 12 2 1 0
	Z_2 2/ 8 1 2 0	Z_2 2/ 12 2 1 0	Z_4 2/ 15 1 1 0
(52,11)	Z_2 3/ 5 7 0 0 1/ 11 5 0 0 3/ 13 2 1 0 1/ 2 3 0 1 1/ 4 5 1 0 1/ 8 1 0 1 3/ 26 0 0 0	Z_2 7/ 5 7 0 0 3/ 9 1 2 0 3/ 11 5 0 0 5/ 13 2 1 0 1/ 4 0 1 1 3/ 4 5 1 0 6/ 6 2 2 0 6/ 8 1 0 1 3/ 14 4 0 0	Z_2 7/ 5 7 0 0 3/ 7 4 1 0 6/ 9 1 2 0 7/ 11 5 0 0 2/ 13 2 1 0 1/ 4 5 1 0 7/ 6 2 2 0 6/ 14 4 0 0

Z_2 2/ 7 4 1 0
4/ 9 1 2 0
2/ 11 5 0 0
1/ 6 2 2 0
2/ 8 1 0 1
3/ 14 4 0 0
5/ 26 0 0 0

Z_2 1/ 7 4 1 0
4/ 9 1 2 0
4/ 11 5 0 0
7/ 13 2 1 0
1/ 14 4 0 0

Z_2 26 0 0 0

(54, 11) Z_2 1/ 5 5 1 0
1/ 2 6 1 0
1/ 6 7 0 0
2/ 12 5 0 0
2/ 14 2 1 0
2/ 24 1 0 0

Z_2 2/ 5 5 1 0
2/ 9 1 0 1
6/ 14 2 1 0
6/ 20 0 1 0

Z_2 2/ 10 1 2 0
2/ 12 5 0 0
2/ 24 1 0 0

Z_4 1/ 5 5 1 0
6/ 11 3 1 0
1/ 6 7 0 0
3/ 12 5 0 0
1/ 14 2 1 0
7/ 24 1 0 0

(56, 11) Z_2 2/ 7 7 0 0
1/ 0 7 1 0
2/ 4 1 3 0
1/ 4 3 0 1
1/ 6 5 1 0
1/ 10 1 0 1
2/ 10 6 0 0
3/ 12 3 1 0
2/ 14 0 2 0
2/ 22 2 0 0

Z_2 6/ 5 7 0 0
4/ 13 2 1 0
2/ 8 1 0 1
7/ 14 4 0 0
5/ 26 0 0 0

Z_2 1/ 9 1 2 0
3/ 26 0 0 0

Z_2 2/ 10 3 1 0

Z_4 2/ 13 2 1 0
1/ 14 4 0 0

Z_2 3/ 5 5 1 0
1/ 9 1 0 1
1/ 4 3 2 0
2/ 6 0 3 0
2/ 6 2 0 1
3/ 6 7 0 0
1/ 10 1 2 0
3/ 12 5 0 0
1/ 20 0 1 0
1/ 24 1 0 0

Z_2 1/ 9 1 0 1
2/ 11 3 1 0
2/ 6 7 0 0
6/ 10 1 2 0
6/ 12 5 0 0
6/ 24 1 0 0

Z_2 2/ 14 2 1 0
2/ 24 1 0 0

Z_4 1/ 6 7 0 0
1/ 14 2 1 0
2/ 24 1 0 0

Z_2 4/ 7 7 0 0
6/ 11 1 2 0
1/ 4 1 3 0
7/ 4 3 0 1
3/ 6 0 1 1
7/ 6 5 1 0
5/ 10 1 0 1
1/ 14 0 2 0
7/ 22 2 0 0

Z_2 2/ 5 7 0 0
2/ 7 4 1 0
4/ 9 1 2 0
2/ 13 2 1 0
3/ 14 4 0 0
7/ 26 0 0 0

Z_2 2/ 11 5 0 0
2/ 13 2 1 0
5/ 14 4 0 0
5/ 26 0 0 0

Z_4 14 4 0 0

Z_2 2/ 5 5 1 0
7/ 9 1 0 1
1/ 6 2 0 1
5/ 6 7 0 0
2/ 10 1 2 0
5/ 14 2 1 0
2/ 24 1 0 0

Z_2 2/ 6 0 3 0
2/ 6 7 0 0

Z_2 2/ 10 1 2 0
2/ 12 5 0 0
3/ 20 0 1 0
2/ 24 1 0 0

Z_4 1/ 6 0 3 0
1/ 6 7 0 0
1/ 14 2 1 0
2/ 24 1 0 0

Z_4 2/ 11 3 1 0

Z_2 3/ 5 3 2 0
1/ 7 0 3 0
4/ 7 7 0 0
3/ 11 1 2 0
2/ 13 5 0 0
3/ 15 2 1 0
1/ 25 1 0 0
1/ 4 3 0 1
5/ 6 0 1 1
1/ 6 5 1 0
3/ 10 1 0 1
1/ 12 3 1 0
7/ 14 0 2 0
3/ 22 2 0 0

Z_2 5/ 5 3 2 0
2/ 7 0 3 0
1/ 7 7 0 0
7/ 11 1 2 0
6/ 13 5 0 0
1/ 6 0 1 1
2/ 6 5 1 0
6/ 10 1 0 1
6/ 10 6 0 0
6/ 12 3 1 0
7/ 14 0 2 0
6/ 22 2 0 0

Z_2 2/ 5 3 2 0
4/ 7 0 3 0
2/ 7 7 0 0
6/ 11 1 2 0
4/ 13 5 0 0
2/ 12 3 1 0
6/ 14 0 2 0

Z_2 4/ 7 0 3 0
2/ 7 7 0 0
4/ 15 2 1 0
4/ 25 1 0 0
6/ 6 5 1 0
6/ 10 1 0 1
5/ 10 6 0 0
2/ 14 0 2 0

Z_2 3/ 7 7 0 0
3/ 15 2 1 0
3/ 25 1 0 0
1/ 6 5 1 0
6/ 10 1 0 1
2/ 10 6 0 0
1/ 12 3 1 0
6/ 14 0 2 0
7/ 22 2 0 0

Z_2 6/ 7 7 0 0
2/ 10 1 0 1
7/ 10 6 0 0
6/ 14 0 2 0
3/ 22 2 0 0

Z_2 2/ 11 1 2 0
1/ 14 0 2 0

Z_2 2/ 7 7 0 0
1/ 10 6 0 0
6/ 12 3 1 0
2/ 14 0 2 0
5/ 22 2 0 0

Z_2 7/ 13 5 0 0
4/ 15 2 1 0
4/ 25 1 0 0
1/ 10 6 0 0
2/ 12 3 1 0
2/ 22 2 0 0

Z_2 7/ 13 5 0 0
2/ 15 2 1 0
5/ 25 1 0 0
2/ 12 3 1 0
1/ 22 2 0 0

Z_2 2/ 13 5 0 0
4/ 15 2 1 0
6/ 25 1 0 0
3/ 22 2 0 0

Z_4 1/ 5 3 2 0
1/ 7 0 3 0
1/ 7 7 0 0
1/ 11 1 2 0
6/ 15 2 1 0
2/ 25 1 0 0
1/ 14 0 2 0
2/ 22 2 0 0

Z_4 1/ 7 0 3 0
1/ 13 5 0 0
4/ 25 1 0 0
3/ 14 0 2 0
5/ 22 2 0 0

Z_4 1/ 13 5 0 0
2/ 25 1 0 0
3/ 22 2 0 0

Z_4 14 0 2 0

(58,11) Z_2 1/ 5 3 0 1
3/ 7 5 1 0
1/ 11 1 0 1
1/ 13 3 1 0
1/ 4 1 1 1
1/ 4 6 1 0
1/ 8 7 0 0
2/ 14 0 0 1
2/ 14 5 0 0
3/ 22 0 1 0
1/ 26 1 0 0

Z_2 2/ 5 3 0 1
3/ 7 0 1 1
6/ 7 5 1 0
2/ 11 1 0 1
2/ 4 6 1 0
2/ 8 7 0 0
6/ 12 1 2 0
7/ 14 0 0 1
7/ 14 5 0 0
6/ 22 0 1 0
2/ 26 1 0 0

Z_2 1/ 7 0 1 1
6/ 7 5 1 0
6/ 11 1 0 1
2/ 13 3 1 0
6/ 8 7 0 0
6/ 14 5 0 0
5/ 20 3 0 0
6/ 22 0 1 0
5/ 26 1 0 0

Z_2 2/ 4 6 1 0
2/ 6 3 2 0
2/ 14 5 0 0
2/ 26 1 0 0

Z_2 2/ 6 3 2 0

Z_2 2/ 7 5 1 0
6/ 13 3 1 0
6/ 14 5 0 0
2/ 22 0 1 0
6/ 26 1 0 0

Z_2 2/ 11 1 0 1
2/ 13 3 1 0
2/ 8 7 0 0
1/ 14 0 0 1
1/ 14 5 0 0
7/ 20 3 0 0
1/ 22 0 1 0
3/ 26 1 0 0

Z_2 6/ 13 3 1 0
1/ 10 4 1 0
2/ 12 1 2 0
1/ 14 5 0 0
3/ 20 3 0 0
2/ 22 0 1 0
1/ 26 1 0 0

Z_2 2/ 8 7 0 0
2/ 12 1 2 0

Z_2 2/ 12 1 2 0
6/ 26 1 0 0

Z_2 2/ 14 5 0 0
3/ 20 3 0 0
6/ 22 0 1 0
5/ 26 1 0 0

Z_4 1/ 5 3 0 1
1/ 11 1 0 1
1/ 8 7 0 0
5/ 22 0 1 0

Z_4 2/ 13 3 1 0
1/ 14 0 0 1
7/ 14 5 0 0
3/ 22 0 1 0

Z_4 2/ 13 3 1 0
5/ 20 3 0 0
5/ 26 1 0 0

Z_4 1/ 7 5 1 0
3/ 13 3 1 0
1/ 4 6 1 0
1/ 6 3 2 0
6/ 14 0 0 1
1/ 14 5 0 0
2/ 20 3 0 0
5/ 22 0 1 0
2/ 26 1 0 0

Z_4 6 3 2 0
26 1 0 0

The leading differential $d^{18}(\beta_1 M_1^{10}) = A[8]C[20]M_1$ determines tentative differentials by assigning the following values to monomials of degree 31 of $Z_8\beta_1 \otimes H_*BP$: $\beta_1 M_1^{10}$ is assigned 1 and all other monomials are assigned 0. The kernel of these tentative differentials is given by the table below. The new β_1-leader ia $2\beta_1 M_1^8 \bar{M}_2$.

DEGREE	GROUP	GENERATOR	DEGREE	GROUP	GENERATOR	DEGREE	GROUP	GENERATOR
(22,11)	Z_2	2/ 8 1 0 0	(24,11)	Z_2	6 2 0 0	(26,11)	Z_2	2/ 10 1 0 0
	Z_2	1/ 6 0 1 0 2/ 10 1 0 0	(28,11)	Z_2	2/ 11 1 0 0 1/ 4 1 1 0 3/ 14 0 0 0		Z_4	2/ 11 1 0 0 2/ 14 0 0 0
(30,11)	Z_2	2/ 12 1 0 0		Z_4	6 3 0 0	(32,11)	Z_2	3/ 7 3 0 0 4/ 13 1 0 0 1/ 6 1 1 0 7/ 10 2 0 0

Z_4 1/ 7 3 0 0
7/ 10 2 0 0

(34,11) Z_2 1/ 7 1 1 0
7/ 10 0 1 0
3/ 14 1 0 0

Z_2 2/ 8 3 0 0

Z_2 2/ 14 1 0 0

(36,11) Z_2 2/ 9 3 0 0
6/ 15 1 0 0
1/ 2 3 1 0
3/ 8 1 1 0
7/ 12 2 0 0

Z_2 2/ 9 3 0 0
2/ 8 1 1 0

Z_4 2/ 15 1 0 0

(38,11) Z_2 2/ 6 2 1 0
2/ 10 3 0 0

Z_2 2/ 10 3 0 0
6/ 12 0 1 0

Z_4 6 2 1 0

(40,11) Z_2 3/ 7 2 1 0
6/ 11 3 0 0
6/ 13 0 1 0
1/ 4 3 1 0
3/ 10 1 1 0
6/ 14 2 0 0

Z_2 6/ 7 2 1 0
4/ 13 0 1 0
1/ 6 0 2 0
6/ 10 1 1 0
6/ 14 2 0 0

Z_2 2/ 7 2 1 0
4/ 11 3 0 0
4/ 13 0 1 0
2/ 10 1 1 0

Z_4 1/ 7 2 1 0
2/ 10 1 1 0
1/ 14 2 0 0

Z_4 2/ 11 3 0 0

(42,11) Z_2 2/ 5 3 1 0
2/ 11 1 1 0
6/ 12 3 0 0

Z_2 6/ 5 3 1 0
6/ 11 1 1 0
1/ 6 0 0 1
6/ 14 0 1 0

Z_2 2/ 12 3 0 0

Z_4 2/ 11 1 1 0

Z_4 5 3 1 0
11 1 1 0
12 3 0 0
14 0 1 0

(44,11) Z_2 6/ 13 3 0 0
1/ 4 1 0 1
6/ 6 3 1 0
7/ 10 4 0 0
6/ 12 1 1 0

Z_2 2/ 6 3 1 0
2/ 12 1 1 0

Z_4 1/ 4 6 0 0
7/ 6 3 1 0
3/ 10 4 0 0

Z_4 2/ 13 3 0 0

(46,11) Z_2 2/ 8 5 0 0
2/ 10 2 1 0
2/ 14 3 0 0

Z_2 2/ 6 1 2 0
6/ 10 2 1 0

Z_2 2/ 14 3 0 0

Z_4 6 1 2 0

Z_8 1/ 7 3 1 0
4/ 13 1 1 0
7/ 10 2 1 0
1/ 14 3 0 0

(48,11) Z_2 1/ 4 2 2 0
6/ 6 1 0 1
3/ 6 6 0 0
6/ 8 3 1 0

Z_2 5/ 7 1 2 0
6/ 9 5 0 0
4/ 11 2 1 0
2/ 15 3 0 0
1/ 6 1 0 1
7/ 10 0 2 0
2/ 14 1 1 0

Z_2 2/ 7 1 2 0
4/ 9 5 0 0
6/ 11 2 1 0
2/ 15 3 0 0
2/ 6 6 0 0
2/ 8 3 1 0
6/ 10 0 2 0
6/ 14 1 1 0

Z_2 2/ 11 2 1 0
2/ 15 3 0 0
6/ 14 1 1 0

Z_4 1/ 7 1 2 0
6/ 15 3 0 0
7/ 10 0 2 0
2/ 14 1 1 0

Z_4 2/ 15 3 0 0

(50,11)	Z_2 3/ 7 1 0 1 4/ 15 1 1 0 1/ 4 2 0 1 6/ 4 7 0 0 6/ 6 4 1 0 2/ 10 0 0 1 3/ 10 5 0 0	Z_2 1/ 7 1 0 1 2/ 4 7 0 0 2/ 6 4 1 0 7/ 10 0 0 1 5/ 10 5 0 0	Z_2 4/ 15 1 1 0 2/ 10 5 0 0 2/ 12 2 1 0 Z_2 2/ 6 4 1 0
	Z_2 2/ 8 1 2 0	Z_2 2/ 12 2 1 0	Z_4 2/ 15 1 1 0
(52,11)	Z_2 3/ 5 7 0 0 6/ 7 4 1 0 4/ 9 1 2 0 7/ 11 5 0 0 3/ 13 2 1 0 1/ 2 3 0 1 1/ 4 5 1 0 7/ 6 2 2 0 7/ 8 1 0 1 5/ 14 4 0 0 6/ 26 0 0 0	Z_2 7/ 5 7 0 0 2/ 9 1 2 0 3/ 11 5 0 0 5/ 13 2 1 0 1/ 4 0 1 1 3/ 4 5 1 0 6/ 6 2 2 0 6/ 8 1 0 1 7/ 14 4 0 0 5/ 26 0 0 0	Z_2 7/ 5 7 0 0 6/ 9 1 2 0 1/ 11 5 0 0 3/ 13 2 1 0 1/ 4 5 1 0 6/ 6 2 2 0 2/ 8 1 0 1 2/ 14 4 0 0 3/ 26 0 0 0
	Z_2 2/ 5 7 0 0 2/ 7 4 1 0 4/ 9 1 2 0 2/ 11 5 0 0 4/ 13 2 1 0 1/ 6 2 2 0 1/ 26 0 0 0	Z_2 6/ 5 7 0 0 6/ 11 5 0 0 2/ 13 2 1 0 2/ 8 1 0 1 2/ 14 4 0 0	Z_2 2/ 5 7 0 0 2/ 7 4 1 0 4/ 9 1 2 0 6/ 11 5 0 0 6/ 14 4 0 0 2/ 26 0 0 0
	Z_2 2/ 10 3 1 0	Z_4 2/ 11 5 0 0	Z_4 2/ 13 2 1 0
(54,11)	Z_2 2/ 5 5 1 0 2/ 9 1 0 1 1/ 2 6 1 0 1/ 4 3 2 0 2/ 6 0 3 0 1/ 6 2 0 1 7/ 6 7 0 0 7/ 10 1 2 0 5/ 12 5 0 0 5/ 14 2 1 0 1/ 20 0 1 0 1/ 24 1 0 0	Z_2 2/ 5 5 1 0 2/ 9 1 0 1 6/ 14 2 1 0 6/ 20 0 1 0 Z_2 2/ 10 1 2 0 2/ 12 5 0 0 2/ 24 1 0 0	Z_2 2/ 10 1 2 0 2/ 12 5 0 0 3/ 20 0 1 0 2/ 24 1 0 0 Z_2 2/ 6 0 3 0 2/ 6 7 0 0 Z_2 2/ 14 2 1 0 2/ 24 1 0 0
	Z_4 6/ 11 3 1 0 1/ 4 3 2 0 1/ 6 2 0 1 1/ 6 7 0 0 1/ 10 1 2 0 2/ 24 1 0 0 Z_4 2/ 11 3 1 0	Z_4 6/ 9 1 0 1 2/ 11 3 1 0 1/ 6 0 3 0 1/ 6 2 0 1 6/ 6 7 0 0 6/ 10 1 2 0 6/ 12 5 0 0 3/ 20 0 1 0 6/ 24 1 0 0	Z_4 6/ 9 1 0 1 2/ 11 3 1 0 1/ 6 2 0 1 6/ 6 7 0 0 6/ 10 1 2 0 6/ 12 5 0 0 3/ 20 0 1 0 6/ 24 1 0 0

(56,11) Z_2
1/ 5 3 2 0
2/ 7 0 3 0
6/ 7 7 0 0
3/ 11 1 2 0
6/ 13 5 0 0
1/ 15 2 1 0
2/ 25 1 0 0
1/ 0 7 1 0
2/ 4 1 3 0
1/ 4 3 0 1
5/ 6 0 1 1
5/ 10 1 0 1
7/ 10 6 0 0
6/ 12 3 1 0
7/ 14 0 2 0

Z_2
3/ 5 3 2 0
6/ 7 0 3 0
3/ 7 7 0 0
7/ 11 1 2 0
2/ 13 5 0 0
1/ 4 1 3 0
7/ 4 3 0 1
2/ 6 0 1 1
5/ 6 5 1 0
7/ 10 1 0 1
6/ 10 6 0 0
2/ 12 3 1 0
2/ 14 0 2 0
1/ 22 2 0 0

Z_2
5/ 5 3 2 0
5/ 7 0 3 0
7/ 7 7 0 0
7/ 11 1 2 0
6/ 13 5 0 0
1/ 4 3 0 1
7/ 6 0 1 1
1/ 10 1 0 1
6/ 14 0 2 0
6/ 22 2 0 0

Z_2
2/ 13 5 0 0
4/ 15 2 1 0
6/ 25 1 0 0
3/ 22 2 0 0

Z_2
5/ 5 3 2 0
2/ 7 0 3 0
7/ 7 7 0 0
7/ 11 1 2 0
6/ 13 5 0 0
1/ 6 0 1 1
2/ 6 5 1 0
6/ 10 1 0 1
5/ 10 6 0 0
5/ 14 0 2 0
1/ 22 2 0 0

Z_2
2/ 5 3 2 0
4/ 7 0 3 0
2/ 7 7 0 0
6/ 11 1 2 0
4/ 13 5 0 0
2/ 12 3 1 0
6/ 14 0 2 0

Z_2
4/ 7 0 3 0
4/ 15 2 1 0
4/ 25 1 0 0
6/ 6 5 1 0
6/ 10 1 0 1
6/ 12 3 1 0
7/ 22 2 0 0

Z_2
4/ 7 7 0 0
2/ 10 1 0 1
6/ 10 6 0 0
2/ 12 3 1 0
6/ 22 2 0 0

Z_4
2/ 11 1 2 0

Z_4
1/ 5 3 2 0
1/ 7 0 3 0
7/ 7 7 0 0
7/ 11 1 2 0
6/ 15 2 1 0
2/ 25 1 0 0
7/ 10 6 0 0
2/ 12 3 1 0
6/ 14 0 2 0
5/ 22 2 0 0

Z_4
1/ 7 0 3 0
4/ 7 7 0 0
6/ 15 2 1 0
6/ 25 1 0 0
6/ 6 5 1 0
2/ 10 6 0 0
6/ 12 3 1 0
5/ 14 0 2 0
5/ 22 2 0 0

Z_4
4/ 7 7 0 0
2/ 15 2 1 0
4/ 25 1 0 0
2/ 6 5 1 0
6/ 10 6 0 0
2/ 12 3 1 0
6/ 14 0 2 0
1/ 22 2 0 0

(58,11) Z_2
4/ 7 0 1 1
2/ 7 5 1 0
4/ 11 1 0 1
2/ 13 3 1 0
2/ 4 1 1 1
6/ 12 1 2 0
6/ 14 0 0 1
2/ 10 3 0 0
3/ 22 0 1 0
6/ 26 1 0 0

Z_2
2/ 5 3 0 1
7/ 7 0 1 1
6/ 7 5 1 0
2/ 11 1 0 1
2/ 13 3 1 0
2/ 4 6 1 0
2/ 8 7 0 0
3/ 10 4 1 0
7/ 14 0 0 1
6/ 14 5 0 0
1/ 20 3 0 0
5/ 26 1 0 0

Z_2	2/ 7 5 1 0
	6/ 13 3 1 0
	6/ 14 5 0 0
	2/ 22 0 1 0
	6/ 26 1 0 0
Z_2	2/ 8 7 0 0
	2/ 12 1 2 0
Z_4	1/ 5 3 0 1
	1/ 11 1 0 1
	2/ 13 3 1 0
	1/ 4 1 1 1
	1/ 6 3 2 0
	1/ 8 7 0 0
	7/ 10 4 1 0
	1/ 20 3 0 0
	6/ 22 0 1 0

Z_2	2/ 4 6 1 0
	2/ 6 3 2 0
	2/ 14 5 0 0
	2/ 26 1 0 0
Z_2	2/ 12 1 2 0
	6/ 26 1 0 0
Z_4	1/ 5 3 0 1
	7/ 11 1 0 1
	6/ 13 3 1 0
	1/ 8 7 0 0
	7/ 14 0 0 1
	1/ 14 5 0 0
	2/ 22 0 1 0
Z_4	2/ 11 1 0 1

Z_2	2/ 14 5 0 0
	3/ 20 3 0 0
	6/ 22 0 1 0
	5/ 26 1 0 0
Z_2	2/ 6 3 2 0
Z_4	2/ 13 3 1 0
	7/ 20 3 0 0
	3/ 26 1 0 0
Z_4	6 3 2 0
	26 1 0 0

DEGREE 14: A[14]

The leading differential $d^{12}(4\nu M_1^3\bar{M}_2) = A[14]$ determines tentative differentials with image $Z_2A[14] \otimes B\langle 4\rangle$. Since $\eta A[14] \neq 0$, the remaining elements are $Z_2A[14]\{M_1^2, \bar{M}_2, M_1^2\bar{M}_2\} \otimes B\langle 4\rangle$, and the A[14]-leader is $A[14]M_1^2$.

The leading differential $d^4(A[14]M_1^2) = \nu A[14]$ determines tentative differentials which are a monomorphism on $Z_2A[14]\{M_1^2, \bar{M}_2, M_1^2\bar{M}_2\} \otimes B\langle 4\rangle$. Thus, there are no remaining elements.

DEGREE 14: σ^2

The leading differential $d^8(\sigma^2\bar{M}_2\langle M_1^4\rangle) = \sigma^3\bar{M}_2$ determines tentative differentials by assigning the following values to monomials of degree 28 of $Z_2\sigma^2 \otimes H_*BP$: $\sigma^2M_1M_2^2$, σ^2M_3 and $\sigma^2M_1^4M_2$ are assigned 1 and $\sigma^2M_1^7$ is assigned 0. The elements of $E^{10}_{*,14}$ in degrees less than 69 with a representative in $Z_2\sigma^2 \otimes H_*BP$ are given by the table below. The new σ^2-leader is $\sigma^2M_1^4\bar{M}_2^2$.

DEGREE	BASIS	DEGREE	BASIS	DEGREE	BASIS
(20,14)	4 2 0 0	(24,14)	2 1 1 0	(36,14)	12 2 0 0
(40,14)	14 2 0 0	(44,14)	4 6 0 0		
(48,14)	2 5 1 0		4 2 2 0	(50,14)	4 2 0 1 4 7 0 0 12 2 1 0
(52,14)	2 1 3 0 2 3 0 1 4 0 1 1 4 5 1 0		20 2 0 0	(54,14)	6 2 0 1 6 7 0 0 14 2 1 0

DEGREE 15: $\eta A[14]$

The leading differential $d^2(A[14]M_1) = \eta A[14]$ determines tentative differentials with cokernel $Z_2\eta A[14]M_1 \otimes B\langle 2\rangle$. The $\eta A[14]$-leader is $\eta A[14]M_1$.

The leading differential $d^8(A[8]M_1^2\bar{M}_2) = \eta A[14]M_1$ determines tentative differentials with image $Z_2\eta A[14]M_1 \otimes B\langle 4\rangle$. The remaining elements are $Z_2\eta A[14]\{M_1^3, M_1\bar{M}_2, M_1^3\bar{M}_2\} \otimes B\langle 4\rangle$, and the new $\eta A[14]$-leader is $\eta A[14]M_1^3$.

The leading differential $d^6(\eta A[14]M_1^3) = 2C[20]$ determines tentative differentials which are a monomorphism on $Z_2\eta A[14]\{M_1^3, M_1\bar{M}_2, M_1^3\bar{M}_2\} \otimes B\langle 4\rangle$. Thus, there are no remaining elements.

DEGREE 17: $\eta A[16]$

Since $\eta^2 A[16] \neq 0$, the only element of $E^4_{*,17}$ with a representative in $Z_2\eta A[16] \otimes H_*BP$ is zero.

DEGREE 17: $\nu A[14]$

Since $\nu^2 A[14] \neq 0$, the only element of $E^6_{*,17}$ with a representative in $Z_2\nu A[14] \otimes H_*BP$ is zero.

DEGREE 18: C[18]

The leading differential $d^2(\eta A[16]M_1) = 4C[18]$ determines tentative differentials with cokernel $[Z_8(C[18]M_1) \otimes B\langle 2\rangle] \oplus [Z_4 C[18] \otimes B\langle 2\rangle]$.

The leading differential $d^4(C[18]M_1^2) = \nu C[18]$ determines tentative differentials with kernel $[Z_8 C[18]\{M_1, M_2\} \otimes B\langle 4\rangle]$

$$\oplus\ [Z_4(2C[18])\{M_1\bar{M}_2, M_1^3\bar{M}_2\} \otimes B\langle 4\rangle] \oplus [Z_2(2C[18]) \otimes B\langle 2\rangle].$$

In Section 3.4 we computed the image of $d^{12}: E^{12}_{*,7} \longrightarrow E^{12}_{*,18}$. However, that computation was done in three stages so that the global image of these d^{12}-differentials is hard to unravel. Therefore, we give the computer calculation of the cokernel of these differentials in the table below. The new C[18]-leader is $2C[18]\bar{M}_2$.

DEGREE	GROUP	GENERATOR	DEGREE	GROUP	GENERATOR	DEGREE	GROUP	GENERATOR
(6,18)	Z_2	2/ 0 1 0 0	(10,18)	Z_2	2/ 2 1 0 0	(12,18)	Z_2	2/ 3 1 0 0 2/ 6 0 0 0
(14,18)	Z_2	2/ 4 1 0 0	(18,18)	Z_2	2/ 0 3 0 0		Z_2	2/ 6 1 0 0
(20,18)	Z_2	2/ 0 1 1 0		Z_2	2/ 7 1 0 0 1/ 4 2 0 0		Z_2	2/ 7 1 0 0 6/ 10 0 0 0
(22,18)	Z_2	2/ 2 3 0 0		Z_2	2/ 4 0 1 0		Z_4	4 0 1 0 8 1 0 0
(24,18)	Z_2	2/ 2 1 1 0		Z_2	2/ 3 3 0 0 6/ 6 2 0 0		Z_2	5 0 1 0 9 1 0 0
(26,18)	Z_2	4/ 3 1 1 0 2/ 0 2 1 0 2/ 4 3 0 0 2/ 10 1 0 0		Z_2 Z_2	2/ 3 1 1 0 6/ 6 0 1 0 2/ 4 3 0 0		Z_4	1/ 0 2 1 0 1/ 4 3 0 0 2/ 6 0 1 0 2/ 10 1 0 0
(28,18)	Z_2	4/ 5 3 0 0 4/ 7 0 1 0 4/ 11 1 0 0 2/ 4 1 1 0		Z_2	2/ 7 0 1 0 4/ 11 1 0 0 2/ 14 0 0 0		Z_8	2/ 5 3 0 0 1/ 7 0 1 0 5/ 11 1 0 0 1/ 4 1 1 0
(30,18)	Z_2	2/ 0 5 0 0		Z_2	2/ 5 1 1 0 6/ 6 3 0 0		Z_2	2/ 6 3 0 0
	Z_2	2/ 12 1 0 0	(32,18)	Z_2	3 2 1 0 7 3 0 0 0 3 1 0		Z_2	2/ 3 2 1 0 6/ 10 2 0 0

Z_2 2/ 6 1 1 0

Z_2 2/ 7 1 1 0
6/ 10 0 1 0
6/ 14 1 0 0

(36,18) Z_2 2/ 3 5 0 0
4/ 5 2 1 0
2/ 15 1 0 0
2/ 6 4 0 0
6/ 18 0 0 0

Z_2 2/ 8 1 1 0

Z_8 1/ 5 2 1 0
7/ 9 3 0 0
2/ 15 1 0 0
2/ 18 0 0 0

Z_2 2/ 10 3 0 0

Z_4 12 0 1 0
16 1 0 0

Z_2 2/ 7 2 1 0
4/ 11 3 0 0
6/ 13 0 1 0
2/ 17 1 0 0
6/ 14 2 0 0

Z_8 1/ 7 2 1 0
1/ 11 3 0 0
1/ 13 0 1 0
3/ 17 1 0 0
1/ 4 3 1 0

Z_2 2/ 3 1 0 1
6/ 5 3 1 0
2/ 11 1 1 0
6/ 0 7 0 0
2/ 4 1 2 0
2/ 6 0 0 1
6/ 8 2 1 0
6/ 12 3 0 0
6/ 14 0 1 0

(44,18) Z_2 6/ 7 5 0 0
4/ 13 3 0 0
6/ 15 0 1 0
2/ 0 5 1 0
6/ 4 6 0 0
6/ 10 4 0 0
2/ 12 1 1 0
2/ 22 0 0 0

Z_2 2/ 7 3 0 0
6/ 10 2 0 0

Z_2 2/ 2 5 0 0

Z_2 2/ 4 2 1 0

Z_2 2/ 5 2 1 0
6/ 9 3 0 0
2/ 15 1 0 0
2/ 18 0 0 0

Z_2 2/ 2 3 1 0

(38,18) Z_2 2/ 3 3 1 0
2/ 6 2 1 0
2/ 12 0 1 0
2/ 16 1 0 0

Z_2 2/ 2 1 2 0

(40,18) Z_2 2/ 2 1 0 1

Z_2 4/ 11 3 0 0
5/ 13 0 1 0
7/ 17 1 0 0
2/ 14 2 0 0

(42,18) Z_2 2/ 5 3 1 0
2/ 11 1 1 0
6/ 12 3 0 0
2/ 14 0 1 0
2/ 18 1 0 0

Z_2 2/ 5 3 1 0
2/ 11 1 1 0
2/ 4 1 2 0
6/ 6 5 0 0
6/ 12 3 0 0
2/ 14 0 1 0
2/ 18 1 0 0

Z_2 2/ 12 3 0 0

Z_2 4/ 7 5 0 0
4/ 13 3 0 0
6/ 19 1 0 0
2/ 4 1 0 1
6/ 6 3 1 0
6/ 12 1 1 0
6/ 22 0 0 0

(34,18) Z_2 2/ 0 1 2 0

Z_2 2/ 14 1 0 0

Z_4 4 2 1 0
8 3 0 0

Z_2 2/ 15 1 0 0
1/ 12 2 0 0
2/ 18 0 0 0

Z_2 2/ 0 1 0 1

Z_2 2/ 4 5 0 0

Z_2 2/ 6 2 1 0

Z_2 2/ 12 0 1 0

Z_2 2/ 3 1 2 0
2/ 6 0 2 0

Z_2 2/ 4 3 1 0

Z_2 2/ 10 1 1 0

Z_2 2/ 8 2 1 0
2/ 12 3 0 0
6/ 18 1 0 0

Z_2 2/ 6 5 0 0

Z_2 2/ 0 7 0 0

Z_4 1/ 8 2 1 0
1/ 12 3 0 0

Z_4 2/ 11 1 1 0
6/ 14 0 1 0

Z_2 2/ 7 5 0 0
4/ 13 3 0 0
6/ 15 0 1 0
2/ 19 1 0 0
6/ 10 4 0 0
6/ 12 1 1 0

Z_2 2/ 12 1 1 0

	Z_2 2/ 7 5 0 0 4/ 15 0 1 0 1/ 4 6 0 0 6/ 22 0 0 0	Z_2 2/ 19 1 0 0 6/ 22 0 0 0 Z_2 2/ 6 3 1 0	Z_8 4/ 13 3 0 0 5/ 15 0 1 0 1/ 19 1 0 0 1/ 12 1 1 0 2/ 22 0 0 0
(46,18)	Z_2 6/ 13 1 1 0 6/ 14 3 0 0	Z_2 6/ 13 1 1 0 2/ 6 1 2 0	Z_2 2/ 0 3 2 0
	Z_2 2/ 4 4 1 0	Z_2 2/ 14 3 0 0	Z_2 2/ 20 1 0 0
	Z_2 2/ 2 7 0 0	Z_4 2/ 7 3 1 0 6/ 10 2 1 0	Z_4 4 4 1 0 8 5 0 0
(48,18)	Z_2 6/ 7 1 2 0 2/ 0 1 3 0 2/ 10 0 2 0	Z_2 2/ 3 7 0 0 6/ 7 1 2 0 2/ 15 3 0 0 2/ 6 6 0 0 2/ 10 0 2 0 2/ 18 2 0 0	Z_2 1/ 5 4 1 0 6/ 7 1 2 0 5/ 9 5 0 0 6/ 15 3 0 0 6/ 8 3 1 0 2/ 14 1 1 0
	Z_2 3/ 11 2 1 0 3/ 15 3 0 0 2/ 21 1 0 0 1/ 8 3 1 0 2/ 18 2 0 0	Z_2 2/ 7 1 2 0 6/ 10 0 2 0	Z_2 2/ 11 2 1 0 2/ 15 3 0 0 4/ 21 1 0 0
	Z_2 2/ 0 3 0 1	Z_2 2/ 2 5 1 0	Z_2 4 2 2 0
	Z_2 2/ 6 1 0 1	Z_2 2/ 14 1 1 0	Z_4 2/ 15 3 0 0 2/ 18 2 0 0
(50,18)	Z_2 2/ 3 5 1 0 2/ 15 1 1 0 2/ 0 1 1 1 2/ 4 0 3 0 2/ 4 7 0 0 2/ 6 4 1 0 2/ 10 5 0 0 2/ 12 2 1 0 6/ 16 3 0 0 6/ 18 0 1 0 6/ 22 1 0 0	Z_2 2/ 3 5 1 0 2/ 15 1 1 0 2/ 4 0 3 0 2/ 4 7 0 0 2/ 6 4 1 0 2/ 10 5 0 0 2/ 12 2 1 0 6/ 16 3 0 0 6/ 18 0 1 0 6/ 22 1 0 0	Z_2 6/ 3 5 1 0 6/ 15 1 1 0 1/ 4 2 0 1 7/ 4 7 0 0 6/ 6 4 1 0 6/ 10 5 0 0 1/ 12 2 1 0 2/ 18 0 1 0 2/ 22 1 0 0
	Z_2 2/ 7 1 0 1 4/ 15 1 1 0 2/ 4 7 0 0 2/ 8 1 2 0 6/ 10 0 0 1 2/ 10 5 0 0 6/ 16 3 0 0 6/ 22 1 0 0	Z_2 4/ 3 5 1 0 2/ 0 6 1 0 2/ 4 7 0 0 2/ 10 5 0 0 Z_2 2/ 8 1 2 0 6/ 12 2 1 0	Z_2 2/ 2 3 2 0 Z_2 2/ 4 7 0 0 Z_2 2/ 12 2 1 0 Z_2 2/ 22 1 0 0

GROUP	GENERATOR
Z_4	2/ 3 5 1 0
	2/ 15 1 1 0
	1/ 4 0 3 0
	2/ 6 4 1 0
	1/ 8 1 2 0
	2/ 10 5 0 0
	6/ 16 3 0 0
	6/ 18 0 1 0
Z_4	2/ 3 5 1 0
	1/ 0 6 1 0
	1/ 4 7 0 0
	2/ 10 5 0 0
Z_4	2/ 15 1 1 0
	6/ 18 0 1 0
Z_4	12 2 1 0
	16 3 0 0

The leading differential $d^6(2C[18]\overline{M}_2) = A[23]$ determines tentative differentials by assigning the following values to monomials of degree 24 of $Z_8C[18] \otimes H_*BP$: $C[18]M_1^3$ is assigned 1 and $C[18]M_2$ is assigned 2. The kernel of these tentative differentials is given by the table below. The new C[18]-leader is $C[18]M_1^4\overline{M}_2^2$.

DEGREE	GROUP	GENERATOR
(20,18)	Z_2	2/ 7 1 0 0
		1/ 4 2 0 0
(22,18)	Z_4	1/ 4 0 1 0
		7/ 8 1 0 0
(24,18)	Z_2	3/ 5 0 1 0
		3/ 9 1 0 0
		2/ 2 1 1 0
(26,18)	Z_4	6/ 3 1 1 0
		1/ 0 2 1 0
		1/ 4 3 0 0
(28,18)	Z_8	2/ 5 3 0 0
		1/ 7 0 1 0
		5/ 11 1 0 0
		1/ 4 1 1 0
(30,18)	Z_2	2/ 5 1 1 0
(32,18)	Z_2	1/ 3 2 1 0
		7/ 7 3 0 0
		1/ 0 3 1 0
		2/ 10 2 0 0
(34,18)	Z_4	6/ 7 1 1 0
		1/ 4 2 1 0
		5/ 8 3 0 0
		2/ 10 0 1 0
		2/ 14 1 0 0
(36,18)	Z_2	2/ 15 1 0 0
		1/ 12 2 0 0
		2/ 18 0 0 0
	Z_8	5/ 5 2 1 0
		3/ 9 3 0 0
		2/ 2 3 1 0
(38,18)	Z_2	2/ 6 2 1 0
		6/ 10 3 0 0
	Z_4	1/ 13 0 1 0
		7/ 16 1 0 0
(40,18)	Z_2	4/ 11 3 0 0
		7/ 13 0 1 0
		5/ 17 1 0 0
		2/ 10 1 1 0
		6/ 14 2 0 0
	Z_8	7/ 7 2 1 0
		1/ 11 3 0 0
		6/ 13 0 1 0
		2/ 17 1 0 0
		1/ 4 3 1 0
(42,18)	Z_4	2/ 5 3 1 0
		2/ 11 1 1 0
		3/ 8 2 1 0
		3/ 12 3 0 0
		2/ 14 0 1 0
		6/ 18 1 0 0
	Z_4	2/ 11 1 1 0
		1/ 8 2 1 0
		1/ 12 3 0 0
		6/ 14 0 1 0
		6/ 18 1 0 0
(44,18)	Z_2	2/ 7 5 0 0
		4/ 15 0 1 0
		1/ 4 6 0 0
		6/ 22 0 0 0
	Z_8	4/ 13 3 0 0
		5/ 15 0 1 0
		1/ 19 1 0 0
		1/ 12 1 1 0
		2/ 22 0 0 0

(46,18)	Z_2 4/ 7 3 1 0 4/ 13 1 1 0	Z_2 2/ 13 1 1 0	Z_4 1/ 4 4 1 0 7/ 8 5 0 0
(48,18)	Z_2 7/ 5 4 1 0 2/ 7 1 2 0 3/ 9 5 0 0 2/ 15 3 0 0 2/ 2 5 1 0 1/ 4 2 2 0 2/ 8 3 1 0	Z_2 6/ 7 1 2 0 3/ 11 2 1 0 3/ 15 3 0 0 2/ 21 1 0 0 1/ 4 2 2 0 5/ 8 3 1 0 2/ 10 0 2 0 6/ 18 2 0 0	Z_4 1/ 11 2 1 0 3/ 15 3 0 0 6/ 21 1 0 0 1/ 8 3 1 0
(50,18)	Z_4 4/ 3 5 1 0 6/ 7 1 0 1 1/ 4 0 3 0 5/ 4 2 0 1 5/ 4 7 0 0 1/ 8 1 2 0 2/ 10 0 0 1 3/ 12 2 1 0 6/ 22 1 0 0	Z_4 4/ 3 5 1 0 6/ 7 1 0 1 1/ 4 2 0 1 5/ 4 7 0 0 2/ 10 0 0 1 2/ 12 2 1 0 3/ 16 3 0 0 6/ 22 1 0 0	Z_4 6/ 15 1 1 0 1/ 12 2 1 0 5/ 16 3 0 0 2/ 18 0 1 0 2/ 22 1 0 0 Z_4 2/ 3 5 1 0 1/ 0 6 1 0 1/ 4 7 0 0

<u>DEGREE 19:</u> A[19]

The leading differential $d^6(\sigma^2\overline{M}_2) = A[19]$ determines tentative differentials with image $Z_2A[19]\{1, M_1, M_2\} \otimes B\langle 4\rangle$. The remaining elements are $Z_2A[19]\{M_1^2, M_1^3, M_1M_2, M_1^2M_2, M_1^3M_2\} \otimes B\langle 4\rangle$, and the A[19]-leader is $A[19]M_1^2$.

The leading differential $d^4(A[19]M_1^2) = \nu A[19]$ determines tentative differentials which are a monomorphism on $Z_2A[19]\{M_1^2, M_1^3, M_1M_2, M_1^2M_2, M_1^3M_2\} \otimes B\langle 4\rangle$. There are no remaining elements.

<u>DEGREE 20:</u> C[20]

The leading differential $d^4(\nu A[14]M_1^2) = 4C[20]$ determines tentative differentials with image $Z_2(4C[20])\{1, M_1, M_1^2, M_2, M_1M_2\} \otimes B\langle 4\rangle$. The leading differential $d^6(\eta A[14]M_1^3) = 2C[20]$ determines tentative differentials with image $Z_2(2C[20])\{1, M_1, M_2\} \otimes B\langle 4\rangle$. The leading differential $d^{12}(\eta A[8]M_1^3\overline{M}_2) = C[20]$ determines tentative differentials with image

$Z_2C[20] \otimes B<4>$. The cokernel of these differentials is $[Z_8C[20]\{M_1^3, M_1^2\bar{M}_2, M_1^3\bar{M}_2\} \otimes B<4>] \oplus [Z_4C[20]\{M_1^2, M_1\bar{M}_2\} \otimes B<4>] \oplus [Z_2C[20]\{M_1, \bar{M}_2\} \otimes B<4>]$. The leading differential $d^2(C[20]M_1) = \eta C[20]$ determines tentative differentials which are a monomorphism on $Z_2(C[20]M_1) \otimes B<2>$. The leading differential $d^4(C[20]M_1^2) = \nu C[20]$ determines tentative differentials which are a monomorphism on $([Z_8C[20]\{\bar{M}_2, M_1^2\bar{M}_2]$ $\oplus [Z_4(2C[20])M_1^3\bar{M}_2] \oplus [Z_4C[20]M_1^2] \oplus [Z_2(2C[20])M_1\bar{M}_2]) \otimes B<4>$. There are no remaining elements.

<u>DEGREE 21:</u> $\eta C[20]$

Since $\eta^2C[20] \neq 0$, the only element of $E^4_{*,21}$ with a representative in $Z_2\eta C[20] \otimes H_*BP$ is zero.

<u>DEGREE 21:</u> $\nu C[18] = \sigma^3$

The leading differential $d^4(C[18]M_1^2) = \nu C[18]$ determines tentative differentials with image $Z_2\nu C[18]\{1, M_1, M_1^2, M_2, M_1M_2\} \otimes B<4>$. The remaining elements are $Z_2\nu C[18]\{M_1^3, M_1^2M_2, M_1^3M_2\} \otimes B<4>$, and the $\nu C[18]$-leader is $\nu C[18]M_1^3$.

The leading differential $d^8(\sigma^2M_1^4\bar{M}_2) = \sigma^3\bar{M}_2$ determines tentative differentials with cokernel given by the table below. The new $\nu C[18]$-leader is $\nu C[18]M_1^3M_2$.

DEGREE	BASIS	DEGREE	BASIS	DEGREE	BASIS
(12,21)	3 1 0 0	(20,21)	7 1 0 0	(24,21)	3 3 0 0
(26,21)	3 1 1 0	(28,21)	11 1 0 0		
(32,21)	6 1 1 0		7 3 0 0	(34,21)	7 1 1 0
(36,21)	3 0 0 1		3 5 0 0	(38,21)	3 3 1 0
(40,21)	3 1 2 0		4 3 1 0		11 3 0 0
(42,21)	3 1 0 1		11 1 1 0		

(44,21)	7 0 0 1		6 3 1 0		7 5 0 0
(46,21)	0 3 2 0		7 3 1 0	(48,21)	14 1 1 0

DEGREE 22: $\eta^2C[20]$

Since $\eta^3C[20] = 4\nu C[20] \neq 0$, the only element of $E^4_{*,22}$ with a representative in $Z_2\eta^2C[20] \otimes H_*BP$ is zero.

DEGREE 22: $\nu A[19]$

The leading differential $d^4(A[19]M_1^2) = \nu A[19]$ determines tentative differentials with image $Z_2\nu A[19]\{1,M_1,M_1^2,M_2,M_1M_2\} \otimes H_*BP$. The remaining elements are $Z_2\nu A[19]\{M_1^3,M_1^2M_2,M_1^3M_2\} \otimes B<4>$, and the $\nu A[19]$-leader is zero.

The leading differential $d^{14}(\eta^2\sigma M_1^7\bar{M}_2) = \nu A[19]\bar{M}_2$ determines tentative differentials with image, in degrees less than 69, of dimension given by the fourth column of the table below.

The leading differntial $d^{10}(\nu A[19]M_1^2\bar{M}_2) = A[31]$ determines tentative differentials which are a monomorphism on $A = Z_2\nu A[19]\{M_1^2\bar{M}_2,M_1^3\bar{M}_2\} \otimes B<4>$. We give the dimensions of A in the fifth column of the table below. As the reader can verify, the sum of the numbers in the last three columns equals the number in the second column except when N = 40. Thus, $E^{16}_{*,22} = E^{16}_{40,22} = Z_2\nu A[19]M_1^7M_2^2<M_3>$ in dimensions less than 69, and the new $\nu A[19]$-leader is $\nu A[19]M_1^7M_2^2<M_3>$.

N	DIM $E^4_{N,22}$	DIM $\nu E^4_{N+4,19}$	DIM $d^{14}(E^{14}_{N+14,9})$	DIM A
0	1	1	0	0
2	1	1	0	0
4	1	1	0	0
6	2	1	1	0
8	2	2	0	0
10	2	1	0	1
12	3	2	0	1
14	4	3	1	0
16	4	4	0	0
18	5	3	1	1
20	6	4	1	1
22	6	4	1	1
24	7	5	0	2
26	8	5	1	2
28	9	7	1	1
30	11	8	2	1
32	12	9	1	2
34	13	8	2	3
36	15	10	2	3
38	16	11	2	3
40	17	12	0	4
42	20	13	3	4
44	22	16	3	3
46	23	16	3	4

DEGREE 23: A[23]

The leading differrential $d^6(2C[18]M_1^3) = A[23]$ determines tentative differentials with image $Z_2A[23]\{1, M_1^2, \bar{M}_2\} \otimes B\langle 4\rangle$. Since $\eta A[23] \neq 0$, the remaining elements are $Z_2A[23]M_1^2\bar{M}_2 \otimes B\langle 4\rangle$, and the A[23]-leader is $A[23]M_1^2\bar{M}_2$.

DEGREE 23: $\nu C[20]$

The leading differential $d^2(\eta^2C[20]M_1) = 4\nu C[20]$ determines tentative differentials with image $Z_2(4\nu C[20]) \otimes B\langle 2\rangle$. The remaining elements are $[Z_8(\nu C[20]M_1) \otimes B\langle 2\rangle] \oplus [Z_4(\nu C[20]) \otimes B\langle 2\rangle]$, and the $\nu C[20]$-leader is $\nu C[20]M_1$.

The leading differential $d^4(C[20]M_1^2) = \nu C[20]$ determines tentative differentials with image

$[Z_8(\nu C[20])\{M_1,M_2\} \oplus Z_4(\nu C[20])\{1,2M_1M_2\} \oplus Z_2(2\nu C[20])M_1] \otimes B\langle 4\rangle$. The remaining elements are

$[Z_2(\nu C[20])\{M_1^2,M_1M_2\} \oplus Z_4(\nu C[20])\{\bar{M}_2,M_1^2\bar{M}_2\} \oplus Z_8(\nu C[20]M_1^3\bar{M}_2)] \otimes B\langle 4\rangle$,

and the new $\nu C[20]$-leader is $\nu C[20]M_1^2$.

The leading differential $d^4(\nu C[20]M_1^2) = \nu^2 C[20]$ determines tentative differentials which are a monomorphism on

$Z_2(\nu C[20])\{M_1^2,M_1M_2,\bar{M}_2,M_1^2\bar{M}_2,M_1^3\bar{M}_2\} \otimes B\langle 4\rangle$. The remaining elements are $[Z_2(\nu C[20])\{2\bar{M}_2,2M_1^2\bar{M}_2\} \oplus Z_4(2\nu C[20]M_1^3\bar{M}_2)] \otimes B\langle 4\rangle$, and the new $\nu C[20]$-leader is $2\nu C[20]\bar{M}_2$.

The leading differential $d^6(2\nu C[20]\bar{M}_2) = A[8]C[20]$ determines tentative differentials which are a monomorphism on $Z_2(\nu C[20])\{2\bar{M}_2,2M_1^2\bar{M}_2,2M_1^3\bar{M}_2\} \otimes B\langle 4\rangle$. The remaining elements are $Z_2(4\nu C[20]M_1^3\bar{M}_2) \otimes B\langle 4\rangle$, and the new $\nu C[20]$-leader is $4\nu C[20]M_1^3\bar{M}_2$.

<u>DEGREE 24:</u> $\eta A[23]$

The leading differential $d^2(A[23]M_1) = \eta A[23]$ determines tentative differentials with image $Z_2\eta A[23] \otimes B\langle 2\rangle$. The remaining elements are $Z_2(\eta A[23]M_1) \otimes B\langle 2\rangle$, and the $\eta A[23]$-leader is $\eta A[23]M_1$.

The leading differential $d^{16}(\eta^2\sigma M_1^9) = \eta A[23]M_1$ determines tentative differentials whose cokernel in degrees less than 69 is $Z_2\eta A[23]\{M_1^{15}\bar{M}_2,M_1^{15}\bar{M}_2^2,M_1^{15}\bar{M}_3\}$. The new $\eta A[23]$-leader is $\eta A[23]M_1^{15}\bar{M}_2$.

<u>DEGREE 26:</u> $\nu^2 C[20]$

The leading differential $d^4(\nu C[20]M_1^2) = \nu^2 C[20]$ determines tentative differentials with image $Z_2\nu^2\{1,M_1,M_1^2,M_2,M_1M_2\} \otimes B\langle 4\rangle$. The remaining elements are $Z_2\nu^2 C[20]\{M_1^3,M_1^2M_2,M_1^3M_2\} \otimes B\langle 4\rangle$, and the $\nu^2 C[20]$-leader is $\nu^2 C[20]M_1^3$.

The leading differential $d^{16}(\beta_1 M_1^{11}) = \nu^2 C[20]M_1^3$ determines tentative differentials whose image in degrees less than 69 has dimensions given by the fourth column in the table below. Since in each row of the table below, the numbers in the third and fourth columns add up to the numbers in the second column, it follows that $E^{18}_{*,26} = 0$ in degrees less than 69.

N	DIM $E^4_{N,26}$	DIM $\nu E^4_{N+4,23}$	DIM $d^{16}(E^{16}_{N+16,11})$
0	1	1	0
2	1	1	0
4	1	1	0
6	2	1	1
8	2	2	0
10	2	1	1
12	3	2	1
14	4	3	1
16	4	4	0
18	5	3	2
20	6	4	2
22	6	4	2
24	7	5	2
26	8	5	3
28	9	7	2
30	11	8	3
32	12	9	3
34	13	8	5
36	15	10	5
38	16	11	5
40	17	12	5
42	20	13	7

<u>DEGREE 28:</u> A[8]C[20]

The leading differential $d^6(2\nu C[20]\bar{M}_2) = A[8]C[20]$ determines tentative differentials with image $Z_2 A[8]C[20]\{1, M_1^2\} \otimes B\langle 4\rangle$, and the A[8]C[20]-leader is $A[8]C[20]M_1$.

The leading differential $d^{18}(\beta_1 M_1^{10}) = A[8]C[20]M_1$ determines tentative differentials whose cokernel in degrees less than 69 is given by the table below. The new A[8]C[20]-leader is $A[8]C[20]M_1\bar{M}_2$.

DEGREE	BASIS	DEGREE	BASIS	DEGREE	BASIS
(8,28)	1 1 0 0	(12,28)	3 1 0 0	(16,28)	5 1 0 0
(20,28)	1 3 0 0		7 1 0 0		
(22,28)	1 1 1 0	(24,28)	3 3 0 0		9 1 0 0
(26,28)	3 1 1 0	(28,28)	5 3 0 0		11 1 0 0
(30,28)	5 1 1 0	(32,28)	7 3 0 0		13 1 0 0
(34,28)	1 3 1 0		7 1 1 0	(36,28)	1 1 2 0
	3 5 0 0		9 3 0 0		15 1 0 0
(38,28)	1 1 0 1		3 3 1 0		9 1 1 0
(40,28	3 1 2 0		5 5 0 0		11 3 0 0
	13 0 1 0		17 1 0 0		

DEGREE 30: A[30]

The leading differential $d^{24}(2\sigma M_1^{12}) = A[30]$ determines tentative differentials with image $Z_2A[30] \otimes Z_2[\langle M_1^4\rangle^2, \langle M_2^2\rangle^2, \langle M_3\rangle^2, \langle M_4\rangle^2, \{M_5\}, \ldots, \{M_n\}, \ldots\}$. Since $\eta A[30] \neq 0$, the remaining elements are $[Z_2A[30]\{M_1^2, \bar{M}_2, M_1^2\bar{M}_2\} \otimes B\langle 4\rangle]$
$\oplus\ [Z_2A[30]\langle M_4\rangle \otimes Z_2[\langle M_1^4\rangle^2, \langle M_2^2\rangle^2, \langle M_3\rangle^2, \langle M_4\rangle^2, \{M_5\}, \ldots, \{M_n\}, \ldots\}] \oplus$

$$\bigoplus_{\alpha,\beta,\gamma} Z_2A[30]\langle M_1^4\rangle^{\alpha}\langle M_2^2\rangle^{\beta}\langle M_3\rangle^{\gamma} \otimes B\langle 8\rangle.$$

The last sum is taken over all $0 \leq \alpha,\ \beta,\ \gamma \leq 1$ with $0 < \alpha\beta\gamma$. The A[30]-leader is $A[30]M_1^2$.

DEGREE 31: A[31]

The leading differential $d^{10}(\nu A[19]M_1^2\bar{M}_2) = A[31]$ determines tentative differentials with image $Z_2A[31]\{1, M_1\} \otimes B\langle 4\rangle$. The remaining elements are $Z_2A[31]\{M_1^2, M_1^3, M_2, M_1M_2, M_1^2M_2, M_1^3M_2\} \otimes B\langle 4\rangle$, and the A[31]-leader is $A[31]M_1^2$.

DEGREE 31: $\eta A[30]$

The leading differential $d^2(A[30]M_1) = \eta A[30]$ determines tentative differentials with image $Z_2\eta A[30] \otimes B\langle 2\rangle$. The cokernel of these differentials is $Z_2\eta A[30]M_1 \otimes B\langle 2\rangle$, and the $\eta A[30]$-leader is $\eta A[30]M_1$.

CHAPTER 6: The Chicago Stems (π^S_N, $32 \le N \le 45$)

1. Introduction

The stems of dimensions 29 to 45 were first computed using the classical Adams spectral sequenceby Barratt, Mahowald and Tangora [10], [37] as corrected by Bruner [16]. Since all three of these authors have connections to the Chicago area we have taken the liberty to name these stems after that city. We continue our calculations from Chapter 5 to recompute these stems from the Atiyah-Hirzebruch spectral sequence. We compute the first seven of these stems in Section 2 and the remainder in Section 3. In Section 4 we collect all the computer computations of tentative differentials. The results of this chapter are summarized in Appendices 1 to 4.

2. Computation of π^S_N, $32 \le N \le 38$

In the tables of leaders below, all leaders of degree greater than 40 will have an asterisk at the left. They will be omitted from all other tables of leaders in this section except for the last one.

Recall from Figure 5.3.7 that there are five leaders of degree 33 and three leaders of degree 34. Now $A[30]M_1^2$ transgresses to $\nu A[30]$. By Lemma 3.3.14, $\eta^2 A[30]$ is divisible by two. As we shall see, the other elements of π^S_{32} all have order two. Thus, $\eta^2 A[30] = 0$ and $\eta A[30]M_1$ must bound. It follows from the following lemma that $d^8(A[16]M_1^6\overline{M}_2) = A[23]M_1^2\overline{M}_2$. Therefore, there is only one possibility: $d^{18}(\sigma^2 M_1^4 \overline{M}_2^2) = \eta A[30]M_1$.

LEMMA 6.2.1 (a) $\eta^2 A[30] = 0$

(b) $\sigma A[16] = A[23]$ and $\nu A[23] = 0$

PROOF. (b) Since $d^{10}(2\sigma M_1^5) = A[16]$, it follows from Theorem 2.4.2 that $A[16] \in \langle\sigma^2, 2, \eta\rangle$. Therefore, $\sigma\cdot A[16] \in \sigma\langle\sigma^2, 2, \eta\rangle \subset \langle\sigma^3, 2, \eta\rangle = \langle\nu C[18], 2, \eta\rangle$. By Theorem 2.4.2, $A[23] = d^6(2C[18]M_1^3)$ is an element of the last triple product. This triple product has indeterminacy $\nu C[18]\cdot\pi_2^S + \eta\cdot\pi_{22}^S$ $= Z_2(4\nu C[20])$ because $\pi_2^S = Z_2\eta^2$ and $\pi_{22}^S = Z_2\nu A[19] \oplus Z_2\eta^2 C[20]$. Thus, $\sigma A[16]$ $= A[23] + 4k\nu C[20]$. However, $A[23]$ was only defined as $d^6(C[18]M_1^3) \in E^6_{0,23}$ $= \pi_{23}^S/Z_8\nu C[20]$. Thus, we can define $A[23] \in \pi_{23}^S$ so that it equals $\sigma A[16]$.∎

We are thus left with three nonbounding leaders of degree 33 which define nonzero elements of π_{32}^S: $d^{24}(\eta^2\sigma M_1^5\bar{M}_3) = A[32,1]$, $d^{22}(2\beta_1 M_1^{11}) = A[32,2]$ and $d^{12}(\nu C[18]M_1^3\bar{M}_2) = A[32,3]$. As we shall see, $\eta A[32,2] \neq 0$ and $\nu A[32,3] \neq 0$. From the former relation it follows that $A[32,2]$ is not divisible by 2. Our computer computations show that $A[32,1]M_1^2$, $A[32,2]M_1^2$ is a d^{24}-boundary, d^{22}-boundary, respectively. Therefore $\nu A[32,1]$ and $\nu A[32,2]$ are divisible by η. It follows that $2\nu A[32,1] = 2\nu A[32,2] = 0$. Hence $A[32,3]$ can not be divisible by 2. Thus, π_{32}^S is an elementary abelian group. We have thus proved the following theorem.

THEOREM 6.2.2 $\pi_{32}^S = Z_2A[32,1] \oplus Z_2A[32,2] \oplus Z_2A[32,3] \oplus Z_2\eta\gamma_3$ and $\eta A[31] = 0$.

The computations of Section 4 show that we have the following leaders.

Row	Degree	Leader	Row	Degree	Leader
9	35	$\eta^2\sigma M_1^7\bar{M}_2^2$	*23	63	$\gamma_2 M_1^{20}$
11	35	$\beta_1 M_1^6\bar{M}_2^2$	23	35	$4\nu C[20]M_1^3\bar{M}_2$
15	37	$\gamma_1 M_1^8\bar{M}_2$	*24	60	$\eta A[23]M_1^{15}\bar{M}_2$
*17	51	$\eta^2\gamma_1 M_1^{17}$	28	36	$A[8]C[20]M_1\bar{M}_2$
18	38	$C[18](M_1^4\bar{M}_2^2+2M_1^7\bar{M}_2)$	30	34	$A[30]M_1^2$

*19	55	$\beta_2M_1^{18}$	31	39	$\eta A[30]M_1\bar{M}_2$
*21	53	$\nu C[18]M_1^6\bar{M}_2\bar{M}_3$	31	35	$A[31]M_1^2$
*22	62	$\nu A[19]M_1^7M_2^2\langle M_3\rangle$	32	34	$A[32,1]M_1, A[32,2]M_1,$
*23	67	$\sigma A[16]M_1^6\bar{M}_2^3\bar{M}_3$			$A[32,3]M_1$

FIGURE 6.2.1: Leaders from Rows 1 to 32 of Degree at Least 34

There are four leaders of degree 34 and four leaders of degree 35. We know that $4\nu C[20]M_1^3\bar{M}_2$ transgresses to $A[14]C[20]$ and $A[31]M_1^2$ transgresses to $\nu A[31]$. Since $A[30] = d^{24}(2\sigma M_1^{12})$, $A[30]M_1^2$ could only bound from below the 7 row, and there are no such leaders of degree 35. Thus, $\nu A[30] = d^4(A[30]M_1^2) \neq 0$. Our computer computations in Section 4 show that $2\beta_1M_1^6\bar{M}_2^2$ in E^{24} is in Image r_{Δ_1}. Therefore it can not hit $A[32,k]M_1$ for $k = 1,2,3$ and must transgress. If $\eta^2A[32,k] \neq 0$ then by Theorem 3.3.14 it must be divisible by two. Assume that $\eta^2\sigma M_1^7\bar{M}_2^2$ transgresses. Then there is one leader of degree 36, $A[8]C[20]M_1\bar{M}_2$, which can hit an $\eta A[32,k_3]M_1$ and there are two leaders of degree 35 to transgress to elements of order four: $d^{26}(\eta^2\sigma M_1^7\bar{M}_2^2) = B_1$, $2B_1 = \eta^2A[32,k_1]$ and $d^{24}(\beta_1M_1^6\bar{M}_2^2) = B_2$, $2B_2 = \eta^2A[32,k_2]$. Since $\eta^3 = 4\nu$, $\eta^3A[32,k_1] = \eta^3A[32,k_2] = 0$. Thus $\eta^2A[32,k_1]M_1$ and $\eta^2A[32,k_2]M_1$ must both bound. However, there is only one leader in degree 37 which can hit such an element: $\gamma_1M_1^8\bar{M}_2$. This contradiction implies that $\eta^2\sigma M_1^7\bar{M}_2^2$ does not transgress and hits one of the $A[32,k]M_1$. Since $A[32,1]$ bounds from the 9 row it follows that if $A[32,1]M_1$ were to bound it would have to bound from below the 9 row. There are no leaders there. Thus, $A[32,1]M_1$ does not bound and $k \neq 1$. Recall from the proof of Theorem 5.3.9 that $d^{18}(\beta_1M_1^{10}) = A[8]C[20]M_1$. Apply r_{Δ_1} to see that $2\beta_1M_1^9$ hits $A[8]C[20]$, and let $R\{2\beta_1M_1^9\}$ represent $2\beta_1M_1^9$ such that $\partial\, R\{2\beta_1M_1^9\} = A[8]C[20]$. Since $2\beta_1M_1^{11}$ transgresses there must be an S in filtration degree 20 such that

$(R\{2\beta_1M_1^9\} \wedge \mu_2) \cup S$ represents $2\beta_1M_1^{11}$ and

$\partial S = A[32,2] \cup (R\{2\beta_1M_1^9\} \wedge \nu) \cup (A[8]C[20] \wedge \mu_2)$. Then

$\partial \{[S \wedge \eta] \cup [R\{2\beta_1M_1^9\} \wedge B_{\eta\nu}] \cup [B_{A[8]C[20]\eta} \wedge \mu_2] \cup [A[8]C[20](\mu_2 \cup_1 \eta)]$
$\cup [R\{2\beta_1M_1^9\}(\nu \cup_1 \eta)]\} = (A[32,2] \wedge \eta) \cup (A[8]C[20]B_{\eta\nu}) \cup (B_{A[8]C[20]\eta} \wedge \nu)$.
Therefore, $\eta A[32,2] \in \langle A[8]C[20],\eta,\nu\rangle$ and $d^6(A[8]C[20]M_1M_2) = \eta A[32,2]M_1$. In E^6, $A[8]M_1M_2$ is homologous to $A[8]M_1\bar{M}_2$. Therefore, $d^6(A[8]C[20]M_1\bar{M}_2) = \eta A[32,2]M_1$. Note that $A[8]C[20]M_1\bar{M}_2$ is the only leader of degree 36 which could hit a leader of the form $\eta A[32,h]M_1$. Thus, $d^6(A[8]C[20]M_1\bar{M}_2)$ must be nonzero. Therefore, $k = 3$ and $d^{24}(\eta^2\sigma M_1^7\bar{M}_2^2) = A[32,3]M_1$. We have thus proved the following theorem.

THEOREM 6.2.3 $\pi_{33}^S = Z_2\nu A[30] \oplus Z_2\eta A[32,1] \oplus Z_2\eta A[32,2] \oplus Z_2\alpha_4 \oplus Z_2\eta^2\gamma_3$ and $\eta A[32,3] = 0$.

The computations of Section 4 show that we have the following leaders.

Row	Degree	Leader	Row	Degree	Leader
*9	63	$\eta^2\sigma M_1^{21}\bar{M}_2^2$	31	39	$\eta A[30]M_1\bar{M}_2$
11	35	$\beta_1M_1^6\bar{M}_2^2$	31	35	$A[31]M_1^2$
15	37	$\gamma_1M_1^8\bar{M}_2$	32	40	$A[32,1]M_1^4$
18	38	$C[18](M_1^4\bar{M}_2^2+2M_1^7\bar{M}_2)$	32	38	$A[32,2]\bar{M}_2$
23	35	$4\nu C[20]M_1^3\bar{M}_2$	32	36	$A[32,3]M_1^2$
28	36	$A[8]C[20]M_1\bar{M}_2$	33	37	$\nu A[30]M_1^2$
30	38	$A[30]M_1^4$	33	35	$\eta A[32,1]M_1$, $\eta A[32,2]M_1$

FIGURE 6.2.2: Leaders from Rows 1 to 33 of Degree at Least 35

There are five leaders of degree 35 and two leaders of degree 36. Now $A[32,3]M_1^2$ transgresses to $\nu A[32,3]$, and we proved above that $A[8]C[20]M_1\bar{M}_2$ hits $\eta A[32,2]M_1$. Thus, the other four leaders of degree 35 transgress to nonzero elements. One of them B[34] must have order four with 2B[34] = η^2A[32,1]. The only possibility is B[34] = $d^{24}(\beta_1 M_1^6 M_2^2)$. By Lemma 3.3.14,

$$B[34] \in \langle \eta, 2, A[32,1] \rangle \qquad [6.1]$$

We have thus proved the following theorem.

THEOREM 6.2.4 $\pi_{34}^S = Z_4 B[34] \oplus Z_2 A[14]C[20] \oplus Z_2 \nu A[31] \oplus Z_2 \eta\alpha_4$ where 2B[34] = η^2A[32,1] and η^2A[32,2] = 0.

The computations of Section 4 show that we have the following leaders.

Row	Degree	Leader	Row	Degree	Leader
11	39	$Z_4(2\beta_1 M_1^{11} M_2)$	32	40	$A[32,1]M_1^4$
15	37	$\gamma_1 M_1^8 \bar{M}_2$	32	38	$A[32,2]\bar{M}_2$
18	38	$C[18](M_1^4 \bar{M}_2^2 + 2M_1^7 \bar{M}_2)$	32	36	$A[32,3]M_1^2$
28	38	$A[8]C[20]M_1^2 \bar{M}_2$	33	37	$\nu A[30]M_1^2$
30	38	$A[30]M_1^4$	*33	41	$\eta A[32,2]M_1 \bar{M}_2$
31	39	$\eta A[30]M_1 \bar{M}_2$	34	40	$\nu A[31]M_1^3$, $B[34]M_1^3$
31	37	$A[31]M_2$	34	36	$A[14]C[20]M_1$, $2B[34]M_1$

FIGURE 6.2.3: Leaders from Rows 1 to 34 of Degree at Least 36

There are three leaders of degree 36 and three leaders of degree 37. Clearly $A[31]M_2$ and $\nu A[30]M_1^2$ transgress. $d^2(2B[34]M_1) = 2\eta B[34] = 0$ and thus $2B[34]M_1$ must bound. The only possibility is $d^{20}(\gamma_1 M_1^8 \bar{M}_2) = 2B[34]M_1$. Since $A[32,1]\bar{M}_2$ is a d^{24}-boundary, νA[32,1] can not be ηA[14]C[20] and νA[32,1] must be zero. We have thus proved the following theorem.

THEOREM 6.2.5 $\pi_{35}^S = Z_2 \eta A[14]C[20] \oplus Z_2 \nu A[32,3] \oplus Z_8 \beta_4$ and ηB[34] = νA[32,1] = 0.

The computations of Section 4 show that we have the following leaders.

Row	Degree	Leader	Row	Degree	Leader
11	39	$Z_4(2\beta_1 M_1^{11} M_2)$	32	38	$A[32,2]\bar{M}_2$, $A[32,3]M_2$
15	39	$Z_4\gamma_1 M_1^{12}$	33	37	$\nu A[30]M_1^2$
18	38	$C[18](M_1^4\bar{M}_2^2+2M_1^7\bar{M}_2)$	34	40	$\nu A[31]M_1^3$, $B[34]M_1^3$
28	38	$A[8]C[20]M_1^2\bar{M}_2$	34	38	$A[14]C[20]M_1^2$
30	38	$A[30]M_1^4$	*34	42	$2B[34]M_1\bar{M}_2$
31	39	$\eta A[30]M_1\bar{M}_2$	*35	41	$\nu A[32,3]M_1^3$
31	37	$A[31]M_2$	35	37	$\eta A[14]C[20]M_1$
32	40	$A[32,1]M_1^4$			

FIGURE 6.2.4: Leaders from Rows 1 to 35 of Degree at Least 37

We will use the following lemma to see that $\nu A[30]M_1^2$ bounds.

LEMMA 6.2.6 (a) $\nu A[30] \in \langle\sigma, C[18], \sigma\rangle = \langle C[18], \sigma, 2\sigma\rangle$.

(b) $\nu^2 A[30] = 0$.

PROOF. (a) $d^{12}(2\sigma M_1^{12}) = A[30]$ so we can represent $2\sigma M_1^{12}$ by

$\mathcal{M} = 2\sigma\langle M^4\rangle^3 \cup (B_{2\sigma 3\sigma} \wedge \langle M_1^4\rangle^2)$ union an element of filtration degree 14.

Then $\nu A[30] = \partial\,[(\nu \wedge \mathcal{M}) \cup \partial\,(B_{\nu 2\sigma} \wedge \langle M_1^4\rangle^3)]$

$= \partial\,[(\nu \wedge B_{2\sigma 3\sigma} \wedge \langle M_1^4\rangle^2) \cup (B_{\nu 2\sigma} \wedge 3\sigma\langle M_1^4\rangle^2) \cup F]$ where F has filtration degree 14;

$= \partial\,[3C[18]\langle M_1^4\rangle^2 \cup F]$ since $C[18] = \langle\nu, 2\sigma, \sigma\rangle$. By Theorems 2.4.2 and 2.3.7(a),

$$\nu A[30] \in \langle\sigma, C[18], \sigma\rangle = \langle C[18], \sigma, 2\sigma\rangle. \qquad [6.2$$

(b) $\nu^2 A[30] \in \nu\langle\sigma, C[18], \sigma\rangle = \langle\nu, \sigma, C[18]\rangle\sigma \in \sigma\cdot\pi_{29}^S = 0.$ ∎

From Figure 6.2.4, we see that there are three leaders of degree 37 and six leaders of degree 38. Clearly $A[30]M_1^4$, $A[32,3]M_2$ and $A[14]C[20]M_1^2$ transgress. Since $d^{18}(\beta_1 M_1^{10}) = A[8]C[20]M_1$, we can apply r_{Δ_1} to see that $2\beta_1 M_1^9$ hits

A[8]C[20]. Using $r_{2\Delta_1+\Delta_2}$, we see that $d^{18}(2\beta_1 M_1^{11} M_2) = A[8]C[20]M_1^2\overline{M}_2$. Since $d^8(A[8]M_1^2\overline{M}_2) = \eta A[14]M_1$, $d^8(A[8]C[20]M_1^2\overline{M}_2) = \eta A[14]C[20]M_1$. Therefore, $\eta A[14]C[20]M_1$ must be zero in E^8. The only possibility is: $d^4(A[32,2]\overline{M}_2) = \eta A[14]C[20]M_1$. It follows that

$$\nu A[32,2] = \eta A[14]C[20]. \qquad [6.3]$$

By Lemma 6.2.6(b), $\nu A[30]M_1^2$ must bound. The only remaining possibility is that $d^{16}(C[18](M_1^4\overline{M}_2^2+2M_1^7\overline{M}_2)) = \nu A[30]M_1^2$. Now there is no possibility for $A[31]M_2$ to bound. Thus, $A[36] = d^6(A[31]M_2)$ defines a nonzero element of π_{36}^S. We have thus proved the following theorem.

THEOREM 6.2.7 $\pi_{36}^S = Z_2A[36]$

The computations of Section 4 show that we have the following leaders.

Row	Degree	Leader	Row	Degree	Leader
11	39	$Z_4(2\beta_1 M_1^{11} M_2)$	32	38	$A[32,3]M_2$
15	39	$Z_4\gamma_1 M_1^{12}$	*33	45	$\nu A[30]M_1^6$
18	40	$2C[18](M_1^4\overline{M}_3+3M_1^8\overline{M}_2)$	*33	41	$\eta A[32,2]M_1\overline{M}_2$
28	38	$A[8]C[20]M_1^2\overline{M}_2$	34	40	$\nu A[31]M_1^3$, $B[34]M_1^3$
30	38	$A[30]M_1^4$	34	38	$A[14]C[20]M_1^2$
31	39	$\eta A[30]M_1\overline{M}_2$	*35	43	$\eta A[14]C[20]M_1\overline{M}_2$
32	40	$A[32,1]M_1^4$	36	38	$A[36]M_1$

FIGURE 6.2.5: Leaders from Rows 1 to 36 of Degree at Least 38

The following relations were discovered by Bruner [16].

LEMMA 6.2.8 $\eta A[36] = \nu^2 A[31] = 0$

PROOF. Since $A[36] = d^6(A[31]M_2)$, Theorem 2.4.4(b) implies that

$$A[36] = \langle\nu, \eta, A[31]\rangle. \qquad [6.4]$$

Therefore, $\eta A[36] = \eta\langle\nu,\eta,A[31]\rangle = \langle\eta,\nu,\eta\rangle A[31] = \nu^2A[31]$. Now $A[31] = d^{10}(\nu A[19]M_1^2\bar{M}_2)$, so $\nu^2A[31]= \eta A[36]$ is hit by $\nu^3A[19]M_1^2\bar{M}_2$ $= \eta A[8]A[19]M_1^2\bar{M}_2$. Thus, A[36] is hit by $A[8]A[19]M_1^2\bar{M}_2$, and $d^{10}(A[8]A[19]M_1^3\bar{M}_2) = A[36]M_1$. Therefore, $\eta A[36] = 0$.∎

There are five leaders of degree 38 and three leaders of degree 39. We have already observed in the proof of Theorem 6.2.7 that $d^{18}(2\beta_1M_1^{11}M_2)$ $= A[8]C[20]M_1^2\bar{M}_2$. Recall that $A[8]\bar{M}_2 = d^6(2\nu M_1^3\bar{M}_2)$ and $A[14] = d^{12}(4\nu M_1^3\bar{M}_2)$. Note that $A[8]\bar{M}_2$ is represented by $(A[8] \wedge \bar{\mu}_2) \cup (B_{A[8]\nu} \wedge \mu_1) \cup B_{\langle A[8],\nu,\eta\rangle}$. Therefore, $2A[8]\bar{M}_2 \cup \partial\, [(B_{2A[8]} \wedge \bar{\mu}_{01}) \cup (B_{\langle 2,A[8],\nu\rangle} \wedge \mu_1)]$ $= (B_{2A[8]} \wedge B_{\nu\eta}) \cup 2B_{\langle A[8],\nu,\eta\rangle} \cup (B_{\langle 2,A[8],\nu\rangle} \wedge \eta)$ represeents A[14]. One conseqeuce is that

$$A[14] \in \langle 2,A[8],\nu,\eta\rangle. \tag{6.5}$$

A second consequence is that $4\beta_1M_1^9M_2$ has a representative with boundary A[14]C[20]. Since $4\beta_1M_1^{11}M_2$ survives to E^{24}, $d^{24}(4\beta_1M_1^{11}M_2) = A[14]C[20]M_1^2$. Observe next that $A[30]M_1^4$ could only be hit from below the 7 row and there are no such leaders of degree 39. Therefore $\sigma A[30] = d^8(A[30]M_1^4) \neq 0$. Since $r_{2\Delta_1}(A[32,3]M_2) = A[32,3]M_1$ which bounds from the 9 row, it follows that $A[32,3]M_2$ can only bound from below the 9 row, and there are no such leaders of degree 39. Thus, $A[37] = d^6(A[32,3]M_2)$ is a nonzero element of π_{37}^S. By the proof of the preceding lemma, $d^{10}(A[8]A[19]M_1^3\bar{M}_2) = A[36]M_1$. Since $A[8]A[19] \in \mathrm{CokJ}_{27} = 0$, $A[36]M_1$ must bound from above the 27 row. There is only one possibility: $d^6(\eta A[30]M_1\bar{M}_2) = A[36]M_1$. (It follows that $(\eta A[30] + A[31])M_1^3$ is a d^{18}-boundary and the image of the leading differential $d^{18}(\sigma^2M_1^4M_2^2)$ should be denoted by $(\eta A[30] + A[31])M_1$ even though $A[31]M_1$ is zero in E^{18}.) We have thus proved the following theorem.

THEOREM 6.2.9 $\pi_{37}^S = Z_2A[37] \oplus Z_2\sigma A[30]$ and $\nu B[34] = \nu A[14]C[20] = 0$.

The computations of Section 4 show that we have the following leaders.

Row	Degree	Leader	Row	Degree	Leader
*11	41	$2\beta_1 M_1^6 \bar{M}_2^3$	*32	42	$A[32,3]M_1^5$
15	39	$Z_4(\gamma_1 M_1^{12})$	34	40	$\nu A[31]M_1^3$, $A[14]C[20]\bar{M}_2$,
18	40	$2C[18](M_1^4\bar{M}_3 + 3M_1^8\bar{M}_2)$			$B[34]M_1^3$
*30	60	$A[30]\langle M_4\rangle$	*34	42	$2B[34]M_1\bar{M}_2$
31	41	$\eta A[30]M_1^5$	*35	41	$\nu A[32,3]M_1^3$
32	40	$A[32,1]M_1^4$	36	40	$A[36]M_1^2$
			37	39	$A[37]M_1$, $\sigma A[30]M_1$

FIGURE 6.2.6: Leaders from Rows 1 to 37 of Degree at Least 39

The proof of the following lemma requires knowledge of the computation of π_N^S for $N \le 50$. Thus, the reader will understand the proof better after reading the rest of this chapter and Section 7.2.

LEMMA 6.2.10 $\eta\sigma A[30] = \eta A[37] = \nu^2 A[32,3]$ and $\sigma A[37] = 0$.

PROOF. Since $A[37] = d^6(A[32,3]M_2)$, Theorem 2.4.4(b) implies that

$$A[37] = \langle \nu, \eta, A[32,3]\rangle. \tag{6.6}$$

Thus, $\sigma A[37] \in \sigma\langle \nu, \eta, A[32,3]\rangle = \langle \sigma, \nu, \eta\rangle A[32,3] = 0$. Also, $\eta A[37] = \eta\langle \nu, \eta, A[32,3]\rangle = \langle \eta, \nu, \eta\rangle A[32,3] = \nu^2 A[32,3]$. We shall see that there is an element $C[44]$ of order eight in π_{44}^S such that $2C[44] = d^{12}(\nu A[30]M_1^6)$ and $4C[44] = \sigma^2 A[30]$. By Theorem 2.4.2,

$$2C[44] \in \langle \nu, \nu A[30], \sigma\rangle = \langle \nu A[30], \nu, \sigma\rangle. \tag{6.7}$$

Then $\sigma^2 A[30] = 4C[44] \in 2\langle \nu A[30], \nu, \sigma\rangle = \langle 2, \nu A[30], \nu\rangle\sigma$ and $\sigma A[30] + kA[37] \in \langle 2, \nu A[30], \nu\rangle$. Therefore, $\eta\sigma A[30] + k\eta A[37] \in \eta\langle 2, \nu A[30], \nu\rangle = \langle \eta, 2, \nu A[30]\rangle\nu = \langle \nu, \eta, 2\rangle A[30] = 0$. Thus, $\eta\sigma A[30] = k\eta A[37]$. We shall see that it is only possible for one of $\eta\sigma A[30]$, $\eta A[37]$ to be zero. Thus, $\eta A[37] \ne 0$. Assume that $\eta\sigma A[30] = 0$. If one were to continue the computation through degree 50 one would have the table of leaders given in Appendix 4 with the following changes: delete $4C[44]M_1^3$, $A[40,1]M_1^2\bar{M}_2$ and add $\eta A[39,3]M_1\bar{M}_2$, $\sigma A[30]M_1^3\bar{M}_2$, $\eta A[37]M_1^3\bar{M}_2$. Then $\eta A[39,3]M_1\bar{M}_2 = \nu A[37]M_1\bar{M}_2$ would transgress to a nonzero element $\xi \in \langle\nu,\eta,\nu A[37],\eta\rangle = \langle\nu,\eta A[37],\nu,\eta\rangle$. Thus, $d^{10}(\eta A[37]M_1^3\bar{M}_2) = \xi M_1$. Now there is no way for $\eta^2 D[45]M_1$ to bound. Thus $0 \neq \eta^3 D[45] = 4\nu D[45]$. However $4\nu D[45]M_1 = d^4(4D[45]\bar{M}_2)$ and $4D[45]\bar{M}_2 = d^8(B[38]\langle M_3\rangle)$, a contradiction. Therefore, $\eta\sigma A[30] \neq 0$, $k = 1$ and $\eta\sigma A[30] = \eta A[37] = \nu^2 A[32,3]$. ▮

In Figure 6.2.6, there are three leaders of degree 39 and six leaders of degree 40. Clearly $A[36]M_1^2$ transgresses to $\nu A[36]$. Since $A[14]C[20]M_1^2$ bounds, $\nu A[14]C[20]$ is divisible by η and is zero by Theorem 6.2.9. Thus, $A[14]C[20]\bar{M}_2$ transgresses. By Lemma 6.2.8, $\nu A[31]M_1^3$ transgresses. By Lemma 6.2.10(a), $B[34]M_1^3$ transgresses. $A[32,1]M_1^4$ transgresses to $\sigma A[32,1]$. By Lemma 6.2.10(b), $(A[37]+\sigma A[30])M_1$ must bound. There is only one possibility: $d^{20}(2C[18](M_1^4\bar{M}_3+3M_1^8\bar{M}_2)) = (A[37]+\sigma A[30])M_1$. All the other leaders of degree 40 transgress. Thus, we have shown that π_{38}^S has a composition series of Z_2 and Z_4.

THEOREM 6.2.11 $\pi_{38}^S = Z_4 B[38] \oplus Z_2\eta\sigma A[30]$ where $\eta\sigma A[30] = \eta A[37]$ and $\sigma A[31]=0$.

PROOF. Let $B[38] = d^{12}(\gamma_1 M_1^{12})$. It remains to show that $4B[38] = 0$, not $\eta\sigma A[30]$. If $4B[38] = \eta\sigma A[30]$ then $4\gamma_1(M_1^{11}\bar{M}_3+2M_1^{15}\bar{M}_2+2M_1^{12}\bar{M}_2^2)$ hits $\eta\sigma A[30]M_1^3\bar{M}_2$ instead of $C[42]\langle M_1^4\rangle$. Now $\eta\sigma A[30]M_1^3\bar{M}_2$ can not hit $\eta^2 D[45]M_1$, $C[42]\langle M_1^4\rangle$ transgresses to $\sigma C[42]$ and there is no leader than can hit $\eta^2 D[45]M_1$. Therefore, $4\nu D[45] = \eta^3 D[45] \neq 0$ which leads to a contradiction as in the proof of Lemma 6.2.10(b). Thus, $4B[38] = 0$. Note that $A[31] = d^{10}(\nu A[19]M_1^2\bar{M}_2)$ is only defined modulo $Z_2\eta A[30]$. Therefore, we can define $A[31]$ so that $\sigma A[31] = 0$. ▮

The computations of Section 4 show that we have the following table of leaders. The leaders of all degrees have been included and those leaders of degree greater than 47 have an asterisk at the left.

Row	Degree	Leader	Row	Degree	Leader
*9	63	$\eta^2\sigma M_1^{21}M_2^2$	32	40	$A[32,1]M_1^4$
11	41	$2\beta_1 M_1^6\bar{M}_2^3$	32	42	$A[32,3]M_1^5$
15	41	$4\gamma_1 M_1^{10}\bar{M}_2$	33	45	$\nu A[30]M_1^6$
*17	51	$\eta^2\gamma_1 M_1^{17}$	33	41	$\eta A[32,2]M_1\bar{M}_2$
18	46	$4C[18](M_1^7\bar{M}_3+M_1^{11}\bar{M}_2)$	34	40	$\nu A[31]M_1^3$, $A[14]C[20]\bar{M}_2$, $B[34]M_1^3$
*19	55	$\beta_2 M_1^{18}$			
*21	53	$\nu C[18]M_1^6\bar{M}_2\bar{M}_3$	34	42	$2B[34]M_1\bar{M}_2$
*22	62	$\nu A[19]M_1^7M_2^2\langle M_3\rangle$	35	41	$\nu A[32,3]M_1^3$
*23	67	$\sigma A[16]M_1^6\bar{M}_2^3\bar{M}_3$	35	43	$\eta A[14]C[20]M_1\bar{M}_2$
*23	63	$\gamma_2 M_1^{20}$	36	40	$A[36]M_1^2$
*24	60	$\eta A[23]M_1^{15}\bar{M}_2$	37	41	$A[37]M_1^2$
*30	60	$A[30]\langle M_4\rangle$	37	45	$\sigma A[30]M_1^4$
31	41	$\eta A[30]M_1^5$	38	40	$\eta\sigma A[30]M_1$, $B[38]M_1$

FIGURE 6.2.7: Leaders from Rows 1 to 38 of Degree at Least 40

3. Computation of π_N^S, $39 \le N \le 45$.

We continue the computations of the preceding section. In the tables of leaders below, all leaders of degree greater than 47 will have an asterisk at the left. They will be omitted from all other tables of leaders except for the last one at the end of this section.

From the table of leaders in Figure 6.2.7, we see that there are seven leaders of degree 40 and six leaders of degree 41. By Lemma 6.2.10(b), $d^4(\nu A[32,3]M_1^3) = \eta A[37]M_1$ and clearly $A[37]M_1^2$ transgresses. Clearly $\eta A[32,2]M_1\bar{M}_2$ survives to E^6 and $d^6(\eta A[32,2]M_1\bar{M}_2)$ can not equal $B[38]M_1$ because $B[38]M_1$ can only bound from below the 15 row. Thus, $\eta A[32,2]M_1\bar{M}_2$ transgresses. Since $d^{24}(2\gamma_1 M_1^{13}) = 2B[38]M_1$ and $4B[38] = 0$, it follows that $4\gamma_1 M_1^{13}$ must transgress. Since

$d^{24}(\beta_1 M_1^6 \overline{M}_2^3) \neq 0$ in the 34 row, it follows that a nonzero differential on $2\beta_1 M_1^6 \overline{M}_2^3$ must land above the 34 row. There are three possibilities for a differential on $2\beta_1 M_1^6 \overline{M}_2^3$: $A[36]M_1^2$, $B[38]M_1$ or an element of π_{40}^S. Assume that $d^{26}(2\beta_1 M_1^6 \overline{M}_2^3) = B[38]M_1$. Then the new B[38]-leader is $B[38]M_1^3$ which is the only leader of degree 44 that can hit an element we will call $\eta A[40,1]M_1$ and which we will show must bound. It follows that $\nu B[38] = \eta A[40,1] \neq 0$. However, the computer calculations in Section 4 show that $B[38]M_1^2$ and $B[38]\overline{M}_2$ are d^{24}-boundaries which implies that $\nu B[38] = 0$, a contradiction. Thus $2\beta_1 M_1^6 \overline{M}_2^3$ must hit $A[36]M_1^2$ or transgress. If it transgresses to ξ then the computer calculation in Section 4 shows that ξM_1 does not bound and thus $\eta\xi \neq 0$. There is no possibility to bound $\eta\xi M_1$ and thus $\eta^2\xi \neq 0$. By Theorem 3.3.14, $\eta^2\xi$ is divisible by two. However, $\eta^2 d^{26}(4\gamma_1 M_1^{13})$ will also be divisible by two. There will be only one element which can be half of one of these elements. Therefore, $2\beta_1 M_1^6 \overline{M}_2^3$ can not transgress and must hit $A[36]M_1^2$. We have thus proved that π_{39}^S has a composition series of Z_2, $3Z_2$, Z_2 and Z_{16} from ImJ_{39}.

THEOREM 6.3.1 $\pi_{39}^S = Z_2\sigma A[32,1] \oplus Z_2\eta B[38] \oplus Z_2 A[39,1] \oplus Z_2 A[39,2] \oplus Z_2 A[39,3] \oplus Z_{16}\gamma_4$

and $\eta^2\sigma A[30] = 0$.

PROOF. Let $A[39,1] = d^6(\nu A[31]M_1^3)$, $A[39,2] = d^6(A[14]C[20]\overline{M}_2)$ and $A[39,3] = d^6(B[34]M_1^3) = d^6(B[34]\overline{M}_2)$. The only possibility for a nontrivial extension is that $2A[39,k]$ could equal $\eta B[38]$ instead of 0 for some $1 \leq k \leq 3$. By Theorem 2.4.2,

$$A[39,1] \in \langle\eta, \nu, \nu A[31]\rangle. \qquad [6.8]$$

Thus, $2A[39,1] \in 2\langle\eta, \nu, \nu A[31]\rangle = \langle 2, \eta, \nu\rangle \nu A[31] = 0$. By Theorem 2.4.4(c),

$$A[39,2] \in \langle\eta, \nu, A[14]C[20]\rangle. \qquad [6.9]$$

Thus $2A[39,2] \in 2\langle\eta, \nu, A[14]C[20]\rangle = \langle 2, \eta, \nu\rangle A[14]C[20] = 0$. By Theorem 2.4.4(c)

$$A[39,3] \in \langle \eta, \nu, B[34] \rangle. \qquad [6.10]$$

Thus, $2A[39,3] \in 2\langle \eta, \nu, B[34] \rangle = \langle 2, \eta, \nu \rangle B[34] = 0$. Therefore, CokJ_{39} is an elementary abelain group. ▮

The computations of Section 4 show that we have the following leaders.

Row	Degree	Leader	Row	Degree	Leader
11	43	$2\beta_1 M_1^7 \overline{M}_2^3$	34	42	$2B[34]M_1\overline{M}_2$
15	41	$4\gamma_1 M_1^{13}$	35	43	$\eta A[14]C[20]M_1\overline{M}_2$
18	46	$4C[18](M_1^7\overline{M}_3 + M_1^{11}\overline{M}_2 + M_1^4\overline{M}_2\overline{M}_3)$	36	42	$A[36]M_1^3$
31	45	$\eta A[30]M_1^5$	37	41	$A[37]M_1^2$
*32	50	$A[32,1]M_1^2\overline{M}_3$	37	45	$\sigma A[30]M_1^4$
32	42	$A[32,3]M_1^5$	*38	52	$\eta\sigma A[30]M_1^3\overline{M}_2$
33	45	$\nu A[30]M_1^6$	*38	50	$B[38]M_1^6$
33	41	$\eta A[32,2]M_1\overline{M}_2$	39	45	$\sigma A[32,1]M_1^3$
34	46	$B[34]M_1^6$	39	41	$\eta B[38]M_1, A[39,1]M_1,$ $A[39,2]M_1, A[39,3]M_1$

FIGURE 6.3.1: Leaders from Rows 1 to 39 of Degree at Least 41

LEMMA 6.3.2 (a) $\eta B[38] = \nu A[36]$.

(b) $\nu A[37] = A[8]A[32,1] = \eta A[39,3] + \eta\sigma A[32,1]$.

PROOF. (a) $\gamma_1 M_1^{10}\overline{M}_2$ hits $2B[34]M_2$ and if it had survived to E^{24} it would have hit $B[38]M_1$. Now $\beta_1 M_1^6 \overline{M}_2^3$ hits $B[34]M_2$. Therefore, if $2\beta_1 M_1^6\overline{M}_2^3$ had survived to E^{28} it would have hit $B[38]M_1$. However, $d^{26}(2\beta_1 M_1^6\overline{M}_2^3) = A[36]M_1^2$. Let $\mathscr{C}$ denote the mapping cone of $A[36]$. Let $'E^r$ denote the Atiyah-Hirzebruch spectral sequence with $'E^2_{n,t} = H_n BP \otimes \pi_t \mathscr{C} \Longrightarrow \mathscr{C}_{n+t} BP$. Then the canonical map $\phi: S \longrightarrow \mathscr{C}$ induces a map of Atiyah-Hirzebruch spectral sequences $\phi_r: E^r \longrightarrow {'E^r}$. Clearly $\phi_2(2\beta_1 M_1^6\overline{M}_2^3)$ survives to $'E^{28}$ and $d^{28}\phi_2(2\beta_1 M_1^6\overline{M}_2^3) = \phi_{28}(B[38]M_1)$. Thus, in π^S_{39}

$\eta B[38] \in A[36]\cdot\pi_3^S$, i.e. $\eta B[38] = k\nu A[36]$. Since $\eta B[38] \neq 0$ and $2\nu A[36] = 0$, k must equal 1.

(b) If $A[37]M_1^2$ is not a boundary then $\nu A[37] \neq 0$ and is not divisible by η. The $\nu A[37]$-leader will be $\nu A[37]M_1^3$. As we shall see, there is no leader of degree 47 below the 40 row. Therefore, $\xi = d^6(\nu A[37]M_1^3)$ is a nonzero element of π_{45}^S. ξM_1 is not a d^6-boundary and as we shall see the one leader of degree 48 below the 40 row bounds a different element. Therefore, $\eta\xi \neq 0$. By Theorem 2.4.2, $\xi \in \langle\eta, A[37], \nu^2\rangle$. Thus, $\eta\xi \in \eta\langle\eta, A[37], \nu^2\rangle \subset \langle\eta^2, A[37], \nu^2\rangle$ $= \langle\langle 2,\eta,2\rangle, A[37], \nu^2\rangle = \langle 2, \langle\eta, 2, A[37]\rangle, \nu^2\rangle + \langle 2, \eta, \langle 2, A[37], \nu^2\rangle\rangle$. Under our present assumptions ($A[37]M_1^2$ does not bound and $\sigma A[32,1]M_1 = d^8(A[32,1]M_1^2\bar{M}_2)$), $\pi_{44}^S = Z_8C[44] \oplus Z_2A[44]$ where $A[44] = d^6(\sigma A[32,1]M_1^3)$ and $\eta C[44] \neq 0$. Then

$\eta\xi \in \langle 2, \langle A[37], 2, \eta\rangle, \nu^2\rangle + \langle 2, \eta, hA[44]+2kC[44]\rangle$

$= \langle 2, \langle A[37], 2, \eta\rangle, \nu^2\rangle + h\langle 2, \eta, A[44]\rangle + kC[44]\langle 2, \eta, 2\rangle$.

By Theorem 2.4.2, $A[44] \in \langle\nu, A[32,1], \eta\sigma\rangle$. Thus, $\eta\xi \in$

$\langle\langle 2, A[37], 2\rangle, \eta, \nu^2\rangle + \langle 2, A[37], \langle 2, \eta, \nu^2\rangle\rangle + h\langle 2, \eta, \langle\nu, A[32,1], \eta\sigma\rangle\rangle + 2k\eta^2C[44]$

$\supset \langle\eta A[37], \eta, \nu^2\rangle + \langle 2, A[37], 0\rangle + h\langle 2, \eta, \nu, A[32,1]\rangle\eta\sigma$ by Theorem 2.3.7(b). (Note that multiplication by η is a monomorphism on π_8^S and $\eta\langle 2, \eta, \nu^2\rangle = \langle\eta, 2, \eta\rangle\nu^2$ $= 2\nu^3 = 0$.) The four-fold Toda bracket above is defined by Theorem 2.2.7(a) because $\langle 2, \eta, \nu\rangle = 0$ and $\langle\eta, \nu, A[32,1]\rangle$ contains $d^6(A[32,1]\bar{M}_2) = 0$. Then $\eta\xi \in \langle 0, \eta, \nu^2\rangle + (2, \eta\sigma, \eta^2) = (\nu^2, 2, \eta\sigma, \eta^2)$. Thus, either 2 divides $\eta\xi$, ν divides $\eta\xi$ or σ divides ξ. We will see that $2\cdot\pi_{46}^S = \nu\cdot\pi_{43}^S = 0$ and thus neither 2 nor ν can divide $\eta\xi$. Therefore, $\xi = \sigma B[38]$. Hence ξM_1 $= d^8(B[38]M_1^2\bar{M}_2)$. However, $B[38]M_1^2\bar{M}_2$ bounds from the 15 row, a contradiction. Therefore, $A[37]M_1^2$ must be a boundary. It can only bound from the 32 row or below. The only such leader is $A[32,3]M_1^5$ which maps to zero under $r_{2\Delta_1}$, and therefore does not hit $A[37]M_1^2$. Thus, $A[37]M_1^2$ must bound by a hidden differential. That is, there is a differential which we thought originated on the 32 row or below and landed in the 39 or 41 row which really hits $A[37]M_1^2$.

There is only one possibility: $d^6(A[32,1]M_1^2\bar{M}_2) = A[37]M_1^2$. Now $\sigma A[32,1]M_1$ becomes a nonbounding leader defining the nonzero element $\eta\sigma A[32,1]$ in π_{40}^S. By Theorem 2.4.4(c),

$$A[37] \in \langle\eta,\nu,A[32,1]\rangle. \qquad [6.10]$$

Thus, $\nu A[37] \in \nu\langle\eta,\nu,A[32,1]\rangle = \langle\nu,\eta,\nu\rangle A[32,1] = A[8]A[32,1] \in \langle\eta,\nu,2\nu\rangle A[32,1]$ $= \eta\langle\nu,2\nu,A[32,1]\rangle = \eta\langle\nu^2,2,A[32,1]\rangle = \eta\langle\langle\eta,\nu,\eta\rangle,2,A[32,1]\rangle$ $= \eta\langle\eta,\langle\nu,\eta,2\rangle,A[32,1]\rangle + \eta\langle\eta,\nu,\langle\eta,2,A[32,1]\rangle\rangle$ $= \eta\langle\eta,0,A[32,1]\rangle + \eta\langle\eta,\nu,\langle\eta,2,A[32,1]\rangle\rangle = \eta\langle\eta,\nu,B[34]\rangle + k\eta\sigma A[32,1]$ since $2\langle\eta,2,A[32,1]\rangle = \langle 2,\eta,2\rangle A[32,1] = \eta^2 A[32,1] = 2B[34]$. ($B[34] = d^{24}(\beta_1 M_1^6\bar{M}_2^2)$ is only defined modulo $Z_2\nu A[31] \oplus Z_2 A[14]C[20] \oplus Z_2(2B[34])$, so we can define $B[34]$ modulo $Z_2(2B[34])$ by insisting that $B[34] \in \langle\eta,2,A[32,1]\rangle$.) By Theorem 2.4.4(c), $A[39,3] = d^6(B[34]\bar{M}_2) \in \langle\eta,\nu,B[34]\rangle$. Thus, $\nu A[37] = \eta A[39,3] + k\eta\sigma A[32,1]$. We shall see that $d^{12}(B[34]\langle M_2^2\rangle) = 2D[45]$ and $d^6(\eta A[39,3]M_1^3) = 4D[45]$. Then $4D[45] \in \langle 2B[34],\sigma,\nu\rangle = \langle\eta^2 A[32,1],\sigma,\nu\rangle$ $= \langle\eta,\eta\sigma A[32,1],\nu\rangle = d^6(\eta A[39,3]M_1^3)$ in E^6. Thus, $\eta A[39,3]M_1^3 = \eta\sigma A[32,1]M_1^3$ in E^6. This can only occur if $d^4(A[37]M_1^2\bar{M}_2) = \eta A[39,3]M_1^3 + \eta\sigma A[32,1]M_1^3$. Therefore, $k = 1$. ∎

In addition to the hidden differential $d^6(A[32,1]M_1^2\bar{M}_2) = A[37]M_1^2$ which we uncovered in the proof of Lemma 6.3.2(b), there are eight leaders of degree 41 and three leaders of degree 42. Since $d^{24}(\beta_1 M_1^7\bar{M}_2^3) = B[34]M_1\bar{M}_2$, $d^{24}(2\beta_1 M_1^7\bar{M}_2^3) = 2B[34]M_1\bar{M}_2$. By Lemma 6.3.2(a), $d^4(A[36]M_1^3) = \eta B[38]M_1$. Since $2\cdot\pi_{41}^S = 0$, $\eta^2 A[39,1] = 0$ by Lemma 3.3.14 . If $\eta A[39,1] \neq 0$ then $\eta A[39,1]M_1$ must bound and the only possibility is $d^{26}(2\gamma_1(M_1^{11}\bar{M}_2+10M_1^{14})) = \eta A[39,1]M_1$. However, this can not occur because $2\gamma_1(M_1^{11}\bar{M}_2+10M_1^{14})) \in \text{Image } r_{\Delta_1}$ and $\eta A[39,1]M_1 \notin \text{Image } r_{\Delta_1}$. It follows that $\eta A[39,1] = 0$. Thus $A[39,1]M_1$ must bound. The only possibility is $d^8(A[32,3]M_1^5) = A[39,1]M_1$ from which it follows that

$$\sigma A[32,3] = A[39,1]. \qquad [6.12]$$

Thus, π^S_{40} has a composition series of $3Z_2$, Z_2, Z_2, Z_2 and a Z_2 from ImJ_{40}.

THEOREM 6.3.3 $\pi^S_{40} = Z_4B[40] \oplus Z_2\eta A[39,3] \oplus Z_2\eta\sigma A[32,1] \oplus Z_2A[40,1]$
$\oplus Z_2A[40,2] \oplus Z_2\eta\gamma_4$.

where $B[40] = C[20]^2$, $2B[40] = \eta A[39,2]$ and $\eta A[39,1] = \eta^2B[38] = 0$.

PROOF. Let $B[40] = d^8(\eta A[32,2]M_1\bar{M}_2)$. In the following argument which identifies $B[40]$ as $C[20]^2$ and $2B[40]$ as $\eta A[39,2]$ we compute modulo ImJ. By Theorem 2.4.5(a), $B[40] \in \langle\eta, \eta A[32,2], \eta, \nu\rangle$. Thus, $2B[40] \in 2\langle\eta, \eta A[32,2], \eta, \nu\rangle \subset \langle\langle 2, \eta, \eta A[32,2]\rangle, \eta, \nu\rangle$. Since $d^6(A[8]C[20]M_1\bar{M}_2) = \eta A[32,2]M_1$, $\eta A[32,2] \in \langle\nu, \eta, A[8]C[20]\rangle$ and $\langle 2, \eta, \eta A[32,2]\rangle \subset \langle 2, \eta, \langle\nu, \eta, A[8]C[20]\rangle\rangle$
$= \langle 2, \langle\eta, \nu, \eta\rangle, A[8]C[20]\rangle + \langle\langle 2, \eta, \nu\rangle, \eta, A[8]C[20]\rangle$
$= \langle 2, \nu^2, A[8]C[20]\rangle + \langle 0, \eta, A[8]C[20]\rangle = \langle 2, \nu^2, A[8]\rangle C[20]$ since $\sigma A[8] = 0$ and $2\mathrm{CokJ}_{35} = 0$. Now $\langle 2, \nu^2, A[8]\rangle = \langle 2, \nu^2, \langle\eta, 2, \nu^2\rangle\rangle$
$= \langle 2, \langle\nu^2, \eta, 2\rangle, \nu^2\rangle + \langle\langle 2, \nu^2, \eta\rangle, 2, \nu^2\rangle = \langle 2, 0, \nu^2\rangle + \langle A[8], 2, \nu^2\rangle = \langle A[8], 2, \nu^2\rangle$
$= \langle A[8], 2, \langle\eta, \nu, \eta\rangle\rangle = \langle A[8], 2, \eta, \nu\rangle\eta$. This four-fold Toda bracket is defined by Theorem 2.2.7(b) since $0 \in \langle A[8], 2, \eta\rangle$ and $0 = \langle 2, \eta, \nu\rangle$. Note that $A[8]^2 \in A[8]\langle\nu, \eta, \nu\rangle = \langle A[8], \nu, \eta\rangle\nu = 0$. Now $A[8]\langle A[8], 2, \eta, \nu\rangle$
$= \langle\langle A[8], A[8], 2\rangle, \eta, \nu\rangle$. However, $\langle A[8], A[8], 2\rangle = \langle\langle\nu, \eta, \nu\rangle, A[8], 2\rangle$
$= \nu\langle\eta, \nu, A[8], 2\rangle = \nu A[14]$. This four-fold Toda bracket is defined by Theorem 2.2.7(a) since $\langle\eta, \nu, A[8]\rangle = 0$ and $\langle\nu, A[8], 2\rangle = 0$. Thus, $A[8]\langle A[8], 2, \eta, \nu\rangle = \langle\nu A[14], \eta, \nu\rangle \supset A[14]\langle\nu, \eta, \nu\rangle = A[14]A[8]$. Therefore, $A[14] + k\sigma^2 \in \langle A[8], 2, \eta, \nu\rangle$ and $\eta A[14] \in \langle 2, \nu^2, A[8]\rangle$. Thus, $2B[40] \in \langle\eta A[14]C[20], \eta, \nu\rangle \supset C[20]\langle\eta A[14], \eta, \nu\rangle = 2C[20]^2$ by 5.6. We shall see that $\eta^2B[40]$ is nonzero and not divisible by two. By Lemma 3.3.14, $2C[20]^2 = 2B[40] \neq 0$. We showed above that $2B[40] \in \langle\langle 2, \eta, \eta A[32,2]\rangle, \eta, \nu\rangle$
$= \langle\langle\eta A[32,2], \eta, 2\rangle, \eta, \nu\rangle \subset \langle\langle\eta, \eta A[32,2], 2\rangle, \eta, \nu\rangle$

$= \eta\langle\eta A[32,2],2,\eta,\nu\rangle$ modulo $Z_2(\nu A[37])$

$\subset \langle\eta A[32,2],2,\langle\eta,\nu,\eta\rangle\rangle = \langle\eta A[32,2],2,\nu^2\rangle = \eta\langle A[32,2],2,\nu^2\rangle$. Now

$\nu\langle A[32,2],2,\nu^2\rangle \subset \langle\nu A[32,2],2,\nu^2\rangle = \langle\eta A[14]C[20],2,\nu^2\rangle = A[14]C[20]\langle\eta,2,\nu^2\rangle$

$= A[14]C[20]A[8] = \eta^2 C[20]^2$ which as we remarked above must be nonzero. Note that the four-fold Toda bracket above is defined by Theorem 2.2.7(b) because $0 \in \langle\eta A[32,2],2,\eta\rangle$ and $0 = \langle 2,\eta,\nu\rangle$. Since $A[39,2] = d^6(A[14]C[20]\bar{M}_2)$, $A[39,2] \in \langle\eta,\nu,A[14]C[20]\rangle$ and $\nu A[39,2] \in \nu\langle\eta,\nu,A[14]C[20]\rangle$

$= \langle\nu,\eta,\nu\rangle A[14]C[20] = A[8]A[14]C[20] = \eta^2 C[20]$. Thus,

$$\nu A[39,2] = \eta^2 C[20]^2 \qquad [6.13]$$

Now $\sigma\langle\eta A[32,2],2,\eta,\nu\rangle \subset \langle\eta A[32,2],2,\langle\eta,\nu,\sigma\rangle\rangle = \langle\eta A[32,2],2,0\rangle \in \eta A[32,2]\cdot\pi_{13}^S$

$= 0$. As we shall see, the only element $\xi \in \pi_{39}^S$ such that $\eta\xi \neq 0$, $\nu\xi = \eta^2 C[20]$ and $\sigma\xi = 0$ is $A[39,2]$. Thus $2C[20]^2 = \eta A[39,2]$ modulo $Z_2(\nu A[37])$. Since $\nu A[37] = \eta A[39,3] + \eta\sigma A[32,1]$, $4C[20]^2 = 0$. Write

$2C[20]^2 = \eta A[39,2] + h\eta A[39,3] + k\eta\sigma A[32,1]$. Redefine $A[39,2]$ as $A[39,2] + hA[39,3] + k\sigma A[32,1]$ so that η times the new $A[39,2]$ equals $2C[20]^2$. Note that $\nu A[39,2]$ and $\sigma A[39,2]$ remain unchanged. Let $A[40,1] = d^6(\eta A[30]M_1^5)$. By Theorem 2.4.2,

$$A[40,1] \in \langle\eta,(\eta A[30],\nu A[32,3]),(\sigma,\nu)^T\rangle. \qquad [6.14]$$

Then $2A[40,1] \in 2\langle\eta,(\eta A[30],\nu A[32,3]),(\sigma,\nu)^T\rangle = \langle 2,\eta,(\eta A[30],\nu A[32,3])\rangle(\sigma,\nu)^T$

$= \langle 2,\eta,\eta A[30]\rangle\sigma + \langle 2,\eta,\nu A[32,3]\rangle\nu = \langle 2,\eta,\eta A[30]\rangle\sigma + \langle 2,\eta,\nu\rangle\nu A[32,3]$

since $2\nu\cdot\pi_{37}^S = 0$

$= \langle 2,\eta,\eta A[30]\rangle\sigma \in \sigma\cdot\pi_{33}^S = \eta\sigma\cdot\pi_{32}^S$. Since $\eta\langle 2,\eta,\eta A[30]\rangle = \langle\eta,2,\eta\rangle\eta A[30]$

$= 2\nu\eta A[30] = 0$, $2A[40,1] \in \{0,\eta\sigma A[32,2]\} = \{0\}$. Let $A[40,2] = d^{26}(4\gamma_1 M_1^{13})$. Since $d^{24}(2\gamma_1 M_1^{12}) = 2B[38]$, it follows from Theorem 2.4.2 that

$$A[40,2] \in \langle\eta,2,2B[38]\rangle. \qquad [6.15]$$

Thus, $2A[40,2] \in 2\langle\eta,2,2B[38]\rangle = \langle 2,\eta,2\rangle 2B[38] = 2\eta^2 B[38] = 0$. ∎

The computations of Section 5 show that we have the following leaders.

Row	Degree	Leader
11	43	$2\beta_1 M_1^7\bar{M}_2^3$
15	43	$Z_8(2\gamma_1(M_1^{11}\bar{M}_2+10M_1^{14}))$
18	46	$4C[18](M_1^7\bar{M}_3+M_1^{11}\bar{M}_2+M_1^4\bar{M}_2\bar{M}_3)$
31	45	$\eta A[30]M_1M_2^2$
*32	50	$A[32,1]M_1^2\bar{M}_3$
33	45	$\nu A[30]M_1^6$
34	46	$B[34]M_1^6$
34	42	$2B[34]M_1\bar{M}_2$
35	43	$\eta A[14]C[20]M_1\bar{M}_2$
36	44	$A[36]M_1M_2$
37	43	$A[37]\bar{M}_2$
37	45	$\sigma A[30]M_1^4$
*39	53	$\sigma A[32,1]M_1^4\bar{M}_2$
*39	51	$\eta B[38]M_1^3\bar{M}_2$
*39	49	$A[39,1]M_1^2M_2$, $A[39,1]M_1^5$, $A[39,3]M_1^2\bar{M}_2$
39	45	$A[39,2]\bar{M}_2$
40	42	$\eta A[39,3]M_1$, $A[40,1]M_1$, $A[40,2]M_1$
40	42	$C[20]^2M_1$, $2C[20]^2M_1$, $\eta\sigma A[32,1]M_1$

FIGURE 6.3.2: Leaders from Rows 1 to 40 of Degree at Least 42

Observe that there are seven leaders of degree 42 and four leaders of degree 43. In the derivation of Theorem 6.3.3, we showed that $d^{24}(2\beta_1M_1^7\bar{M}_2^3) = 2B[34]M_1\bar{M}_2$. By Lemma 6.3.2(b), $d^4(A[37]\bar{M}_2) = \eta A[39,3]M_1 + \eta\sigma A[32,1]M_1$. In addition, $d^6(\eta A[14]C[20]M_1\bar{M}_2) = 2C[20]^2M_1$ because $d^6(\eta A[14]M_1^3) = 2C[20]$. Since $2\pi_{41}^S = 0$, $\eta^2\sigma A[32,1] = 0$ by Lemma 3.3.14. Thus, $\eta\sigma A[32,1]M_1$ must bound. There is only one possibility: $d^{26}(2\gamma_1(M_1^{11}\bar{M}_2+10M_1^{14})) = \eta\sigma A[32,1]M_1$. Now the remaining three leaders of degree 42 must transgress to define nonzero elements of π_{41}^S. All of these elements have order two because they are divisible by η. Note that $\nu B[38] = 0$ because $B[38]\bar{M}_2$ is a d^{24}-boundary. We have thus proved the following theorem.

THEOREM 6.3.4 $\pi_{41}^S = Z_2\eta C[20]^2 \oplus Z_2\eta A[40,1] \oplus Z_2\eta A[40,2] \oplus Z_2\alpha_S \oplus Z_2\eta^2\gamma_4$ and $\eta^2\sigma A[32,1] = \eta^2A[39,2] = \eta^2A[39,3] = \nu B[38] = 0$.

The computations of Section 4 show that we have the following leaders.

Row	Degree	Leader	Row	Degree	Leader
*11	57	$4\beta_1(M_1^7\bar{M}_2^3\bar{M}_3+M_1^{10}\bar{M}_2^2\bar{M}_3+M_1^{14}\bar{M}_2^3)$	37	45	$\sigma A[30]M_1^4$
15	43	$Z_4(4\gamma_1)(M_1^{11}\bar{M}_2+2M_1^{14})$	39	45	$A[39,2]\bar{M}_2$
18	46	$4C[18](M_1^7\bar{M}_3+M_1^{11}\bar{M}_2+M_1^4\bar{M}_2\bar{M}_3)$	*40	52	$\eta A[39,3]M_1^3\bar{M}_2, A[40,1]M_1^6$
31	45	$\eta A[30]M_1M_2^2$	*40	66	$A[40,2]M_1^6\bar{M}_3$
33	45	$\nu A[30]M_1^6$	40	46	$C[20]^2\bar{M}_2,\ \eta\sigma A[32,1]M_1^3$
34	46	$B[34]M_1^6,\ 2B[34]M_1^3\bar{M}_2$	*40	48	$2C[20]^2M_1\bar{M}_2$
36	44	$A[36]M_1M_2$	41	43	$\eta C[20]^2M_1, \eta A[40,1]M_1,$
*37	51	$(A[37]+\sigma A[30])M_1^4M_2$			$\eta A[40,2]M_1$

FIGURE 6.3.3: Leaders from Rows 1 to 41 of Degree at Least 43

There are four leaders of degree 43 and one leader of degree 44. Since $A[40,1] \in \langle\eta,(\eta A[30],\nu A[32,3]),(\sigma,\nu)^T\rangle$, $\eta^2A[40,1] \in \eta^2\langle\eta,(\eta A[30],\nu A[32,3]),(\sigma,\nu)^T\rangle \subset \langle\eta^3,(\eta A[30],\nu A[32,3]),(\sigma,\nu)^T\rangle$ $= \langle 4\nu,(\eta A[30],\nu A[32,3]),(\sigma,\nu)^T\rangle \subset \langle\nu,(4\eta A[30],4\nu A[32,3]),(\sigma,\nu)^T\rangle$ $= \langle\nu,0,(\sigma,\nu)^T\rangle = \sigma\cdot\pi_{35}^S + \nu\cdot\pi_{39}^S = \nu\cdot\pi_{39}^S$. Thus, if $\eta^2A[40,1] \neq 0$ then there is $\xi \in \pi_{39}^S$ such that either ξM_1^3 or $\xi\bar{M}_2$ is a leader and $\nu\xi = \eta^2A[40,1]$. There is only one possibility: $\xi = A[39,2]$. In the proof of Theorem 6.3.3 we showed that $\nu A[39,2] = \eta^2C[20]$. Thus $\eta^2A[40,1] = 0$ and $\eta A[40,1]M_1$ must bound. There is only one possibility: $d^6(A[36]M_1M_2) = \eta A[40,1]M_1$. By Theorem 2.4.4(b),

$$\eta A[40,1] \in \langle\nu,\eta,A[36]\rangle \qquad [6.16]$$

In addition, we have shown that

$$\nu A[39,1] = \nu A[39,3] = 0 \qquad [6.17]$$

Now the other three leaders of degree 43 must transgress to nonzero elements of π_{42}^S. Thus, π_{42}^S has a composition series of $2Z_2$, Z_4 and a Z_2 from $\mathrm{Im}J_{42}$.

THEOREM 6.3.5 $\pi_{42}^S = Z_8C[42] \oplus Z_2\eta^2C[20]^2 \oplus Z_2\eta\alpha_5$, $4C[42] = \eta^2A[40,2]$ and $\eta^2A[40,1] = 0$.

PROOF. By Lemma 3.3.14, $\eta^2 A[40,2]$ must be divisible by two. There is only one possibility: $4d^{28}(4\gamma_1 M_1^{11} M_2) = \eta^2 A[40,2]$. Thus, $C[42] = d^{28}(4\gamma_1(M_1^{11}\bar{M}_2 + 2M_1^{14}))$ has order eight and $4C[42] = \eta^2 A[40,2]$. ∎

The computations of Section 4 show that we have the following leaders.

Row	Degree	Leader	Row	Degree	Leader
15	45	$16\gamma_1 M_1^8 \langle M_3 \rangle$	39	45	$A[39,2]\bar{M}_2$
18	46	$4C[18](M_1^7\bar{M}_3 + M_1^{11}\bar{M}_2 + M_1^4\bar{M}_2\bar{M}_3)$	40	46	$\eta\sigma A[32,1]M_1^3, C[20]^2\bar{M}_2$
31	45	$\eta A[30]M_1M_2^2$	41	47	$\eta A[40,1]M_1^3$
33	45	$\nu A[30]M_1^6$	42	46	$C[42]M_1^2$
34	46	$B[34]M_1^6,\ 2B[34]M_1^3\bar{M}_2$	*42	48	$2C[42]M_1^3$
*36	54	$A[36]M_1^2\langle M_3 \rangle$	42	44	$4C[42]M_1,\ \eta^2 C[20]^2 M_1$
37	45	$\sigma A[30]M_1^4$			

FIGURE 6.3.4: Leaders from Rows 1 to 42 of Degree at Least 44

There are two leaders of degree 44 and five leaders of degree 45. In the proof of Theorem 6.3.3 we showed that $\nu A[39,2] = \eta^2 C[20]$. Thus, $d^4(A[39,2]\bar{M}_2) = \eta^2 C[20]^2 M_1$. Since $r_{\Delta_1}(16\gamma_1 M_1^8 \langle M_3 \rangle)$ is homologous to $16\gamma_1 M_1^{11} M_2$ in E^{16}, it follows that $d^{28}(16\gamma_1 M_1^8 \bar{M}_3) = 4C[42]M_1$. Thus all the leaders of degree 44 bound and we have proved the following theorem.

THEOREM 6.3.6 $\pi_{43}^S = Z_8 \beta_5$

The computations of Section 4 show that we have the following leaders.

Row	Degree	Leader	Row	Degree	Leader
*15	67	$7\gamma_1(M_1^{26} + 2e)$	37	45	$\sigma A[30]M_1^4$
18	46	$4C[18](M_1^7\bar{M}_3 + M_1^{11}\bar{M}_2 + M_1^4\bar{M}_2\bar{M}_3)$	40	46	$\eta\sigma A[32,1]M_1^3, C[20]^2\bar{M}_2$

31	45	$\eta A[30]M_1M_2^2$	41	47	$\eta A[40,1]M_1^3$
33	45	$\nu A[30]M_1^6$	42	46	$C[42]M_1^2$
34	46	$B[34]M_1^6$, $2B[34]M_1^3\bar{M}_2$	*42	50	$\eta^2C[20]^2M_1\bar{M}_2$

FIGURE 6.3.5: Leaders from Rows 1 to 43 of Degree at Least 45

There are three leaders of degree 45 and six leaders of degree 46. If $\eta A[30]M_1M_2^2$, $\nu A[30]M_1^6$ or $\sigma A[30]M_1^4$ were to bound it would have to be hit by a leader of degree 46 which is below the 30 row. The only such leader is $4C[18](M_1^7\bar{M}_3+M_1^{11}\bar{M}_2+M_1^4\bar{M}_2\bar{M}_3)$. However, $d^{20}(2C[18](M_1^7\bar{M}_3+M_1^{11}\bar{M}_2+M_1^4\bar{M}_2\bar{M}_3))$ $= (\sigma A[30]+A[37])M_1M_2$. Therefore, $4C[18](M_1^7\bar{M}_3+M_1^{11}\bar{M}_2+M_1^4\bar{M}_2\bar{M}_3)$ survives to E^{22} and can not hit $\eta A[30]M_1M_2^2$, $\nu A[30]M_1^6$ or $\sigma A[30]M_1^4$. Thus, we have shown that π_{44}^S has a composition series of Z_2, Z_2, Z_2.

THEOREM 6.3.7 $\pi_{44}^S = Z_8C[44]$ where $4C[44] = \sigma^2A[30]$.

PROOF. Let $C[44] = d^{14}(\eta A[30]M_1M_2^2)$. Note that $\mu_1 \wedge \mu_1$ has boundary $(\eta \wedge \mu_1) \cup (\mu_1 \wedge \eta)$, and $d^4(M_1^2) = \nu$. Therefore, $2d^{14}(\eta A[30]M_1M_2^2)$ is represented by $d^{12}(\nu A[30]M_2^2)$. Since $\nu A[30]M_2^2$ is homologous in E^4 to $\nu A[30]M_1^6$, $2C[44] = d^{12}(\nu A[30]M_2^2)$. By Theorem 2.4.2,

$$2C[44] \in \langle\nu,\nu A[30],\sigma\rangle = \langle\nu A[30],\nu,\sigma\rangle. \qquad [6.18]$$

Note that $\mu_2 \wedge \mu_2$ has boundary $(\nu \wedge \mu_2) \cup (\mu_2 \wedge \nu)$, and $d^8\langle M_1^4\rangle = \sigma$. Therefore, $2d^{12}(\nu A[30]M_1^6)$ is represented by $d^8(\sigma A[30]M_1^4) = \sigma^2A[30]$. Thus, $4C[44] = \sigma^2A[30]$, and $C[44]$ has order eight. ∎

The computations of Section 4 show that we have the following leaders.

Row	Degree	Leader	Row	Degree	Leader
18	46	$4C[18](M_1^7\bar{M}_3+M_1^{11}\bar{M}_2+M_1^4\bar{M}_2\bar{M}_3)$	42	46	$C[42]M_1^2$
34	46	$B[34]M_1^6$, $2B[34]M_1^3\bar{M}_2$	44	46	$C[44]M_1$

40	46	$\eta\sigma A[32,1]M_1^3$, $C[20]^2\bar{M}_2$	*44	56	$2C[44]M_1^6$
41	47	$\eta A[40,1]M_1^3$	*44	58	$4C[44]M_1^7$

FIGURE 6.3.6: Leaders from Rows 1 to 44 of Degree at Least 46

There are seven leaders of degree 46 and one leader of degree 47. Clearly $\eta A[40,1]M_1^3$ transgresses. Thus π_{45}^S has a composition series of

Z_2, Z_2, $2Z_2$, $2Z_2$, Z_2.

THEOREM 6.3.8 $\pi_{45}^S = Z_{16}D[45] \oplus Z_2A[45,1] \oplus Z_2A[45,2] \oplus Z_2\eta C[44]$ where $8D[45] = \nu C[42]$.

PROOF. Note that $2\gamma_1M_1^{11}M_2$ hits $\eta\sigma A[32,1]M_1$ and $4\gamma_1M_1^{11}M_2$ hits $C[42]$. It follows that

$$C[42] \in \langle 2,\eta,\eta\sigma A[32,1]\rangle. \qquad [6.19]$$

Thus, $\nu C[42] \in \nu\langle 2,\eta,\eta\sigma A[32,1]\rangle = 2\langle\eta,\eta\sigma A[32,1],\nu\rangle$. By Theorem 2.4.4(a), $\langle\eta,\eta\sigma A[32,1],\nu\rangle$ contains $d^6(\eta\sigma A[32,1]M_1^3)$. Therefore, $\nu C[42] = d^4(C[42]M_1^2)$ is twice $d^6(\eta\sigma A[32,1]M_1^3)$. Since $\eta^2A[40,2] = 4C[42]$, $4C[42] \in 2\langle\eta,2,A[40,2]\rangle$ and

$$2C[42] \in \langle\eta,2,A[40,2]\rangle. \qquad [6.20]$$

Thus, $2\nu C[42] \in \nu\langle\eta,2,A[40,2]\rangle = \langle\nu,\eta,2\rangle A[40,2] = 0$. Let $\xi = d^{12}(B[34]M_1^6)$. We shall see in Chapter 7 that $\sigma B[34] = \eta A[40,1]$. Therefore by Theorem 2.4.2, $\xi \in \langle\nu,(B[34],A[40,1]),(\sigma,\eta)^T\rangle$. Thus, $2\xi \in 2\langle\nu,(B[34],A[40,1]),(\sigma,\eta)^T\rangle$
$\subset \langle\nu,(2B[34],2A[40,1]),(\sigma,\eta)^T\rangle = \langle\nu,(\eta^2A[32,1],0),(\sigma,\eta)^T\rangle$
$= \langle\nu,\eta^2A[32,1],\sigma\rangle$ modulo $\eta\cdot\pi_{44}^S$
$\supset \langle\nu,\eta,\eta\sigma A[32,1]\rangle$ modulo $\eta\cdot\pi_{44}^S$ and $\langle\nu,\eta,\eta\sigma A[32,1]\rangle$contains $d^6(\eta\sigma A[32,1]M_2)$. Thus, 2ξ projects in E^6 to $d^6(\eta\sigma A[32,1]M_1^3)$. Let $A[45,1] = d^{12}(2B[34]M_1^3\bar{M}_2)$. Since $A[45,1]M_1$ can not bound, $\eta A[45,1] \neq 0$ and $A[45,1]$ is not divisible by two. Let $A[45,2] = d^6(C[20]^2\bar{M}_2)$. By Theorem 2.4.4(c),

$$A[45,2] \in \langle\eta,\nu,C[20]^2\rangle. \qquad [6.21]$$

Thus, $2A[45,2] \in 2\langle\eta,\nu,C[20]^2\rangle = \langle 2,\eta,\nu\rangle C[20]^2 = 0$, and $2A[45,2] = 0$. Since

$A[45,2]M_1$ can not bound, $\eta A[45,2] \neq 0$ and $A[45,2]$ can not be divisible by two. Now there is no possibility for $2A[45,1]$ to be nonzero. Let $D[45] = d^{28}(4C[18](M_1^7\bar{M}_3+M_1^{11}\bar{M}_2+M_1^4\bar{M}_2\bar{M}_3))$. We shall see in Chapter 7 that $\eta^2D[45]$ is nonzero and not divisible by two. By Lemma 3.3.14, $2D[45] \neq 0$. The only possibility is that $2D[45] = \xi$ which we already know has order eight. In addition, we showed above that

$$2D[45] \in \langle \nu, (B[34], A[40,1]), (\sigma, \eta)^T \rangle \qquad [6.22]$$

Since $4D[45] = d^6(\eta\sigma A[32,1]M_1^3)$, it follows from that Theorem 2.4.2 that

$$4D[45] \in \langle \eta, \eta\sigma A[32,1], \nu \rangle \qquad [6.23]$$ ∎

The computations of Section 4 show that we have the following table of leaders. This table contains the leaders of all degrees.

Row	Degree	Leader	Row	Degree	Leader
9	63	$\eta^2\sigma M_1^{21}M_2^2$	39	51	$\eta B[38]M_1^3\bar{M}_2$
11	57	$4\beta_1(M_1^7\bar{M}_2^3\bar{M}_3+M_1^{10}\bar{M}_2^2\bar{M}_3+M_1^{14}\bar{M}_2^3)$	39	49	$A[39,1]M_1^2M_2$, $A[39,1]M_1^5$,
15	67	$\gamma_1(M_1^{26}+2e)$			$A[39,3]M_1^2\bar{M}_2$
17	51	$\alpha_2M_1^{14}\bar{M}_2$	40	52	$A[40,1]M_1^6$,
18	64	$4C[18]M_1^7\bar{M}_2M_2^2\bar{M}_3$			$(\eta A[39,3]+\eta\sigma A[32,1])M_1^3\bar{M}_2$
19	55	$\beta_2M_1^{18}$	40	66	$A[40,2]M_1^6\bar{M}_3$
21	53	$\nu C[18]M_1^6\bar{M}_2\bar{M}_3$	40	48	$2C[20]^2M_1\bar{M}_2$
22	62	$\nu A[19]M_1^7M_2^2\langle M_3\rangle$	41	47	$\eta A[40,1]M_1^3$
23	67	$\sigma A[16]M_1^6\bar{M}_2^3\bar{M}_3$	42	48	$2C[42]\bar{M}_2$
23	63	$\gamma_2M_1^{20}$	42	50	$\eta^2C[20]^2M_1\bar{M}_2$
24	60	$\eta A[23]M_1^{15}\bar{M}_2$	44	48	$C[44]M_1^2$
30	60	$A[30]\langle M_4\rangle$	44	56	$2C[44]M_1^6$
32	50	$A[32,1]M_1^2\bar{M}_3$	44	58	$4C[44]M_1^7$
34	48	$B[34]M_1^4M_2$	45	47	$D[45]M_1$
34	64	$2B[34]M_1^5\bar{M}_2\bar{M}_3$	45	51	$2D[45]M_1^3$, $8D[45]M_1^3$

36	54	$A[36]M_1^2\langle M_3\rangle$	45	49	$4D[45]M_1^2$
38	50	$\eta\sigma A[30]M_1^3\bar{M}_2$, $B[38]M_1^6$	45	47	$\eta C[44]M_1$, $A[45,1]M_1$,
39	53	$\sigma A[32,1]M_1^4\bar{M}_2$			$A[45,2]M_1$, $D[45]M_1$

FIGURE 6.3.7: Leaders from Rows 1 to 45 of Degree at Least 47

4. Tentative Differentials

In this section we give the computer calculations of the tenative differentials which are determined by the leading differentials on leaders of degrees 33 through 46 which we established in Sections 2 and 3. We use the same notation and conventions as in Sections 4.4 and 5.4. These tables are ordered by the row numbers of the leaders.

<u>DEGREE 9:</u> $\eta^2\sigma$ and α_1

The leading differential $d^{24}(\eta^2\sigma M_1^5\bar{M}_3) = A[32,1]$ determines tentative differentials by making the following assignments to monomials of $[Z_2\eta^2\sigma M_1 \otimes B\langle 2\rangle] \oplus [Z_2\alpha_1 \otimes B\langle 2\rangle]$ of degree 33: $\eta^2\sigma M_1^3\bar{M}_2^3$ and $\eta^2\sigma M_1^5\bar{M}_3$ are assigned 1 and all other monomials are assigned 0. The kernel of these tentaive differentials in degrees less than 70 is given by the table below, and the new $\eta^2\sigma$-leader is $\eta^2\sigma M_1^7\bar{M}_2^2$.

DEGREE	BASIS				DEGREE	BASIS				DEGREE	BASIS			
(26,9)	7	2	0	0	(42,9)	7	0	2	0	(50,9)	5	2	2	0
											7	6	0	0
											11	0	2	0
(54,9)	7	2	2	0		21	2	0	0	(56,9)	3	1	1	1
	21	2	0	0		27	0	0	0		13	5	0	0
											21	0	1	0
											6	0	1	1
(58,9)	23	2	0	0										

The leading differential $d^{24}(\eta^2\sigma M_1^{13}) = A[32,3]M_1$ determines tentative differentials by making the following assignments to monomials of $[Z_2\eta^2\sigma M_1 \otimes B\langle 2\rangle] \oplus [Z_2\alpha_1 \otimes B\langle 2\rangle]$ of degree 35: $\eta^2\sigma M_1^3\overline{M}_2\overline{M}_3$, $\eta^2\sigma M_1^7\overline{M}_2^2$ are assigned 1 and all other monomials are assigned 0. The only remaining element in degrees less than 70 is $\eta^2\sigma(M_1^{21}\overline{M}_2^2+M_1^{27})$.

DEGREE 11: β_1

The leading differential $d^{18}(2\beta_1 M_1^{14}) = A[8]C[20]M_1^2\overline{M}_2$ determines tentative differentials by making the following assignments to monomials of $Z_8\beta_1 \otimes H_*BP$ of degree 39: $\beta_1M_1^{11}M_2$ is assigned 1; $\beta_1M_1^5M_2^3$, $\beta_1M_1^8M_2^2$ are assigned 2; $\beta_1M_1^{14}$ is assigned 6; $\beta_1M_1^7M_3$, $\beta_1M_1^4M_2M_3$ are assigned 4 and all the other monomials are assigned 0. The kernel of these tentative differentials in degrees less than 70 is given by the table below, and the β_1-leader remains $2\beta_1M_1^8\overline{M}_2$.

DEGREE	GROUP	BASIS	DEGREE	GROUP	BASIS	DEGREE	GROUP	BASIS
(22,11)	Z_2	2/ 8 1 0 0	(24,11)	Z_2	6 2 0 0	(26,11)	Z_2	1/ 6 0 1 0 2/ 10 1 0 0
	Z_2	2/ 10 1 0 0	(28,11)	Z_2	2/ 11 1 0 0 1/ 4 1 1 0 3/ 14 0 0 0		Z_2	4/ 11 1 0 0
(30,11)	Z_2	2/ 12 1 0 0	(32,11)	Z_2	3/ 7 3 0 0 4/ 13 1 0 0 1/ 6 1 1 0 7/ 10 2 0 0		Z_4	1/ 7 3 0 0 7/ 10 2 0 0
	Z_4	6 3 0 0						
(34,11)	Z_2	1/ 7 1 1 0 7/ 10 0 1 0 3/ 14 1 0 0		Z_2	2/ 8 3 0 0	(36,11)	Z_2	2/ 9 3 0 0 6/ 15 1 0 0 1/ 2 3 1 0 3/ 8 1 1 0 7/ 12 2 0 0
				Z_2	2/ 14 1 0 0			
	Z_2	2/ 9 3 0 0 2/ 8 1 1 0		Z_2	4/ 15 1 0 0	(38,11)	Z_2	2/ 6 2 1 0 2/ 10 3 0 0
	Z_2	2/ 10 3 0 0 6/ 12 0 1 0	(40,11)	Z_2	3/ 7 2 1 0 6/ 11 3 0 0 6/ 13 0 1 0 1/ 4 3 1 0 3/ 10 1 1 0 6/ 14 2 0 0		Z_2	6/ 7 2 1 0 4/ 13 0 1 0 1/ 6 0 2 0 6/ 10 1 1 0 6/ 14 2 0 0
	Z_4	6 2 1 0						

Z_2 2/ 7 2 1 0
4/ 11 3 0 0
4/ 13 0 1 0
2/ 10 1 1 0

Z_2 4/ 11 3 0 0

Z_4 1/ 7 2 1 0
2/ 10 1 1 0
1/ 14 2 0 0

(42, 11) Z_2 6/ 5 3 1 0
6/ 11 1 1 0
1/ 6 0 0 1
6/ 14 0 1 0

Z_2 2/ 5 3 1 0
2/ 11 1 1 0
6/ 12 3 0 0

Z_2 2/ 12 3 0 0

Z_2 2/ 14 0 1 0

Z_4 1/ 5 3 1 0
3/ 11 1 1 0
1/ 12 3 0 0
1/ 14 0 1 0

(44, 11) Z_2 6/ 13 3 0 0
1/ 4 1 0 1
6/ 6 3 1 0
7/ 10 4 0 0
6/ 12 1 1 0

Z_2 2/ 6 3 1 0
2/ 12 1 1 0

Z_2 4/ 13 3 0 0

Z_4 1/ 4 6 0 0
1/ 6 3 1 0
7/ 10 4 0 0
2/ 12 1 1 0

(46, 11) Z_2 2/ 6 1 2 0
6/ 10 2 1 0

Z_2 2/ 14 3 0 0

Z_2 2/ 8 5 0 0
2/ 10 2 1 0
2/ 14 3 0 0

Z_4 6 1 2 0

Z_8 1/ 7 3 1 0
4/ 13 1 1 0
7/ 10 2 1 0
1/ 14 3 0 0

(48, 11) Z_2 1/ 4 2 2 0
6/ 6 1 0 1
3/ 6 6 0 0
6/ 8 3 1 0
1/ 10 0 2 0

Z_2 5/ 7 1 2 0
6/ 9 5 0 0
4/ 11 2 1 0
2/ 15 3 0 0
1/ 6 1 0 1
7/ 10 0 2 0
2/ 14 1 1 0

Z_2 2/ 7 1 2 0
4/ 9 5 0 0
6/ 11 2 1 0
2/ 15 3 0 0
2/ 6 6 0 0
2/ 8 3 1 0
6/ 10 0 2 0
6/ 14 1 1 0

Z_2 4/ 9 5 0 0
6/ 11 2 1 0
6/ 15 3 0 0
2/ 6 6 0 0
2/ 8 3 1 0
2/ 14 1 1 0

Z_2 2/ 11 2 1 0
2/ 15 3 0 0
6/ 14 1 1 0

Z_4 1/ 7 1 2 0
7/ 10 0 2 0
2/ 14 1 1 0

(50, 11) Z_2 3/ 7 1 0 1
4/ 15 1 1 0
1/ 4 2 0 1
6/ 4 7 0 0
6/ 6 4 1 0
2/ 10 0 0 1
3/ 10 5 0 0

Z_2 1/ 7 1 0 1
2/ 4 7 0 1
2/ 6 4 1 0
7/ 10 0 0 1
5/ 10 5 0 0

Z_2 2/ 6 4 1 0

Z_2 2/ 8 1 2 0

Z_2 2/ 12 2 1 0

Z_2 4/ 15 1 1 0

Z_2 4/ 15 1 1 0
2/ 10 5 0 0
2/ 12 2 1 0

Z_2 3/ 5 7 0 0
6/ 7 4 1 0
4/ 9 1 2 0
7/ 11 5 0 0
3/ 13 2 1 0
1/ 2 3 0 1
1/ 4 5 1 0
7/ 6 2 2 0
7/ 8 1 0 1
5/ 14 4 0 0
6/ 26 0 0 0

Z_2 7/ 5 7 0 0
2/ 9 1 2 0
3/ 11 5 0 0
5/ 13 2 1 0
1/ 4 0 1 1
3/ 4 5 1 0
6/ 6 2 2 0
6/ 8 1 0 1
7/ 14 4 0 0
5/ 26 0 0 0

Z_2 7/ 5 7 0 0
6/ 9 1 2 0
1/ 11 5 0 0
3/ 13 2 1 0
1/ 4 5 1 0
6/ 6 2 2 0
2/ 8 1 0 1
2/ 14 4 0 0
3/ 26 0 0 0

Z_2 2/ 10 3 1 0

Z_2 2/ 5 7 0 0
2/ 7 4 1 0
4/ 9 1 2 0
2/ 11 5 0 0
4/ 13 2 1 0
1/ 6 2 2 0
1/ 26 0 0 0

Z_2 4/ 11 5 0 0

Z_2 4/ 13 2 1 0

Z_2 6/ 5 7 0 0
6/ 11 5 0 0
2/ 13 2 1 0
2/ 8 1 0 1
2/ 14 4 0 0

Z_2 2/ 5 7 0 0
2/ 7 4 1 0
4/ 9 1 2 0
6/ 11 5 0 0
6/ 14 4 0 0
2/ 26 0 0 0

(54,11) Z_2 2/ 5 5 1 0
2/ 9 1 0 1
1/ 2 6 1 0
1/ 4 3 2 0
2/ 6 0 3 0
1/ 6 2 0 1
7/ 6 7 0 0
7/ 10 1 2 0
5/ 12 5 0 0
5/ 14 2 1 0
1/ 20 0 1 0
1/ 24 1 0 0

Z_4 6/ 9 1 0 1
1/ 4 3 2 0
1/ 6 0 3 0
2/ 6 2 0 1
3/ 6 7 0 0
1/ 10 1 2 0
2/ 24 1 0 0

Z_2 4/ 11 3 1 0
2/ 4 3 2 0
2/ 6 2 0 1
2/ 6 7 0 0
2/ 10 1 2 0

Z_2 2/ 6 0 3 0
2/ 6 7 0 0

Z_2 2/ 10 1 2 0
2/ 12 5 0 0
2/ 24 1 0 0

Z_4 6 0 3 0

Z_2 2/ 5 5 1 0
2/ 9 1 0 1
6/ 14 2 1 0
6/ 20 0 1 0

Z_2 2/ 10 1 2 0
2/ 12 5 0 0
3/ 20 0 1 0
2/ 24 1 0 0

Z_2 2/ 14 2 1 0
2/ 24 1 0 0

Z_4 6/ 9 1 0 1
1/ 6 2 0 1
2/ 6 7 0 0

(56,11) Z_2 1/ 5 3 2 0
2/ 7 0 3 0
6/ 7 7 0 0
3/ 11 1 2 0
6/ 13 5 0 0
1/ 15 2 1 0
2/ 25 1 0 0
1/ 0 7 1 0
2/ 4 1 3 0
1/ 4 3 0 1
5/ 6 0 1 1
5/ 10 1 0 1
7/ 10 6 0 0
6/ 12 3 1 0
7/ 14 0 2 0

Z_2 3/ 5 3 2 0
6/ 7 0 3 0
3/ 7 7 0 0
7/ 11 1 2 0
2/ 13 5 0 0
1/ 4 1 3 0
7/ 4 3 0 1
2/ 6 0 1 1
5/ 6 5 1 0
7/ 10 1 0 1
6/ 10 6 0 0
2/ 12 3 1 0
2/ 14 0 2 0
1/ 22 2 0 0

Z_2 5/ 5 3 2 0
5/ 7 0 3 0
7/ 7 7 0 0
7/ 11 1 2 0
6/ 13 5 0 0
1/ 4 3 0 1
7/ 6 0 1 1
1/ 10 1 0 1
6/ 14 0 2 0
6/ 22 2 0 0

Z_2 2/ 13 5 0 0
4/ 15 2 1 0
6/ 25 1 0 0
3/ 22 2 0 0

Z_2 5/ 5 3 2 0
2/ 7 0 3 0
7/ 7 7 0 0
7/ 11 1 2 0
6/ 13 5 0 0
1/ 6 0 1 1
2/ 6 5 1 0
6/ 10 1 0 1
5/ 10 6 0 0
5/ 14 0 2 0
1/ 22 2 0 0

Z_2 2/ 5 3 2 0
4/ 7 0 3 0
2/ 7 7 0 0
6/ 11 1 2 0
4/ 13 5 0 0
2/ 12 3 1 0
6/ 14 0 2 0

Z_2 4/ 11 1 2 0

Z_2 2/ 7 0 3 0
4/ 15 2 1 0
4/ 25 1 0 0
2/ 14 0 2 0
2/ 22 2 0 0

Z_2 4/ 7 7 0 0
2/ 10 1 0 1
6/ 10 6 0 0
2/ 12 3 1 0
6/ 22 2 0 0

Z_2 4/ 15 2 1 0
4/ 25 1 0 0
2/ 6 5 1 0
2/ 10 1 0 1
2/ 12 3 1 0
1/ 22 2 0 0

Z_4 1/ 5 3 2 0
1/ 7 0 3 0
7/ 7 7 0 0
3/ 11 1 2 0
6/ 15 2 1 0
2/ 25 1 0 0
7/ 10 6 0 0
2/ 12 3 1 0
6/ 14 0 2 0
5/ 22 2 0 0

Z_4 1/ 7 0 3 0
2/ 25 1 0 0
3/ 14 0 2 0
6/ 22 2 0 0

(58,11) Z_2 4/ 7 0 1 1
2/ 7 5 1 0
4/ 11 1 0 1
2/ 13 3 1 0
2/ 4 1 1 1
6/ 12 1 2 0
6/ 14 0 0 1
2/ 20 3 0 0
3/ 22 0 1 0
6/ 26 1 0 0

Z_2 2/ 7 5 1 0
6/ 13 3 1 0
6/ 14 5 0 0
2/ 22 0 1 0
6/ 26 1 0 0

Z_2 2/ 12 1 2 0
6/ 26 1 0 0

Z_4 6 3 2 0
26 1 0 0

Z_2 2/ 5 3 0 1
6/ 11 1 0 1
4/ 13 3 1 0
2/ 8 7 0 0
6/ 14 0 0 1
2/ 14 5 0 0

Z_2 2/ 4 6 1 0
2/ 6 3 2 0
2/ 14 5 0 0
2/ 26 1 0 0

Z_2 4/ 11 1 0 1

Z_2 2/ 8 7 0 0
2/ 12 1 2 0

Z_4 1/ 5 3 0 1
1/ 11 1 0 1
2/ 13 3 1 0
1/ 4 1 1 1
1/ 6 3 2 0
1/ 8 7 0 0
7/ 10 4 1 0
4/ 14 0 0 1
1/ 20 3 0 0
6/ 22 0 1 0

Z_2 1/ 7 0 1 1
2/ 7 5 1 0
4/ 11 1 0 1
2/ 13 3 1 0
6/ 4 6 1 0
5/ 10 4 1 0
7/ 14 0 0 1
7/ 20 3 0 0
3/ 26 1 0 0

Z_2 2/ 6 3 2 0

Z_2 2/ 14 5 0 0
3/ 20 3 0 0
6/ 22 0 1 0
5/ 26 1 0 0

Z_4 1/ 5 3 0 1
3/ 11 1 0 1
1/ 8 7 0 0
7/ 14 0 0 1
1/ 14 5 0 0
7/ 20 3 0 0
2/ 22 0 1 0
3/ 26 1 0 0

The leading differential $d^{22}(2\beta_1 M_1^8 \overline{M}_2) = A[32,2]$ determines tentative differentials by making the following assignments to monomials of $Z_8\beta_1 \otimes H_*BP$ of degree 33: $\beta_1 M_1^5 M_2^2$ is assigned 2, $\beta_1 M_1^{11}$ is assigned 1 and all the other monomials are assigned 0. The following table gives the kernel of these tentative differentials in degrees less than 70, and the new β_1-leader is $\beta_1 M_1^6 \overline{M}_2^2$.

DEGREE	GROUP	BASIS	DEGREE	GROUP	BASIS	DEGREE	GROUP	BASIS
(24,11)	Z_2	6 2 0 0	(26,11)	Z_2	1/ 6 0 1 0	(28,11)	Z_2	4/ 11 1 0 0
					2/ 10 1 0 0			
	Z_2	2/ 11 1 0 0	(30,11)	Z_4	6 3 0 0	(32,11)	Z_2	3/ 7 3 0 0
		1/ 4 1 1 0						4/ 13 1 0 0
		3/ 14 0 0 0						1/ 6 1 1 0
								7/ 10 2 0 0
	Z_4	1/ 7 3 0 0	(34,11)	Z_2	1/ 7 1 1 0	(36,11)	Z_2	6/ 15 1 0 0
		7/ 10 2 0 0			7/ 10 0 1 0			1/ 2 3 1 0
					1/ 14 1 0 0			1/ 8 1 1 0
								3/ 12 2 0 0
	Z_2	4/ 15 1 0 0	(38,11)	Z_4	1/ 6 2 1 0	(40,11)	Z_2	4/ 11 3 0 0
					2/ 10 3 0 0			
	Z_2	3/ 7 2 1 0		Z_2	4/ 7 2 1 0		Z_4	1/ 7 2 1 0
		6/ 11 3 0 0			4/ 11 3 0 0			2/ 10 1 1 0
		6/ 13 0 1 0			1/ 6 0 2 0			1/ 14 2 0 0
		1/ 4 3 1 0			6/ 14 2 0 0			
		3/ 10 1 1 0						
		6/ 14 2 0 0						
(42,11)	Z_2	1/ 6 0 0 1		Z_2	2/ 5 3 1 0		Z_4	1/ 5 3 1 0
		6/ 14 0 1 0			2/ 11 1 1 0			3/ 11 1 1 0
					6/ 12 3 0 0			7/ 12 3 0 0
					6/ 14 0 1 0			1/ 14 0 1 0
(44,11)	Z_2	6/ 13 3 0 0		Z_2	4/ 13 3 0 0	(46,11)	Z_4	6 1 2 0
		1/ 4 1 0 1						
		4/ 6 3 1 0		Z_4	1/ 4 6 0 0		Z_8	1/ 7 3 1 0
		3/ 10 4 0 0			7/ 6 3 1 0			4/ 13 1 1 0
					3/ 10 4 0 0			1/ 10 2 1 0
								1/ 14 3 0 0
(48,11)	Z_2	5/ 7 1 2 0		Z_2	5/ 7 1 2 0		Z_2	2/ 7 1 2 0
		6/ 9 5 0 0			6/ 9 5 0 0			4/ 15 3 0 0
		2/ 11 2 1 0			2/ 11 2 1 0			6/ 10 0 2 0
		1/ 4 2 2 0			1/ 6 1 0 1			
		5/ 6 1 0 1			7/ 10 0 2 0		Z_4	1/ 7 1 2 0
		5/ 6 6 0 0						7/ 10 0 2 0
								2/ 14 1 1 0
(50,11)	Z_2	3/ 7 1 0 1		Z_2	1/ 7 1 0 1		Z_2	4/ 15 1 1 0
		4/ 15 1 1 0			2/ 4 7 0 0			
		1/ 4 2 0 1			7/ 10 0 0 1			
		6/ 4 7 0 0			5/ 10 5 0 0			
		2/ 10 0 0 1			6/ 12 2 1 0			
		3/ 10 5 0 0						

(52,11) Z_2
6/ 5 7 0 0
2/ 9 1 2 0
4/ 13 2 1 0
1/ 2 3 0 1
2/ 6 2 2 0
5/ 8 1 0 1
7/ 14 4 0 0

Z_2
7/ 5 7 0 0
2/ 9 1 2 0
3/ 11 5 0 0
5/ 13 2 1 0
1/ 4 0 1 1
3/ 4 5 1 0
6/ 6 2 2 0
6/ 8 1 0 1
7/ 14 4 0 0
5/ 26 0 0 0

Z_2
1/ 5 7 0 0
6/ 9 1 2 0
3/ 11 5 0 0
1/ 13 2 1 0
1/ 4 5 1 0
6/ 6 2 2 0
3/ 26 0 0 0

Z_2
4/ 11 5 0 0
4/ 13 2 1 0
1/ 6 2 2 0
6/ 14 4 0 0
7/ 26 0 0 0

Z_2
4/ 11 5 0 0

Z_2
4/ 13 2 1 0

(54,11) Z_2
4/ 5 5 1 0
4/ 9 1 0 1
1/ 2 6 1 0
1/ 4 3 2 0
4/ 6 0 3 0
1/ 6 2 0 1
5/ 6 7 0 0
1/ 10 1 2 0
3/ 12 5 0 0
3/ 14 2 1 0
3/ 24 1 0 0

Z_2
4/ 11 3 1 0
2/ 4 3 2 0
2/ 6 2 0 1
2/ 6 7 0 0
2/ 10 1 2 0

Z_2
5/ 20 0 1 0

Z_4
2/ 5 5 1 0
1/ 4 3 2 0
3/ 6 0 3 0
2/ 6 2 0 1
1/ 6 7 0 0
3/ 10 1 2 0
2/ 12 5 0 0
6/ 14 2 1 0
7/ 20 0 1 0

Z_4
6/ 5 5 1 0
6/ 9 1 0 1
1/ 6 0 3 0
6/ 6 7 0 0
6/ 10 1 2 0
2/ 12 5 0 0
2/ 14 2 1 0
2/ 20 0 1 0
6/ 24 1 0 0

Z_4
4/ 5 5 1 0
2/ 9 1 0 1
1/ 6 2 0 1
2/ 14 2 1 0
7/ 20 0 1 0
2/ 24 1 0 0

(56,11) Z_2
4/ 5 3 2 0
1/ 7 7 0 0
2/ 11 1 2 0
5/ 15 2 1 0
6/ 25 1 0 0
1/ 0 7 1 0
3/ 4 1 3 0
7/ 6 0 1 1
3/ 6 5 1 0
2/ 10 1 0 1
1/ 10 6 0 0
6/ 12 3 1 0
5/ 14 0 2 0

Z_2
1/ 5 3 2 0
2/ 7 0 3 0
1/ 7 7 0 0
1/ 11 1 2 0
6/ 13 5 0 0
1/ 4 1 3 0
7/ 4 3 0 1
2/ 6 0 1 1
1/ 6 5 1 0
3/ 10 1 0 1
2/ 10 6 0 0
5/ 22 2 0 0

Z_2
4/ 11 1 2 0

Z_2
5/ 5 3 2 0
5/ 7 0 3 0
7/ 7 7 0 0
7/ 11 1 2 0
6/ 13 5 0 0
4/ 15 2 1 0
4/ 25 1 0 0
1/ 4 3 0 1
7/ 6 0 1 1
6/ 6 5 1 0
7/ 10 1 0 1
6/ 12 3 1 0
2/ 14 0 2 0
5/ 22 2 0 0

DEGREE	GROUP	BASIS
	Z_2	5/ 5 3 2 0
		2/ 7 0 3 0
		7/ 7 7 0 0
		7/ 11 1 2 0
		6/ 13 5 0 0
		1/ 6 0 1 1
		2/ 6 5 1 0
		6/ 10 1 0 1
		5/ 10 6 0 0
		5/ 14 0 2 0
		1/ 22 2 0 0
	Z_2	2/ 7 0 3 0
		4/ 15 2 1 0
		4/ 25 1 0 0
		2/ 14 0 2 0
		2/ 22 2 0 0
	Z_2	2/ 13 5 0 0
		4/ 15 2 1 0
		6/ 25 1 0 0
		3/ 22 2 0 0
	Z_4	1/ 5 3 2 0
		7/ 7 0 3 0
		7/ 7 7 0 0
		7/ 11 1 2 0
		6/ 15 2 1 0
		6/ 25 1 0 0
		2/ 6 5 1 0
		7/ 10 6 0 0
		6/ 22 2 0 0
	Z_2	1/ 7 0 3 0
		2/ 25 1 0 0
		3/ 14 0 2 0
		6/ 22 2 0 0
(58,11)	Z	4/ 7 0 1 1
		2/ 7 5 1 0
		4/ 11 1 0 1
		2/ 13 3 1 0
		2/ 4 1 1 1
		2/ 4 6 1 0
		2/ 6 3 2 0
		6/ 14 0 0 1
		2/ 14 5 0 0
		2/ 20 3 0 0
		7/ 22 0 1 0
		2/ 26 1 0 0
	Z_4	1/ 5 3 0 1
		1/ 11 1 0 1
		2/ 13 3 1 0
		1/ 4 1 1 1
		1/ 6 3 2 0
		7/ 8 7 0 0
		7/ 10 4 1 0
		1/ 20 3 0 0
		6/ 22 0 1 0
		2/ 26 1 0 0
	Z_2	2/ 5 3 0 1
		6/ 11 1 0 1
		4/ 13 3 1 0
		2/ 8 7 0 0
		6/ 14 0 0 1
		2/ 14 5 0 0
	Z_2	2/ 7 5 1 0
		6/ 13 3 1 0
		2/ 6 3 2 0
		1/ 20 3 0 0
		5/ 26 1 0 0
	Z_2	4/ 11 1 0 1
	Z_4	1/ 5 3 0 1
		3/ 11 1 0 1
		7/ 8 7 0 0
		7/ 14 0 0 1
		1/ 14 5 0 0
		7/ 20 3 0 0
		2/ 22 0 1 0
		5/ 26 1 0 0
	Z_2	1/ 7 0 1 1
		4/ 11 1 0 1
		4/ 13 3 1 0
		6/ 4 6 1 0
		5/ 10 4 1 0
		3/ 14 0 0 1
		6/ 14 5 0 0
		7/ 20 3 0 0
		6/ 22 0 1 0
		5/ 26 1 0 0
	Z_4	6 3 2 0
		26 1 0 0

The leading differential $d^{24}(\beta_1 M_1^{12}) = B[34]$ determines tentative differntials by making the following assignments to monomials of $Z_8\beta_1 \otimes H_*BP$ of degree 35: $\beta_1 M_1^6 M_2^2$ is assigned 1 and all other monomials are assigned 0. The kernel of these tentative differentials is given by the table below, and the new β_1-leader is $4\beta_1 M_1^{11}\overline{M}_2$.

DEGREE	GROUP	BASIS
(28,11)	Z_2	4/ 11 1 0 0
(30,11)	Z_2	2/ 6 3 0 0
(32,11)	Z_2	2/ 7 3 0 0
		6/ 10 2 0 0

(36,11)	Z_2	4/ 15	1 0 0
	Z_2	4/ 11	3 0 0
(38,11)	Z_2	2/ 6	2 1 0
(40,11)	Z_2	2/ 7	2 1 0
		2/ 14	2 0 0
	Z_2	4/ 11	1 1 0
(42,11)	Z_2	2/ 5	3 1 0
		6/ 11	1 1 1
		6/ 12	3 0 0
		2/ 14	0 1 0
(44,11)	Z_2	4/ 13	3 0 0
		2/ 4	1 0 1
		6/ 4	6 0 0
		2/ 6	3 1 0
	Z_2	4/ 13	3 0 0
(46,11)	Z_2	2/ 6	1 2 0
	Z_4	2/ 7	3 1 0
		2/ 10	2 1 0
		2/ 14	3 0 0
(48,11)	Z_2	2/ 7	1 2 0
		6/ 10	0 2 0
	Z_2	4/ 15	3 0 0
(50,11)	Z_2	4/ 15	1 1 0
(52,11)	Z_2	5/ 5	7 0 0
		4/ 9	1 2 0
		5/ 11	5 0 0
		3/ 13	2 1 0
		1/ 2	3 0 1
		7/ 4	5 1 0
		5/ 8	1 0 1
		7/ 14	4 0 0
		1/ 26	0 0 0
	Z_2	4/ 11	5 0 0
	Z_2	4/ 13	2 1 0
	Z_2	4/ 5	5 1 0
		2/ 4	3 2 0
		6/ 6	0 3 0
		2/ 6	7 0 0
		6/ 10	1 2 0
		6/ 20	0 1 0
	Z_2	4/ 5	5 1 0
		4/ 11	3 1 0
		2/ 6	0 3 0
		6/ 6	2 0 1
		2/ 20	0 1 0
(54,11)	Z_2	4/ 5	5 1 0
		4/ 9	1 0 1
		2/ 2	6 1 0
		2/ 4	3 2 0
		6/ 6	0 3 0
		2/ 6	2 0 1
		6/ 6	7 0 0
		6/ 10	1 2 0
		2/ 12	5 0 0
		2/ 14	2 1 0
		2/ 24	1 0 0
	Z_2	4/ 9	1 0 1
		2/ 6	2 0 1
		6/ 20	0 1 0
	Z_2	5/ 20	0 1 0
(56,11)	Z_2	6/ 5	3 2 0
		6/ 7	0 3 0
		2/ 7	7 0 0
		6/ 11	1 2 0
		4/ 13	5 0 0
		2/ 4	1 3 0
		2/ 4	3 0 1
		6/ 6	0 1 1
		2/ 6	5 1 0
		6/ 10	1 0 1
		2/ 14	0 2 0
		2/ 22	2 0 0
	Z_2	2/ 7	0 3 0
		4/ 25	1 0 0
		6/ 14	0 2 0
	Z_2	4/ 15	2 1 0
		6/ 22	2 0 0
	Z_2	4/ 11	1 2 0
(58,11)	Z_2	2/ 5	3 0 1
		2/ 11	1 0 1
		4/ 13	3 1 0
		2/ 4	1 1 1
		2/ 6	3 2 0
		6/ 8	7 0 0
		6/ 10	4 1 0
		2/ 20	3 0 0
	Z_2	2/ 5	3 0 1
		6/ 11	1 0 1
		6/ 8	7 0 0
		6/ 14	0 0 1
		2/ 14	5 0 0
		6/ 20	3 0 0
		2/ 26	1 0 0
	Z_2	4/ 7	0 1 1
		6/ 4	6 1 0
		2/ 6	3 2 0
		6/ 10	4 1 0
		3/ 20	3 0 0
		5/ 22	0 1 0
		5/ 26	1 0 0

DEGREE	GROUP	BASIS	DEGREE	GROUP	BASIS	DEGREE	GROUP	BASIS
	Z_2	2/ 6 3 2 0 2/ 26 1 0 0		Z_2	4/ 11 1 0 1		Z_2	4/ 13 3 1 0 2/ 20 3 0 0 6/ 26 1 0 0

The leading differential $d^{24}(2\beta_1 M_1^7 \overline{M}_2^3) = 2B[34]M_1\overline{M}_2$ determines tentative differentials by making the following assignments to monomials of $Z_8\beta_1 \otimes H_*BP$ of degree 43: $\beta_1 M_1^7 M_2^3$ is assigned 1; $\beta_1 M_1^{13} M_2$, $\beta_1 M_1^6 M_2 M_3$, $\beta_1 M_1^{10} M_2^2$ are assigned 2; $\beta_1 M_1 M_4$, $\beta_1 M_1^3 M_2^2 M_3$, $\beta_1 M_1^4 M_2^4$ are assigned 4 and all other monomials are assigned 0. The kernel of these tentative differntials is given by the table below, and the β_1-leader remains $4\beta_1 M_1^{11}\overline{M}_2$.

DEGREE	GROUP	BASIS	DEGREE	GROUP	BASIS	DEGREE	GROUP	BASIS
(28,11)	Z_2	4/ 11 1 0 0	(30,11)	Z_2	2/ 6 3 0 0	(36,11)	Z_2	4/ 15 1 0 0
(38,11)	Z_2	2/ 6 2 1 0	(40,11)	Z_2	4/ 11 3 0 0	(42,11)	Z_2	4/ 11 1 1 0
	Z_2	2/ 5 3 1 0 6/ 11 1 1 0 6/ 12 3 0 0 2/ 14 0 1 0	(44,11)	Z_2	4/ 13 3 0 0	(46,11)	Z_2	2/ 6 1 2 0
							Z_2	4/ 7 3 1 0
(48,11)	Z_2	4/ 15 3 0 0	(50,11)	Z_2	4/ 15 1 1 0	(52,11)	Z_2	4/ 11 5 0 0
							Z_2	4/ 13 2 1 0
(54,11)	Z_2	4/ 5 5 1 0 4/ 9 1 0 1 2/ 2 6 1 0 2/ 4 3 2 0 6/ 6 0 3 0 2/ 6 2 0 1 6/ 6 7 0 0 6/ 10 1 2 0 2/ 12 5 0 0 2/ 14 2 1 0 2/ 24 1 0 0		Z_2	4/ 5 5 1 0 2/ 4 3 2 0 6/ 6 0 3 0 2/ 6 7 0 0 6/ 10 1 2 0 6/ 20 0 1 0		Z_2	4/ 5 5 1 0 4/ 11 3 1 0 2/ 6 0 3 0 6/ 6 2 0 1 2/ 20 0 1 0
				Z_2	5/ 20 0 1 0		Z_2	4/ 9 1 0 1 2/ 6 2 0 1 6/ 20 0 1 0
(56,11)	Z_2	4/ 15 2 1 0 6/ 22 2 0 0		Z_2	4/ 11 1 2 0			
(58,11)	Z_2	2/ 5 3 0 1 2/ 11 1 0 1 4/ 13 3 1 0 2/ 4 1 1 1 2/ 6 3 2 0 6/ 8 7 0 0 6/ 10 4 1 0 2/ 20 3 0 0		Z_2	2/ 5 3 0 1 6/ 11 1 0 1 6/ 8 7 0 0 6/ 14 0 0 1 2/ 14 5 0 0 6/ 20 3 0 0 2/ 26 1 0 0		Z_2	4/ 7 0 1 1 6/ 4 6 1 0 2/ 6 3 2 0 6/ 10 4 1 0 3/ 20 3 0 0 5/ 22 0 1 0 5/ 26 1 0 0

	Z_2	2/ 6 3 2 0		Z_2	4/ 11 1 0 1		Z_2	4/ 13 3 1 0
		2/ 26 1 0 0						2/ 20 3 0 0
								6/ 26 1 0 0

The tentative differential $d^{24}(4\beta_1 M_1^{11}\overline{M}_2) = A[14]C[20]M_1^2$ determines tentative differentials by making the following assignments to monomials of $Z_8\beta_1 \otimes H_*BP$ of degree 39: $\beta_1 M_1^{11} M_2$ is assigned 1; $\beta_1 M_1^5 M_2^3$, $\beta_1 M_1^8 M_2^2$ are assigned 2; $\beta_1 M_1^{14}$ is assigned 6; $\beta_1 M_1^7 M_3$, $\beta_1 M_1^4 M_2 M_3$ are assigned 4 and all other monomials are assigned 0. The kernel of these tentative differentials is given by the table below, and the new β_1-leader is $2\beta_1 M_1^6 \overline{M}_2^3$.

DEGREE	GROUP	BASIS	DEGREE	GROUP	BASIS	DEGREE	GROUP	BASIS
(30,11)	Z_2	2/ 6 3 0 0	(38,11)	Z_2	2/ 6 2 1 0	(42,11)	Z_2	2/ 5 3 1 0
								6/ 11 1 1 0
								6/ 12 3 0 0
								2/ 14 0 1 0
(46,11)	Z_2	2/ 6 1 2 0		Z_2	4/ 7 3 1 0			
(54,11)	Z_2	4/ 5 5 1 0		Z_2	4/ 5 5 1 0		Z_2	5/ 20 0 1 0
		4/ 9 1 0 1			2/ 4 3 2 0			
		2/ 2 6 1 0			6/ 6 0 3 0			
		2/ 4 3 2 0			2/ 6 7 0 0			
		6/ 6 0 3 0			6/ 10 1 2 0			
		2/ 6 2 0 1			6/ 20 0 1 0			
		6/ 6 7 0 0						
		6/ 10 1 2 0		Z_2	4/ 9 1 0 1			
		2/ 12 5 0 0			2/ 6 2 0 1			
		2/ 14 2 1 0			6/ 20 0 1 0			
		2/ 24 1 0 0						
(58,11)	Z_2	2/ 5 3 0 1		Z_2	2/ 5 3 0 1		Z_2	4/ 7 0 1 1
		2/ 11 1 0 1			6/ 11 1 0 1			4/ 13 3 1 0
		4/ 13 3 1 0			6/ 8 7 0 0			2/ 4 6 1 0
		2/ 4 1 1 1			6/ 14 0 0 1			6/ 6 3 2 0
		2/ 6 3 2 0			2/ 14 5 0 0			2/ 10 4 1 0
		6/ 8 7 0 0			6/ 20 3 0 0			7/ 20 3 0 0
		6/ 10 4 1 0			2/ 26 1 0 0			3/ 22 0 1 0
		2/ 20 3 0 0						1/ 26 1 0 0
	Z_2	2/ 6 3 2 0						
		2/ 26 1 0 0						

The leading differential $d^{26}(2\beta_1 M_1^6 M_2^3\ ) = A[36]M_1^2$ determines tentative differentials by making the following assignments to monomials of $Z_8\beta_1 \otimes H_*BP$ of degree 41: $\beta_1 M_1^6 M_2^3$ is assigned 1; $\beta_1 M_1^8 M_3$ is assigned 2; $\beta_1 M_1^5 M_2 M_3$ $\beta_1 M_1^9 M_2^2$,

$\beta_1 M_1^{12} M_2$ are assigned 6; $\beta_1 M_1 M_3^2$, $\beta_1 M_1^3 M_2^4$, $\beta_1 M_4$ are assigned 4 and all other monomials are assigned 0. The kernel of these tentative differentials is given by the table below, and the new β_1-leader is $4\beta_1 M_1^7 \overline{M}_2^3 \overline{M}_3$.

DEGREE	GROUP	BASIS	DEGREE	GROUP	BASIS	DEGREE	GROUP	BASIS
(46,11)	Z_2	4/ 7 3 1 0	(54,11)	Z_2	4/ 5 5 1 0		Z_2	5/ 20 0 1 0
					4/ 9 1 0 1			
					2/ 4 3 2 0			
					6/ 6 0 3 0			
					2/ 6 2 0 1			
					2/ 6 7 0 0			
					6/ 10 1 2 0			
(58,11)	Z_2	4/ 5 3 0 1		Z_2	4/ 7 0 1 1			
		4/ 7 0 1 1			4/ 13 3 1 0			
		2/ 4 1 1 1			6/ 4 6 1 0			
		6/ 4 6 1 0			6/ 10 4 1 0			
		6/ 14 0 0 1			1/ 20 3 0 0			
		2/ 14 5 0 0			5/ 22 0 1 0			
		5/ 20 3 0 0			1/ 26 1 0 0			
		1/ 22 0 1 0						
		1/ 26 1 0 0						

DEGREE 14: σ^2

The leading differential $d^{18}(\sigma^2 M_1^4 M_2^2) = \eta A[30]M_1 + A[31]M_1$ determines tentative differentials which are a monomorphism on $E^{18}_{*,14}$ in degrees less than 69. There are no remaining elements.

DEGREE 15: γ_1

The leading differential $d^{20}(\gamma_1 M_1^8 \overline{M}_2) = \eta^2 A[32,1]M_1$ determines tentative differentials by making the following assignments to monomials of $Z_{32}\gamma_1 \otimes H_*BP$ of degree 37: $\gamma_1 M_1^{11}$ is assigned 1 and all other monomials are assigned 0. The kernel of these tentative differentials is given by the table below, and the γ_1-leader is $\gamma_1 M_1^{12}$.

DEGREE	GROUP	GENERATOR	DEGREE	GROUP	GENERATOR	DEGREE	GROUP	GENERATOR
(24,15)	Z_4	12 0 0 0	(26,15)	Z_4	2/ 10 1 0 0	(28,15)	Z_2	2/ 11 1 0 0
								22/ 14 0 0 0

Z_{16} 14 0 0 0

(30,15) Z_4 26/ 15 0 0 0
16/ 8 0 1 0
30/ 12 1 0 0

Z_{32} 14/ 15 0 0 0
1/ 8 0 1 0
7/ 12 1 0 0

(32,15) Z_4 6/ 13 1 0 0
1/ 10 2 0 0

Z_8 2/ 13 1 0 0

(34,15) Z_4 2/ 10 0 1 0

Z_{32} 10 0 1 0
14 1 0 0

(36,15) Z_2 20/ 15 1 0 0
2/ 8 1 1 0
6/ 12 2 0 0

Z_2 2/ 11 0 1 0
28/ 15 1 0 0
22/ 12 2 0 0

Z_{16} 14/ 15 1 0 0
1/ 12 2 0 0

Z_{16} 2/ 15 1 0 0

(38,15) Z_4 20/ 13 2 0 0
2/ 10 3 0 0
22/ 12 0 1 0

Z_4 2/ 13 2 0 0
2/ 6 2 1 0
8/ 10 3 0 0
20/ 12 0 1 0

Z_{32} 6 2 1 0

(40,15) Z_2 10/ 11 3 0 0
18/ 13 0 1 0
4/ 8 4 0 0
30/ 10 1 1 0
10/ 14 2 0 0

Z_4 26/ 11 3 0 0
14/ 13 0 1 0
1/ 8 4 0 0
2/ 10 1 1 0
24/ 14 2 0 0

Z_8 2/ 10 1 1 0
2/ 14 2 0 0

Z_{16} 14 2 0 0

Z_8 2/ 11 3 0 0

(42,15) Z_4 18/ 11 1 1 0
6/ 15 2 0 0
2/ 8 2 1 0
6/ 12 3 0 0
24/ 14 0 1 0

Z_4 2/ 6 5 0 0
4/ 12 3 0 0

Z_{32} 1/ 6 5 0 0
2/ 12 3 0 0
1/ 14 0 1 0

Z_8 2/ 11 1 1 0
28/ 12 3 0 0
28/ 14 0 1 0

Z_{32} 10/ 15 2 0 0
1/ 8 2 1 0
23/ 12 3 0 0
27/ 14 0 1 0

(44,15) Z_2 24/ 13 3 0 0
4/ 15 0 1 0
2/ 6 3 1 0
22/ 10 4 0 0
26/ 12 1 1 0

Z_2 2/ 9 2 1 0
2/ 15 0 1 0
10/ 10 4 0 0
12/ 12 1 1 0

Z_8 2/ 13 3 0 0
30/ 15 0 1 0
4/ 10 4 0 0
14/ 12 1 1 0

Z_{16} 10 4 0 0

Z_{16} 18/ 13 3 0 0
20/ 15 0 1 0
4/ 12 1 1 0

Z_{32} 6 3 1 0

(46,15) Z_4 16/ 11 4 0 0
2/ 4 4 1 0
2/ 8 0 0 1
22/ 8 5 0 0
14/ 10 2 1 0
4/ 14 3 0 0

Z_4 2/ 11 4 0 0
12/ 13 1 1 0
2/ 8 5 0 0
24/ 10 2 1 0
28/ 14 3 0 0

Z_8 26/ 11 4 0 0
26/ 13 1 1 0
4/ 8 0 0 1
4/ 10 2 1 0

Z_{32} 30/ 11 4 0 0
4/ 13 1 1 0
1/ 4 4 1 0
1/ 8 0 0 1
1/ 8 5 0 0
15/ 10 2 1 0
7/ 14 3 0 0

Z_{32} 12/ 11 4 0 0
2/ 13 1 1 0
1/ 8 0 0 1
1/ 8 5 0 0
10/ 10 2 1 0
22/ 14 3 0 0

(48,15)	Z_2	2/ 9	5 0 0
		10/ 11	2 1 0
		4/ 15	3 0 0
		2/ 6	6 0 0
		6/ 8	3 1 0
		30/ 10	0 2 0
		30/ 12	4 0 0
		10/ 14	1 1 0
	Z_4	6/ 11	2 1 0
		1/ 6	6 0 0
		2/ 8	3 1 0
		7/ 12	4 0 0
		6/ 14	1 1 0
	Z_4	8/ 9	5 0 0
		18/ 11	2 1 0
		12/ 15	3 0 0
		16/ 8	3 1 0
		23/ 10	0 2 0
		5/ 12	4 0 0
		20/ 14	1 1 1
	Z_8	2/ 9	5 0 0
		16/ 15	3 0 0
		1/ 10	0 2 0
		2/ 12	4 0 0
		28/ 14	1 1 0
	Z_8	2/ 8	3 1 0
		2/ 14	1 1 0
	Z_8	2/ 11	2 1 0
	Z_{16}	16/ 9	5 0 0
		6/ 15	3 0 0
		16/ 12	4 0 0
		12/ 14	1 1 0
	Z_{16}	2/ 14	1 1 0
(50,15)	Z_4	22/ 9	3 1 0
		14/ 13	4 0 0
		26/ 15	1 1 0
		4/ 4	7 0 0
		12/ 6	4 1 0
		4/ 8	1 2 0
		18/ 10	0 0 1
		22/ 10	5 0 0
		22/ 12	2 1 0
	Z_4	2/ 6	4 1 0
		12/ 12	2 1 0
	Z_4	8/ 9	3 1 0
		12/ 13	4 0 0
		4/ 8	1 2 0
		10/ 10	0 0 1
		28/ 10	5 0 0
		16/ 12	2 1 0
	Z_4	14/ 9	3 1 0
		8/ 13	4 0 0
		2/ 15	1 1 0
		2/ 10	0 0 1
		18/ 10	5 0 0
		18/ 12	2 1 0
	Z_{16}	8/ 9	3 1 0
		30/ 13	4 0 0
		10/ 15	1 1 0
		2/ 8	1 2 0
		18/ 10	5 0 0
		20/ 12	2 1 0
	Z_{32}	26/ 9	3 1 0
		30/ 13	4 0 0
		30/ 15	1 1 0
		1/ 4	7 0 0
		17/ 10	5 0 0
		7/ 12	2 1 0
	Z_{32}	26/ 9	3 1 0
		22/ 13	4 0 0
		16/ 15	1 1 0
		1/ 10	0 0 1
		25/ 10	5 0 0
		20/ 12	2 1 0
(52,15)	Z_2	28/ 11	5 0 0
		12/ 13	2 1 0
		2/ 4	5 1 0
		10/ 8	1 0 1
		26/ 8	6 0 0
		20/ 10	3 1 0
		2/ 12	0 2 0
		4/ 14	4 0 0
		16/ 26	0 0 0
	Z_2	8/ 11	5 0 0
		28/ 13	2 1 0
		2/ 8	1 0 1
		28/ 8	6 0 0
		8/ 10	3 1 0
		18/ 12	0 2 0
		26/ 14	4 0 0
		20/ 26	0 0 0
	Z_2	2/ 7	4 1 0
		4/ 11	0 0 1
		28/ 11	5 0 0
		12/ 13	2 1 0
		14/ 8	6 0 0
		30/ 14	4 0 0
		17/ 26	0 0 0
	Z_2	2/ 11	0 0 1
		8/ 11	5 0 0
		12/ 13	2 1 0
		12/ 8	6 0 0
		16/ 10	3 1 0
		14/ 12	0 2 0
		10/ 14	4 0 0
		21/ 26	0 0 0
	Z_2	24/ 11	5 0 0
		16/ 13	2 1 0
		16/ 8	6 0 0
		8/ 10	3 1 0
		12/ 12	0 2 0
		24/ 14	4 0 0
		31/ 26	0 0 0
	Z_8	28/ 13	2 1 0
		1/ 8	6 0 0
		10/ 10	3 1 0
		27/ 12	0 2 0
		29/ 14	4 0 0
		27/ 26	0 0 0
	Z_8	30/ 11	5 0 0
		18/ 13	2 1 0
		2/ 10	3 1 0
		8/ 12	0 2 0
		18/ 14	4 0 0
		23/ 26	0 0 0
	Z_{16}	30/ 11	5 0 0
		28/ 13	2 1 0
		1/ 12	0 2 0
		18/ 14	4 0 0
		4/ 26	0 0 0
	Z_{16}	2/ 11	5 0 0
		30/ 13	2 1 0
		14/ 14	4 0 0
		1/ 26	0 0 0

	Z_{16}	2/ 13	2 1 0
		3/ 14	4 0 0
		12/ 26	0 0 0

Z_{16}	1/ 14	4 0 0
	22/ 26	0 0 0

(54,15)	Z_2	16/ 9	6 0 0
		16/ 11	3 1 0
		20/ 15	4 0 0
		4/ 6	2 0 1
		28/ 6	7 0 0
		16/ 8	4 1 0
		20/ 10	1 2 0
		28/ 12	0 0 1
		24/ 12	5 0 0
		8/ 14	2 1 0
		10/ 24	1 0 0

Z_4	8/ 9	6 0 0
	8/ 11	3 1 0
	20/ 13	0 2 0
	8/ 15	4 0 0
	2/ 6	2 0 1
	18/ 6	7 0 0
	24/ 8	4 1 0
	6/ 10	1 2 0
	30/ 12	0 0 1
	18/ 12	5 0 0

Z_4	20/ 11	3 1 0
	2/ 6	7 0 0
	30/ 8	4 1 0
	2/ 12	5 0 0
	20/ 14	2 1 0

Z_4	24/ 9	6 0 0
	12/ 11	3 1 0
	20/ 13	0 2 0
	18/ 15	4 0 0
	2/ 8	4 1 0
	26/ 10	1 2 0
	12/ 12	0 0 1
	18/ 12	5 0 0
	8/ 14	2 1 0
	6/ 24	1 0 0

Z_4	26/ 9	6 0 0
	20/ 11	3 1 0
	8/ 13	0 2 0
	4/ 15	4 0 0
	2/ 10	1 2 0
	24/ 12	0 0 1
	22/ 12	5 0 0
	28/ 14	2 1 0
	12/ 24	1 0 0

Z_4	26/ 9	6 0 0
	16/ 11	3 1 0
	14/ 13	0 2 0
	22/ 15	4 0 0
	4/ 12	0 0 1
	24/ 12	5 0 0
	4/ 24	1 0 0

Z_8	8/ 9	6 0 0
	24/ 11	3 1 0
	16/ 13	0 2 0
	20/ 15	4 0 0
	1/ 6	2 0 1
	1/ 8	4 1 0
	19/ 12	5 0 0
	14/ 14	2 1 0
	18/ 24	1 0 0

Z_8	2/ 9	6 0 0
	30/ 11	3 1 0
	28/ 13	0 2 0
	16/ 15	4 0 0
	10/ 12	5 0 0
	4/ 14	2 1 0
	12/ 24	1 0 0

Z_{32}	30/ 11	3 1 0
	30/ 13	0 2 0
	20/ 15	4 0 0
	1/ 6	7 0 0
	1/ 8	4 1 0
	5/ 12	5 0 0
	7/ 14	2 1 0

Z_{32}	30/ 11	3 1 0
	20/ 15	4 0 0
	1/ 8	4 1 0
	11/ 12	5 0 0
	30/ 14	2 1 0
	2/ 24	1 0 0

Z_{32}	12/ 11	3 1 0
	24/ 13	0 2 0
	14/ 15	4 0 0
	1/ 12	0 0 1
	1/ 12	5 0 0
	20/ 14	2 1 0
	12/ 24	1 0 0

The leading differential $d^{24}(\gamma_1 M_1^{12}) = B[38]$ determines tentative differentials by making the following assignments to monomials of $Z_{32}\gamma_1 \otimes H_*BP$ of degree 39: $\gamma_1 M_1^{12}$ is assigned 1 and all other monomials are assigned 0. The kernel of these tentative differentials is given by the table below, and the new γ_1-leader is $4\gamma_1 M_1^{10}\overline{M}_2$.

DEGREE	GROUP	GENERATOR	DEGREE	GROUP	GENERATOR	DEGREE	GROUP	GENERATOR
(26,15)	Z_2	4/ 10 1 0 0	(28,15)	Z_8	2/ 11 1 0 0	(30,15)	Z_2	20/ 15 0 0 0
					20/ 14 0 0 0			32/ 8 0 1 0
								28/ 12 1 0 0
	Z_8	16/ 15 0 0 0	(32,15)	Z_4	20/ 13 1 0 0	(34,15)	Z_2	4/ 10 0 1 0
		4/ 8 0 1 0			4/ 10 2 0 0			
		4/ 12 1 0 0						
	Z_8	4/ 14 1 0 0	(36,15)	Z_8	2/ 11 0 1 0		Z_8	4/ 15 1 0 0
					20/ 12 2 0 0			
(38,15)	Z_2	28/ 13 2 0 0		Z_2	8/ 13 2 0 0		Z_8	28/ 13 2 0 0
		4/ 6 2 1 0			4/ 10 3 0 0			8/ 12 0 1 0
		12/ 10 3 0 0			12/ 12 0 1 0			
		28/ 12 0 1 0						
(40,15)	Z_4	8/ 11 3 0 0		Z_4	4/ 11 3 0 0		Z_8	30/ 11 3 0 0
		24/ 13 0 1 0			4/ 10 1 1 0			2/ 13 0 1 0
		4/ 8 4 0 0						2/ 8 4 0 0
		12/ 10 1 1 0						2/ 10 1 1 0
								8/ 14 2 0 0
(42,15)	Z_2	4/ 6 5 0 0		Z_2	4/ 11 1 1 0		Z_8	2/ 11 1 1 0
		8/ 12 3 0 0			12/ 15 2 0 0			2/ 6 5 0 0
					4/ 8 2 1 0			28/ 14 0 1 0
					12/ 12 3 0 0			
					16/ 14 0 1 0			
	Z_8	4/ 11 1 1 0		Z_8	4/ 14 0 1 0	(44,15)	Z_4	12/ 13 3 0 0
		4/ 15 2 0 0						12/ 15 0 1 0
		16/ 12 3 0 0						4/ 6 3 1 0
		4/ 14 0 1 0						8/ 10 4 0 0
								24/ 12 1 1 0
	Z_8	2/ 9 2 1 0		Z_8	16/ 13 3 0 0		Z_8	16/ 13 3 0 0
		20/ 13 3 0 0			8/ 15 0 1 0			20/ 15 0 1 0
		10/ 15 0 1 0			4/ 10 4 0 0			16/ 12 1 1 0
		2/ 6 3 1 0			4/ 12 1 1 0			
		30/ 10 4 0 0						
		10/ 12 1 1 0						
(46,15)	Z_2	4/ 4 4 1 0		Z_2	4/ 11 4 0 0		Z_4	20/ 11 4 0 0
		4/ 8 0 0 1			24/ 13 1 1 0			20/ 13 1 1 0
		12/ 8 5 0 0			4/ 8 5 0 0			8/ 8 0 0 1
		28/ 10 2 1 0			16/ 10 2 1 0			8/ 10 2 1 0
		8/ 14 3 0 0			24/ 14 3 0 0			
	Z_8	12/ 11 4 0 0		Z_8	4/ 14 3 0 0	(48,15)	Z_4	4/ 9 5 0 0
		16/ 13 1 1 0						16/ 11 2 1 0
		4/ 8 0 0 1						24/ 15 3 0 0
		24/ 10 2 1 0						4/ 6 6 0 0
								8/ 8 3 1 0
								8/ 12 4 0 0

Z_4 12/ 9 5 0 0
4/ 11 2 1 0
8/ 15 3 0 0
8/ 10 0 2 0
28/ 12 4 0 0
24/ 14 1 1 0

Z_8 14/ 9 5 0 0
14/ 11 2 1 0
8/ 15 3 0 0
1/ 6 6 0 0
12/ 8 3 1 0
13/ 10 0 2 0
27/ 12 4 0 0
12/ 14 1 1 0

Z_8 32/ 9 5 0 0
12/ 15 3 0 0
24/ 14 1 1 0

Z_8 4/ 14 1 1 0

(50,15) Z_2 16/ 9 3 1 0
12/ 13 4 0 0
16/ 15 1 1 0
8/ 4 7 0 0
24/ 6 4 1 0
8/ 8 1 2 0
8/ 10 5 0 0
8/ 12 2 1 0

Z_2 16/ 9 3 1 0
24/ 13 4 0 0
8/ 8 1 2 0
20/ 10 0 0 1
24/ 10 5 0 0

Z_2 28/ 9 3 1 0
16/ 13 4 0 0
4/ 15 1 1 0
4/ 10 0 0 1
4/ 10 5 0 0
4/ 12 2 1 0

Z_4 4/ 9 3 1 0
26/ 13 4 0 0
20/ 15 1 1 0
4/ 4 7 0 0
14/ 6 4 1 0
2/ 10 0 0 1
4/ 10 5 0 0
28/ 12 2 1 0

Z_8 4/ 9 3 1 0
14/ 13 4 0 0
4/ 15 1 1 0
2/ 6 4 1 0
14/ 10 0 0 1

Z_8 4/ 9 3 1 0
24/ 13 4 0 0
20/ 15 1 1 0
4/ 12 2 1 0

Z_8 28/ 13 4 0 0
4/ 15 1 1 0
4/ 8 1 2 0
20/ 10 5 0 0
24/ 12 2 1 0

(52,15) Z_2 24/ 11 5 0 0
16/ 13 2 1 0
16/ 8 6 0 0
8/ 10 3 1 0
12/ 12 0 2 0
24/ 14 4 0 0
31/ 26 0 0 0

Z_4 28/ 11 5 0 0
20/ 13 2 1 0
2/ 4 5 1 0
10/ 8 1 0 1
24/ 8 6 0 0
12/ 12 0 2 0
10/ 14 4 0 0
26/ 26 0 0 0

Z_4 28/ 7 4 1 0
24/ 11 0 0 1
4/ 11 5 0 0
4/ 13 2 1 0
4/ 8 1 0 1
16/ 8 6 0 0
12/ 10 3 1 0
4/ 12 0 2 0
28/ 14 4 0 0
25/ 26 0 0 0

Z_8 2/ 7 4 1 0
2/ 11 0 0 1
20/ 11 5 0 0
12/ 13 2 1 0
28/ 10 3 1 0
28/ 12 0 2 0
30/ 26 0 0 0

Z_8 2/ 11 0 0 1
12/ 11 5 0 0
16/ 13 2 1 0
20/ 8 6 0 0
4/ 12 0 2 0
8/ 14 4 0 0
29/ 26 0 0 0

Z_8 16/ 11 5 0 0
8/ 13 2 1 0
4/ 8 6 0 0
28/ 10 3 1 0
16/ 12 0 2 0
4/ 14 4 0 0
24/ 26 0 0 0

Z_8 4/ 11 5 0 0
14/ 26 0 0 0

(54,15) Z_2 4/ 9 6 0 0
4/ 13 0 2 0
12/ 15 4 0 0
8/ 12 5 0 0
20/ 24 1 0 0

Z_2 16/ 15 4 0 0
4/ 6 2 0 1
4/ 8 4 1 0
12/ 12 5 0 0
24/ 14 2 1 0
8/ 24 1 0 0

Z_2	12/ 9	6 0 0	Z_2	16/ 9	6 0 0	Z_2	20/ 9	6 0 0
	16/ 13	0 2 0		24/ 11	3 1 0		8/ 11	3 1 0
	24/ 15	4 0 0		8/ 13	0 2 0		16/ 13	0 2 0
	4/ 6	7 0 0		4/ 15	4 0 0		8/ 15	4 0 0
	28/ 8	4 1 0		4/ 8	4 1 0		4/ 10	1 2 0
	28/ 10	1 2 0		20/ 10	1 2 0		16/ 12	0 0 1
	16/ 12	0 0 1		24/ 12	0 0 1		12/ 12	5 0 0
	24/ 12	5 0 0		4/ 12	5 0 0		24/ 14	2 1 0
	16/ 14	2 1 0		16/ 14	2 1 0		24/ 24	1 0 0
	8/ 24	1 0 0		12/ 24	1 0 0			

Z_2	12/ 9	6 0 0	Z_4	2/ 9	6 0 0	Z_4	4/ 11	3 1 0
	16/ 11	3 1 0		16/ 11	3 1 0		12/ 13	0 2 0
	28/ 15	4 0 0		14/ 13	0 2 0		12/ 15	4 0 0
	4/ 12	0 0 1		2/ 15	4 0 0		20/ 12	5 0 0
	28/ 12	5 0 0		2/ 6	2 0 1		24/ 14	2 1 0
	16/ 14	2 1 0		2/ 6	7 0 0		28/ 24	1 0 0
	14/ 24	1 0 0		2/ 10	1 2 0			
				2/ 12	0 0 1			
				22/ 12	5 0 0			
				24/ 14	2 1 0			
				16/ 24	1 0 0			

Z_8	4/ 13	0 2 0	Z_8	16/ 15	4 0 0	Z_8	16/ 15	4 0 0
	16/ 15	4 0 0		4/ 14	2 1 0		4/ 12	5 0 0
	28/ 12	5 0 0		16/ 24	1 0 0		16/ 14	2 1 0
	16/ 14	2 1 0					14/ 24	1 0 0
	12/ 24	1 0 0						

The leading differential $d^{26}(4\gamma_1 M_1^{10}\bar{M}_2) = A[40,2]$ determines tentative differentials by making the following assignments to monomials of $Z_{32}\gamma_1 \otimes H_*BP$ of degree 41: $\gamma_1 M_1^{13}$ is assigned 1; $\gamma_1 M_1^{10} M_2$ is assigned 2; $\gamma_1 M_1^{7} M_2^{2}$ is assigned 4 and all other monomials are assigned 0. The kernel of these tentative differentials is given by the table below, and the new γ_1-leader is $2\gamma_1(M_1^{11}\bar{M}_2+10M_1^{14})$.

DEGREE	GROUP	GENERATOR		DEGREE	GROUP	GENERATOR		DEGREE	GROUP	GENERATOR	
(28,15)	Z_8	2/ 11	1 0 0	(30,15)	Z_8	16/ 15	0 0 0	(32,15)	Z_2	8/ 13	1 0 0
		20/ 14	0 0 0			4/ 8	0 1 0			8/ 10	2 0 0
						4/ 12	1 0 0				
(34,15)	Z_8	4/ 10	0 1 0	(36,15)	Z_4	4/ 11	0 1 0		Z_8	2/ 11	0 1 0
		28/ 14	1 0 0			8/ 12	2 0 0			4/ 15	1 0 0
										20/ 12	2 0 0
(38,15)	Z_8	8/ 13	2 0 0	(40,15)	Z_2	16/ 11	3 0 0		Z_2	8/ 11	3 0 0
		4/ 16	2 1 0			16/ 13	0 1 0			8/ 10	1 1 0
		16/ 10	3 0 0			8/ 8	4 0 0				
						24/ 10	1 1 0				

	Z_8	10/ 11	3 0 0
		30/ 13	0 1 0
		2/ 8	4 0 0
		2/ 10	1 1 0
		24/ 14	2 0 0
(42,15)	Z_8	2/ 11	1 1 0
		12/ 15	2 0 0
		2/ 6	5 0 0
		4/ 8	2 1 0
		20/ 12	3 0 0
		20/ 14	0 1 0
	Z_8	8/ 15	2 0 0
		4/ 8	2 1 0
		28/ 12	3 0 0
		12/ 14	0 1 0
	Z_8	4/ 11	1 1 0
		8/ 15	2 0 0
		8/ 12	3 0 0
		20/ 14	0 1 0
(44,15)	Z_2	24/ 13	3 0 0
		24/ 15	0 1 0
		8/ 6	3 1 0
		16/ 10	4 0 0
		16/ 12	1 1 0
	Z_4	8/ 15	0 1 0
	Z_8	22/ 9	2 1 0
		18/ 15	0 1 0
		2/ 6	3 1 0
		2/ 10	4 0 0
		22/ 12	1 1 0
	Z_8	16/ 13	3 0 0
		8/ 15	0 1 0
		4/ 10	4 0 0
		4/ 12	1 1 0
(46,15)	Z_2	8/ 11	4 0 0
		8/ 13	1 1 0
		16/ 8	0 0 1
		16/ 10	2 1 0
	Z_8	8/ 11	4 0 0
		8/ 13	1 1 0
		4/ 4	4 1 0
		20/ 8	0 0 1
		12/ 8	5 0 0
		12/ 10	2 1 0
		4/ 14	3 0 0
	Z_8	8/ 11	4 0 0
		24/ 13	1 1 0
		4/ 8	0 0 1
		28/ 8	5 0 0
		8/ 10	2 1 0
		8/ 14	3 0 0
(48,15)	Z_2	8/ 9	5 0 0
		16/ 15	3 0 0
		8/ 6	6 0 0
		16/ 8	3 1 0
		16/ 12	4 0 0
	Z_2	24/ 9	5 0 0
		8/ 11	2 1 0
		16/ 15	3 0 0
		16/ 10	0 2 0
		24/ 12	4 0 0
		16/ 14	1 1 0
	Z_4	8/ 15	3 0 0
		16/ 14	1 1 0
	Z_8	28/ 9	5 0 0
		20/ 11	2 1 0
		8/ 10	0 2 0
		12/ 12	4 0 0
		12/ 14	1 1 0
	Z_8	14/ 9	5 0 0
		14/ 11	2 1 0
		8/ 15	3 0 0
		1/ 6	6 0 0
		12/ 8	3 1 0
		13/ 10	0 2 0
		27/ 12	4 0 0
		12/ 14	1 1 0
(50,15)	Z_4	24/ 9	3 1 0
		18/ 13	4 0 0
		16/ 15	1 1 0
		4/ 4	7 0 0
		14/ 6	4 1 0
		24/ 8	1 2 0
		10/ 10	0 0 1
		8/ 10	5 0 0
		24/ 12	2 1 0
	Z_4	8/ 13	4 0 0
		24/ 15	1 1 0
		8/ 6	4 1 0
		24/ 8	1 2 0
		16/ 10	0 0 1
		24/ 10	5 0 0
	Z_8	24/ 9	3 1 0
		6/ 13	4 0 0
		2/ 6	4 1 0
		24/ 8	1 2 0
		22/ 10	0 0 1
		4/ 10	5 0 0
		28/ 12	2 1 0
	Z_8	24/ 9	3 1 0
		24/ 13	4 0 0
		16/ 15	1 1 0
		4/ 10	0 0 1
		4/ 10	5 0 0
(52,15)	Z_2	24/ 11	5 0 0
		8/ 13	2 1 0
		4/ 4	5 1 0
		20/ 8	1 0 1
		16/ 8	6 0 0
		24/ 12	0 2 0
		20/ 14	4 0 0
		20/ 26	0 0 0
	Z_2	24/ 7	4 1 0
		16/ 11	0 0 1
		8/ 11	5 0 0
		8/ 13	2 1 0
		8/ 8	1 0 1
		24/ 10	3 1 0
		8/ 12	0 2 0
		24/ 14	4 0 0
		18/ 26	0 0 0

DEGREE	GROUP	GENERATOR
	Z_2	24/ 11 5 0 0
		16/ 13 2 1 0
		16/ 8 6 0 0
		8/ 10 3 1 0
		12/ 12 0 2 0
		24/ 14 4 0 0
		31/ 26 0 0 0
	Z_8	14/ 11 0 0 1
		24/ 11 5 0 0
		4/ 13 2 1 0
		4/ 8 1 0 1
		24/ 10 3 1 0
		24/ 12 0 2 0
		16/ 14 4 0 0
		12/ 26 0 0 0
	Z_2	4/ 9 6 0 0
		28/ 13 0 2 0
		12/ 15 4 0 0
		4/ 6 7 0 0
		4/ 8 4 1 0
		4/ 10 1 2 0
		12/ 12 0 0 1
		24/ 14 2 1 0
		8/ 24 1 0 0
	Z_8	4/ 9 6 0 0
		24/ 11 3 1 0
		4/ 13 0 2 0
		4/ 15 4 0 0
		4/ 10 1 2 0
		12/ 12 0 0 1
		20/ 12 5 0 0
		20/ 14 2 1 0
		4/ 24 1 0 0
	Z_4	4/ 7 4 1 0
		4/ 11 0 0 1
		24/ 11 5 0 0
		24/ 13 2 1 0
		8/ 10 3 1 0
		16/ 14 4 0 0
		14/ 26 0 0 0
	Z_8	2/ 7 4 1 0
		16/ 11 0 0 1
		20/ 11 5 0 0
		20/ 13 2 1 0
		4/ 12 0 2 0
		20/ 14 4 0 0
		10/ 26 0 0 0
	Z_2	8/ 9 6 0 0
		16/ 11 3 1 0
		16/ 13 0 2 0
		24/ 15 4 0 0
		4/ 8 4 1 0
		8/ 10 1 2 0
		12/ 12 0 0 1
		20/ 12 5 0 0
		8/ 14 2 1 0
		14/ 24 1 0 0
	Z_8	16/ 9 6 0 0
		16/ 11 3 1 0
		24/ 15 4 0 0
		4/ 12 0 0 1
		8/ 12 5 0 0
		6/ 24 1 0 0
	Z_8	6/ 11 0 0 1
		20/ 11 5 0 0
		24/ 13 2 1 0
		2/ 4 5 1 0
		6/ 8 1 0 1
		12/ 8 6 0 0
		4/ 10 3 1 0
		14/ 14 4 0 0
		17/ 26 0 0 0
(54,15)	Z_2	28/ 9 6 0 0
		4/ 13 0 2 0
		4/ 15 4 0 0
		4/ 6 2 0 1
		28/ 6 7 0 0
		28/ 10 1 2 0
		20/ 12 0 0 1
		12/ 12 5 0 0
	Z_4	6/ 9 6 0 0
		20/ 11 3 1 0
		22/ 13 0 2 0
		10/ 15 4 0 0
		2/ 6 2 0 1
		2/ 6 7 0 0
		2/ 10 1 2 0
		26/ 12 0 0 1
		10/ 12 5 0 0
		16/ 14 2 1 0
		28/ 24 1 0 0
	Z_8	16/ 15 4 0 0
		4/ 12 5 0 0
		16/ 14 2 1 0
		14/ 24 1 0 0

The leading differential $d^{26}(2\gamma_1(M_1^{11}\overline{M}_2+10M_1^{14})) = \eta\sigma A[32,1]M_1$ determines tentative differentials by making the following assignments to monomials of $Z_{32}\gamma_1 \otimes H_*BP$ of degree 43: M_1^{14} is assigned 1; $M_1^{11}M_2$ is assigned 2; $M_1^8M_2^2$ is assigned 12; $M_1^4M_2M_3$ is assigned 8 and all other monomials are assigned 0. The kernel of these tentative differentials is given by the table below, and the new γ_1-leader is $4\gamma_1(M_1^{11}\overline{M}_2+2M_1^{14})$.

DEGREE	GROUP	GENERATOR
(28,15)	Z_4	4/ 11 1 0 0
		8/ 14 0 0 0
(30,15)	Z_8	16/ 15 0 0 0
		4/ 8 0 1 0
		4/ 12 1 0 0
(32,15)	Z_2	8/ 13 1 0 0
		8/ 10 2 0 0

(34,15)	Z_8	4/ 10	0 1 0
		28/ 14	1 0 0
(36,15)	Z_4	4/ 11	0 1 0
		8/ 15	1 0 0
		8/ 12	2 0 0
	Z_4	8/ 15	1 0 0
(38,15)	Z_8	8/ 13	2 0 0
		4/ 6	2 1 0
		16/ 10	3 0 0
(40,15)	Z_2	16/ 11	3 0 0
		16/ 13	0 1 0
		8/ 8	4 0 0
		24/ 10	1 1 0
	Z_2	8/ 11	3 0 0
		8/ 10	1 1 0
	Z_4	20/ 11	3 0 0
		28/ 13	0 1 0
		4/ 8	4 0 0
		4/ 10	1 1 0
		16/ 14	2 0 0
(42,15)	Z_4	4/ 11	1 1 0
		24/ 15	2 0 0
		4/ 6	5 0 0
		8/ 8	2 1 0
		8/ 12	3 0 0
		8/ 14	0 1 0
	Z_8	8/ 15	2 0 0
		4/ 8	2 1 0
		28/ 12	3 0 0
		12/ 14	0 1 0
	Z_8	4/ 11	1 1 0
		8/ 15	2 0 0
		8/ 12	3 0 0
		20/ 14	0 1 0
(44,15)	Z_2	24/ 13	3 0 0
		24/ 15	0 1 0
		8/ 6	3 1 0
		16/ 10	4 0 0
		16/ 12	1 1 0
	Z_4	12/ 9	2 1 0
		4/ 15	0 1 0
		4/ 6	3 1 0
		4/ 10	4 0 0
		12/ 12	1 1 0
	Z_4	8/ 15	0 1 0
	Z_8	16/ 13	3 0 0
		4/ 10	4 0 0
		4/ 12	1 1 0
(46,15)	Z_2	8/ 11	4 0 0
		8/ 13	1 1 0
		16/ 8	0 0 1
		16/ 10	2 1 0
	Z_8	4/ 4	4 1 0
		4/ 8	0 0 1
		12/ 8	5 0 0
		28/ 10	2 1 0
		4/ 14	3 0 0
	Z_8	8/ 11	4 0 0
		24/ 13	1 1 0
		4/ 8	0 0 1
		28/ 8	5 0 0
		8/ 10	2 1 0
		8/ 14	3 0 0
(48,15)	Z_2	8/ 9	5 0 0
		16/ 15	3 0 0
		8/ 6	6 0 0
		16/ 8	3 1 0
		16/ 12	4 0 0
	Z_2	24/ 9	5 0 0
		8/ 11	2 1 0
		16/ 15	3 0 0
		16/ 10	0 2 0
		24/ 12	4 0 0
		16/ 14	1 1 0
	Z_4	4/ 9	5 0 0
		20/ 11	2 1 0
		2/ 6	6 0 0
		24/ 8	3 1 0
		10/ 10	0 2 0
		30/ 12	4 0 0
		8/ 14	1 1 0
	Z_4	8/ 15	3 0 0
		16/ 14	1 1 0
	Z_8	28/ 9	5 0 0
		20/ 11	2 1 0
		8/ 10	0 2 0
		12/ 12	4 0 0
		12/ 14	1 1 0
(50,15)	Z_4	8/ 13	4 0 0
		24/ 15	1 1 0
		8/ 6	4 1 0
		24/ 8	1 2 0
		16/ 10	0 0 1
		24/ 10	5 0 0
	Z_4	24/ 9	3 1 0
		18/ 13	4 0 0
		16/ 15	1 1 0
		4/ 4	7 0 0
		14/ 6	4 1 0
		24/ 8	1 2 0
		10/ 10	0 0 1
		8/ 10	5 0 0
		24/ 12	2 1 0
	Z_8	24/ 9	3 1 0
		6/ 13	4 0 0
		2/ 6	4 1 0
		24/ 8	1 2 0
		22/ 10	0 0 1
		4/ 10	5 0 0
		28/ 12	2 1 0
	Z_8	24/ 9	3 1 0
		24/ 13	4 0 0
		16/ 15	1 1 0
		4/ 10	0 0 1
		4/ 10	5 0 0

(52,15)

Z_2	
24/ 11	5 0 0
8/ 13	2 1 0
4/ 4	5 1 0
20/ 8	1 0 1
16/ 8	6 0 0
24/ 12	0 2 0
20/ 14	4 0 0
20/ 26	0 0 0

Z_2	
24/ 7	4 1 0
16/ 11	0 0 1
8/ 11	5 0 0
8/ 13	2 1 0
8/ 8	1 0 1
24/ 10	3 1 0
8/ 12	0 2 0
24/ 14	4 0 0
18/ 26	0 0 0

Z_2	
24/ 11	5 0 0
16/ 13	2 1 0
16/ 8	6 0 0
8/ 10	3 1 0
12/ 12	0 2 0
24/ 14	4 0 0
31/ 26	0 0 0

Z_4	
4/ 11	0 0 1
24/ 11	5 0 0
24/ 8	6 0 0
16/ 14	4 0 0

Z_4	
4/ 7	4 1 0
8/ 11	5 0 0
8/ 13	2 1 0
8/ 12	0 2 0
8/ 14	4 0 0
20/ 26	0 0 0

Z_4	
24/ 11	5 0 0
16/ 13	2 1 0
8/ 8	6 0 0
8/ 10	3 1 0
24/ 12	0 2 0
24/ 14	4 0 0
26/ 26	0 0 0

Z_8	
12/ 11	5 0 0
4/ 13	2 1 0
2/ 4	5 1 0
2/ 8	1 0 1
4/ 8	6 0 0
20/ 10	3 1 0
16/ 12	0 2 0
22/ 14	4 0 0
7/ 26	0 0 0

(54,15)

Z_2	
28/ 9	6 0 0
4/ 13	0 2 0
4/ 15	4 0 0
4/ 6	2 0 1
28/ 6	7 0 0
28/ 10	1 2 0
20/ 12	0 0 1
12/ 12	5 0 0

Z_2	
4/ 9	6 0 0
28/ 13	0 2 0
12/ 15	4 0 0
4/ 6	7 0 0
4/ 8	4 1 0
4/ 10	1 2 0
12/ 12	0 0 1
24/ 14	2 1 0
8/ 24	1 0 0

Z_2	
8/ 9	6 0 0
16/ 11	3 1 0
16/ 13	0 2 0
24/ 15	4 0 0
4/ 8	4 1 0
8/ 10	1 2 0
12/ 12	0 0 1
20/ 12	5 0 0
8/ 14	2 1 0
14/ 24	1 0 0

Z_4	
6/ 9	6 0 0
20/ 11	3 1 0
22/ 13	0 2 0
10/ 15	4 0 0
2/ 6	2 0 1
2/ 6	7 0 0
2/ 10	1 2 0
26/ 12	0 0 1
10/ 12	5 0 0
16/ 14	2 1 0
28/ 24	1 0 0

Z_8	
4/ 9	6 0 0
24/ 11	3 1 0
4/ 13	0 2 0
4/ 15	4 0 0
4/ 10	1 2 0
12/ 12	0 0 1
20/ 12	5 0 0
20/ 14	2 1 0
4/ 24	1 0 0

Z_8	
16/ 9	6 0 0
16/ 11	3 1 0
24/ 15	4 0 0
4/ 12	0 0 1
8/ 12	5 0 0
6/ 24	1 0 0

Z_8	
16/ 15	4 0 0
4/ 12	5 0 0
16/ 14	2 1 0
14/ 24	1 0 0

The leading differential $d^{28}(4\gamma_1(M_1^{11}\bar{M}_2+2M_1^{14}) = C[42]$ determines tentative differentials by making the following assignments to monomials of $Z_{32}\gamma_1 \otimes H_*BP$ of degree 43: M_1^{14} is assigned 1; $M_1^{11}M_2$ is assigned 2; $M_1^8M_2^2$ is assigned 12; $M_1^4M_2M_3$ is assigned 8 and all other monomials are assigned 0. The kernel of these differentials is given by the table below, and the new γ_1-leader is $16\gamma_1(M_1^8\bar{M}_3+M_1^{12}\bar{M}_2)$.

DEGREE	GROUP	GENERATOR	DEGREE	GROUP	GENERATOR	DEGREE	GROUP	GENERATOR
(30,15)	Z_2	16/ 8 0 1 0	(34,15)	Z_2	16/ 10 0 1 0	(36,15)	Z_2	16/ 15 1 0 0
		16/ 12 1 0 0			16/ 14 1 0 0			
(38,15)	Z_2	16/ 6 2 1 0	(40,15)	Z_2	16/ 11 3 0 0	(42,15)	Z_2	16/ 11 1 1 0
					16/ 13 0 1 0			32/ 6 5 0 0
					8/ 8 4 0 0			16/ 8 2 1 0
					24/ 10 1 1 0			16/ 12 3 0 0
	Z_2	16/ 8 2 1 0	(44,15)	Z_2	16/ 10 4 0 0		Z_2	16/ 15 0 1 0
		16/ 12 3 0 0			16/ 12 1 1 0			
		16/ 14 0 1 0						
(46,15)	Z_2	16/ 4 4 1 0		Z_2	16/ 8 0 0 1			
		16/ 8 0 0 1			16/ 8 5 0 0			
		16/ 8 5 0 0						
		16/ 10 2 1 0						
		16/ 14 3 0 0						
(48,15)	Z_2	16/ 9 5 0 0		Z_2	16/ 14 1 1 0	(50,15)	Z_2	16/ 15 1 1 0
		16/ 15 3 0 0						16/ 8 1 2 0
		4/ 6 6 0 0		Z_2	16/ 15 3 0 0			16/ 10 0 0 1
		16/ 8 3 1 0						16/ 10 5 0 0
		4/ 10 0 2 0						
		4/ 12 4 0 0						
	Z_2	16/ 13 4 0 0		Z_2	24/ 13 4 0 0	(52,15)	Z_2	8/ 11 5 0 0
		32/ 4 7 0 0			8/ 6 4 1 0			16/ 13 2 1 0
		16/ 6 4 1 0			24/ 10 0 0 1			8/ 10 3 1 0
		16/ 10 5 0 0			16/ 10 5 0 0			20/ 12 0 2 0
					16/ 12 2 1 0			24/ 14 4 0 0
								17/ 26 0 0 0
	Z_2	24/ 11 5 0 0		Z_2	16/ 7 4 1 0		Z_2	16/ 8 6 0 0
		8/ 13 2 1 0			16/ 11 0 0 1			16/ 10 3 1 0
		4/ 4 5 1 0			16/ 11 5 0 0			16/ 12 0 2 0
		20/ 8 1 0 1			16/ 8 6 0 0			16/ 14 4 0 0
		16/ 8 6 0 0			16/ 10 3 1 0			12/ 26 0 0 0
		24/ 12 0 2 0			16/ 12 0 2 0			
		20/ 14 4 0 0			16/ 14 4 0 0			
		20/ 26 0 0 0			16/ 26 0 0 0			
(54,15)	Z_2	8/ 9 6 0 0		Z_2	24/ 9 6 0 0		Z_2	16/ 9 6 0 0
		8/ 11 3 1 0			8/ 13 0 2 0			8/ 11 3 1 0
		16/ 13 0 2 0			8/ 15 4 0 0			16/ 15 4 0 0
		8/ 15 4 0 0			8/ 6 7 0 0			4/ 8 4 1 0
		4/ 6 2 0 1			8/ 8 4 1 0			24/ 10 1 2 0
		28/ 8 4 1 0			24/ 10 1 2 0			12/ 12 0 0 1
		8/ 12 0 0 1			8/ 12 0 0 1			4/ 12 5 0 0
		20/ 12 5 0 0			16/ 12 5 0 0			8/ 14 2 1 0
		8/ 14 2 1 0						10/ 24 1 0 0
		16/ 24 1 0 0						
	Z_2	32/ 10 1 2 0		Z_2	16/ 12 5 0 0			
		16/ 12 0 0 1			8/ 24 1 0 0			
		8/ 24 1 0 0						

The leading differential $d^{28}(16\gamma_1 M_1^8 \langle M_3 \rangle) = 4C[42]M_1$ determines tentative differentials by making the following assignments to monomials of $Z_{32}\gamma_1 \otimes H_*BP$ of degree 45: M_1^{15} is assigned 1; $M_1^{12}M_2$ is assigned 2; $M_1^9 M_2^2$ is assigned 12; $M_1^6 M_2^3$ is assigned 8 and all other monomials are assigned 0. The table below gives the kernel of these tentative differentials, and the new γ_1-leader is $\gamma_1(M_1^{26}+\cdots)$.

DEGREE	GROUP	GENERATOR
(52,15)	Z_2	16/ 11 5 0 0
		24/ 13 2 1 0
		4/ 4 5 1 0
		20/ 8 1 0 1
		8/ 10 3 1 0
		20/ 12 0 2 0
		12/ 14 4 0 0
		7/ 26 0 0 0

DEGREE	GROUP	GENERATOR
(54,15)	Z_2	16/ 9 6 0 0
		16/ 11 3 1 0
		24/ 13 0 2 0
		4/ 6 2 0 1
		8/ 6 7 0 0
		8/ 8 4 1 0
		16/ 10 1 2 0
		12/ 12 0 0 1
		8/ 12 5 0 0
		16/ 14 2 1 0
		2/ 24 1 0 0

DEGREE 16: A[16]

The leading differential $d^8(A[16]M_1^6\bar{M}_2) = A[23]M_1^2\bar{M}_2$ determines tentative differentials which are a monomorphism on $Z_2A[16] \otimes H_*BP \,/\, d^{10}(E^{10}_{*,7})$ in degrees less than 69. Thus there are no remaining elements.

DEGREE 18: C[18]

The leading differential $d^{16}(C[18](M_1^4\bar{M}_2^2+2M_1^7\bar{M}_2)) = \nu A[30]M_1^2$ determines tentative differentials by making the following assignments to monomials of $Z_8C[18] \otimes H_*BP$ of degree 38: $C[18]M_1^{10}$ is assigned 1 and all other monomials are assigned 0. The kernel of these tentative differentials is given by the table below, and the new C[18]-leader is $2C[18](M_1^4\bar{M}_3+3M_1^8\bar{M}_2)$.

DEGREE	GROUP	GENERATOR
(22,18)	Z_2	2/ 4 0 1 0
		6/ 8 1 0 0

DEGREE	GROUP	GENERATOR
(26,18)	Z_2	4/ 3 1 1 0
		2/ 0 2 1 0
		2/ 4 3 0 0

DEGREE	GROUP	GENERATOR
(28,18)	Z_4	4/ 5 3 0 0
		2/ 7 0 1 0
		2/ 11 1 0 0
		2/ 4 1 1 0

DEGREE	GROUP	GENERATOR
(30,18)	Z_2	2/ 5 1 1 0
(34,18)	Z_2	4/ 7 1 1 0
		2/ 4 2 1 0
		2/ 8 3 0 0
(36,18)	Z_4	2/ 5 2 1 0
		6/ 9 3 0 0
		4/ 15 1 0 0
		6/ 12 2 0 0
(38,18)	Z_2	2/ 6 2 1 0
		6/ 10 3 0 0
	Z_2	2/ 12 0 1 0
		6/ 16 1 0 0
(40,18)	Z_4	6/ 7 2 1 0
		2/ 11 3 0 0
		6/ 13 0 1 0
		2/ 17 1 0 0
		2/ 4 3 1 0
(42,18)	Z_2	4/ 5 3 1 0
		4/ 11 1 1 0
		6/ 8 2 1 0
		6/ 12 3 0 0
	Z_4	2/ 5 3 1 0
		4/ 11 1 1 0
(44,18)	Z_4	2/ 15 0 1 0
		2/ 19 1 0 0
		2/ 12 1 1 0
(46,18)	Z_2	2/ 4 4 1 0
		6/ 8 5 0 0
	Z_2	4/ 7 3 1 0
		4/ 13 1 1 0
	Z_2	2/ 13 1 1 0
(48,18)	Z_2	2/ 11 2 1 0
		6/ 15 3 0 0
		4/ 21 1 0 0
		2/ 8 3 1 0
	Z_2	4/ 7 1 0 1
		2/ 4 2 0 1
		2/ 4 7 0 0
		6/ 16 3 0 0
(50,18)	Z_2	4/ 3 5 1 0
		2/ 0 6 1 0
		2/ 4 7 0 0
	Z_2	4/ 7 1 0 1
		2/ 4 0 3 0
		2/ 4 2 0 1
		2/ 4 7 0 0
		2/ 8 1 2 0
		6/ 12 2 1 0
	Z_2	8/ 3 5 1 0
		4/ 15 1 1 0
		2/ 12 2 1 0
		2/ 16 3 0 0

The leading differential $d^{20}(2C[18](M_1^4\bar{M}_3+3M_1^8\bar{M}_2)) = (\sigma A[30]+A[37])M_1$ determines tentative differentials by making the following assignments to monomials of $Z_8C[18] \otimes H_*BP$ of degree 38: $C[18]M_1^{11}$ is assigned 1 and all other monomials are assigned 0. The kernel of these tentative differentials is given by the table below, and the new C[18]-leader is $4C[18](M_1^7\bar{M}_3+M_1^{11}\bar{M}_2)$.

DEGREE	GROUP	GENERATOR
(28,18)	Z_2	4/ 7 0 1 0
		4/ 11 1 0 0
(36,18)	Z_2	4/ 5 2 1 0
		4/ 9 3 0 0
(40,18)	Z_2	4/ 7 2 1 0
		4/ 11 3 0 0
		4/ 13 0 1 0
		4/ 17 1 0 0
(42,18)	Z_2	4/ 5 3 1 0
(44,18)	Z_2	4/ 15 0 1 0
		4/ 19 1 0 0
(46,18)	Z_2	4/ 7 3 1 0
		4/ 13 1 1 0
(48,18)	Z_2	2/ 11 2 1 0
		6/ 15 3 0 0
		4/ 21 1 0 0
		2/ 8 3 1 0
(50,18)	Z_2	4/ 7 1 0 1
		4/ 15 1 1 0
		2/ 4 2 0 1
		2/ 4 7 0 0
		2/ 12 2 1 0

The leading differential $d^{28}(4C[18](M_1^7\overline{M}_3+M_1^{11}\overline{M}_2)) = D[45]$ determines tentative differentials by making the following assignments to monomials of $Z_8C[18] \otimes H_*BP$ of degree 46: $C[18]M_1^{11}M_2$ is assigned 1 and all other monomials are assigned 0. The only element of degree less than 69 in the kernel of these differntials is $4C[18](M_1^7\overline{M}_2^3M_3+M_1^{13}\overline{M}_2\overline{M}_3)$ which is the new C[18]-leader.

<u>DEGREE 21:</u> $\nu C[18]$

The leading differential $d^{12}(\nu C[18]M_1^3\overline{M}_2) = A[32,3]$ determines tentative differentials by making the following assignments to monomials of $Z_2\nu C[18] \otimes H_*BP$ of degree 12: $\nu C[18]M_1^3M_2$ is assigned 1 and the other two monomials are asssigned 0. The kernel of these differentials in degrees less than 68 is given by the table below, and the new $\nu C[18]$-leader is $\nu C[18]M_1^6\overline{M}_2\overline{M}_3$.

DEGREE	BASIS	DEGREE	BASIS	DEGREE	BASIS
(32,21)	6 1 1 0	(40,21)	4 3 1 0	(44,21)	6 3 1 0
(48,21)	14 1 1 0				

<u>DEGREE 23:</u> $A[23] = \sigma A[16]$

A[23] is defined by the leading differential $A[23] = d^6(2C[18]\overline{M}_2)$. In addition $\eta A[23] \neq 0$ and $d^8(A[16]M_1^6\overline{M}_2) = A[23]M_1^2\overline{M}_2$. Let $A(23,4) = Z_2A[23] \otimes B\langle 2\rangle$. The following table is constructed from computer printouts. In each row of this table, except for the last row, the sum of the numbers in the third and fourth columns equals the number in the second column. Thus, the only nonzero element of $E^{10}_{*,23}$ in degrees less than 68 with a representative in $Z_2A[23] \otimes H_*BP$ is $A[23]M_1^6\overline{M}_2^3\overline{M}_3$ which is the new A[23]-leader.

DEGREE	DIM A(23,4)	DIM $d^6(E^{18})$	DIM $d^8(E^{16})$
(0,23)	1	1	0
(2,23)	0	0	0
(4,23)	1	1	0
(6,23)	1	1	0
(8,23)	1	1	0
(10,23)	1	0	1
(12,23)	2	2	0
(14,23)	2	2	0
(16,23)	2	2	0
(18,23)	3	2	1
(20,23)	3	3	0
(22,23)	3	2	1
(24,23)	4	3	1
(26,23)	4	3	1
(28,23)	5	5	0
(30,23)	6	5	1
(32,23)	6	5	1
(34,23)	7	5	2
(36,23)	8	7	1
(38,23)	8	6	2
(40,23)	9	7	2
(42,23)	11	9	2
(44,23)	11	10	0

DEGREE 23: $\nu C[20]$

The leading differential $d^{12}(4\nu C[20]M_1^3M_2) = A[14]C[20]$ determines tentative differentials which are a monomorphism on $Z_2(4\nu C[20]M_1^3M_2) \otimes B\langle 4\rangle$. There are no remaining elements.

DEGREE 28: $A[8]C[20]$

The leading differential $d^6(A[8]C[20]M_1\bar{M}_2) = \eta A[32,2]M_1$ determines tentative differentials which are a monomorphism on $Z_2A[8]C[20]\{M_1\bar{M}_2, M_1^3\bar{M}_2\} \otimes B\langle 4\rangle$. The remaining elements in degrees less than 69 are

$Z_2A[8]C[20]M_1^{13}\bar{M}_3 \oplus (Z_2A[8]C[20]M_1^2\bar{M}_2 \otimes B\langle 4\rangle)$, and the new $A[8]C[20]$-leader is $A[8]C[20]M_1^2\bar{M}_2$.

The leading differential $d^{18}(2\beta_1M_1^{11}M_2) = A[8]C[20]M_1^2\bar{M}_2$ determines tentative differentials with image $Z_2A[8]C[20]M_1^2\bar{M}_2 \otimes H_*BP$ in degrees less than 69. The only remaining element in degrees less than 69 is $A[8]C[20]M_1^{13}\bar{M}_3$.

DEGREE 30: A[30]

The leading differential $d^4(A[30]M_1^2) = \nu A[30]$ determines tentative differentials which are a monomorphism on $Z_2A[30]\{M_1^2, \bar{M}_2, M_1^2\bar{M}_2\} \otimes B\langle 4\rangle$. The remaining elements are

$$(Z_2A[30]\langle M_4\rangle \otimes Z_2[\langle M_1^4\rangle^2, \langle M_2^2\rangle^2, \langle M_3\rangle^2, \langle M_4\rangle^2, \{M_5\}, \ldots, \{M_n\}, \ldots]) \oplus \bigoplus_{\alpha,\beta,\gamma} Z_2A[30]\langle M_1^4\rangle^{\alpha}\langle M_2^2\rangle^{\beta}\langle M_3\rangle^{\gamma} \otimes B\langle 8\rangle$$

where the sum is taken over all $0 \le \alpha, \beta, \gamma \le 1$ with $0 < \alpha\beta\gamma$. The new A[30]-leader is $A[30]\langle M_1^4\rangle$.

The leading differential $d^8(A[30]\langle M_1^4\rangle) = \sigma A[30]$ determines tentative differentials which are a monomorphism on $\bigoplus_{\alpha,\beta,\gamma} Z_2A[30]\langle M_1^4\rangle^{\alpha}\langle M_2^2\rangle^{\beta}\langle M_3\rangle^{\gamma} \otimes B\langle 8\rangle$ where the sum is taken over all $0 \le \alpha, \beta, \gamma \le 1$ with $0 < \alpha\beta\gamma$. The remaining elements are $Z_2A[30]\langle M_4\rangle \otimes Z_2[\langle M_1^4\rangle^2, \langle M_2^2\rangle^2, \langle M_3\rangle^2, \langle M_4\rangle^2, \{M_5\}, \ldots, \{M_n\}, \ldots]$, and the new A[30]-leader is $A[30]\langle M_4\rangle$.

DEGREE 31: A[31]

The leading differential $d^4(A[31]M_1^2) = \nu A[31]$ determines tentative differentials which are a monomorphism on $Z_2A[31]\{M_1^2, M_1^3, M_1M_2, M_1^2M_2, M_1^3M_2\} \otimes B\langle 4\rangle$. The remaining elements are $Z_2A[31]M_2 \otimes B\langle 4\rangle$, and the new A[31]-leader is $A[31]M_2$.

The leading differential $d^6(A[31]M_2) = A[36]$ determines tentative differentials which are a monomorphism on $Z_2A[36]M_2 \otimes B\langle 4\rangle$. There are no remaining elements.

DEGREE 31: $\eta A[30]$

The leading differential $d^{18}(\sigma^2M_1^4M_2^2) = \eta A[30]M_1 + A[31]M_1$ determines tentative differentials with cokernel in degrees less than 68 given by the table below. The new $\eta A[30]$-leader is $\eta A[30]M_1\bar{M}_2$.

DEGREE	BASIS	DEGREE	BASIS	DEGREE	BASIS
(8,31)	1 1 0 0	(10,31)	5 0 0 0	(12,31)	3 1 0 0
(14,31)	1 2 0 0		7 0 0 0	(16,31)	1 0 1 0
	5 1 0 0	(18,31)	3 2 0 0	(20,31)	1 3 0 0
	3 0 1 0		7 1 0 0	(22,31)	1 1 1 0
	5 2 0 0	(24,31)	3 3 0 0		5 0 1 0
	9 1 0 0	(26,31)	3 1 1 0		7 2 0 0
	13 0 0 0	(28,31)	1 2 1 0		5 3 0 0
	7 0 1 0		11 1 0 0	(30,31)	5 1 1 0
	9 2 0 0		15 0 0 0	(32,31)	1 5 0 0
	3 2 1 0		7 3 0 0		9 0 1 0
	13 1 0 0	(34,31)	1 3 1 0		5 4 0 0
	7 1 1 0		11 2 0 0		
(36.31)	1 1 2 0		3 5 0 0		5 2 1 0
	9 3 0 0		11 0 1 0		15 1 0 0

The leading differential $d^6(\eta A[30]M_1\bar{M}_2) = A[36]M_1$ determines tentative differentials which are a monomorphism on $Z_2\eta A[30]\{M_1\bar{M}_2, M_1^3\bar{M}_2\} \otimes B\langle 4\rangle$. The remaining elements of $Z_2\eta A[30] \otimes H_*BP$ in degrees less than 68 are given by the table below, and the new $\eta A[30]$-leader is $\eta A[30]M_1^5$.

DEGREE	BASIS	DEGREE	BASIS	DEGREE	BASIS
(10,31)	5 0 0 0	(14,31)	1 2 0 0		7 0 0 0
(16,31)	1 0 1 0 5 1 0 0	(18,31)	3 2 0 0	(20,31)	3 0 1 0 7 1 0 0
(22,31)	5 2 0 0	(24,31)	5 0 1 0 9 1 0 0	(26,31)	7 2 0 0
	13 0 0 0	(28,31)	1 2 1 0 5 3 0 0		7 0 1 0 11 1 0 0

(30,31)	9 2 0 0		15 0 0 0	(32,31)	3 2 1 0
					7 3 0 0
	9 0 1 0	(34,31)	5 4 0 0		11 2 0 0
	13 1 0 0				
(36,31)	5 2 1 0		11 0 1 0		
	9 3 0 0		15 1 0 0		

The leading differential $d^{10}(\eta A[30]M_1^5) = A[40,1]$ determines tentative differntials with kernel in degrees less than 68 equal to $Z_2\eta A[30]\{M_1M_2^2, M_1^9M_2^2\}$, and the new $\eta A[30]$-leader is $\eta A[30]M_1M_2^2$.

The leading differential $d^{14}(\eta A[30]M_1M_2^2) = C[44]$ determines the tentative differential $d^{14}(\eta A[30]M_1^9M_2^2) = C[44]M_1^8$. There are no remaining elements in degrees less than 68.

DEGREE 32: A[32,1]

The leading differential $d^{24}(\eta^2\sigma M_1^5\overline{M}_3) = A[32,1]$ determines tentative differentials with cokernel in degrees less than 69 given by the table below. This table takes into account that $\eta A[32,1] \neq 0$. The A[32,1]-leader is $A[32,1]M_1^4$.

DEGREE	BASIS	DEGREE	BASIS	DEGREE	BASIS
(8,32)	4 0 0 0	(10,32)	2 1 0 0	(12,32)	6 0 0 0
(14,32)	4 1 0 0	(16,32)	2 2 0 0	(18,32)	0 3 0 0
	2 0 1 0		6 1 0 0	(20,32)	0 1 1 0
	4 2 0 0	(22,32)	2 3 0 0		4 0 1 0
(24,32)	2 1 1 0		6 2 0 0		12 0 0 0
(26,32)	0 2 1 0		4 3 0 0		6 0 1 0
	10 1 0 0	(28,32)	4 1 1 0		14 0 0 0
(30,32)	2 2 1 0		6 3 0 0		8 0 1 0
	12 1 0 0	(32,32)	0 3 1 0		4 4 0 0
	6 1 1 0		10 2 0 0		

(34,32)	2 0 0 1		2 5 0 0		4 2 1 0
	8 3 0 0		10 0 1 0		14 1 0 0
(36,32)	2 3 1 0		4 0 2 0		6 4 0 0
	8 1 1 0		12 2 0 0		

The leading differential $d^6(A[32,1]M_1^2\bar{M}_2) = A[37]M_1^2$ determines tentative differentials which are a monomorphism on $Z_2A[32,1]M_1^2\bar{M}_2$ ⊗ B<4>. The remaining elements of $Z_2A[32,1]$ ⊗ B<4> in degrees less than 69 are given by the table below, and the A[32,1]-leader remains $A[32,1]M_1^4$.

DEGREE	BASIS	DEGREE	BASIS	DEGREE	BASIS
(8,32)	4 0 0 0	(12,32)	6 0 0 0	(14,32)	4 1 0 0
(16,32)	2 2 0 0	(18,32)	0 3 0 0		2 0 1 0 6 1 0 0
(20,32)	0 1 1 0		4 2 0 0	(22,32)	4 0 1 0
(24,32)	6 2 0 0		12 0 0 0	(26,32)	0 2 1 0
	4 3 0 0		6 0 1 0 10 1 0 0	(28,32)	4 1 1 0
	14 0 0 0	(30,32)	2 2 1 0 6 3 0 0		8 0 1 0
	12 1 0 0	(32,32)	0 3 1 0		4 4 0 0
	10 2 0 0	(34,32)	2 0 0 1 2 5 0 0 10 0 1 0		4 2 1 0
	8 3 0 0		10 0 1 0 14 1 0 0	(36,32)	4 0 2 0
	6 4 0 0		8 1 1 0		12 2 0 0

The leading differential $d^8(A[32,1]M_1^4) = \sigma A[32,1]$ determines tentative differentials with kernel given by the table below. The new A[32,1]-leader is $A[32,1](\bar{M}_2^3+M_1^2\bar{M}_3+M_1^8\bar{M}_2)$.

DEGREE	BASIS	DEGREE	BASIS	DEGREE	BASIS
(18,32)	0 3 0 0 2 0 1 0 6 1 0 0	(26,32)	0 2 1 0 6 0 1 0 10 1 0 0	(30,32)	8 0 1 0

(34,32)	2 0 0 1	8 3 0 0
	2 5 0 0	10 0 1 0
	10 0 1 0	14 1 0 0

DEGREE 32: A[32,2]

The leading differential $d^{22}(2\beta_1 M_1^{11}) = A[32,2]$ determines tentative differentials with image in degrees less than 69 equal to $Z_2A[32,2]\{1,M_1^2\} \otimes B\langle 2\rangle$. Since $\eta A[32,2] \neq 0$, the remaining elements are $Z_2A[32,2]\{\bar{M}_2, M_1^2\bar{M}_2\} \otimes B\langle 4\rangle$, and the A[32,2]-leader is $A[32,2]\bar{M}_2$.

The leading differential $d^4(A[32,2]\bar{M}_2) = \nu A[32,2]M_1 = \eta A[14]C[20]M_1$ determines tentative differential which are a monomorphism on $Z_2A[32,2]\{\bar{M}_2, M_1^2\bar{M}_2\} \otimes B\langle 4\rangle$. There are no remaining elements of $Z_2A[32,2] \otimes H_*BP$ in degrees less than 69.

DEGREE 32: A[32,3]

The leading differential $d^{12}(\nu C[18]M_1^3\bar{M}_2) = A[32,3]$ determines tentative differentials with image $Z_2A[32,3] \otimes B\langle 4\rangle$. The remaining elements are $Z_2A[32,3]\{M_1, M_1^2, M_1^3, M_2, M_1M_2, M_1^2M_2, M_1^3M_2\} \otimes B\langle 4\rangle$, and the A[32,3]-leader is $A[32,3]M_1$.

The leading differential $d^4(A[32,3]M_1^2) = \nu A[32,3]$ determines tentative differentials which are a monomorphism on $Z_2A[32,3]\{M_1^2, M_1^3, M_1M_2, M_1^2M_2, M_1^3M_2\} \otimes B\langle 4\rangle$. The remaining elements are $Z_2A[32,3]\{M_1, M_2\} \otimes B\langle 4\rangle$, and the A[32,3]-leader remains $A[32,3]M_1$.

The leading differential $d^6(A[32,3]M_2) = A[37]$ determines tentative differentials which are a monomorphism on $Z_2A[32,3]M_2 \otimes B\langle 4\rangle$. The remaining elements are $Z_2A[32,3]M_1 \otimes B\langle 4\rangle$, and the A[32,3]-leader remains $A[32,3]M_1$.

The leading differential $d^8(A[32,3]M_1^5) = A[39,1]M_1$ determines tentaive differentials with kernel in degrees less than 69 given by the table below. The A[32,3]-leader remains $A[32,3]M_1$.

DEGREE	BASIS	DEGREE	BASIS	DEGREE	BASIS
(2,32)	1 0 0 0	(18,32)	9 0 0 0	(26,32)	1 4 0 0 13 0 0 0
(30,32)	1 0 2 0 9 2 0 0	(32,32)	1 0 0 1 1 5 0 0 9 0 1 0	(34,32)	17 0 0 0

The leading differential $d^{24}(\eta^2\sigma M_1^{13}) = A[32,3]M_1$ determines tentative differentials with image in degrees less than 69 equal to all the elements in the above table. Thus, there are no remaining elements.

<u>DEGREE 33:</u> $\eta A[32,1]$

Since $\eta^2 A[32,1]$ is nonzero, the only element of $E^4_{*,33}$ with a representative in $Z_2\eta A[32,1] \otimes H_*BP$ is zero.

<u>DEGREE 33:</u> $\eta A[32,2]$

The leading differential $d^2(A[32,2]M_1) = \eta A[32,2]$ determines tentative differentials with image $Z_2\eta A[32,1] \otimes B\langle 2\rangle$. The remaining elements are $Z_2\eta A[32,2]M_1 \otimes B\langle 2\rangle$, and the $\eta A[32,2]$-leader is $\eta A[32,2]M_1$.

The leading differential $d^6(A[8]C[20]M_1\bar{M}_2) = \eta A[32,2]M_1$ determines tentative differentials with image $Z_2\eta A[32,2]\{M_1, M_1^3\} \otimes B\langle 4\rangle$. The remaining elements are $Z_2\eta A[32,2]\{M_1\bar{M}_2, M_1^3\bar{M}_2\} \otimes B\langle 4\rangle$, and the new $\eta A[32,2]$-leader is $\eta A[32,2]M_1\bar{M}_2$.

The leading differential $d^8(\eta A[32,2]M_1\bar{M}_2) = C[20]^2$ determines tentative differentials which are a monomorphism on $Z_2\eta A[32,2]\{M_1\bar{M}_2, M_1^3\bar{M}_2\} \otimes B\langle 4\rangle$. Thus, there are no remaining elements.

<u>DEGREE 33:</u> $\nu A[30]$

The leading differential $d^4(A[30]M_1^2) = \nu A[30]$ determines tentative differentials with image $Z_2\nu A[30]\{1,M_1,M_2\} \otimes B\langle 4\rangle$. There are also tentative d^{16}-differentials determined by the leading differential $d^{16}(C[18](M_1^4\overline{M}_2^2+2M_1^7\overline{M}_2)) = \nu A[30]M_1^2$. In addition the leading differential $d^{12}(\nu A[30]M_1^6) = 2C[44]$ determines tentative differentials with image K. These tentative differentials are determined by assigning 1 to every monomial of degree 45 of $Z_2\nu A[30] \otimes H_*BP$. Let $'E^2_{*,33} = Z_2\nu A[30] \otimes H_*BP$. In the following table, the numbers in the last three columns add up to the numbers in the second column. Therefore, the only element of degree less than 68 in $E^{18}_{*,33}$ with a representative in $Z_2\nu A[30] \otimes H_*BP$ is zero.

DEGREE	DIM $'E^2_{*,33}$	DIM $d^4(E^4_{*,30})$	DIM $d^{16}(E^{16}_{*,18})$	DIM K
(0,33)	1	1	0	0
(2,33)	1	1	0	0
(4,33)	1	0	1	0
(6,33)	2	1	1	0
(8,33)	2	1	1	0
(10,33)	2	1	1	0
(12,33)	3	1	1	1
(14,33)	4	3	0	1
(16,33)	4	2	1	1
(18,33)	5	2	1	2
(20,33)	6	2	2	2
(22,33)	6	3	1	2
(24,33)	7	3	2	2
(26,33)	8	4	1	3
(28,33)	9	4	2	3
(30,33)	11	6	1	4
(32,33)	12	5	3	4
(34,33)	13	5	4	4

DEGREE 34: B[34]

The leading differential $d^{24}(\beta_1 M_1^6\overline{M}_2^2) = B[34]$ determines tentative differentials which have cokernel in $Z_2 \otimes [Z_4B[34] \otimes H_*BP]$ in degrees less than 69 given by the table below. The B[34]-leader is $B[34]\overline{M}_2$.

DEGREE	GENERATOR	DEGREE	GENERATOR	DEGREE	GENERATOR
(6,34)	0 1 0 0	(10,34)	2 1 0 0	(12,34)	3 1 0 0
	6 0 0 0	(14,34)	0 0 1 0		4 1 0 0

	7 0 0 0	(16,34)	5 1 0 0	(18,34)	0 3 0 0
	2 0 1 0		6 1 0 0	(20,34)	0 1 1 0
	1 3 0 0		4 2 0 0		7 1 0 0
(22,34)	1 1 1 0		2 3 0 0		4 0 1 0
	8 1 0 0	(24,34)	2 1 1 0		3 3 0 0
	5 0 1 0		6 2 0 0	(26,34)	0 2 1 0
	3 1 1 0		4 3 0 0		6 0 1 0
	7 2 0 0		10 1 0 0	(28,34)	1 2 1 0
	4 1 1 0		5 3 0 0		7 0 1 0
	11 1 0 0		14 0 0 0	(30,34)	0 5 0 0
	2 2 1 0		5 1 1 0		6 3 0 0
	8 0 1 0		12 1 0 0		15 0 0 0
(32,34)	0 3 1 0		3 2 1 0		6 1 1 0
	7 3 0 0		13 1 0 0	(34,34)	0 1 2 0
	1 3 1 0		2 5 0 0		4 2 1 0
	7 1 1 0		8 3 0 0		10 0 1 0
	14 1 0 0				

The leading differential $d^6(B[34]M_1^3) = A[39,3]$ determines tentative differentials by assigning 1 to $B[34]M_1^3$ and 0 to $B[34]M_2$. The kernel of these tentative differentials on the remaining elements of $Z_2 \otimes (Z_4B[34] \otimes H_*BP)$ in degrees less than 69 is given by the table below. The new B[34]-leader is $B[34]M_1^6$.

DEGREE	GENERATOR	DEGREE	GENERATOR	DEGREE	GENERATOR
(12,34)	6 0 0 0	(14,34)	0 0 1 0 4 1 0 0		7 0 0 0 4 1 0 0
(16,34)	5 1 0 0	(18,34)	2 0 1 0 6 1 0 0	(20,34)	1 3 0 0
	4 2 0 0	(22,34)	1 1 1 0		4 0 1 0 8 1 0 0
(24,34)	5 0 1 0		6 2 0 0	(26,34)	0 2 1 0 4 3 0 0

	7 2 0 0	6 0 1 0	(28,34)	1 2 1 0
	4 3 0 0	10 1 0 0		
	7 0 1 0	5 3 0 0	(30,34)	2 2 1 0
	11 1 0 0			6 3 0 0
	4 1 1 0	14 0 0 0		
	5 1 1 0	8 0 1 0		15 0 0 0
		12 1 0 0		12 1 0 0
(32,34)	3 2 1 0	13 1 0 0		
	7 3 0 0			
	0 3 1 0			
(34,34)	1 3 1 0	4 2 1 0		10 0 1 0
		8 3 0 0		14 1 0 0

The leading differential $d^{12}(B[34]M_1^6) = 2D[45]$ determines tentative differentials by assigning 1 to $B[34]M_1^6$ and 0 to all other monomials of $Z_2 \otimes (Z_4 \otimes B[34] \otimes H_{12}BP)$. There is no point to computing the kernel of these tentative differentials now because there will be a hidden differential $d^8(B[34]M_1^4M_2) = \eta A[40,1]M_1^3$ whose tentative differentials must be computed first. Thus, we postpone both of these computations to Section 7.6. The new B[34]-leader is $B[34]M_1^4M_2$.

<u>DEGREE 34:</u> $2B[34] = \eta^2 A[32,1]$

The leading differentials $d^2(\eta A[32,1]M_1) = 2B[34]$, $d^{20}(\gamma_1 M_1^{11}) = 2B[34]M_1$ and $d^{24}(2\beta_1 M_1^7\bar{M}_2^3) = 2B[34]M_1\bar{M}_2$ determine tentative differentials with image in $Z_2(2B[34]) \otimes H_*BP$. In addition the leading differential $d^{12}(2B[34]M_1^3\bar{M}_2) = A[45,1]$ determines tentative differentials which are a monomorphism on $Z_2(2B[34]M_1^3\bar{M}_2) \otimes B$<4>. The table below shows that the only nonzero elements of $E_{*,34}^{26}$ with representatives in $Z_2(2B[34]) \otimes H_*BP$ of degree less than 69 are $2B[34]M_1^5\bar{M}_2\bar{M}_3$ and $2B[34]M_1\bar{M}_2^3\bar{M}_3$. The new 2B[34]-leader is $2B[34]M_1^5\bar{M}_2\cdot\bar{M}_3$.

DEGREE	DIM $2E^4_{*,34}$	DIM $d^{20}(E^{20}_{*,15})$	DIM $d^{24}(E^{24}_{*,11})$	DIM $d^{12}(2E^{12}_{*,34})$
(0,34)	0	0	0	0
(2,34)	1	1	0	0
(4,34)	0	0	0	0
(6,34)	1	1	0	0
(8,34)	1	0	1	0
(10,34)	1	1	0	0
(12,34)	1	0	0	1
(14,34)	2	2	0	0
(16,34)	2	1	1	0
(18,34)	2	2	0	0
(20,34)	3	1	1	1
(22,34)	3	2	1	0
(24,34)	3	1	1	1
(26,34)	4	3	0	1
(28,34)	4	2	1	1
(30,34)	5	4	0	0
(32,34)	6	3	2	1
(34,34)	6	4	0	1

<u>DEGREE 34:</u> $\nu A[31]$

The leading differential $d^4(A[31]M_1^2) = \nu A[31]$ determines tentative differentials with image $Z_2\nu A[31]\{1, M_1M_1^2, M_2, M_1M_2\} \otimes B\langle 4\rangle$ and cokernel $Z_2\nu A[31]\{M_1^3, M_1^2M_2, M_1^3M_2\}$. The $\nu A[31]$-leader is $\nu A[31]M_1^3$.

The leading differntial $d^6(\nu A[31]M_1^3) = A[39,1]$ determines tentative differentials which are a monomorphism on $Z_2\nu A[31]\{M_1^3, M_1^2M_2, M_1^3M_2\}$. Thus, there are no remaining elements.

<u>DEGREE 34:</u> $A[14]C[20]$

The leading differential $d^{12}(4\nu C[20]M_1^3M_2) = A[14]C[20]$ determines tentative differentials with image $Z_2A[14]C[20] \otimes B\langle 4\rangle$. Since $\eta A[14]C[20] \neq 0$, the remaining elements are $Z_2A[14]C[20]\{M_1^2, \bar{M}_2, M_1^2\bar{M}_2\} \otimes B\langle 4\rangle$, and the $A[14]C[20]$-leader is $A[14]C[20]M_1^2$.

The leading differential $d^{24}(4\beta_1M_1^{11}M_2) = A[14]C[20]M_1^2$ determines tentative differentials with image equal to $Z_2A[14]C[20]\{M_1^2, \bar{M}_2, M_1^2\bar{M}_2\} \otimes B\langle 4\rangle$ in degrees less than 69. Thus, there are no remaining elements.

DEGREE 35: $\nu A[32,3]$

The leading differential $d^4(A[32,3]M_1^2) = \nu A[32,3]$ determines tentative differentials with image $Z_2\nu A[32,3]\{1,M_1,M_1^2,M_2,M_1M_2\} \otimes B\langle 4\rangle$. The remaining elements are $Z_2\nu A[32,3]\{M_1^3,M_1^2M_2,M_1^3M_2\} \otimes B\langle 4\rangle$, and the $\nu A[32,3]$-leader is $\nu A[32,3]M_1^3$.

The leading differential $d^4(\nu A[32,3]M_1^3) = \eta\sigma A[30]M_1$ determines tentative differentials which are a monomorphism on $Z_2\nu A[32,3]\{M_1^3,M_1^2M_2,M_1^3M_2\} \otimes B\langle 4\rangle$. Thus, there are no remaining elements.

DEGREE 35: $\eta A[14]C[20]$

The leading differential $d^2(A[14]C[20]M_1) = \eta A[14]C[20]$ determines tentative differentials with image $Z_2\eta A[14]C[20] \otimes B\langle 2\rangle$ and cokernel $Z_2\eta A[14]C[20]M_1 \otimes B\langle 2\rangle$. The $\eta A[14]C[20]$-leader is $\eta A[14]C[20]M_1$.

The leading differential $d^4(A[32,2]\bar{M}_2) = \eta A[14]C[20]M_1$ determines tentative differentials with image $Z_2\eta A[14]C[20]\{M_1,M_1^3\} \otimes B\langle 4\rangle$. The remaining elements are $Z_2\eta A[14]C[20]\{M_1\bar{M}_2,M_1^3\bar{M}_2\} \otimes B\langle 4\rangle$, and the new $\eta A[14]C[20]$-leader is $\eta A[14]C[20]M_1\bar{M}_2$.

The leading differential $d^6(\eta A[14]C[20]M_1\bar{M}_2) = 2C[20]^2M_1$ determines tentative differentials which are a monomorphism on $Z_2\eta A[14]C[20]\{M_1\bar{M}_2,M_1^3\bar{M}_2\} \otimes B\langle 4\rangle$. There are no remaining elements.

DEGREE 36: $A[36]$

The leading differential $d^6(A[31]M_2) = A[36]$ determines tentative differentials with image $Z_2A[36] \otimes B\langle 4\rangle$. The remaining elements are $Z_2A[36]\{M_1,M_1^2,M_1^3,M_2,M_1M_2,M_1^2M_2,M_1^3M_2\} \otimes B\langle 4\rangle$, and the $A[36]$-leader is $A[36]M_1$.

The leading differential $d^6(\eta A[30]M_1\bar{M}_2) = A[36]M_1$ determines tentative differentials with image $Z_2A[36]\{M_1, M_2\} \otimes B\langle 4\rangle$. The remaining elements are $Z_2A[36]\{M_1^2, M_1^3, M_1M_2, M_1^2M_2, M_1^3M_2\} \otimes B\langle 4\rangle$, and the A[36]-leader is $A[36]M_1^2$.

The leading differential $d^4(A[36]M_1^3) = \eta B[38]M_1$ determines tentative differentials which are a monomorphism on $Z_2A[36]\{M_1^3, M_1^2M_2, M_1^3M_2\} \otimes B\langle 4\rangle$. The remaining elements are $Z_2A[36]\{M_1^2, M_1M_2\} \otimes B\langle 4\rangle$, and the A[36]-leader remains $A[36]M_1^2$.

The leading differential $d^6(A[36]M_1M_2) = \eta A[40,1]M_1$ determines tentative differentials which are a monomorphism on $Z_2A[36]M_1M_2 \otimes B\langle 4\rangle$. The remaining elements are $Z_2A[36]M_1^2 \otimes B\langle 4\rangle$, and the A[36]-leader remains $A[36]M_1^2$.

The leading differential $d^{25}(2\beta_1M_1^6\bar{M}_2^3) = A[36]M_1^2$ determines tentative differentials with cokernel in degrees less than 69 given by the table below. The new A[36]-leader is $A[36]M_1^2\bar{M}_3$.

DEGREE	GENERATOR	DEGREE	GENERATOR	DEGREE	GENERATOR
(18,36)	2 0 1 0	(24,36)	6 2 0 0	(26,36)	6 0 1 0
(30,36)	2 2 1 0				

DEGREE 37: A[37] and σA[30]

The leading differential $d^6(A[32,3]M_2) = A[37]$ determines tentative differentials with image $Z_2A[37] \otimes B\langle 4\rangle$. The remaining elements are $(Z_2A[37]M_1 \otimes B\langle 2\rangle) \oplus (Z_2A[37]\{M_1^2, \bar{M}_2, M_1^2\bar{M}_2\} \otimes B\langle 4\rangle)$, and the A[37]-leader is $A[37]M_1$.

The leading differential $d^8(A[30]M_1^4) = \sigma A[30]$ determines tentative differentials with image in degrees less than 68 equal to

$$Z_2\sigma A[30]\{1, M_1^2, \bar{M}_2, M_2^2, \bar{M}_3, M_1^8, M_1^2\bar{M}_3+\bar{M}_2M_2^2, M_1^{10}, M_1^8\bar{M}_2, M_1^{12}+M_2^4, M_2^2\bar{M}_3+M_1^{10}\bar{M}_2, \langle M_3\rangle^2, M_1^2M_2^4+M_1^{14}, M_1^8M_2^2, \bar{M}_2M_2^4+\bar{M}_2M_1^{12}, \langle M_4\rangle, M_1^8\bar{M}_3\}.$$

The remaining elements in degrees less than 68 are

$(Z_2\sigma A[30]M_1 \otimes B\langle 2\rangle) \oplus (Z_2\sigma A[30]\{\langle M_1^4\rangle, M_1^2\bar{M}_2, \langle M_2^2\rangle, \langle M_3\rangle, M_1^2\langle M_2^2\rangle, M_1^2\langle M_3\rangle, M_1^6\bar{M}_2, \bar{M}_2\langle M_3\rangle,$
$M_1^2\bar{M}_2^3, M_1^4\langle M_3\rangle, M_1^6M_2^2, M_1^{12}, M_1^6\langle M_3\rangle, M_1^2\bar{M}_2\langle M_3\rangle, M_1^{10}\bar{M}_2, M_1^4\bar{M}_2^3, M_1^4\bar{M}_2\bar{M}_3, M_1^{14}, M_1^2M_2^2\bar{M}_3, M_1^6\bar{M}_2^3, \bar{M}_2^5\})$.

The $\sigma A[30]$-leader is $\sigma A[30]M_1$.

Note that $\eta\sigma A[30] = \eta A[37] \neq 0$. In addition the leading differential $d^{18}(2C[18](M_1^4\bar{M}_3+3M_1^8\bar{M}_2) = (\sigma A[30]+A[37])M_1$ determines tentative differentials with image in $Z_2(\sigma A[30]+A[37])M_1 \otimes B\langle 2\rangle$. The cokernel of these tentative differentials in degrees less than 68 is given by $Z_2(A[37]+\sigma A[30])M_1^3\bar{M}_2 \otimes B\langle 4\rangle$ as well as the elements listed in the table below. The remaining elements in $(Z_2A[37] \oplus Z_2\sigma A[30]) \otimes B\langle 2\rangle$ of degrees less than 68 are

$(Z_2A[37]\{M_1^2, \bar{M}_2, M_1^2\bar{M}_2\} \otimes B\langle 4\rangle)$
$\oplus (Z_2\sigma A[30]\{\langle M_1^4\rangle, M_1^2\bar{M}_2, \langle M_2^2\rangle, \langle M_3\rangle, M_1^2\langle M_2^2\rangle, M_1^2\langle M_3\rangle, M_1^6\bar{M}_2, \bar{M}_2\langle M_3\rangle, M_1^2\bar{M}_2^3, M_1^4\langle M_3\rangle, M_1^6M_2^2,$
$M_1^{12}, M_1^6\langle M_3\rangle, M_1^2\bar{M}_2\langle M_3\rangle, M_1^{10}\bar{M}_2, M_1^4\bar{M}_2^3, M_1^4\bar{M}_2\bar{M}_3, M_1^{14}, M_1^2M_2^2\bar{M}_3, M_1^6\bar{M}_2^3, \bar{M}_2^5\})$.

The new $A[37]$-leader is $A[37]M_1^2$, and the new $\sigma A[30]$-leader is $\sigma A[30]M_1^4$.

DEGREE	BASIS	DEGREE	BASIS
(14,37)	$(\sigma A[30]+A[37])M_1^7$	(16,37)	$(\sigma A[30]+A[37])M_1^5\bar{M}_2$
(20,37)	$(\sigma A[30]+A[37])M_1\bar{M}_2^3$	(22,37)	$(\sigma A[30]+A[37])M_1^5\bar{M}_2^2$
(24,37)	$(\sigma A[30]+A[37])M_1^5\bar{M}_3$	(26,37)	$(\sigma A[30]+A[37])M_1^7\langle M_2^2\rangle$
(28,37)	$(\sigma A[30]+A[37])M_1^5\bar{M}_2^3$		$(\sigma A[30]+A[37])M_1^7\langle M_3\rangle$
	$(\sigma A[30]+A[37])M_1\bar{M}_2^2\langle M_3\rangle$		
(30,37)	$(\sigma A[30]+A[37])M_1^{15}$		$(\sigma A[30]+A[37])M_1^5\bar{M}_2\langle M_3\rangle$

The leading differential $d^6(A[32,1]M_1^3\bar{M}_2) = A[37]M_1^2$ determines tentative differentials with image $Z_2A[37]M_1^2 \otimes B\langle 4\rangle$. The remaining elements from $Z_2A[37] \otimes B\langle 2\rangle$ are $Z_2A[37]\{\bar{M}_2, M_1^2\bar{M}_2\} \otimes B\langle 4\rangle$, and the new $A[37]$-leader is $A[37]\bar{M}_2$. The leading differential $d^4(A[37]\bar{M}_2) = \eta A[39,3]M_1 + \eta\sigma A[32,1]M_1$ determines tentative differentials which are a monomorphism on $(Z_2A[37]\{\bar{M}_2, M_1^2\bar{M}_2\} \otimes B\langle 4\rangle)$ $\oplus (Z_2(A[37]+\sigma A[30])M_1^3\bar{M}_2$. There are no remaining elements from $Z_2A[37] \otimes B\langle 2\rangle$. The leading differentials $d^8(\sigma A[30]M_1^4) = \sigma^2A[30] = 4C[44]$ and $d^8(A[37]M_1^4) = 0$

determine tentative differentials which are a monomorphism on the remaining elements of $Z_2(A[37]+\sigma A[30])M_1 \otimes B<2>$ given by the table above and on

$$(Z_2\sigma A[30]\{<M_1^4>, M_1^2\bar{M}_2, <M_2^2>, <M_3>, M_1^2<M_2^2>, M_1^2<M_3>, M_1^6\bar{M}_2, \bar{M}_2<M_3>, M_1^2\bar{M}_2^3, M_1^4<M_3>, M_1^6M_2^2,$$
$$M_1^{12}, M_1^6<M_3>, M_1^2\bar{M}_2<M_3>, M_1^{10}\bar{M}_2, M_1^4\bar{M}_2^3, M_1^4\bar{M}_2\bar{M}_3, M_1^{14}, M_1^2M_2^2\bar{M}_3, M_1^6\bar{M}_2^3, \bar{M}_2^5\}).$$

There are no remaining elements in degrees less than 68 from

$(Z_2(A[37]+\sigma A[30])M_1 \otimes B<2>) \oplus (Z_2\sigma A[30] \otimes B<2>)$.

<u>DEGREE 38:</u> B[38]

The leading differential $d^{24}(\gamma_1 M_1^{12}) = B[38]$ defines B[38], and $\eta B[38] \neq 0$. Therefore the tentative differentials defined by this leading d^{24}-differential have image in $(Z_4 B[38] \otimes B<2>) \oplus (Z_2(2B[38]M_1) \otimes B<2>)$. The cokernel of these tentative differntials is given by the table below. The B[38]-leader is $B[38]M_1^6$.

DEGREE	GROUP	GENERATOR	DEGREE	GROUP	GENERATOR	DEGREE	GROUP	GENERATOR
(12,38)	Z_2	6 0 0 0	(14,38)	Z_4	4 1 0 0	(18,38)	Z_4	6 1 0 0
(20,38)	Z_2	0 1 1 1		Z_2 2/	7 1 0 0	(22,38)	Z_4	4 0 1 0
(24,38)	Z_2	2 1 1 0		Z_2	6 2 0 0	(26,38)	Z_4	4 3 0 0
	Z_4	6 0 1 0	(28,38)	Z_2 2/	7 0 1 0		Z_2	14 0 0 0
	Z_4	4 1 1 0	(30,38)	Z_4	6 3 0 0		Z_4	12 1 0 0

<u>DEGREE 38:</u> $\eta A[37] = \eta\sigma A[30]$

The leading differential $d^2(A[37]M_1) = \eta A[37]$ determines tentative differentials with image $Z_2\eta A[37] \otimes B<2>$ and cokernel $Z_2\eta A[37]M_1 \otimes B<2>$. The $\eta A[37]$-leader is $\eta A[37]M_1$.

The leading differential $d^4(\nu A[32,3]M_1^3) = \eta A[37]M_1$ determines tentative differentials with image $Z_2\eta A[37]\{M_1, M_1^3, M_1\bar{M}_2\} \otimes B<4>$. The remainng elements are $Z_2\eta A[37]M_1^3\bar{M}_2 \otimes B<4>$, and the new $\eta A[37]$-leader is $\eta A[37]M_1^3\bar{M}_2$.

DEGREE 39: A[39,1]

The leading differential $d^6(\nu A[31]M_1^3) = A[39,1]$ determines tentative differentials with image $Z_2A[39,1]\{1,M_1^2,\bar{M}_2\} \otimes B\langle 4\rangle$. The cokernel of these tentative differentials is $Z_2A[39,1]\{M_1,M_1^3,M_1\bar{M}_2,M_1^2\bar{M}_2,M_1^3\bar{M}_2\} \otimes B\langle 4\rangle$, and the A[39,1]-leader is $A[39,1]M_1$.

The leading differential $d^8(A[32,3]M_1^5) = A[39,1]M_1$ determines tentative differentials. The remaining elements in degrees less than 70 are $Z_2A[39,1]\{M_1^2M_2,M_1^3M_2\} \otimes B\langle 4\rangle$ as well as the elements listed in the table below. The new A[39,1]-leaders are $A[39,1]M_1^5$ and $A[39,1]M_1^2M_2$.

DEGREE	GENERATOR	DEGREE	GENERATOR	DEGREE	GENERATOR
(10,39)	5 0 0 0	(14,39)	7 0 0 0	(16,39)	5 1 0 0
(18,39)	3 2 0 0	(20,39)	3 0 1 0	(22,39)	5 2 0 0
	1 1 1 0	(24,39)	5 0 1 0		3 3 0 0
(26,39)	7 2 0 0		13 0 0 0	(28,39)	7 0 1 0
	5 3 0 0	(30,39)	5 1 1 0		15 0 0 0
	6 3 0 0				

DEGREE 39: A[39,2]

The leading differential $d^6(A[14]C[20]\bar{M}_2) = A[39,2]$ determines tentative differentials with image $Z_2A[39,2]\{1,M_1^2\} \otimes B\langle 4\rangle$. Since $\eta A[39,2] \neq 0$, the remaining elements are $Z_2A[39,2]\{\bar{M}_2,M_1^2\bar{M}_2\} \otimes B\langle 4\rangle$ and the A[39,2]-leader is $A[39,2]\bar{M}_2$. The leading differential $d^4(A[39,2]\bar{M}_2) = \eta^2C[20]^2M_1$ determines tentative differentials which are a monomorphism on $Z_2A[39,2]\{\bar{M}_2,M_1^2\bar{M}_2\} \otimes B\langle 4\rangle$. There are no remaining elements.

DEGREE 39: $A[39,3]$

The leading differential $d^6(B[34]M_1^3) = A[39,3]$ determines tentative differentials with image $Z_2A[39,3]\{1, M_1^2, \bar{M}_2\} \otimes B<4>$. Since $\eta A[39,3] \neq 0$, the remaining elements are $Z_2A[39,3]M_1^2\bar{M}_2 \otimes B<4>$, and the $A[39,3]$-leader is $A[39,3]M_1^2\bar{M}_2$.

DEGREE 39: $\eta B[38]$

The leading differential $d^2(B[38]M_1) = \eta B[38]$ determines tentative differntials with image $Z_2\eta B[38] \otimes B<2>$. The remaining elements are $Z_2\eta B[38]M_1 \otimes B<2>$, and the $\eta B[38]$-leader is $\eta B[38]M_1$.

The leading differential $d^4(A[36]M_1^3) = \eta B[38]M_1$ determines tentative differentials with image $Z_2\eta B[38]\{M_1, M_1^3, M_1\bar{M}_2\} \otimes B<4>$. The remaining elements are $Z_2\eta B[38]M_1^3\bar{M}_2 \otimes B<4>$, and the new $\eta B[38$-leader is $\eta B[38]M_1^3\bar{M}_2$.

DEGREE 39: $\sigma A[32,1]$

The leading differential $d^8(A[32,1]M_1^4) = \sigma A[32,1]$ determines tentative differntials with image in $Z_2\sigma A[32,1] \otimes B<2>$ since $\eta\sigma A[32,1] \neq 0$. The cokernel of these tentative differentials in degrees less than 68 is given by the table below. The $\sigma A[32,1]$-leader is $\sigma A[32,1]M_1^4\bar{M}_2$.

DEGREE	GENERATOR	DEGREE	GENERATOR	DEGREE	GENERATOR
(14,39)	4 1 0 0	(18,39)	6 1 0 0	(20,39)	4 2 0 0
(22,39)	4 0 1 0	(24,39)	6 2 0 0	(26,39)	4 3 0 0
	6 0 1 0	(28,39)	4 1 1 0		

DEGREE 40: $A[40,1]$

The leading differential $d^{10}(\eta A[30]M_1^5) = A[40,1]$ determines tentative differentials with image in $Z_2A[40,1] \otimes B\langle 2\rangle$ because $\eta A[40,1] \neq 0$. The cokernel of these differentials in degrees less than 69 is given by the table below, and the A[40,1]-leader is $A[40,1]M_1^6$.

DEGREE	GENERATOR	DEGREE	GENERATOR	DEGREE	GENERATOR
(12,40)	6 0 0 0	(14,40)	4 1 0 0	(18,40)	6 1 0 0
(20,40)	4 2 0 0		0 1 1 0	(22,40)	2 3 0 0
(24,40)	6 2 0 0		2 1 1 0	(26,40)	6 0 1 0
	4 3 0 0	(28,40)	14 0 0 0		4 1 1 0

DEGREE 40: A[40,2]

The leading differential $d^{26}(4\gamma_1 M_1^{13}) = A[40,2]$ determines tentative differentials with image in $Z_2A[40,2] \otimes B\langle 2\rangle$ because $\eta A[40,2] \neq 0$. The cokernel of these tentative differentials in degrees less than 69 equals $Z_2A[40,2]\{M_1^6\bar{M}_2, M_1^6\bar{M}_3\}$. The A[40,2]-leader is $A[40,2]M_1^6\bar{M}_2$.

DEGREE 40: $C[20]^2$

The leading differential $d^8(\eta A[32,2]M_1\bar{M}_2) = C[20]^2$ determines tentative differentials with image $Z_2C[20]^2\{1,M_1^2\} \otimes B\langle 4\rangle$. Since $\eta C[20]^2 \neq 0$, the remaining elements are $Z_2C[20]^2\{\bar{M}_2, M_1^2\bar{M}_2\} \otimes B\langle 4\rangle$, and the $C[20]^2$-leader is $C[20]^2\bar{M}_2$.

The leading differential $d^6(C[20]^2\bar{M}_2) = A[45,2]$ determines tentative differentials which are a monomorphism on $Z_2C[20]^2\{\bar{M}_2, M_1^2\bar{M}_2\} \otimes B\langle 4\rangle$. There are no remaining elements.

DEGREE 40: $2C[20]^2 = \eta A[39,2]$

The leading differential $d^2(A[39,2]M_1) = \eta A[39,2]$ determines tentative differentials with image $Z_2(2C[20]^2) \otimes B<2>$. The remaining elements are $Z_2(2C[20]^2M_1) \otimes B<2>$, and the $2C[20]^2$-leader is $2C[20]^2M_1$.

The leading differential $d^6(\eta A[14]C[20]M_1\bar{M}_2) = 2C[20]^2M_1$ determines tentative differentials with image $Z_2(2C[20]^2)\{M_1,M_1^3\} \otimes B<4>$. The remainig elements are $Z_2(2C[20]^2)\{M_1\bar{M}_2,M_1^3\bar{M}_2\} \otimes B<4>$, and the new $2C[20]^2$-leader is $2C[20]^2M_1\bar{M}_2$.

DEGREE 40: $\eta A[39,3]$

The leading differential $d^2(A[39,3]M_1) = \eta A[39,3]$ determines tentative differntials with image $Z_2\eta A[39,3] \otimes B<2>$. The remaining elements ae $Z_2\eta A[39,3]M_1 \otimes B<2>$, and the $\eta A[39,3]$-leader $\eta A[39,3]M_1$.

The leading differential $d^4(A[37]\bar{M}_2) = \eta A[39,3]M_1 + \eta\sigma A[32,1]M_1$ determines tentative differentials with image $Z_2\eta(A[39,3]+\sigma A[32,1])\{M_1,M_1^3,M_1\bar{M}_2\} \otimes B<4>$. The remaining elements are $Z_2(\eta A[39,3]+\eta\sigma A[32,1])M_1^3\bar{M}_2 \otimes B<4>$, and the new $\eta A[39,3]$-leader is $(\eta A[39,3]+\eta\sigma A[32,1])M_1^3\bar{M}_2$.

DEGREE 40: $\eta\sigma A[32,1]$

The leading differential $d^2(\sigma A[32,1]M_1) = \eta\sigma A[32,1]$ determines tentative differentials with image $Z_2\eta\sigma A[32,1] \otimes B<2>$. The remaining elements are $Z_2\eta\sigma A[32,1]M_1 \otimes B<2>$, and the $\eta\sigma A[32,1]$-leader is $\eta\sigma A[32,1]M_1$.

The leading differential $d^{26}(2\gamma_1(M_1^{11}\bar{M}_2+2M_1^{14})) = \eta\sigma A[32,1]M_1$ determines tentative differentials with image in degrees less than 60 euqal to all the elements of $Z_2\eta\sigma A[32,1]M_1 \otimes B<4>$ except $Z_2\eta\sigma A[32,1]\{M_1^5<M_3>,M_1<M_2^2><M_3>\}$. Thus, the remaining elements in degrees less than 69 are $(Z_2\eta\sigma A[32,1]\{M_1^3,M_1\bar{M}_2,M_1^3\bar{M}_2\} \otimes B<4>) \oplus (Z_2\eta\sigma A[32,1]\{M_1^5<M_3>,M_1<M_2^2><M_3>\})$. The new $\eta\sigma A[32,1]$-leader is $\eta\sigma A[32,1]M_1^3$.

The leading differential $d^6(\eta\sigma A[32,1]M_1^3) = 4D[45]$ determines tentative differentials which are a monomorphism on $Z_2\eta\sigma A[32,1]\{M_1^3, M_1\bar{M}_2, M_1^3\bar{M}_2\} \otimes B\langle 4\rangle$. The remaining elements in degrees less than 69 are $Z_2\eta\sigma A[32,1]\{M_1^5\langle M_3\rangle, M_1\langle M_2^2\rangle\langle M_3\rangle\}$, and the new $\eta\sigma A[32,1]$-leader is $\eta\sigma A[32,1]M_1^5\langle M_3\rangle$.

<u>DEGREE 41:</u> $\eta A[40,1]$

The leading differential $d^2(A[40,1]M_1) = \eta A[40,1]$ determines tentative differentials with image $Z_2\eta A[40,1] \otimes B\langle 2\rangle$. The remaining elements are $Z_2\eta A[40,1]M_1 \otimes B\langle 2\rangle$, and the $\eta A[40,1]$-leader is $\eta A[40,1]M_1$.

The leading differential $d^6(A[36]M_1M_2) = \eta A[40,1]M_1$ determines tentative differentials with image $Z_2\eta A[40,1]M_1 \otimes B\langle 4\rangle$. The remaining elements are $Z_2\eta A[40,1]\{M_1^3, M_1\bar{M}_2, M_1^3\bar{M}_2\} \otimes B\langle 4\rangle$, and the new $\eta A[40,1]$-leader is $\eta A[40,1]M_1^3$.

<u>DEGREE 41:</u> $\eta A[40,2]$ and $\eta C[20]^2$

Since $\eta^2 A[40,2]$ and $\eta^2 C[20]^2$ are nonzero, the only element of $E^4_{*,41}$ with a representative in $(Z_2\eta A[40,2] \oplus Z_2\eta C[20]^2) \otimes H_*BP$ is zero.

<u>DEGREE 42:</u> $C[42]$

The leading differential $d^{28}(4\gamma_1(M_1^{11} \,|\, \bar{M}_2 \,|\, 2M_1^{14})) = C[42]$ defines an element of order four in E^{28} and determines tentative differentials whose cokernel in $Z_4 \otimes (Z_8C[42] \otimes H_*BP)$ in degrees less than 69 is given by the table below. The $C[42]$-leader is $C[42]M_1^2$.

<u>DEGREE</u>	<u>GROUP</u>	<u>GENERATOR</u>	<u>DEGREE</u>	<u>GROUP</u>	<u>GENERATOR</u>	<u>DEGREE</u>	<u>GROUP</u>	<u>GENERATOR</u>
(4,42)	Z_2	2 0 0 0	(6,42)	Z_4	0 1 0 0	(8,42)	Z_2	1 1 0 0
(10,42)	Z_4	2 1 0 0	(12,42)	Z_2	6 0 0 0		Z_4	3 1 0 0

(14,42)	Z_4	4 1 0 0	(16,42)	Z_2	1 0 1 0		Z_2	2 2 0 0
(18,42)	Z_2	2 0 1 0 6 1 0 0		Z_4	0 3 0 0		Z_4	6 1 0 0
(20,42)	Z_2	1 3 0 0		Z_2	10 0 0 0		Z_4	0 1 1 0
	Z_4	7 1 0 0	(22,42)	Z_2	1 1 1 0		Z_4	2 3 0 0
	Z_4	8 1 0 0	(24,42)	Z_2	5 0 1 0		Z_2	6 2 0 0
	Z_4	2 1 1 0		Z_4	3 3 0 0	(26,42)	Z_2	6 0 1 0 10 1 0 0
	Z_4	3 1 1 0		Z_4	4 3 0 0		Z_4	10 1 0 0

The leading differential $d^4(C[42]M_1^2) = \nu C[42] = 8D[45]$ determines tentative differentials which are a monomorphism on $Z_2 \otimes ((Z_8C[42])\{M_1^2, M_1^3, M_1M_2, M_1^2M_2, M_1^3M_2\} \otimes B\langle 4\rangle)$. The kernel of these tentative differentials in degrees less than 69 on elements with representatives in $Z_4 \otimes (Z_8C[42] \otimes H_*BP)$ is given by the table below. The new C[42]-leader is $2C[42]\bar{M}_2$.

DEGREE	GROUP	GENERATOR	DEGREE	GROUP	GENERATOR	DEGREE	GROUP	GENERATOR
(6,42)	Z_2	2/ 0 1 0 0	(10,42)	Z_2	2/ 2 1 0 0	(12,42)	Z_2	2/ 3 1 0 0
(14,42)	Z_2	2/ 4 1 0 0	(18,42)	Z_2	2/ 0 3 0 0		Z_2	2/ 6 1 0 0
(20,42)	Z_2	2/ 0 1 1 0		Z_2	2/ 7 1 0 0	(22,42)	Z_2	2/ 2 3 0 0
	Z_2	2/ 8 1 0 0	(24,42)	Z_2	2/ 2 1 1 0		Z_2	2/ 3 3 0 0
(26,42)	Z_2	2/ 3 1 1 0		Z_2	2/ 4 3 0 0		Z_2	2/ 10 1 0 0

<u>DEGREE 42:</u> $4C[42] = \eta^2 A[40,2]$

The leading differential $d^2(\eta A[40,2]M_1) = 4C[42]$ determines tentative differentials with image $Z_2(4C[42]) \otimes B\langle 2\rangle$. The remaining elements are $Z_2(4C[42]M_1) \otimes B\langle 2\rangle$, and the 4C[42]-leader is $4C[42]M_1$.

The leading differential $d^{28}(16\gamma_1 M_1^8\bar{M}_3) = 4C[42]M_1$ determines tentative differentials with image in degrees less than 69 equal to $Z_2(4C[42])M_1 \otimes B\langle 2\rangle$. Thus, there are no remaining elements.

DEGREE 42: $\eta^2 C[20]^2$

The leading differential $d^2(\eta C[20]^2 M_1) = \eta^2 C[20]^2$ determines tentative differentials with image $Z_2\eta^2 C[20]^2 \otimes B\langle 2\rangle$. The remaining elements are $Z_2\eta^2 C[20]^2 M_1 \otimes B\langle 2\rangle$, and the $\eta^2 C[20]^2$-leader is $\eta^2 C[20]^2 M_1$.
The leading differntial $d^4(A[39,2]\bar{M}_2) = \eta^2 C[20]^2 M_1$ determines tentative differentials with image $Z_2(\eta^2 C[20]^2)\{M_1, M_1^3\} \otimes B\langle 4\rangle$. The remaining elements are $Z_2(\eta^2 C[20]^2)\{M_1\bar{M}_2, M_1^3\bar{M}_2\} \otimes B\langle 4\rangle$, and the new $\eta^2 C[20]^2$-leader is $\eta^2 C[20]^2 M_1\bar{M}_2$.

DEGREE 44: C[44]

The leading differential $d^{14}(\eta A[30]M_1M_2^2) = C[44]$ determines tentative differentials with image in $Z_2 \otimes (Z_8 C[44] \otimes B\langle 2\rangle)$ since $\eta C[44] \neq 0$. The cokernel of these tentative differentials in degrees less than 69 is given by the table below. The C[44]-leader is $C[44]M_1^2$.

DEGREE	GENERATOR	DEGREE	GENERATOR	DEGREE	GENERATOR
(4,44)	2 0 0 0	(6,44)	0 1 0 0	(8,44)	4 0 0 0
(10,44)	2 1 0 0	(12,44)	0 2 0 0		6 0 0 0
(14,44)	0 0 1 0		4 1 0 0	(16,44)	2 2 0 0
(18,44)	0 3 0 0		2 0 1 0		6 1 0 0
(20,44)	0 1 1 0		4 2 0 0		10 0 0 0
(22,44)	2 3 0 0		4 0 1 0		8 1 0 0
(34,44)	2 1 1 0		12 0 0 0		0 4 0 0

DEGREE 44: 2C[44]

The leading differential $d^{12}(\nu A[30]M_1^6) = 2C[44]$ determines tentative differentials with cokernel in degrees less than 69 given by the table below. The 2C[44]-leader is $2C[44]M_1^6$.

DEGREE	GENERATOR	DEGREE	GENERATOR	DEGREE	GENERATOR
(12,44)	6 0 0 0	(14,44)	7 0 0 0	(16,44)	5 1 0 0
(18,44)	6 1 0 0	(20,44)	7 1 0 0		4 2 0 0
(22,44)	5 2 0 0		2 3 0 0	(24,44)	5 0 1 0
	6 2 0 0		3 3 0 0		

DEGREE 44: $4C[44] = \sigma^2 A[30]$

The leading differential $d^8(\sigma A[30]M_1^4) = 4C[44]$ determines tentative differentials whose cokernel in degrees less than 69 is given by the table below. The 4C[44]-leader is $4C[44]M_1^7$.

DEGREE	GENERATOR	DEGREE	GENERATOR	DEGREE	GENERATOR
(14,44)	7 0 0 0	(18,44)	6 1 0 0	(20,44)	7 1 0 0
(22,44)	4 0 1 0				

DEGREE 45: A[45,1]

The leading differential $d^{12}(2B[34]M_1^3\bar{M}_2) = A[45,1]$ determines tentative differentials with image $Z_2A[45,1] \otimes B\langle 4\rangle$. Since $\eta A[45,1] \neq 0$, the remaining elements are $Z_2A[45,1]\{M_1^2, \bar{M}_2, M_1^2\bar{M}_2\} \otimes B\langle 4\rangle$. The A[45,1]-leader is $A[45,1]M_1^2$.

DEGREE 45: A[45,2]

The leading differential $d^6(C[20]^2\bar{M}_2) = A[45,2]$ determines tentative differentials with image $Z_2A[45,2]\{1, M_1^2\} \otimes B\langle 4\rangle$. Since $\eta A[45,2] \neq 0$, the remaining elements are $Z_2A[45,2]\{\bar{M}_2, M_1^2\bar{M}_2\} \otimes B\langle 4\rangle$, and the A[45,2]-leader is $A[45,2]\bar{M}_2$.

DEGREE 45: D[45]

The leading differential $d^{28}(4C[18](M_1^7\bar{M}_3+M_1^{11}\bar{M}_2)) = D[45]$ determines tentative differentials with image in $Z_2 \otimes (Z_{16}D[45] \otimes B\langle 2\rangle)$ since $\eta D[45] \neq 0$. The cokernel of these tentative differentials in degrees less than 68 is given by the table below. The D[45]-leader is $D[45]M_1^2$.

DEGREE	GENERATOR	DEGREE	GENERATOR	DEGREE	GENERATOR
(4,45)	0 2 0 0	(6,45)	0 1 0 0	(10,45)	2 1 0 0
(12,45)	6 0 0 0	(14,45)	4 1 0 0	(16,45)	2 2 0 0
(18,45)	0 3 0 0		6 1 0 0	(20,45)	0 1 1 0
	10 0 0 0	(22,45)	2 3 0 0		8 1 0 0

DEGREE 45: 2D[45]

The leading differential $d^{12}(B[34]M_1^6) = 2D[45]$ determines tentative differentials whose cokernel in degrees less than 67 will be computed in Section 7.6 because of the hidden differential mentioned in the discussion of $d^{12}(B[34]M_1^6)$ above. The 2D[45]-leader is $2D[45]M_1^2$.

DEGREE 45: 4D[45]

The leading differential $d^6(\eta\sigma A[32,1]M_1^3) = 4D[45]$ determines tentative differentials with image $Z_2 \otimes (Z_4(4D[45])\{1,M_1,M_2\} \otimes B\langle 4\rangle)$. The remainig elements are $Z_2 \otimes (Z_4D[45]\{M_1^2,M_1^3,M_1M_2,M_1^2M_2,M_1^3M_2\} \otimes B\langle 4\rangle)$, and the 4D[45]-leader is $4D[45]M_1^2$.

DEGREE 45: $8D[45] = \nu C[42]$

The leading differential $d^4(C[42]M_1^2) = 8D[45]$ determines tentative differentials with image $Z_2(8D[45])\{1,M_1,M_1^2,M_2,M_1M_2\} \otimes B\langle 4\rangle$. The remaining elements are $Z_2(8D[45])\{M_1^3,M_1^2M_2,M_1^3M_2\} \otimes B\langle 4\rangle$, and the 8D[45]-leader is $8D[45]M_1^3$.

DEGREE 45: $\eta C[44]$

Since $\eta^2C[44] \neq 0$, the only element of $E^4_{*,45}$ with a representative in $Z_2\eta C[44] \otimes H_*BP$ is zero.

CHAPTER 7: THE NEW STEMS (π_N^S, $46 \le N \le 64$)

1. Introduction

In Sections 2 through 5 we continue the computations of Chapter 6 to determine the next 19 stable stems. The tables of leaders in this chapter only include those leaders of degree less than 67. We are unable to determine three group extensions in degrees 54, 62 and 63. In particular we can not determine whether $2A[54,1]$ equals zero or $\eta A[8]D[45]$. Whenever we deal with $A[54,1]$ we take into account both possibilities for $2A[54,1]$. In Section 6 we collect all the computations of tentative differentials which are determined by the leading differentials discovered in the previous sections. The results of this chapter are summarized in Appendices 1 to 4.

The computations of π_N^S, $46 \le N \le 59$, agree with the tentative computation of these stems by Mark Mahowald using the Adams spectral sequence [55]. Tables of the Adams spectral sequence through degree 64 are given in Appendix 6.

2. Computation of π_N^S, $46 \le N \le 50$.

In the tables of leaders below, all leaders of degree greater than 52 will have an asterisk at the left. They will be omitted from all other tables of leaders in this section except for the last one.

Recall from the table of leaders in Figure 6.3.7 that there are five leaders of degree 47 and four leaders of degree 48. Let $\phi = d^6(\eta A[40,1]M_1^3)$. We show that $\phi = 0$. By Theorem 2.4.2, $\phi \in \langle\eta^2, A[40,1], \nu\rangle = \langle\langle 2,\eta,2\rangle, A[40,1],\nu\rangle$ $= \langle 2,\langle\eta,2,A[40,1]\rangle,\nu\rangle + \langle 2,\eta,\langle 2,A[40,1],\nu\rangle\rangle$. Now $2\langle\eta,2,A[40,1]\rangle$ $= \langle 2,\eta,2\rangle A[40,1] = \eta^2 A[40,1] = 0$. Hence η^2 divides $\langle\eta,2,A[40,1]\rangle$, $0 \in \langle 2,\langle\eta,2,A[40,1]\rangle,\nu\rangle$ and $\phi \in \langle 2,\eta,\langle 2,A[40,1],\nu\rangle\rangle$. ($\phi$ is not divisible by ν

while the other elements of π^S_{46} are divisible by η and thus have order two.) Now $\eta\langle 2, A[40,1], \nu\rangle = \langle \eta, 2, A[40,1]\rangle\nu = 0$. Thus, $\langle 2, A[40,1], \nu\rangle = 2kC[44]$. Therefore, $\phi \in \langle 2, \eta, 2kC[44]\rangle = k\langle 2, \eta, 2\rangle C[44] = k\eta^2C[44]$, a contradiction, since ϕ is not divisible by η. Thus, $\phi = 0$ and $\eta A[40,1]M_1^3$ must bound. There is only one possibility: $d^8(B[34]M_1^4M_2) = \eta A[40,1]M_1^3$ and

$$\sigma B[34] = \eta A[40,1]. \qquad [7.1]$$

Clearly $C[44]M_1^2$ transgresses. Note that $2C[42]\bar{M}_2$ transgresses because $2\nu C[42] = 16D[45] = 0$. Observe that $2C[20]^2M_1\bar{M}_2$ equals $C[20]^2\bar{M}_2$ times the infinite cycle $2M_1$, and $d^6(C[20]^2\bar{M}_2) = A[45,2]$. Therefore $d^6(2C[20]^2M_1\bar{M}_2) = 2A[45,2]M_1 = 0$, and $2C[20]^2M_1\bar{M}_2$ transgresses. Thus, $\eta C[44]M_1$, $D[45]M_1$, $A[45,1]M_1$ and $A[45,2]M_1$ can not be boundaries. We have thus proved the following theorem.

THEOREM 7.2.1 $\pi^S_{46} = Z_2\eta^2C[44] \oplus Z_2\eta D[45] \oplus Z_2\eta A[45,1] \oplus Z_2\eta A[45,2]$.

The computations in Section 6 show that we have the following leaders.

Row	Degree	Leader	Row	Degree	Leader
17	51	$\eta^2\gamma_1M_1^{17}$	40	48	$2C[20]^2M_1\bar{M}_2$
32	50	$A[32,1](M_1^2M_3+M_2^3)$	42	48	$2C[42]M_1^3$
38	50	$\eta\sigma A[30]M_1^3\bar{M}_2$, $B[38]M_1^6$	42	50	$\eta^2C[20]^2M_1\bar{M}_2$
38	52	$2B[38]M_1^4\bar{M}_2$	44	48	$C[44]M_1^2$
39	51	$\eta B[38]M_1^3\bar{M}_2$	45	49	$A[45,1]M_1^2$, $Z_8D[45]M_1^2$
39	49	$A[39,1]M_1^2M_2$, $A[39,1]M_1^5$, $A[39,3]M_1^2\bar{M}_2$	45	51	$8D[45]M_1^3$, $A[45,2]\bar{M}_2$
40	52	$(\eta A[39,3]+\eta\sigma A[32,1])M_1^3\bar{M}_2$, $A[40,1]M_1^6$	46	48	$\eta^2C[44]M_1$, $\eta D[45]M_1$, $\eta A[45,1]M_1$, $\eta A[45,2]M_1$

FIGURE 7.2.1: Leaders from Rows 1 to 46 of Degree at Least 48

LEMMA 7.2.2 (a) $\sigma^2A[32,1] = 0$

(b) $\sigma A[39,3] = 0$

PROOF. (a) $\sigma^2A[32,1] \in A[32,1]\langle\nu,\sigma,\nu\rangle = \langle A[32,1],\nu,\sigma\rangle\nu \subset \nu\cdot\pi_{43}^S = 0$.

(b) Since $A[39,3] \in \langle\nu,B[34],\eta\rangle$, $\sigma A[39,3] \in \sigma\langle\nu,B[34],\eta\rangle$ $= \langle\sigma,\nu,B[34]\rangle\eta = \langle\sigma,\nu,\langle\eta,2,A[32,1]\rangle\rangle\eta \supset \langle\sigma,\nu,\eta,2\rangle\eta A[32,1] = 0$ because $\langle\sigma,\nu,\eta,2\rangle \in \pi_{13}^S = 0$. Note that the four-fold Toda bracket is defined by Theorem 2.2.7(a) because $\langle\sigma,\nu,\eta\rangle = 0$ and $\langle\nu,\eta,2\rangle = 0$. Thus, $\sigma A[39,3] \in \eta(\text{Indet }\langle\sigma,\nu,B[34]\rangle) = \{0,\eta\sigma B[38]\}$ since $\eta B[34] = 0$ and $\eta^2\sigma^2A[30] = 0$. Therefore, $\sigma(A[39,3] + k\eta B[38]) = 0$. Note that A[39,3] is defined as $d^6(B[34]M_1^3)$ and is thus only defined modulo $Z_2\eta B[38]$. Therefore, we can define A[39,3] so that $\sigma A[39,3] = 0$. Alternatively, we shall see that $\sigma B[38] = 4D[45]$, and thus $0 = \eta\sigma B[38] = \sigma A[39,3]$.∎

There are seven leaders of degree 48 and six leaders of degree 49. By Lemma 3.3.14, if $\eta^2A[45,1]$ or $\eta^2A[45,2]$ is nonzero then it must be divisible by two. Let $d^8(2C[20]^2M_1\bar{M}_2) = B[47]$. Since $A[45,2] = d^6(C[20]^2\bar{M}_2)$, it follows from Theorem 2.4.2 that

$$B[47] \in \langle\eta,2,A[45,2]\rangle. \qquad [7.2]$$

Thus $2B[47] \in 2\langle\eta,2,A[45,2]\rangle = \langle 2,\eta,2\rangle A[45,2] = \eta^2A[45,2]$. Let $A[47] = d^6(2C[42]\bar{M}_2)$. By Theorem 2.4.4(c),

$$A[47] \in \langle\eta,\nu,2C[42]\rangle. \qquad [7.3.]$$

Thus, $2A[47] \in 2\langle\eta,\nu,2C[42]\rangle = \langle 2,\eta,\nu\rangle 2C[42] = 0$. By 6.18, $2C[44] \in \langle\sigma,\nu,\nu A[30]\rangle$. Thus, $2\nu C[44] \in \nu\langle\sigma,\nu,\nu A[30]\rangle = \langle\nu,\sigma,\nu\rangle\nu A[30]$ $= \sigma^2\nu A[30] = 0$. Therefore, B[47] is the only elements of CokJ_{47} which may not have order two. Thus, $\eta^2A[45,1] = 0$, and $\eta A[45,1]M_1$ must be a boundary. In addition, $\eta^2C[44]M_1$ must bound because $\eta^3C[44] = 4\nu C[44] = \nu\sigma^2A[30] = 0$. There are only three leaders of degree 49 which do not clearly transgress: $A[39,1]M_1^2M_2$, $A[39,1]M_1^5$ and $A[39,3]M_1^2\bar{M}_2$. If $d^8(A[39,1]M_1^2M_2)$ equals $\eta A[45,1]M_1$ or $\eta^2C[44]M_1$ then r_{Δ_1} applied to $d^8(A[39,1]M_1^3M_2)$ produces a contradiction because there is no possibility for a hidden differential on $A[39,1]M_1^3M_2$.

Thus, $A[39,1]M_1^2M_2$ transgresses. Therefore, $2C[20]^2M_1\bar{M}_2$, $2C[42]\bar{M}_2$, $C[44]M_1^2$, $\eta A[45,2]M_1$ and $\eta D[45]M_1$ can not bound.

The only possibility for $\eta^2C[44]M_1$ and $\eta A[45,1]M_1$ to be boundaries is $\{d^8(A[39,1]M_1^5), d^8(A[39,3]M_1^2\bar{M}_2\} = \{\eta^2C[44]M_1, \ \eta A[45,1]M_1\}$. Now $\sigma A[39,1] \in \sigma\langle\nu A[31],\nu,\eta\rangle = \langle\sigma,\nu A[31],\nu\rangle\eta$ and $\nu\langle\sigma,\nu A[31],\nu\rangle = \langle\nu,\sigma,\nu A[31]\rangle\nu$ $= \langle\nu,\sigma,\nu\rangle\nu A[31] = (\sigma^2)\nu A[31] = 0$. There is only one leader of degree 50 below the 34 row: $A[32,1](M_1^2M_3+M_2^3)$. As we shall see in the proof of Theorem 7.2.4, $A[39,1]M_1^2M_2$ must bound and the only possibility is $d^8(A[32,1](M_1^2M_3+M_2^3))$ $= A[39,1]M_1^2M_2$. Thus, $A[45,1]M_1^2$ can not bound and $\nu A[45,1]$ is nonzero. Therefore, $\langle\sigma,\nu A[31],\nu\rangle \neq A[45,1]$, $\sigma A[39,1] \neq \eta A[45,1]$ and $d^8(A[39,1]M_1^5) \neq$ $\eta A[45,1]M_1$. Thus, $d^8(A[39,1]M_1^5) = \eta^2C[44]M_1$, $d^8(A[39,3]M_1^2\bar{M}_2) = \eta A[45,1]M_1$ and

$$\sigma A[39,1] = \eta^2C[44]. \qquad [7.4]$$

We have thus proved the following theorem.

THEOREM 7.2.3 $\pi^S_{47} = Z_4B[47] \oplus Z_2A[47] \oplus Z_2\eta^2D[45] \oplus Z_2\nu C[44] \oplus Z_{32}\gamma_5$ where $2B[47] = \eta^2A[45,2]$.

The computations in Section 6 show that we have the following leaders.

Row	Degree	Leader
17	51	$\eta^2\gamma_1M_1^{17}$
32	50	$A[32,1](M_1^2M_3+M_2^3)$
38	50	$\eta\sigma A[30]M_1^3\bar{M}_2$, $B[38]M_1^6$
38	52	$2B[38]M_1^4\bar{M}_2$
39	51	$\eta B[38]M_1^3\bar{M}_2$
39	49	$A[39,1]M_1^2M_2$
40	52	$(\eta A[39,3]+\eta\sigma A[32,1])M_1^3\bar{M}_2$, $A[40,1]M_1^6$
42	50	$\eta^2C[20]^2M_1\bar{M}_2$
44	52	$C[44]M_1^4$
45	49	$A[45,1]M_1^2$, $Z_8(D[45]M_1^2)$
45	51	$8D[45]M_1^3$, $A[45,2]\bar{M}_2$
46	52	$\eta A[45,1]M_1^3$
*46	66	$\eta^2C[44]M_1^7\bar{M}_2$
47	49	$\eta^2D[45]M_1$, $\eta^2A[45,2]M_1$, $A[47]M_1$, $B[47]M_1$
47	51	$\nu C[44]M_1^2$

FIGURE 7.2.2: Leaders from Rows 1 to 47 of Degree at Least 49

There are nine leaders of degree 49 and four leaders of degree 50. Since $d^6(C[20]^2\overline{M}_2) = A[45,2]$, $d^6(\eta^2C[20]^2M_1\overline{M}_2) = \eta^2A[45,2]M_1$. Note that $\xi = d^{10}(A[39,1]M_1^2M_2) = d^{10}(\sigma A[32,3]M_1^2M_2) \in \langle A[37],\sigma,\nu\rangle$ by Theorem 2.4.2. Assume that ξ is nonzero. Then ξM_1^2 can not bound, $\nu\xi \neq 0$ and $\nu\xi \notin (\eta)$. Thus, $\xi \in \langle\langle A[32,3],\eta,\nu\rangle,\sigma,\nu\rangle = \langle A[32,3],\langle\eta,\nu,\sigma\rangle,\nu\rangle + \langle A[32,3],\eta,\langle\nu,\sigma,\nu\rangle\rangle =$ $\langle A[32,3],0,\nu\rangle + \langle A[32,3].\eta,\sigma^2\rangle = \langle A[32,3],\eta,\sigma^2\rangle$ since ξ is not divisible by ν. Since $A[32,3] = d^{12}(\sigma^3M_1^3\overline{M}_2) = d^{12}[(\sigma^2M_1^3)\sigma\overline{M}_2]$, Theorem 2.4.6(c) implies that

$$A[32,3] \in \langle A[19],\sigma,\nu,\eta\rangle. \qquad [7.5]$$

Then $\nu\xi \in \nu\langle A[32,3],\eta,\sigma^2\rangle = A[32,3]\langle\eta,\sigma^2,\nu\rangle = A[32,3]A[19]$

$\in A[19]\langle A[19],\sigma,\nu,\eta\rangle \subset \langle A[19]^2,\sigma,\nu,\eta\rangle = k\langle\eta\sigma A[30],\sigma,\nu,\eta\rangle$ since $\eta A[19]^2 = 0$;

$= k[d^8(\sigma^2A[30]M_1\overline{M}_2) + \text{Indet}\ \langle\eta\sigma A[30],\sigma,\nu,\eta\rangle]$ by Theorem 2.4.6(c);

$= k[d^8(4C[44]M_1\overline{M}_2) + \text{Indet}\ \langle\eta\sigma A[30],\sigma,\nu,\eta\rangle]$;

$= \text{Indet}\ \langle\eta\sigma A[30],\sigma,\nu,\eta\rangle]$ since $4C[44]M_1\overline{M}_2$ is a d^8-boundary.

By Theorem 2.3.1(b), there is $\eta\zeta \in \pi^S_{46}$ such that $\nu\xi \in \langle\eta\sigma A[30],\beta_1,\eta\rangle+\langle\eta\zeta,\nu,\eta\rangle$. Then $\nu\xi \in \sigma A[30]\langle\eta,\beta_1,\eta\rangle + \nu^2\zeta = \sigma A[30](\nu\beta_1) + \nu^2\zeta = \nu^2\zeta$. Therefore, ν divides ξ, a contradiction. Thus, $A[39,1]M_1^2M_2$ must be a boundary. The only possibility is: $d^8(A[32,1](M_1^2M_3+M_2^3)) = A[39,1]M_1^2M_2$. Since $d^4(4D[45]M_1^2) =$ $4\nu D[45] = \eta^3D[45] = 0$ in E^4, $4D[45]M_1^2$ must bound from below the 40 row. There is only one possibility: $d^8(B[38]M_1^6) = 4D[45]M_1^2$ and

$$\sigma B[38] = 4D[45]. \qquad [7.6]$$

Now $\eta^3D[45] = 4\nu D[45] = \nu\sigma B[38] = 0$, and $\eta^2D[45]M_1$ must be a boundary. There is only one possibility: $d^8(\eta\sigma A[30]M_1^3\overline{M}_2) = \eta^2D[45]M_1$. We have thus proved the following theorem.

THEOREM 7.2.4 $\pi^S_{48} = Z_4\nu D[45] \oplus Z_2\nu A[45,1] \oplus Z_2\eta A[47] \oplus Z_2\eta B[47] \oplus Z_2\eta\gamma_5$.

The computations in Section 6 show that we have the following leaders.

Row	Degree	Leader
17	51	$\eta^2\gamma_1M_1^{17}$
*32	62	$A[32,1]M_1^8\bar{M}_3$
*38	62	$B[38]M_1^2\bar{M}_2\bar{M}_3$
38	52	$2B[38]M_1^4\bar{M}_2$
39	51	$\eta B[38]M_1^3\bar{M}_2$, $A[39,1]M_1^3M_2$
40	52	$(\eta A[39,3]+\eta\sigma A[32,1])M_1^3\bar{M}_2$,
		$A[40,1]M_1^6$
44	52	$C[44]M_1^4$
*45	59	$2D[45]M_1\langle M_2^2\rangle$
*45	53	$4D[45]M_1M_2$
45	51	$8D[45]M_1^3$, $A[45,2]\bar{M}_2$
46	52	$\eta A[45,1]M_1^3$
*47	53	$\eta^2D[45]M_1^3$, $B[47]\bar{M}_2$
*47	55	$\eta^2A[45,2]M_1\bar{M}_2$
*47	57	$A[47]M_1^2\bar{M}_2$
47	51	$\nu C[44]M_1^2$
48	52	$\nu A[45,1]M_1^2$, $\nu D[45]M_1^2$
*48	54	$2\nu D[45]M_1^3$
48	50	$\eta A[47]M_1$, $\eta B[47]M_1$

FIGURE 7.2.3: Leaders from Rows 1 to 48 of Degree at Least 50

There are two leaders of degree 50 and six leaders of degree 51. Note that $B[47] \in \langle\eta,2,A[45,2]\rangle = \langle\eta,2,\langle\eta,\nu,C[20]^2\rangle\rangle$
$= \langle\eta,\langle 2,\eta,\nu\rangle,C[20]^2\rangle + \langle\langle\eta,2,\eta\rangle,\nu,C[20]^2\rangle = \langle\eta,0,C[20]^2\rangle + \langle 2\nu,\nu,C[20]^2\rangle$.
Since $\sigma C[20] = 0$, $B[47] \in \langle 2\nu,\nu,C[20]^2\rangle$. Thus, $\eta B[47] \in \eta\langle 2\nu,\nu,C[20]^2\rangle$
$= \langle\eta,2\nu,\nu\rangle C[20]^2 = A[8]C[20]^2 = \langle\nu,\eta,\nu\rangle C[20]^2 = \nu\langle\eta,\nu,C[20]^2\rangle = \nu A[45,2]$ and

$$\eta B[47] = \nu A[45,2]. \tag{7.7}$$

Therefore, $d^4(A[45,2]\bar{M}_2) = \eta B[47]M_1$. If $\eta^2A[47]$ were nonzero, it would be divisible by two. Since there is no other possibility for a nonzero element of $\mathrm{Cok}J_{49}$, $\eta^2A[47] = 0$ and $\eta A[47]M_1$ must bound. Since $\eta^2\gamma_1M_1^{17} \in \mathrm{Image}\ r_{\Delta_1}$, it can not hit $\eta A[47]M_1$. Let X represents $A[39,1]M_1^3M_2$ as an element of E^8 with $\partial X = (A[39,1] \wedge \sigma \wedge \mu_2) \cup (A[39,1] \wedge B_{\sigma\nu})$. Since $\sigma A[39,1] = \eta^2C[44]$, we can represent $A[39,1]M_1^3M_2$ by $X \cup (\eta C[44] \wedge \mu_{01})$ which transgresses. There remains only one possibility: $d^{10}(\eta B[38]M_1^3\bar{M}_2) = \eta A[47]M_1$. We have thus proved the following theorem.

THEOREM 7.2.5 $\pi_{49}^S = Z_2\alpha_6 \oplus Z_2\eta^2\gamma_5$.

The computations in Section 6 show that we have the following leaders.

Row	Degree	Leader	Row	Degree	Leader
17	51	$\eta^2\gamma_1 M_1^{17}$	45	51	$8D[45]M_1^3$
38	52	$2B[38]M_1^4\bar{M}_2$	46	52	$\eta A[45,1]M_1^3$
39	51	$A[39,1]M_1^3\bar{M}_2$	47	51	$\nu C[44]M_1^2$
40	52	$(\eta A[39,3]+\eta\sigma A[32,1])M_1^3\bar{M}_2$,	48	52	$\nu D[45]M_1^2$, $\nu A[45,1]M_1^2$
		$A[40,1]M_1^6$	*48	54	$\eta A[47]M_1^3$
44	52	$C[44]<M_1^4>$	*48	56	$\eta B[47]M_1\bar{M}_2$

FIGURE 7.2.4: Leaders from Rows 1 to 49 of Degree at Least 51

We have four leaders of degree 51 and seven leaders of degree 52. Since $d^{12}(B[38]M_1^4\bar{M}_2) = 4D[45]M_1^3$, it follows that $d^{12}(2B[38]M_1^4\bar{M}_2) = 8D[45]M_1^3$. Clearly $A[50,1] = d^{34}(\eta^2\gamma_1 M_1^{17})$ and $A[50,2] = d^{12}(A[39,1]M_1^3\bar{M}_2)$ are nonzero. Recall that $C[44] = d^{14}(\eta A[30]M_1M_2^2)$ and note that $\eta A[30]M_1M_2^2$ is represented by
$\mathcal{R} = (\mu_{02}\wedge \eta \wedge A[30] \wedge \mu_1) \cup (\mu_{02}\wedge B_{\eta A[30]\eta}) \cup (\mu_4\wedge B_{\nu\eta}\wedge A[30] \wedge \mu_1)$
$\cup\ (B_{<\sigma,\nu,\eta>}\wedge A[30] \wedge \mu_1) \cup (\mu_4\wedge A[31] \wedge \mu_{01}) \cup (\mu_4\wedge B_{A[31]\eta} \wedge \mu_2)$
$\cup\ (\mu_4 \wedge B_{<\nu,\eta,A[30]\eta>\cup<A[31],\eta,\nu>}) \cup (B_{\sigma A[31]}\wedge \mu_{01}) \cup (B_{<\sigma,A[31],\eta>}\wedge \mu_1)$.
Observe that $\mathcal{R}$, without the last term, shows that $0 = d^{12}(\eta A[30]M_1M_2^2)$ $= <\sigma,A[31],\eta>M_1$. Thus, $<\sigma,A[31],\eta>M_1$ is zero in E^{12}. Therefore, $<\sigma,A[31],\eta> \in Z_2\eta A[39,2] \oplus Z_2\eta A[39,3] \oplus Z_2\eta\gamma_4 \subset \mathrm{Indet}<\sigma,A[31],\eta>$, and $0 \in <\sigma,A[31],\eta>$. Now $\partial\mathcal{R} = (B_{\sigma\nu}\wedge B_{\eta A[30]\eta}) \cup (B_{<\sigma,\nu,\eta>}\wedge A[30] \wedge \eta)$ $\cup\ (\sigma \wedge B_{<\nu,\eta,A[30]\eta>\cup<A[31],\eta,\nu>}) \cup (B_{\sigma A[31]}\wedge B_{\eta\nu}) \cup (B_{<\sigma,A[31],\eta>}\wedge \nu)$ which represents an element of $<\sigma, \left[A[31],\nu\right], \begin{bmatrix}\eta & 0\\ 0 & \eta\end{bmatrix}, \begin{bmatrix}\nu\\ A[30]\eta\end{bmatrix}>$. Thus,

$$C[44] \in <\sigma, \left[A[31],\nu\right], \begin{bmatrix}\eta & 0\\ 0 & \eta\end{bmatrix}, \begin{bmatrix}\nu\\ \eta A[30]\end{bmatrix}>. \qquad [7.8]$$

Let $\mathfrak{C}$ denote the mapping cone of $A[40,1]$ with $\rho:S \longrightarrow \mathfrak{C}$ the canonical map. In the Atiyah-Hirzebruch spectral sequence for $\mathfrak{C}_*BP$, $\rho(\eta A[30]M_1^3M_2^2)$ survives to

E^{12}. Since $\eta A[40,1] \neq 0$ and $\eta^2 A[40,1] = 0$, $\rho: \pi^S_{42} \longrightarrow \mathfrak{C}_{42}$ is an isomorphism. The only possibility for $d^{12}(\rho(\eta A[30]M_1^3M_2^2))$ to be nonzero is $d^{12}(\rho(\eta A[30]M_1^3M_2^2)) = \rho(2C[42]M_1^3)$ which implies that $d^{12}(\eta A[30]M_1M_2^2) = 2C[42]M_1$ which contradicts that $\eta A[30]M_1M_2^2$ survives to E^{14}. Thus, $\rho(\eta A[30]M_1^3M_2^2)$ survives to E^{14} and $r_{2\Delta_1}$ shows that $d^{14}(\rho(\eta A[30]M_1^3M_2^2)) = \rho(C[44]M_1^2)$. Therefore $\rho(\nu C[44]) = 0$ and $A[40,1]$ divides $\nu C[44]$. The only possibility is

$$\sigma A[40,1] = \nu C[44]. \tag{7.9}$$

Thus, $d^8(A[40,1]M_1^6) = \nu C[44]M_1^2$. Clearly $2A[50,2] = 0$. We shall see in Section 7.3 that $\eta A[50,2] \neq 0$. Therefore $2A[50,1] = 0$. We have thus proved the following theorem.

THEOREM 7.2.6 $\pi^S_{50} = Z_2 A[50,1] \oplus Z_2 A[50,2] \oplus Z_2 \eta\alpha_6$.

The computations in Section 6 show that we have the following leaders. Since this is the final table of leaders of this section, we include the leaders of all degrees.

Row	Degree	Leader
9	63	$\eta^2\sigma M_1^{21}M_2^2$
11	57	$4\beta_1(M_1^7\bar{M}_2^3\bar{M}_3+M_1^{10}\bar{M}_2^2\bar{M}_3+M_1^{14}\bar{M}_2^3)$
17	59	$\eta^2\gamma_1M_1^{15}\bar{M}_2^2$
18	64	$4C[18]M_1^7\bar{M}_2M_2^2\bar{M}_3$
19	55	$\beta_2M_1^{18}$
21	53	$\nu C[18]M_1^6\bar{M}_2\bar{M}_3$
22	62	$\nu A[19]M_1^7M_2^2\langle M_3\rangle$
23	63	$\gamma_2M_1^{20}$
24	60	$\eta A[23]M_1^{15}\bar{M}_2$
30	60	$A[30]\langle M_4\rangle$
40	58	$A[40,2]M_1^6\bar{M}_2$
40	52	$(\eta A[39,3]+\eta\sigma A[32,1])M_1^3\bar{M}_2$
44	52	$C[44]\langle M_1^4\rangle$
44	56	$2C[44]M_1^6$
44	58	$4C[44]M_1^7$
45	59	$2D[45]M_1\langle M_2^2\rangle$
45	53	$4D[45]M_1M_2$
46	52	$\eta A[45,1]M_1^3$
46	66	$\eta^2C[44]M_1^7\bar{M}_2$
47	55	$\nu C[44]M_1M_2$, $\eta^2A[45,2]M_1\bar{M}_2$

32	62	$A[32,1]M_1^8\bar{M}_3$
34	64	$2B[34]M_1^5\bar{M}_2\bar{M}_3$
36	54	$A[36]M_1^2\langle M_3\rangle$
38	62	$B[38]M_1^2\bar{M}_2\bar{M}_3$
39	53	$\sigma A[32,1]M_1^4\bar{M}_2$
39	61	$A[39,1]M_1M_2M_3$
40	64	$\eta\sigma A[32,1]M_1^5\bar{M}_3$
40	60	$A[40,1]\bar{M}_2\bar{M}_3$
47	53	$\eta^2D[45]M_1^3$, $B[47]\bar{M}_2$
47	57	$A[47]M_1^2\bar{M}_2$
48	52	$\nu D[45]M_1^2$, $\nu A[45,1]M_1^2$
48	54	$2\nu D[45]M_1^3$, $\eta A[47]M_1^3$
48	56	$\eta B[47]M_1\bar{M}_2$
50	56	$A[50,1]M_1^3$, $A[50,1]M_2$
50	52	$A[50,2]M_1$

FIGURE 7.2.5: Leaders from Rows 1 to 50 of Degree at Least 52

3. Computation of π_N^S, $51 \le N \le 55$.

We continue the computations of Section 2. In the tables of leaders below, all leaders of degree greater than 57 will have an asterisk at the left. They will be omitted from all tables of leaders in this section except for the last one.

From Figure 7.2.5, we see that there are six leaders of degree 52 and five leaders of degree 53. Observe that η, ν times $4D[45]$ and η, ν times $\eta^2D[45]$ and ν times $B[47]$ are zero. Therefore, $4D[45]M_1M_2$, $\eta^2D[45]M_1^3$ and $B[47]\bar{M}_2$ must transgress. Assume that $\nu A[45,1]M_1^2$ is not a boundary. Then $\nu^2A[45,1]$ is nonzero and not divisible by η. In particular, there is no way for $\eta A[50,2]M_1$ to bound. Therefore, $\eta^2A[50,2]$ is nonzero. By Lemma 3.3.14, $\eta^2A[50,2]$ must be divisible by two. This is impossible because all the other elements of π_{52}^S have order two. Thus, $\nu A[45,1]M_1^2$ must be a boundary.

We show that $(\eta A[39,3]+\eta\sigma A[32,1])M_1^3\bar{M}_2$ must be a boundary. Assume that $\xi = d^{12}((\eta A[39,3]+\eta\sigma A[32,1])M_1^3\bar{M}_2)$ is nonzero. The only leader of degree 54 or 56 below the 40 row is $A[36]M_1^2\bar{M}_3$ which we shall see must bound $4D[45]M_1M_2$. Thus, ξM_1 and ξM_1^2 can not bound. Therefore $\eta\xi$, $\nu\xi$ are nonzero and $\nu\xi$ is not divisible by η. If $\eta^2\xi$ is nonzero then $\eta^2\xi M_1$ must be a boundary because $\eta^3\xi = 4\nu\xi = 0$. There are only four leaders of degree 56 below row 52: $A[50,1]M_2$, $C[44]\langle M_2^2\rangle$, $2C[44]\langle M_2^2\rangle$ and $\eta B[47]M_1\bar{M}_2$. Now $A[50,1]M_2$ transgresses. and $d^6(\eta B[47]M_1\bar{M}_2) = \eta A[52,2]M_1$. Since $\langle M_2^2\rangle$ has a representative with boundary $(\sigma \wedge \mu_2) \cup B_{\sigma\nu}$, $C[44]\langle M_2^2\rangle$ hits $\sigma C[44]M_1^2$ if $\sigma C[44] \neq 0$ or transgresses if $\sigma C[44] = 0$. Then $2C[44]\langle M_2^2\rangle$ hits $2\sigma C[44]M_1^2$ if $2\sigma C[44] \neq 0$ or transgresses to an element of $\langle 2,\sigma C[44],\nu\rangle$ if $2\sigma C[44] = 0$. Thus, $\eta^2\xi M_1$ can not be a boundary, $\eta^2\xi = 0$ and $\eta\xi M_1$ must be a boundary. Assume that $d^6(\eta^2 A[45,2]M_1\bar{M}_2) = \eta\xi M_1$. Then $\eta\xi \in \langle\nu,\eta,\eta^2 A[45,2]\rangle \subset \langle\nu,\eta^3,A[45,2]\rangle = \langle\nu,4\nu,A[45,2]\rangle \supset 2\langle\nu,2\nu,A[45,2]\rangle = 0$ and $\eta\xi \in \text{Indet}\ \langle\nu,4\nu,A[45,2]\rangle = \nu\cdot\pi_{49}^S + A[45,2]\cdot\pi_7^S = \sigma A[45,2] \in \sigma\langle C[20]^2,\nu,\eta\rangle = \langle\sigma,C[20]^2,\nu\rangle\eta$ and $\xi \in \langle\sigma,C[20]^2,\nu\rangle$. Then $\nu\xi \in \nu\langle\sigma,C[20]^2,\nu\rangle = \langle\nu,\sigma,C[20]^2\rangle\nu$, $\xi \in \langle\nu,\sigma,C[20]^2\rangle$ and $\eta\xi \in \eta\langle\nu,\sigma,C[20]^2\rangle = \langle\eta,\nu,\sigma\rangle C[20]^2 = 0$, a contradicition. Thus, $d^6(\eta^2 A[45,2]M_1\bar{M}_2)$ can not equal $\eta\xi M_1$. If $\beta_2 M_1^{18}$ hits $\eta\xi M_1$ then there is a hidden differential on $\beta_2 M_1^{19}$ which can only hit $2C[44]M_1^6$. However, there is no possibility for a hidden differential on $\beta_2 M_1^{19} M_2$, $r_{3\Delta_1}(\beta_2 M_1^{19} M_2) = \beta_2 M_1^{19}$ and $2C[44]M_1^6 \notin \text{Image}\ r_{3\Delta_1}$. There remains only one possibility: $d^6(\nu C[44]M_1 M_2) = \eta\xi M_1$. Then $\eta\xi \in \langle\nu,\eta,\nu C[44]\rangle = \langle\nu,\eta,\nu\rangle C[44] = A[8]C[44] \in \langle\eta,\nu,2\nu\rangle C[44] = \eta\langle\nu,2\nu,C[44]\rangle$ and $\xi \in \langle\nu,2\nu,C[44]\rangle$. Thus, $\nu\xi \in \nu\langle\nu,2\nu,C[44]\rangle \subset \langle\nu^2,2\nu,C[44]\rangle \supset \langle\nu^2,2,\nu C[44]\rangle \supset \langle\langle\eta,\nu,\eta\rangle,2,\nu C[44]\rangle \supset \eta\langle\nu,\eta,2,\nu C[44]\rangle$. This Toda bracket is defined by Theorem 2.2.7(a) because $\langle\nu,\eta,2\rangle = 0$ and $\langle\eta,2,\nu C[44]\rangle$ contains an element of $\text{CokJ}_{49} = 0$. Since $\nu\xi$ is not divisible by η, $\nu\xi \in \text{Indet}\ \langle\nu^2,2\nu,C[44]\rangle = \nu^2\cdot\pi_{48}^S + \sigma\cdot\pi_{47}^S$. Thus, $\nu\xi \in \{\sigma A[47],\sigma B[47]\}$, and $\nu\xi M_1^2 \in \{d^8(A[47]\langle M_2^2\rangle),d^8(B[47]\langle M_2^2\rangle)\}$. This contradicts the fact that both $A[47]\langle M_2^2\rangle$ and $B[47]\langle M_2^2\rangle$ are boundaries. There remains no possiblity for $\eta\xi M_1$ to bound. This contradicition implies that ξ must be zero, and $(\eta A[39,3]+\eta\sigma A[32,1])M_1^3\bar{M}_2$ must bound from below the 37 row. There is only one possibility: $d^{20}(\nu C[18]M_1^6\bar{M}_2\bar{M}_3) = (\eta A[39,3]+\eta\sigma A[32,1])M_1^3\bar{M}_2$.

Now the only possibility for $\nu A[45,1]M_1^2$ to be a boundary is $d^{10}(\sigma A[32,1]M_1^4\bar{M}_2) = \nu A[45,1]M_1^2$. Then $\eta A[45,1]M_1^3$, $C[44]M_1^4$, $\nu D[45]M_1^2$ and $A[50,2]M_1$ can not be boundaries. Therefore $\lambda = d^6(\eta A[45,1]M_1^3)$, $\sigma C[44]$, $\nu^2 D[45]$ and $\eta A[50,2]$ are nonzero. We shall see that $d^8(2C[44]M_1^6) = \lambda M_1^2$. Therefore, $2\sigma C[44] = \lambda$ and

$$2\sigma C[44] \in \langle \eta, \eta A[45,1], \nu \rangle. \qquad [7.10]$$

If $2\lambda = k\eta A[50,2] + \nu^2 D[45]$ then $d^8(4C[44]M_1^7) = \nu^2 D[45]M_1^3$ since $\eta A[50,2]M_1^3 = d^4(\nu A[45,1]M_1^2 M_2)$. This contradicts $d^4(\nu^2 D[45]M_1^3) = \eta A[8]D[45]M_1 \neq 0$. Thus, $2\lambda = k\eta A[50,2]$. Assume that $k = 1$. We will see that $\nu^2 A[50,2]$ is nonzero and not divisible by two. Then $\nu^2 A[50,2] \in \langle \eta, \nu, \eta \rangle A[50,2] \subset \langle \eta, \nu, \eta A[50,2] \rangle = \langle \eta, \nu, 4\sigma C[44] \rangle \supset 4C[44]\langle \eta, \nu, \sigma \rangle = 0$. Thus, $\nu^2 A[50,2] \in \mathrm{Indet}\, \langle \eta, \nu, 4\sigma C[44] \rangle = = (\eta)$, a contradiction. Therefore, $k = 0$ and $4\sigma C[44] = 0$. We have thus proved the following theorem.

THEOREM 7.3.1 $\pi_{51}^S = Z_4 \sigma C[44] \oplus Z_2 \eta A[50,2] \oplus Z_2 \nu^2 D[45] \oplus Z_8 \beta_6$

The computations in Section 6 show that we have the following leaders.

Row	Degree	Leader	Row	Degree	Leader
11	57	$4\beta_1(M_1^7\bar{M}_2^3\bar{M}_3 + M_1^{10}\bar{M}_2^2\bar{M}_3 + M_1^{14}\bar{M}_2^3)$	47	57	$A[47]M_1^2\bar{M}_2$
19	55	$\beta_2 M_1^{18}$	48	54	$2\nu D[45]M_1^3$, $\nu A[45,1]M_1^3$,
36	54	$A[36]M_1^2\bar{M}_3$			$\eta A[47]M_1^3$
*39	59	$\sigma A[32,1]M_1^4 M_2^2$	48	56	$\eta B[47]M_1\bar{M}_2$
*40	66	$\eta A[39,3]M_1^3\bar{M}_2\langle M_3\rangle$	50	56	$A[50,1]M_1^3$, $A[50,1]M_2$
44	56	$2C[44]M_1^6$	50	54	$A[50,2]M_1^2$
45	53	$4D[45]M_1M_2$	51	53	$\sigma C[44]M_1$, $\eta A[50,2]M_1$
47	55	$\nu C[44]M_1M_2$, $\eta^2 A[45,2]M_1\bar{M}_2$	51	55	$2\sigma C[44]M_1^2$
47	53	$\eta^2 D[45]M_1^3$, $B[47]\bar{M}_2$	51	57	$\nu^2 D[45]M_1^3$

FIGURE 7.3.1: Leaders from Rows 1 to 51 of Degree at Least 53

There are five leaders of degree 53 and five leaders of degree 54. Let $\langle\mu_{001}\rangle$ represent $\langle M_3\rangle$ such that $\partial\langle\mu_{001}\rangle = (\sigma \wedge \bar{\mu}_{01}) \cup (B_{\sigma\nu} \wedge \mu_1) \cup B_{\langle\nu,\sigma,\eta\rangle}$. Since $\sigma A[36] = 0$ and $\nu A[36] = \eta B[38]$, $A[36]M_1^2\langle M_3\rangle$ can be represented by $[\mu_2 \wedge B_{A[36]\sigma} \wedge \bar{\mu}_{01}] \cup [((\mu_2 \wedge A[36]) \cup (\mu_1 \wedge B[38])) \wedge \langle\mu_{001}\rangle]$ which has boundary $(\mu_1 \wedge B[38]] \wedge \sigma \wedge \bar{\mu}_{01})$ union elements of filtration degree six. Thus, $d^{10}(A[36]M_1^2\langle M_3\rangle) = \sigma B[38]M_1\bar{M}_2 = 4D[45]M_1\bar{M}_2$. Let $A[52,1] = d^6(\eta^2D[45]M_1^3)$. Then

$$A[52,1] \in \langle\eta,\eta^2D[45],\nu\rangle. \qquad [7.11]$$

Thus, $2A[52,1] \in 2\langle\eta,\eta^2D[45],\nu\rangle = \langle 2,\eta,\eta^2D[45]\rangle\nu \in \nu\cdot\pi_{49}^S = 0$. Let $A[52,2] = d^6(B[47]\bar{M}_2)$. Then

$$A[52,2] \in \langle\eta,\nu,B[47]\rangle. \qquad [7.12]$$

Thus, $2A[52,2] \in 2\langle\eta,\nu,B[47]\rangle = \langle 2,\eta,\nu\rangle B[47] = 0$. By Lemma 3.3.14, $\eta^2A[50,2]$ must be divisible by two. However, there is no possibility for an element of order four in π_{52}^S. Thus, $\eta^2A[50,2] = 0$, and $\eta A[50,2]M_1$ must bound. Clearly $A[50,2]M_1^2$, $\eta A[47]M_1^3$ and $2\nu D[45]M_1^3$ transgress. Thus, $d^4(\nu A[45,1]M_1^3) = \eta A[50,2]M_1$ and

$$\nu^2A[45,1] = \eta A[50,2]. \qquad [7.13]$$

Now $\eta\sigma C[44]$, $A[52,1]$ and $A[52,2]$ are nonzero. Observe that $A[52,1]$ can not be divisible by σ because we shall see that $\nu A[52,1] \neq 0$. If σ times $A[45,1]$, $A[45,2]$ or $D[45]$ were $A[52,2]$ then $A[52,2]\bar{M}_2$ would be $d^8(A[45,1]\langle M_3\rangle)$, $d^8(A[45,2]\langle M_3\rangle)$ or $d^8(D[45]\langle M_3\rangle)$. This would contradict that $A[45,1]\langle M_3\rangle$, $A[45,2]\langle M_3\rangle$ and $D[45]\langle M_3\rangle$ are boundaries. Note that $A[45,1]$, $A[45,2]$ and $D[45]$ are only defined modulo $\eta C[44]$. Thus, we can define $A[45,1]$, $A[45,2]$ and $D[45]$ such that

$$\sigma A[45,1] = \sigma A[45,2] = \sigma D[45] = 0. \qquad [7.14]$$

We have thus proved the following theorem.

THEOREM 7.3.2 $\pi_{52}^S = Z_2A[52,1] \oplus Z_2A[52,2] \oplus Z_2\eta\sigma C[44]$

The computations in Section 6 show that we have the following leaders.

Row	Degree	Leader	Row	Degree	Leader
11	57	$4\beta_1(M_1^7\bar{M}_2^3\bar{M}_3+M_1^{10}\bar{M}_2^2\bar{M}_3+M_1^{14}\bar{M}_2^3)$	50	56	$A[50,1]M_1^3$, $A[50,1]M_2$
19	55	$\beta_2M_1^{18}$	50	54	$A[50,2]M_1^2$
*36	60	$A[36]M_1^6\bar{M}_2^2$	*51	59	$\sigma C[44]M_1^4$
44	56	$2C[44]M_1^6$	51	55	$2\sigma C[44]M_1^2$
45	57	$4D[45]M_1^3M_2$	*51	63	$\eta A[50,2]M_1^3\bar{M}_2$
47	55	$\nu C[44]M_1M_2$, $\eta^2A[45,2]M_1\bar{M}_2$	51	57	$\nu^2D[45M_1^3$
47	57	$A[47]M_1^2\bar{M}_2$	52	54	$\eta\sigma C[44]M_1$, $A[52,2]M_1$
48	54	$2\nu D[45]M_1^3$, $\eta A[47]M_1^3$	52	56	$A[52,1]M_1^2$
48	56	$\eta B[47]M_1\bar{M}_2$			

FIGURE 7.3.2: Leaders from Rows 1 to 52 of Degree at Least 54

There are five leaders of degree 54 and five leaders of degree 55. Clearly $\sigma C[44]M_1^2$ and $2\sigma C[44]M_1^2$ transgress. We show that $\eta\sigma C[44]M_1$ must be a boundary. Observe that $\eta^3\sigma C[44] = 4\nu\sigma C[44] = 0$. If $\eta^2\sigma C[44] \neq 0$ then $\eta^2\sigma C[44]M_1$ must be a boundary. There are only three leaders of degree 56 below row 52: $A[50,1]M_2$, $2C[44]\langle M_2^2\rangle$ and $\eta B[47]M_1\bar{M}_2$. Now $A[50,1]M_2$ transgresses, $d^8(C[44]\langle M_2^2\rangle) = 2\sigma C[44]M_1^2$ and $d^6(\eta B[47]M_1\bar{M}_2) = \eta A[52,2]M_1$ since $A[52,2] = d^6(B[47]\bar{M}_2)$. (If $\eta A[52,2] = 0$ then $\eta B[47]M_1\bar{M}_2$ transgresses.) Thus, $\eta^2\sigma C[44]M_1$ can not be a boundary, $\eta^2\sigma C[44] = 0$ and $\eta\sigma C[44]M_1$ must be a boundary.

Assume that $\xi = d^6(\eta A[47]M_1^3)$ is nonzero. By Theorem 2.4.2, $\xi \in \langle\eta^2, A[47], \nu\rangle$

$= \langle\langle 2,\eta,2\rangle, A[47],\nu\rangle = \langle 2,\langle\eta,2,A[47]\rangle,\nu\rangle + \langle 2,\eta,\langle 2.A[47],\nu\rangle\rangle$

$= \langle 2,0,\nu\rangle + \langle 2,\eta,\langle\nu,A[47],2\rangle\rangle$ since $\mathrm{Cok}J_{49} = 0$;

$\equiv \langle 2,\langle\eta,\nu,A[47]\rangle,2\rangle + \langle\langle 2,\eta,\nu\rangle,A[47],2\rangle$ modulo $(2,\nu)$;

$\equiv k\eta\langle\eta,\nu,A[47]\rangle + \langle 0,A[47],2\rangle$ modulo $(2,\nu)$;

$\subset \langle 2,\eta,\nu\rangle$. Since ξ is nonzero in E^6, ξ must divisible by two. However, there is no possibility for ξ to be divisible by two. Thus, $\xi = 0$ and $\eta A[47]M_1^3$ must bound. If $d^{30}(\beta_2 M_1^{18}) = \eta A[47]M_1^3$ then $d^{30}(\beta_2 M_1^{18} M_2) = \eta A[47]M_1^3\bar{M}_2$. Since $r_{3\Delta_1}(\beta_2 M_1^{18} M_2) = \beta_2 M_1^{18}$, $r_{3\Delta_1}(\eta A[47]M_1^3\bar{M}_2) = \eta A[47]M_1^3$ not $\eta A[47]\bar{M}_2$, a contradiction. Therefore the only way for $\eta A[47]M_1^3$ to bound is by a hidden differential which replaces a tentative differential which originates below the 39 row and lands above the 48 row. The only possibility is $d^{32}(\eta^2\gamma_1 M_1^{19}) = \eta A[47]M_1^3$.

Now $A[50,1]M_1^2$ becomes a new leader of degree 54 which can not bound since there are no leaders of degree 55 below the 17 row. Since $A[50,2]M_1^2$ can only bound from below the 39 row, only $\beta_2 M_1^{18}$ could hit it. In that case we can argue as above to show that $r_{3\Delta_1}(A[50,2]M_1^2 M_2) = A[50,2]M_1^2$, a contradicition. Thus, $A[50,2]M_1^2$ can not bound. If $\beta_2 M_1^{18}$ hits $\eta\sigma C[44]M_1$, $A[52,2]M_1$ or $2\nu D[45]M_1^3$ then there must be a hidden differential on $\beta_2 M_1^{16}\bar{M}_2$ which lands below the 52 row. There is no possibility for such a hidden differential because $A[50,1]M_2$ can only bound from below the 17 row. Thus, $\beta_2 M_1^{18}$ must transgress. Assume that $d^6(\eta^2 A[45,2]M_1\bar{M}_2) = \eta\sigma C[44]M_1$. Then $\eta\sigma C[44] \in \langle \nu,\eta,\eta^2 A[45,2]\rangle = \langle \nu,\eta,2B[47]\rangle \supset \langle \nu,\eta,2\rangle B[47] = 0$, and $\eta\sigma C[44]$ is divisible by ν, a contradiction. There remains only one possibility: $d^6(\nu C[44]M_1 M_2) = \eta\sigma C[44]M_1$. Now $\eta A[52,2]$, $\nu A[50,1]$, $\nu A[50,2]$ and $d^6(2\nu D[45]M_1^3) = A[8]D[45]$ must be nonzero. We have thus proved the following theorem.

THEOREM 7.3.3 $\pi_{53}^S = Z_2 A[8]D[45] \oplus Z_2\nu A[50,1] \oplus Z_2\nu A[50,2] \oplus Z_2\eta A[52,2]$

The computations in Section 6 show that we have the following leaders.

Row	Degree	Leader	Row	Degree	Leader
11	57	$4\beta_1(M_1^7\bar{M}_2^3\bar{M}_3+M_1^{10}\bar{M}_2^2\bar{M}_3+M_1^{14}\bar{M}_2^3)$	50	56	$A[50,1]M_2$
*17	59	$\eta^2\gamma_1M_1^{15}\bar{M}_2^2$	51	55	$2\sigma C[44]M_1^2$
19	55	$\beta_2M_1^{18}$	51	57	$\nu^2D[45]M_1^3$
44	56	$2C[44]M_1^6$	52	56	$A[52,1]M_1^2$
45	57	$4D[45]M_1^3M_2$	*52	58	$A[52,2]\bar{M}_2$
47	55	$\eta^2A[45,2]M_1\bar{M}_2$ or $\nu C[44]M_1M_2$	*52	60	$\eta B[51]M_1\bar{M}_2$
*47	65	$\nu C[44]M_1^3\langle M_2^2\rangle$	53	55	$\eta A[52,2]M_1, A[8]D[45]M_1$
47	57	$A[47]M_1^2\bar{M}_2$	53	59	$\nu A[50,1]M_1^3$
48	56	$\eta B[47]M_1\bar{M}_2$	53	57	$\nu A[50,2]M_1^2$

FIGURE 7.3.3: Leaders from Rows 1 to 53 of Degree at Least 55

There are four leaders of degree 55 and four leaders of degree 56. Clearly $A[52,1]M_1^2$ and $A[50,1]M_2$ transgress. Since $A[52,2] = d^6(B[47]\bar{M}_2)$, $\eta A[52,2]M_1 = d^6(\eta B[47]M_1\bar{M}_2)$. Clearly $d^8(2C[44]\langle M_2^2\rangle) = 2\sigma C[44]M_1^2$. Now $A[54,1] = d^{36}(\beta_2M_1^{18})$, $A[54,2] = d^8(\eta^2A[45,2]M_1\bar{M}_2)$ and $\eta A[8]D[45] = d^2(A[8]D[45]M_1)$ must be nonzero. By Theorem 2.4.5(a),

$$A[54,2] \in \langle\nu,\eta,\eta^2A[45,2],\eta\rangle. \tag{7.15}$$

Note that $d^6(B[47]\bar{M}_2) = A[52,2]$ and $d^8(2B[47]M_1\bar{M}_2) = d^8(\eta^2A[45,2]M_1\bar{M}_2) = A[54,2]$. By Theorem 2.4.2,

$$A[54,2] \in \langle\eta,2,A[52,2]\rangle. \tag{7.16}$$

Therefore, $2A[54,2] \in 2\langle\eta,2,A[52,2]\rangle = \langle 2,\eta,2,\rangle A[52,2] = \eta^2A[52,2] = 0$.

We show that $2A[54,1]$ can only be a multiple of $\eta A[8]D[45]$. Assume that $2A[54,1] = A[54,2] + \lambda\eta A[8]D[45]$. Then $d^{36}(2\beta_2M_1^{16}\bar{M}_2) = A[54,2]M_1$, $\nu A[54,1] = 0$ and $A[54,1]\bar{M}_2$ will transgress to define an element $\xi \in \pi_{59}^S$ such that $2\xi = d^6(A[54,2]\bar{M}_2) = A[59,1]$. By Theorem 2.4.4(c), $\xi \in \langle\eta,\nu,A[54,1]\rangle$ and $2\xi \in 2\langle\eta,\nu,A[54,1]\rangle = \langle 2,\eta,\nu\rangle A[54,1] = 0$, a contradiction. Thus, $2A[54,1] \neq A[54,2]$. Thus, $2A[54,1]$

$= \lambda\eta A[8]D[45]$. We are unable to determine whether or not λ is zero because, as we shall see, the value of λ makes no significant difference in the computations through degree 64 in our spectral sequence.

Now $\sigma A[47] \in \sigma\langle\eta, 2C[42], \nu\rangle \supset \sigma\langle\eta, 2, \nu C[42]\rangle = \sigma\langle\eta, 2, 8D[45]\rangle = \sigma\langle\eta, 2, 2\sigma B[38]\rangle$ $\supset \sigma^2\langle\eta, 2, 2B[38]\rangle = \langle\sigma^2, \eta, 2\rangle 2B[38] = 0$ since $\eta\langle\sigma^2, \eta, 2\rangle = 2\langle\eta, \sigma^2, \eta\rangle = 2\nu\sigma^2 = 0$ while $\eta A[16] \neq 0$. Thus, $\sigma A[47] \in \text{Indet}\ \sigma\langle\eta, 2C[42], \nu\rangle = \eta\sigma\cdot\pi^S_{46} = \eta^2\sigma\cdot\pi^S_{45} = 0$ and

$$\sigma A[47] = 0. \qquad [7.17]$$

Since $B[47] = d^8(2C[20]^2M_1\bar{M}_2)$, Theorem 2.4.5(a) implies that

$$B[47] \in \langle\nu, \eta, 2C[20]^2, \eta\rangle. \qquad [7.18]$$

Then $\sigma B[47] \in \sigma\langle\nu, \eta, 2C[20]^2, \eta\rangle \supset \sigma\langle\nu, \eta, 2, \eta C[20]^2\rangle \subset \langle\langle\sigma, \nu, \eta\rangle, 2, \eta C[20]^2\rangle$ $= \langle 0, 2, \eta C[20]^2\rangle = \eta C[20]^2\cdot\pi^S_{13} = 0$ and $\sigma B[47] \in \text{Indet}\ \sigma\langle\nu, \eta, 2C[20]^2, \eta\rangle$ $= \sigma\langle\nu, \eta, \pi^S_{42}\rangle + \sigma\langle\nu, X, \eta\rangle$ with $X = 2hC[42] + k\eta^2C[20]^2$ by Theorem 2.3.1(b); $= \langle\sigma, \nu, \eta\rangle\cdot\pi^S_{42} + h\sigma\langle\nu, 2C[42], \eta\rangle + k\sigma\langle\nu, \eta^2C[20]^2, \eta\rangle = h\sigma A[47] + k\sigma\langle\nu, \eta, \eta^2C[20]^2\rangle$ $= k\langle\sigma, \nu, \eta\rangle\eta^2C[20]^2 = 0$. Thus,

$$\sigma B[47] = 0. \qquad [7.19]$$

We have thus proved the following theorem.

THEOREM 7.3.4 π^S_{54} has a composition series

$$Z_2A[54,1] \oplus Z_2A[54,2],\ Z_2\eta A[8]D[45]$$

where $2A[54,1] = \lambda\eta A[8]D[45]$ and $2A[54,2] = 0$.

We now have the following table of leaders.

Row	Degree	Leader	Row	Degree	Leader
11	57	$4\beta_1(M_1^7\bar{M}_2^3\bar{M}_3 + M_1^{10}\bar{M}_2^2\bar{M}_3 + M_1^{14}\bar{M}_2^3)$	52	56	$A[52,1]M_1^2$
19	57	$2\beta_2M_1^{16}\bar{M}_2$	*53	61	$\eta A[52,2]M_1\bar{M}_2$
*44	66	$2C[44]M_1^2M_2^3$	53	57	$\nu A[50,2]M_1^2$

45	57	$4D[45]M_1^3M_2$	*53	63	$A[8]D[45]M_1^2\overline{M}_2$
47	57	$A[47]M_1^2\overline{M}_2$	*54	60	$A[54,1]M_1^3$
51	57	$\nu^2D[45]M_1^3$	54	56	$A[54,2]M_1, \eta A[8]D[45]M_2$

FIGURE 7.3.4: Leaders from Rows 1 to 54 of Degree at Least 56

There are four leaders of degree 56 and six leaders of degree 57. Since $\nu^3 = \eta A[8]$, $d^4(\nu^2D[45]M_1^3) = \eta A[8]D[45]M_1$. Assume that $A[52,1]M_1^2$ is not a boundary. Then $\nu A[52,1]$ is nonzero and not divisible by η. Since $A[52,1] = d^6(\eta^2D[45]M_2)$, $A[52,1] \in \langle\nu,\eta,\eta^2D[45]\rangle \subset \langle\nu,\eta^3,D[45]\rangle = \langle\nu,4\nu,D[45]\rangle$. Then $\nu A[52,1] \in \nu\langle\nu,4\nu,D[45]\rangle \subset \langle\nu^2,4\nu,D[45]\rangle = \langle\langle\eta,\nu,\eta\rangle,4\nu,D[45]\rangle \supset \eta\langle\nu,\eta,4\nu,D[45]\rangle$. Note that this Toda bracket is defined by Theorem 2.2.7(a) because $\langle\nu,\eta,4\nu\rangle = 4A[8] + \nu\cdot\pi_5^S = 0$ and $\langle\eta,4\nu,D[45]\rangle \supset \langle\eta,\nu,4D[45]\rangle = \langle\eta,\nu,\sigma B[38]\rangle \supset \langle\eta,\nu,\sigma\rangle B[38] = 0$. Thus, $\nu A[52,1] \in \mathrm{Indet}\ \langle\nu^2,4\nu,D[45]\rangle = \nu^2\cdot\pi_{49}^S + D[45]\cdot\pi_{10}^S = \eta\alpha_1D[45]$, a contradiction. Thus, $A[52,1]M_1^2$ must be a boundary. Now $A[52,1]M_1^2$ can only bound from below the 47 row, and we shall see that $A[52,1]M_1M_2$ must also bound. There is only one possibility: $d^8(4D[45]M_1^3M_2) = A[52,1]M_1^2$ and $A[52,1]M_1M_2$ bounds from below the 45 row.

Assume that $\xi = d^6(A[50,1]M_2)$ is nonzero. Then ξM_1 can not bound because we will show that the only leader of degree 58 below the 50 row must hit $A[47]M_1^2\overline{M}_2$. Thus, $\eta\xi \neq 0$. By Theorem 2.4.4(b), $\xi \in \langle\nu,\eta,A[50,1]\rangle$ and $\eta\xi \in \eta\langle\nu,\eta,A[50,1]\rangle = \langle\eta,\nu,\eta\rangle A[50,1] = \nu^2A[50,1]$. Thus, $\eta\xi M_1 = d^4(\nu A[50,1]M_1^3)$. However, we shall see in the derivation of π_{58}^S that $\nu A[50,1]M_1^3$ must be a boundary. Thus, $\xi = 0$ and $A[50,1]M_2$ must bound from below the 17 row. There is only one possibility: $A[50,1]M_2 = d^{40}(4\beta_1M_1^7\overline{M}_2^3\overline{M}_3)$.

Clearly $2\beta_2M_1^{19} = 2M_1\cdot 2\beta_2M_1^{18}$ survives to E^{36} and hits $2A[54,1]M_1 = 0$, i.e. $2\beta_2M_1^{19}$ transgresses. Since $A[54,2]M_1$ can only bound from below the 47 row, $\eta A[54,2]$ is nonzero. We have thus proved the following theorem.

THEOREM 7.3.5 $\pi_{55}^S = Z_2\eta A[54,2] \oplus Z_{16}\gamma_6$

The computations of Section 6 show that we have the following leaders. This table contains the leaders of all degrees.

Row	Degree	Leader
9	63	$\eta^2\sigma M_1^{21}M_2^2$
11	65	$\beta_1(5M_1^{20}\bar{M}_3+4M_1^6\bar{M}_2^7)$
17	59	$\eta^2\gamma_1 M_1^{15}\bar{M}_2^2$
18	64	$4C[18]M_1^7\bar{M}_2M_2^2\bar{M}_3$
19	57	$2\beta_2M_1^{16}\bar{M}_2$
22	62	$\nu A[19]M_1^7M_2^2\langle M_3\rangle$
23	63	$\gamma_2M_1^{20}$
24	60	$\eta A[23]M_1^{15}\bar{M}_2$
30	60	$A[30]\langle M_4\rangle$
32	62	$A[32,1]M_1^8\bar{M}_3$
34	64	$2B[34]M_1^5\bar{M}_2\bar{M}_3$
36	60	$A[36]M_1^6\bar{M}_2^2$
38	62	$B[38]M_1^2\bar{M}_2\bar{M}_3$
39	61	$A[39,1]M_1M_2M_3$
39	66	$\eta A[39,3]M_1^3\bar{M}_2\langle M_3\rangle$
39	59	$\sigma A[32,1]M_1^4M_3^2$
40	64	$\eta\sigma A[32,1]M_1^5\bar{M}_3$
40	60	$A[40,1]\bar{M}_2\bar{M}_3$
40	58	$A[40,2]M_1^6\bar{M}_2$
44	66	$2C[44]M_1^2M_2^3$
44	58	$4C[44]M_1^7$
45	59	$2D[45]M_1\langle M_2^2\rangle$
45	63	$4D[45M_1^6M_2$
46	66	$\eta^2C[44]M_1^7\bar{M}_2$
47	65	$\nu C[44]M_1^3\langle M_2^2\rangle$
47	57	$A[47]M_1^2\bar{M}_2$
51	63	$\eta A[50,2]M_1^3\bar{M}_2$
51	59	$\sigma C[44]M_1^4$
52	58	$A[52,1]M_1^3$, $A[52,2]\bar{M}_2$
52	60	$\eta\sigma C[44]M_1\bar{M}_2$
53	61	$\eta A[52,2]M_1\bar{M}_2$
53	59	$\nu A[50,1]M_1^3$
53	57	$\nu A[50,2]M_1^2$
53	63	$A[8]D[45]M_1^2\bar{M}_2$
54	60	$A[54,1]M_1^3$, $A[54,2]\bar{M}_2$
54	66	$\eta A[8]D[45]M_1^3\bar{M}_2$
55	57	$\eta A[54,2]M_1$

FIGURE 7.3.5: Leaders from Rows 1 to 55 of Degree at Least 57

4. Computation of π_N^S, $56 \le N \le 60$.

We continue the computations of Section 3. In the tables of leaders below, all leaders of degree greater than 62 will have an asterisk at the left. They will be omitted from all tables of leaders in this section except for the last one.

From Figure 7.3.5, we see that there are four leaders of degree 57 and four leaders of degree 58. Note that $\nu A[52,2] \in \nu\langle\eta,\nu,B[47]\rangle = \langle\nu,\eta,\nu\rangle B[47]$ $= A[8]B[47] = \langle\eta,2\nu,\nu\rangle B[47] = \eta\langle 2\nu,\nu,B[47]\rangle = \eta\langle\langle\eta,2,\eta\rangle,\nu,B[47]\rangle$ $= \eta\langle\eta,\langle 2,\eta,\nu\rangle,B[47]\rangle + \eta\langle\eta,2,\langle\eta,\nu,B[47]\rangle\rangle = \eta\langle\eta,0,B[47]\rangle + \eta\langle\eta,2,A[52,2]\rangle$ $= \eta A[54,2]$ since $\eta^2\cdot\pi_{53}^S = 0$ and $\eta\sigma B[47] = 0$. We will show in the derivation of π_{61}^S that $\nu A[52,1]$ equals 0 and does not equal $\nu A[52,2]$. Therefore,

$$\nu A[52,2] = \eta A[54,2] \text{ and } \nu A[52,1] = 0. \qquad [7.20]$$

Thus, $d^4(A[52,2]\bar{M}_2) = \eta A[54,2]M_1$. Observe that $4C[44]M_1^7$ survives to E^{10} since $4\sigma C[44] = 0$. Assume that $d^{10}(4C[44]M_1^7) = \nu A[50,2]M_1^2$. Then $\nu A[50,2] \in \langle\eta,4C[44],\sigma\rangle = \langle\eta,\sigma^2A[30],\sigma\rangle$ and $\nu A[50,2] \in \langle\eta,\sigma^2,\sigma A[30]\rangle$ because $\sigma\cdot\pi_{46}^S = \eta\sigma\cdot\pi_{45}^S$ which can not contain $\nu A[50,2]$. Thus, $\nu A[50,2] \in \langle\eta,\sigma^3,A[30]\rangle$ $= \langle\eta,\nu C[18],A[30]\rangle$ and $\nu A[50,2] \in \langle\eta,\nu,C[18]A[30]\rangle$. Then $\nu A[50,2]M_1^2 =$ $d^6(C[18]A[30]M_1^2M_2)$, a contradiction. Therefore, $4C[44]M_1^7$ transgresses. (Note that $A[30]\cdot\pi_{23}^S = \{\sigma A[16]A[30], \nu C[20]A[30], A[30]\gamma_2\}$. Now $A[16]A[30]$ is divisible by η. Also, $A[30]\gamma_2 \in A[30]\langle 2\gamma_1,16,\sigma\rangle = \langle A[30],2\gamma_1,16\rangle\sigma$ which is divisible by η. In addition, $A[30]A[14] \in A[30]\langle 2,A[8],\nu,\eta\rangle$ $\subset \langle\langle A[30],2,A[8]\rangle,\nu,\eta\rangle$ which projects to zero in E^8. However, π_{44}^S projects monomorphicly into E^8. Thus, $A[30]A[14] = 0$. Now $C[20]A[30] \in$ $A[30]\langle A[14],2,\eta,\nu\rangle \subset \langle\langle A[30],A[14],2\rangle,\eta,\nu\rangle$. Thus, $C[20]A[30]$ is zero in E^8. Since π_{50}^S projects monomorphically into E^8, $C[20]A[30] = 0$.) We will use the following lemma to continue our analysis of the leaders of degree 57.

LEMMA 7.4.1 (a) $\sigma A[40,2] = A[47]$

(b) $\nu^2 A[50,1] = 0$

(c) $\nu^2 A[50,2]$ is not divisible by η.

PROOOF. (a) By 6.20, $2C[42] \in \langle\eta,2,A[40,2]\rangle$. Thus, $A[47] \in \langle\eta,\nu,2C[42]\rangle$ $\subset \langle\eta,\nu,\langle\eta,2,A[40,2]\rangle\rangle = \langle\langle\eta,\nu,\eta\rangle,2,A[40,2]\rangle + \langle\eta,\langle\nu,\eta,2\rangle,A[40,2]\rangle$ $= \langle\nu^2,2,A[40,2]\rangle + \langle\eta,0,A[40,2]\rangle$ and $A[47] \in \langle\nu^2,2,A[40,2]\rangle \subset \langle\nu,2\nu,A[40,2]\rangle$ $\supset \langle 2\nu,\nu,A[40,2]\rangle = \langle\langle\eta,2,\eta\rangle,\nu,A[40,2]\rangle \supset \eta\langle 2,\eta,\nu,A[40,2]\rangle$. Note that this four-fold Toda bracket is defined by Theorem 2.2.7(a) because $\langle 2,\eta,\nu\rangle = 0$ and $0 = d^6(A[40,2]\bar{M}_2) \in \langle\eta,\nu,A[40,2]\rangle$. Therefore, $A[47] \in$ Indet $\langle\nu^2,2,A[40,2]\rangle$ $= \nu^2\cdot\pi^S_{41} + A[40,2]\cdot\pi^S_7 = \{\sigma A[40,2]\}$ because $\nu\cdot\pi^S_{41} = 0$. Thus, $\sigma A[40,2] = A[47]$.

(b) The hidden differential from the 17 row which hits $\eta A[47]M_1\bar{M}_2$ instead of $A[50,1]\bar{M}_2$ shows that these elements have homologous representatives. Therefore, representatives of $0 = \nu\eta A[47]M_1\bar{M}_2$ and $\nu A[50,1]\bar{M}_2$ are also homologous. Hence $\nu A[50,1]\bar{M}_2$ must bound. In addition, $d^4(A[50,1]M_1M_2) = \nu A[50,1]M_1^2$. Thus, $\nu^2 A[50,1] = 0$.

(c) By (a), $d^8(A[40,2]M_1^6\bar{M}_2) = A[47]M_1^2\bar{M}_2$. There remains no possibility for $\nu A[50,2]M_1^2$ to be a boundary. Thus, $\nu^2 A[50,2]$ is nonzero and is not divisible by η. ∎

By Lemma 7.4.1(a), $A[47]M_1^2\bar{M}_2 = d^8(A[40,2]M_1^6\bar{M}_2)$. Thus, $\nu A[50,2]M_1^2$, $2\beta_2 M_1^{19}$ can not bound and $\nu^2 A[50,2]$, $A[56] = d^{38}(2\beta_2 M_1^{19})$ are nonzero. Since $2A[54,1] = \lambda\eta A[8]D[45]$ and $d^{36}(\beta_2 M_1^{18}) = A[54,1]$, Theorem 2.4.2 implies that

$$A[56] \in \langle\eta,(2,\eta A[8]),(A[54,1],\lambda D[45])^T\rangle. \qquad [7.21]$$

Then $2A[56] \in 2\langle\eta,(2,\eta A[8]),(A[54,1],\lambda D[45])^T\rangle$ $= \langle 2,\eta,(2,\eta A[8])\rangle(A[54,1],\lambda D[45])^T = \langle 2,\eta,2\rangle A[54,1] + \lambda\langle 2,\eta,\eta A[8]\rangle D[45]$ $= \eta^2 A[54,1] + 2\lambda'\beta_1 D[45]$ since $\langle 2,\eta,\eta A[8]\rangle = \langle 2,\eta,\nu^3\rangle \supset \langle 2,\eta,\nu\rangle\nu^2 = 0$. Now $\eta^2 A[54,1] = 0$. Moreover, $\lambda' = 0$ because $\eta A[56] \neq 0$ so that $A[56]$ $\neq \lambda'\beta_1 D[45]$. Thus, $2A[56] = 0$. We have now proved the following theorem.

THEOREM 7.4.2 $\pi^S_{56} = Z_2A[56] \oplus Z_2\nu^2A[50,2] \oplus Z_2\eta\gamma_6$

The computations in Section 6 show that we have the following leaders.

Row	Degree	Leader	Row	Degree	Leader
17	59	$\eta^2\gamma_1M_1^{15}\overline{M}_2^2$	44	58	$4C[44]M_1^7$
19	61	$\beta_2M_1^{14}\overline{M}_3$	45	59	$2D[45]M_1\langle M_2^2\rangle$
22	62	$\nu A[19]M_1^7M_2^2\langle M_3\rangle$	51	59	$\sigma C[44]M_1^4$
24	60	$\eta A[23]M_1^{15}\overline{M}_2$	52	58	$A[52,1]M_1^3$
30	60	$A[30]\langle M_4\rangle$	52	60	$\eta\sigma C[44]M_1\overline{M}_2$
32	62	$A[32,1]M_1^8$	53	61	$\eta A[52,2]M_1\overline{M}_2$
36	60	$A[36]M_1^6\overline{M}_2^2$	53	59	$\nu A[50,1]M_1^3$
38	62	$B[38]M_1^2\overline{M}_2\overline{M}_3$	54	60	$A[54,1]M_1^3$, $A[54,2]\overline{M}_2$
39	61	$A[39,1]M_1M_2M_3$	*55	63	$\eta A[54,2]M_1\overline{M}_2$
39	59	$\sigma A[32,1]M_1^4M_2^2$	56	58	$A[56]M_1$
40	60	$A[40,1]\overline{M}_2\overline{M}_3$	56	62	$\nu^2A[50,2]M_1^3$

FIGURE 7.4.1: Leaders from Rows 1 to 56 of Degree at Least 58

There are three leaders of degree 58 and five leaders of degree 59. Clearly $\sigma C[44]M_1^4$ transgresses. Since $D[45]\langle M_1^4\rangle$ bounds from the 18 row, $2D[45]$ bounds from the 34 row and $2M_2$ is an infinite cycle, $2D[45]M_2\langle M_1^4\rangle$ must bound from the r-row for some $18 \le r < 34$. There are two such leader of degree 60: $A[30]\langle M_4\rangle$ and $\eta A[23]M_1^{15}\overline{M}_2$. In the derivation of π^S_{58}, we will show that $d^{16}(A[30]\langle M_4\rangle) = \sigma A[32,1]M_1^4M_2^2$. Thus, we must have $d^{22}(\eta A[23]M_1^{15}\overline{M}_2) = 2D[45]M_2\langle M_1^4\rangle$. By Lemma 7.4.1(b), $\nu A[50,1]M_1^3$ transgresses. Since $\sigma^2A[32,1] = 0$, $\sigma A[32,1]\langle M_1^4\rangle\langle M_2^2\rangle$ survives to E^{16}. Since $A[56]M_1$ can only bound from below the 19 row, $\sigma A[32,1]\langle M_1^4\rangle\langle M_2^2\rangle$ transgresses.

We show that $A[52,1]M_1^3$ is a boundary. Assume that $\xi = d^6(A[52,1]M_1^3) \neq 0$. There is no possibility for ξM_1 to bound from below the 52 row, and thus

$\eta\xi \neq 0$. (We shall see that $A[40,1]\bar{M}_2\bar{M}_3$, $A[36]M_1^6\bar{M}_2^2$, $A[30]<M_4>$ and $\eta A[23]M_1^{15}\bar{M}_2$ are needed to bound other elements.) By Theorem 2.4.4(a), $\xi \in <\eta, A[52,1], \nu>$ and $\eta\xi \in \eta<\eta, A[52,1], \nu> \subset <\eta^2, A[52,1], \nu> = <<2,\eta,2>, A[52,1], \nu>$ $\supset 2<\eta, 2, A[52,1], \nu> \subset 2\cdot\pi_{58}^S$ which, as we shall see, must be zero. Thus, $\eta\xi \in \text{Indet } <\eta^2, A[52,1], \nu>$ and $\eta\xi \in \nu\cdot\pi_{55}^S = 0$. Therefore, $\xi = 0$ and $A[52,1]M_1^3$ must be a boundary. The only possibility is $d^{36}(\eta^2\gamma_1 M_1^{15}\bar{M}_2^2) = A[52,1]M_1^3$. (Use Theorem 2.2.7(e) to show that the four-fold Toda bracket is defined. By Theorem 2.4.2, $<\eta, 2, A[52,1]>$ contains $d^8(2\eta^2 D[45]M_1M_2) = 0$. Therefore, $<\eta, 2, A[52,1]>$ is zero in E^8, and the only possibility is $<\eta, 2, A[52,1]>$ $= \eta A[8]D[45] \in \text{Indet } <\eta, 2, A[52,1]>$. Thus, $<\eta, 2, A[52,1]>$ contains 0. Now $\eta<2, A[52,1], \nu> = \nu<\eta, 2, A[52,1]> = <\nu, \eta, 2>A[52,1] = 0$. Observe that if $\eta A[56] = 0$ then $A[56]M_1 = d^{40}(\eta^2\gamma_1 M_1^{15}\bar{M}_2^2)$, $\xi \neq 0$ and $\eta\xi \neq 0$ as we remarked above. Since $2\xi \in 2<\eta, \nu, A[52,1]> = <2, \eta, \nu>A[52,1] = 0$, $\eta^2\xi$ is divisible by two which we shall see is impossible. Thus, $\eta^2\xi = 0$ and $\eta\xi M_1$ must bound. The only possibility is $d^6(\eta A[52,2]M_1\bar{M}_2) = \eta\xi M_1$, and $\pi_{60}^S = Z_2\eta A[59,2]$. However, we shall see that $\eta A[59,2]$ is divisible by two, a contradiciton. Therefore, $\eta A[56] \neq 0$ and $<2, A[52,1], \nu> \subset Z_2\nu^2 A[50,2] \subset \text{Indet } <2, A[52,1], \nu>$. Thus, $0 \in <2, A[52,1], \nu>$. Finally, π_{53}^S is generated by $A[8]D[45]$, $\eta A[52,2]$, $\nu A[50,1]$ and $\nu A[50,2]$. Note that ν times $A[8]D[45]$ and $\eta A[52,2]$ are zero while η times $\nu A[50,1]$ and $\nu A[50,2]$ are zero.)

Now $A[56]M_1$ and $4C[44]M_1^7$ can not be boundaries. Thus, $\eta A[56]$ and $A[57] = d^{14}(4C[44]M_1^7)$ are nonzero. Since $<\sigma, 4C[44], \nu> \in \text{CokJ}_{55} = \nu\cdot\pi_{52}^S$, $<\sigma, 4C[44], \nu>$ contains 0. Also $0 \in <4C[44], \nu, \eta>$ because $\text{CokJ}_{49} = 0$. Since $\sigma\pi_{48}^S = 0$, $<\sigma, 4C[44], \nu, \eta>$ is defined by Theorem 2.2.7(c). By Theorem 2.4.6(c),

$$A[57] \in <\sigma, 4C[44], \nu, \eta>. \qquad [7.22]$$

Then $A[57] \in <\sigma, \sigma^2 A[30], \nu, \eta> \supset <\sigma^2 A[30], \sigma, \nu, \eta> = <4C[44], \sigma, \nu, \eta>$ and

$2A[57] \in 2<\eta, \nu, \sigma, 4C[44]> + 2\cdot\text{Indet } <\sigma, \sigma^2 A[30], \nu, \eta>$;

$\subset <<2, \eta, \nu>, \sigma, 4C[44]> + 2[<\sigma, X, \eta> + <Y, \nu, \eta>]$ by Theorem 2.3.1(b);

$= \langle 0,\sigma,4C[44]\rangle + \sigma\langle X,\eta,2\rangle + Y\langle \nu,\eta,2\rangle = \sigma\langle X,\eta,2\rangle$. If $\nu X = 0$ then $\nu\langle X,\eta,2\rangle = \langle \nu,X,\eta\rangle 2 = 0$ and $\langle X,\eta,2\rangle = 0$ since multiplication by ν is a monomorphism on CokJ_{50}. Thus, $X = h\nu D[45]+k\nu A[45,1]$ and $2A[57] = \sigma\langle h\nu D[45]+k\nu A[45,1],\eta,2\rangle = \sigma(hD[45]+kA[45,1])\langle \nu,\eta,2\rangle = 0$. Since A[57] bounds from the 44 row which is below the 50 row, A[57] can not be divisible by σ. Thus, $\sigma A[50,2] = k\eta A[56]$ and

$$\eta\sigma A[50,2] = 0. \qquad [7.23]$$

We have thus proved the following theorem.

THEOREM 7.4.4 $\pi^S_{57} = Z_2\eta A[56] \oplus Z_2 A[57] \oplus Z_2\alpha_7 \oplus Z_2\eta^2\gamma_6$

We now have the following table of leaders.

Row	Degree	Leader	Row	Degree	Leader
19	61	$\beta_2 M_1^{14}\bar{M}_3$	40	60	$A[40,1]\bar{M}_2\bar{M}_3$
22	62	$\nu A[19]M_1^7 M_2^2\langle M_3\rangle$	45	59	$2D[45]M_1\langle M_2^2\rangle$
24	60	$\eta A[23]M_1^{15}\bar{M}_2$	51	59	$\sigma C[44]M_1^4$
30	60	$A[30]\langle M_4\rangle$	52	60	$A[52,1]M_1M_2$, $\eta\sigma C[44]M_1\bar{M}_2$
32	62	$A[32,1]M_1^8\bar{M}_3$	53	61	$\eta A[52,2]M_1\bar{M}_2$
36	60	$A[36]M_1^6\bar{M}_2^2$	53	59	$\nu A[50,1]M_1^3$
38	62	$B[38]M_1^2\bar{M}_2\bar{M}_3$	54	60	$A[54,1]M_1^3$, $A[54,2]\bar{M}_2$
39	61	$A[39,1]M_1M_2M_3$	56	62	$\nu^2 A[50,2]M_1^3$
39	59	$\sigma A[32,1]M_1^4M_2^2$	57	59	$A[57]M_1$, $\eta A[56]M_1$

FIGURE 7.4.2: Leaders from Rows 1 to 57 of Degree at Least 59

There are six leaders of degree 59 and eight leaders of degree 60. We digress to show that $A[32,1] \in \langle \eta,2,A[30]\rangle$. Note that $d^8(4M_1^2\langle M_1^4\rangle^2\bar{M}_3) \equiv 4\sigma M_1^{13}$ modulo (8σ). In addition, if $4M_1^2\langle M_1^4\rangle^2\bar{M}_3$ survived to E^{10} it would hit $\eta^2\sigma M_1^5\bar{M}_3$. Thus, A[32,1] is represented by the boundary of a representative of

$\eta^2\sigma M_1^5\bar{M}_3$ which is homologous to the boundary of a representative of $4\sigma M_1^{13}$. By Theorem 2.4.2, the boundary of a representative of $4\sigma M_1^{13} = M_1\cdot 2\cdot 2\sigma M_1^{12}$ is an element of $<\eta,2,A[30]>$ and

$$A[32,1] \in <\eta,2,A[30]>. \qquad [7.24]$$

Recall that $d^8<M_4> = 2\sigma M_1\bar{M}_2<M_3>$, and μ_1, $\bar{\mu}_{01}$, $<\mu_{001}>$ is a representative of M_1, $\bar{M}_2$, $<M_3>$, respectively. Then $A[30]<M_4>$ is represented by $A[30]<M_4> \cup (\sigma \wedge B_{A[30]\cdot 2} \wedge \mu_1 \wedge \bar{\mu}_{01} \wedge <\mu_{001}>)$ which has boundary $(\sigma A[30] \wedge B_{2\eta} \wedge \bar{\mu}_{01} \wedge <\mu_{001}>) \cup (\sigma \wedge B_{A[30]\cdot 2} \wedge \eta \wedge \bar{\mu}_{01} \wedge <\mu_{001}>)$ $= \sigma<A[30],2,\eta> \wedge \bar{\mu}_{01} \wedge <\mu_{001}>$ modulo elements of filtration degree 18. Thus, $d^{10}(A[30]<M_4>) = \sigma A[32,1]\bar{M}_2<M_3> = \sigma A[32,1]M_1^4M_2^2 + d^8(A[32,1]M_1^4\bar{M}_2\bar{M}_3)$ $= \sigma A[32,1]M_1^4M_2^2$ in E^{10}.

Since $\nu A[54,2] \in \nu<\eta,2,A[52,2]> = <\nu,\eta,2>A[52,2] = 0$, $A[54,2]\bar{M}_2$ must transgress. If $d^6(\eta\sigma C[44]M_1\bar{M}_2) = \eta A[56]M_1$ then $d^6(\sigma C[44]\bar{M}_2) = A[56]$, a contradiction. Since $A[57]M_1$ can only bound from below the 44 row, $\eta\sigma C[44]M_1\bar{M}_2$ must transgress. In the derivation of π_{59}^S, we show that $A[52,1]M_1M_2$ is a boundary. Since there is no possibility for an element of order four in π_{58}^S, $\eta A[56]M_1$ must be a boundary. Since $\sigma A[40,1] = \nu C[44]$, a representative $\mathcal{R}$ of $A[40,1]\bar{M}_2<M_3>$ has boundary $(C[44] \wedge \nu \wedge <\mu_{02}>) \cup (C[44] \wedge B_{\nu\sigma} \wedge \mu_2)$. Then $\mathcal{R} \cup (C[44] \wedge \mu_2 \wedge <\mu_{02}>)$ has boundary $C[44] \wedge \mu_2 \wedge \sigma \wedge \mu_2$ union elements of filtration degree 4. Thus, $d^{18}(A[40,1]\bar{M}_2<M_3>) = \sigma C[44]M_1^4$. Recall that in the derivation of π_{57}^S we showed that $d^{22}(\eta A[23]M_1^{15}\bar{M}_2) = 2D[45]<M_1^4>M_2$. In the proof of Lemma 7.4.1(b) we showed that $\nu A[50,1]\bar{M}_2$ must be a boundary. Since $\nu A[50,1]\bar{M}_2 = \nu A[50,1]M_1^3$ in E^4, $\nu A[50,1]M_1^3$ must bound from below the 50 row. The only possibility is $d^{18}(A[36]M_1^6\bar{M}_2^2) = \nu A[50,1]M_1^3$. There remains only one way for $\eta A[56]M_1$ to bound: $d^4(A[54,1]M_1^3) = \eta A[56]M_1$. Thus,

$$\nu A[54,1] = \eta A[56] \text{ and } \nu A[54,2] = 0. \qquad [7.25]$$

Now $A[57]M_1$ can not be a boundary. We have thus proved the following theorem.

THEOREM 7.4.5 $\pi_{58}^S = Z_2\eta A[57] \oplus Z_2\eta\alpha_7$

The computations in Section 6 show that we have the following leaders.

Row	Degree	Leader	Row	Degree	Leader
19	61	$\beta_2 M_1^{14}\bar{M}_3$	51	61	$\sigma C[44]M_1^2\bar{M}_2$
22	62	$\nu A[19]M_1^7 M_2^2 <M_3>$	52	60	$A[52,1]M_1M_2$, $\eta\sigma C[44]M_1\bar{M}_2$
32	62	$A[32,1]M_1^8\bar{M}_3$	53	61	$\eta A[52,2]M_1\bar{M}_2$
38	62	$B[38]M_1^2\bar{M}_2\bar{M}_3$	*53	63	$\nu A[50,1]M_1^2 M_2$
*39	63	$\sigma A[32,1]M_1^6 M_2^2$	54	60	$A[54,2]\bar{M}_2$
39	61	$A[39,1]M_1M_2M_3$	56	62	$\nu^2 A[50,2]M_1^3$
45	61	$2D[45]M_1<M_3>$	58	60	$\eta A[57]M_1$

FIGURE 7.4.3: Leaders from Rows 1 to 58 of Degree at Least 60

There are four leaders of degree 60 and five leaders of degree 61. Since $\eta A[56] = \nu A[54,1]$, $d^4(A[54,1]M_1^3) = \eta A[56]M_1$. Since $D[45]<M_3>$ is a d^{28}-boundary and $2M_1$ is an infinite cycle, $2M_1D[45]<M_3>$ must be a boundary. Let $A[59,2] = d^6(A[54,2]\bar{M}_2)$. By Theorem 2.4.4(c),

$$A[59,2] \in <\eta,\nu,A[54,2]>. \qquad [7.26]$$

Thus, $2A[59,2] \in 2<\eta,\nu,A[54,2]> = <2,\eta,\nu>A[54,2] = 0$. Let $A[59,1] = d^6(\eta\sigma C[44]M_1\bar{M}_2)$. By Theorem 2.4.5(a),

$$A[59,1] \in <\eta,\eta\sigma C[44],\eta,\nu>. \qquad [7.27]$$

Thus, $2A[59,1] \in 2<\eta,\eta\sigma C[44],\eta,\nu> \subset <<2,\eta,\eta\sigma C[44]>,\eta,\nu> = <<\eta\sigma C[44],\eta,2>,\eta,\nu>$ $= <\eta\sigma C[44],<\eta,2,\eta>,\nu> + <\eta\sigma C[44],\eta,<2,\eta,\nu>> = <\eta\sigma C[44],2\nu,\nu> + <\eta\sigma C[44],\eta,0>$ $= A[8](\sigma C[44]) + \nu\cdot\pi_{56}^S = \{\nu^3 A[50,2],\nu A[56]\} = \{\eta A[8]A[50,2]\}$ because $\nu A[56] \in \nu<\eta,(2,\eta A[8]),(A[54,1],\lambda D[45])^T> = <\nu,\eta,(2,\eta A[8])>(A[54,1],\lambda D[45])^T$ $= <\nu,\eta,2>A[54,1] + \lambda<\nu,\eta,\eta A[8]>D[45] = \lambda<\nu,\eta,\nu^3>D[45] = \lambda\nu^2 A[8]D[45] = 0$. Observe that $<\eta^2,\nu,\eta,2>$ is defined by Theorem 2.2.7(a) since $<\eta^2,\nu,\eta>$ $= <\nu,\eta,2> = 0$. Then $\eta<\eta^2,\nu,\eta,2> = <\eta^2,\nu,<\eta,2,\eta>> = <\eta^2,\nu,2\nu> = \eta A[8]$. Since multiplication by η is a monomorphism on π_8^S,

$$A[8] \in \langle\eta^2,\nu,\eta,2\rangle. \tag{7.28}$$

Thus, $A[50,2]A[8] = A[50,2]\langle\eta^2,\nu,\eta,2\rangle \subset \langle\langle A[50,2],\eta^2,\nu\rangle,\eta,2\rangle$. Now $\langle A[50,2],\eta^2,\nu\rangle \supset \langle\eta A[50,2],\eta,\nu\rangle$ which contains $d^6(\eta A[50,2]M_2) = 0$. Therefore, $A[50,2]A[8] = k\langle\nu^2 A[50,2],\eta,2\rangle = k\nu A[50,2]\langle\nu,\eta,2\rangle = 0$ since $2\cdot\pi_{58}^S = 0$. Thus, $2A[59,1] = 0$ and

$$\nu^3 A[50,2] = \nu A[56] = 0. \tag{7.29}$$

We show that $A[52,1]M_1M_2$ must be a boundary. Assume that $\xi = d^8(A[52,1]M_1M_2)$ is not zero. Let $\mathfrak{C}$ denote the mapping cone of $2\nu D[45]$ with $\rho: S \longrightarrow \mathfrak{C}$ the canonical map. In the Atiyah-Hirzebruch spectral sequence for $\mathfrak{C}_*BP$, $\rho(2D[45])M_1^5M_2$ clearly survives to E^6. The only possibility for $d^6(\rho(2D[45])M_1^5M_2)$ is $\partial^{-1}(\eta)M_1^5$. In that case, $d^6(\rho(2D[45])M_1M_2) = \partial^{-1}(\eta)M_1$ and $0 = \eta\partial^{-1}(\eta) = \partial^{-1}(\eta^2)$, a contradiction. Thus, $\rho(2D[45])M_1^5M_2$ survives to E^8 and $d^8(\rho(2D[45])M_1^5M_2) = \rho(A[52,1])M_1M_2$. Then $\rho(\xi)$ must be zero in E^8. If $\rho(\xi) = d^2(\partial^{-1}(h\eta^2\sigma+k\eta A[8])M_1)$ then in π_*^S, $\xi \in \langle\eta,h\eta^2\sigma+k\eta A[8],2\nu D[45]\rangle \supset \langle\eta,h\eta^2\sigma+k\eta A[8],2\rangle\nu D[45] = 0$, a contradiction. If $\rho(\xi) = d^4(\partial^{-1}(\sigma)M_1^2)$ then $\xi \in \langle\nu,\sigma,2\nu D[45]\rangle \supset \langle\nu,\sigma,\nu\rangle 2D[45] = \sigma^2(2D[45]) = 0$, a contradiction. Note that $\rho(\xi)$ can not be in Image d^6 because $\pi_5^S = 0$. Thus, $\xi \in 2\nu D[45]\cdot\pi_{11}^S = 0$, and $A[52,1]M_1M_2$ must bound from below the 47 row. Note that if $d^{34}(\beta_2M_1^{14}\bar{M}_3) = A[52,1]M_1M_2$ then $r_{\Delta_2}\circ d^{34}(\beta_2M_1^{15}M_2^3) = A[52,1]M_1M_2$, an impossibility. There is is only one remaining possibility: $d^{14}(A[39,1]M_1M_2M_3) = A[52,1]M_1M_2$.

Since $\sigma C[44]M_1^2\bar{M}_2$ survives to E^6, it must transgress. Observe that the argument above that $\beta_2M_1^{14}\bar{M}_3$ can not hit $A[52,1]M_1M_2$ also shows that $\beta_2M_1^{14}\bar{M}_3$ can not hit $A[54,2]\bar{M}_2$ nor $\eta\sigma C[44]M_1\bar{M}_2$. Thus, $A[54,2]\bar{M}_2$, $\eta\sigma C[44]M_1\bar{M}_2$ can not be boundaries, and $A[59,1]$, $A[59,2]$ are nonzero. By Lemma 3.3.14, $\eta^2 A[57]$ must be zero since it can not be divisible by two. Thus, $\eta A[57]M_1$ must be a boundary. In Lemma 7.4.7(c) we shall see that $\eta A[52,2] = \eta C[20]A[32,2]$, and thus $\eta A[52,2]M_1\bar{M}_2 = C[20](\eta A[32,2]M_1\bar{M}_2)$ transgresses. If $d^8(\sigma C[44]M_1^2\bar{M}_2) = \eta A[57]M_1$ then $\beta_2 M_1^{14}\bar{M}_3$ transgresses to a nonzero element B which is indecomposable by Lemma 1.3.10. However, in the proof of Lemma 7.4.7(e) we shall see that $2B = \eta A[59,2]$ and $B = C[20]^3$, a contradiction. Thus, $d^{40}(\beta_2 M_1^{14}\bar{M}_3) = \eta A[57]M_1$. Since $\sigma C[44]M_1^2\bar{M}_2$ transgresses,

$$\sigma^2 C[44] = 0. \qquad [7.30]$$

Since $A[52,1] \in \langle \nu, \eta, \eta^2 D[45]\rangle$, $\sigma A[52,1] \in \sigma\langle \nu, \eta, \eta^2 D[45]\rangle = \langle \sigma, \nu, \eta\rangle \eta^2 D[45] = 0$. Since $A[52,2] \in \langle B[47], \nu, \eta\rangle$, $\sigma A[52,2] \in \langle \sigma, B[47], \nu\rangle \eta \subset \eta \cdot \mathrm{CokJ}_{58} = 0$. Thus,

$$\sigma A[52,1] = \sigma A[52,2] = 0. \qquad [7.31]$$

We have thus proved the following theorem.

THEOREM 7.4.6 $\pi_{59}^S = Z_2 A[59,1] \oplus Z_2 A[59,2] \oplus Z_8 \beta_7$

The computations in Section 6 show that we have the following leaders.

Row	Degree	Leader	Row	Degree	Leader
*19	63	$4\beta_2 M_1^{19}\bar{M}_2$	51	61	$\sigma C[44]M_1^2\bar{M}_2$
22	62	$\nu A[19]M_1^7 M_2^2\langle M_3\rangle$	*52	64	$A[52,1]M_1^3 M_2$
32	62	$A[32,1]M_1^8\bar{M}_3$	53	61	$\eta A[52,2]M_1\bar{M}_2$
38	62	$B[38]M_1^2\bar{M}_2\bar{M}_3$	56	62	$\nu^2 A[50,2]M_1^3$
45	61	$2D[45]M_1\langle M_3\rangle$	59	61	$A[59,1]M_1$, $A[59,2]M_1$

FIGURE 7.4.4: Leaders from Rows 1 to 59 of Degree at Least 61

There are five leaders of degree 61 and four leaders of degree 62. Since $A[59,1]M_1$ and $A[59,2]M_1$ can only bound from below the 54 row, $\nu^2 A[50,2]M_1^3$ must transgress. Recall that in the derivation of π_{59}^S we showed that $2D[45]M_1\langle M_3\rangle$ must bound from below the 34 row. Moreover, $2D[45]M_2\langle M_3\rangle$ equals $2M_2$ times the boundary $D[45]\langle M_3\rangle$ and must bound from below the 34 row. The lowest row of such an element of degree 66 is the 24 row. Therefore, $d^{14}(A[32,1]M_1^8\bar{M}_3)$ $= 2D[45]M_1\langle M_3\rangle$.

Let $\xi = d^{10}(\sigma C[44]M_1^2\bar{M}_2)$. Then $\sigma C[44]M_1^2\bar{M}_2$ has a representative $\mathcal{R} = (C[44] \wedge \mu_2 \wedge \sigma \wedge \bar{\mu}_{01}) \cup (C[44] \wedge \mu_2 \wedge B_{\sigma\nu} \wedge \mu_1) \cup (C[44] \wedge B_{\nu\sigma} \wedge \bar{\mu}_{01}) \cup (C[44] \wedge \mu_2 \wedge B_{\langle\sigma,\nu,\eta\rangle}) \cup (B_{C[44]\langle\nu,\sigma,\nu\rangle} \wedge \mu_1)$ with $\partial\,\mathcal{R} = C[44] \wedge [(B_{\nu\sigma} \wedge B_{\nu\eta}) \cup (\nu \wedge B_{\langle\sigma,\nu,\eta\rangle})] \cup [B_{C[44]\langle\nu,\sigma,\nu\rangle} \wedge \eta]$. Observe that $\partial\,[(B_{\nu\sigma} \wedge B_{\nu\eta}) \cup (\nu \wedge B_{\langle\sigma,\nu,\eta\rangle})] = \langle\nu,\sigma,\nu\rangle \wedge \eta$. Thus, $\partial\,\mathcal{R}$ represents an element of $\langle C[44],\langle\nu,\sigma,\nu\rangle,\eta\rangle = \langle C[44],\sigma^2,\eta\rangle$. By Theorem 2.4.2, $\xi M_1 = d^{16}(\eta\sigma^2A[30]M_1^3M_2^2) = 0$, and ξM_1 must bound from between the 45 and 51 rows, i.e. from the 47 row. There is no leader of degree 63 in the 47 row. Thus, $\xi = 0$ and $\sigma C[44]M_1^2\bar{M}_2$ must bound from below the 40 row. Assume that $d^{14}(B[38]M_1^2\bar{M}_2\bar{M}_3) = \sigma C[44]M_1^2\bar{M}_2$. Since $d^{12}(A[40,1]\bar{M}_2\langle M_3\rangle) = \sigma C[44]M_1^4$, r_{Δ_1} shows that a representative of $B[38]M_1^4\bar{M}_3$ is homologous to a representative of $A[40,1]\bar{M}_2\langle M_3\rangle$. Then r_{Δ_3} shows that a representative of $B[38]M_1^4$ is homologous to a representative of $A[40,1]\bar{M}_2$. Therefore, $4D[45] = \sigma B[38] = d^6(A[40,1]\bar{M}_2)$, a contradiction. The only other possibility for $\sigma C[44]M_1^2\bar{M}_2$ to bound from below the 40 row is $d^{30}(\nu A[19]M_1^7M_2^2\langle M_3\rangle) = \sigma C[44]M_1^2\bar{M}_2$.

We show that $\eta A[59,1] = 0$. Assume that $\eta A[59,1] \neq 0$. Since we shall see that there is no possibility for an element of π_{61}^S to have order 4, $\eta^2A[59,1] = 0$ by Lemma 3.3.14. Thus, $\eta A[59,1]M_1$ must be a boundary. Clearly $d^6(\eta A[54,2]M_1\bar{M}_2) = \eta A[59,2]M_1$. The arguments in the derivation of π_{61}^S that show that $\eta^2\sigma M_1^{21}M_2^2$, $4\beta_2M_1^{19}\bar{M}_2$ and $\gamma_2M_1^{20}$ can not bound $B[60]M_1$ also apply to show that these elements do not bound $\eta A[59,1]M_1$. (The argument there shows that if $d^{42}(4\beta_2M_1^{19}\bar{M}_2) = \eta A[59,1]M_1$ then $d^6(\eta A[54,2]M_1\bar{M}_2) = \eta A[59,1]M_1$. Hence $d^6(A[54,2]\bar{M}_2) = A[59,1]$, a contradiction.) In the derivation of π_{61}^S we will also show that $d^6(\eta A[50,2]M_1^3\bar{M}_2) = \nu^2A[50,2]M_1^3$ and that $4D[45]M_1^6M_2$, $\sigma A[32,1]M_1^6M_2^2$ are boundaries. Since $\nu A[50,1]M_1^2M_2 = r_{\Delta_1}(\nu A[50,1]M_1^3M_2)$ and there is no possibility for a hidden differential on $\nu A[50,1]M_1^3M_2$, $\nu A[50,1]M_1^2M_2$ must transgress. Note that $d^8(A[8]D[45]M_1^2\bar{M}_2) = D[45]d^8(A[8]M_1^2\bar{M}_2) = D[45]\eta A[14]M_1$. If $d^8(A[8]D[45]M_1^2\bar{M}_2) = \eta A[59,1]M_1$ then $A[59,1] = A[14]D[45]$. Since $A[59,1] = d^8(\eta\sigma C[44]M_1\bar{M}_2)$ and $A[14] = d^{12}(4\nu M_1^3\bar{M}_2)$, it would follow that twice a representative of $2\nu D[45]M_1^3\bar{M}_2$ union $\partial\ [(B_{4\nu D[45]} \wedge \mu_1 \wedge \mu_2 \wedge \bar{\mu}_{01}) \cup (2\nu D[45] \wedge B_{2\eta} \wedge \mu_2 \wedge \bar{\mu}_{01})]$ represents $\eta\sigma C[44]M_1\bar{M}_2$. Thus, $\eta\sigma C[44] \in \langle 2D[45],2\nu,\nu\rangle \subset \langle D[45],4\nu,\nu\rangle = \langle D[45],\eta^3,\nu\rangle = \langle\eta^2D[45],\eta,\nu\rangle$. Therefore, $\eta\sigma C[44]M_1 = d^6(\eta^2D[45]M_1\bar{M}_2)$, a contradiction. Hence $d^8(A[8]D[45]M_1^2\bar{M}_2) \neq \eta A[59,1]M_1$. Thus, $\eta A[59,1]M_1$ can not be a boundary a contradiction. Hence $\eta A[59,1] = 0$ and $A[59,1]M_1$ must bound from below the 52 row. The only possibility is $d^{22}(B[38]M_1^2\bar{M}_2\bar{M}_3) = A[59,1]M_1$.

Now $\eta A[59,2]$ and $B[60] = d^8(\eta A[52,2]M_1\bar{M}_2)$ are nonzero. To identify $2B[60]$ as $\eta A[59,2]$, we derive several relations.

LEMMA 7.4.7 (a) $\eta^2A[45,2] = \eta A[14]A[32,2]$ and $\eta A[45,2] \equiv A[14]A[32,2]$ modulo $(\eta^2C[44],\eta A[45,1])$.

(b) $A[54,2] = A[14]C[20]^2$

(c) $\nu A[52,2] = \nu C[20]A[32,2]$ and $A[52,2] = C[20]A[32,2]$ modulo $(A[52,1],\eta\sigma C[44])$.

(d) $A[59,2] = A[14]A[45,2]$

(e) $B[60] = C[20]^3$ and $2B[60] = \eta A[59,2]$.

(f) $\eta B[47] = A[8]C[20]^2$

(g) $\eta^2C[20]^3 = \nu A[14]A[45,2]$

PROOF. (a) $d^6(A[8]C[20]M_1\bar{M}_2) = \eta A[32,2]M_1$. Thus, $\eta^2A[45,2]M_1 = d^6(\eta^2C[20]^2M_1\bar{M}_2) = d^6(A[8]A[14]C[20]M_1\bar{M}_2) = A[14]d^6(A[8]C[20]M_1\bar{M}_2) = \eta A[14]A[32,2]M_1$.

(b) $A[14]C[20]^2 \in A[14]\langle\eta,\eta A[32,2],\eta,\nu\rangle \subset \langle\eta,\eta A[14]A[32,2],\eta,\nu\rangle = \langle\eta,\eta^2A[45,2],\eta,\nu\rangle = A[54,2]$.

(c) $\nu A[52,2] = \eta A[54,2] = \eta A[14]C[20]^2 = \nu A[32,2]C[20]$.

(d) $A[59,2] \in \langle \eta, \nu, A[54,2] \rangle = \langle \eta, \nu, A[14]C[20]^2 \rangle = A[14]\langle \eta, \nu, C[20]^2 \rangle$
$= A[14]A[45,2]$.

(e) $2C[20]^3 = C[20]^2 \langle \nu, \eta, \eta A[14] \rangle = \langle C[20]^2, \nu, \eta \rangle \eta A[14] = A[45,2]\eta A[14]$
$= \eta A[59,2] \neq 0$. The only possibility for $C[20]^3$ is $B[60]$.

(f) $\eta B[47] = \nu A[45,2] = \nu \langle \eta, \nu, C[20]^2 \rangle = \langle \nu, \eta, \nu \rangle C[20]^2 = A[8]C[20]^2$.

(g) $\nu A[14]A[45,2] = \eta A[14]B[47] = A[14]A[8]C[20]^2 = \eta^2 C[20]^3$. ∎

The above relations were motivated by relations (d) and (e) which were observed by Mark Mahowald from the Adams spectral sequence. We have now proved the following theorem.

THEOREM 7.4.8 $\pi_{60}^S = Z_4 B[60]$ and $2B[60] = \eta A[59,2]$.

The computations of Section 6 show that we have the following leaders. Since this is the final table of leaders of this section, we inlcude the leaders of all degrees.

Row	Degree	Leader	Row	Degree	Leader
9	63	$\eta^2 \sigma M_1^{21} M_2^2$	46	66	$\eta^2 C[44] M_1^7 \bar{M}_2$
11	65	$\beta_1 M_1^{20} \bar{M}_3$	47	65	$\nu C[44] M_1^3 \langle M_2^2 \rangle$
18	64	$4C[18] M_1^7 \bar{M}_2 M_2^2 \bar{M}_3$	51	65	$\sigma C[44] M_1^4 \bar{M}_2$
19	63	$4\beta_2 M_1^{19} \bar{M}_2$	51	63	$\eta A[50,2] M_1^3 \bar{M}_2$
23	63	$\gamma_2 M_1^{20}$	52	64	$A[52,1] M_1^3 M_2$
32	66	$A[32,1] M_1^2 \langle M_4 \rangle$	53	63	$\nu A[50,1] M_1^2 M_2$, $A[8]D[45] M_1^2 \bar{M}_2$
34	64	$2B[34] M_1^5 \bar{M}_2 \bar{M}_3$	54	66	$\eta A[8]D[45] M_1^3 \bar{M}_2$
39	63	$\sigma A[32,1] M_1^6 M_2^2$	55	63	$\eta A[54,2] M_1 \bar{M}_2$
40	66	$\eta A[39,3] M_1^3 \bar{M}_2 \langle M_3 \rangle$	56	62	$\nu^2 A[50,2] M_1^3$
40	64	$\eta\sigma A[32,1] M_1^5 \langle M_3 \rangle$	56	66	$A[56] M_1^2 \bar{M}_2$
44	66	$2C[44] M_1^2 M_2^3$	59	65	$A[59,1] M_2$, $A[59,1] \bar{M}_2$,

45	65	$2D[45]\langle M_1^4\rangle\langle M_2^2\rangle$			$A[59,2]\bar{M}_2$
45	63	$4D[45]M_1^6M_2$	60	62	$Z_4B[60]M_1$

FIGURE 7.4.5: Leaders from Rows 1 to 60 of Degree at Least 62

5. Computation of π_N^S, $61 \le N \le 64$.

We continue the computations of Section 4. In the tables of leaders we include all leaders of degree less than 67. This will suffice to compute through the 64 stem modulo group extension problems which we can not resolve in the 62 and 63 stems. We use Tangora's computation [59] of E_2 of the Adams spectral sequence through degree 70 to see that A[61] is not divisible by σ, to eliminate one group extension in degree 62 and to eliminate several possible differentials in degree 64. Diagrams summarizing his computation are given in Appendix 6.

From Figure 7.4.5, we see that there are three leaders of degree 62 and nine leaders of degree 63. Let $\mathfrak{C}$ denote the mapping cone of η^2 with $\rho: S \longrightarrow \mathfrak{C}$ the canonical map. If $d^6(\eta A[50,2]M_1^3\bar{M}_2) = \nu^2A[50,2]M_1^3$ then in the Atiyah-Hirzebruch spectral sequence for $\mathfrak{C}_*BP$, $d^6(\rho(\eta)M_1^3\bar{M}_2) = \rho(\nu)^2M_1^3$. Thus, $d^4(\rho(\nu)^2M_1^3) = \rho(\eta A[8])M_1$ is zero, and $\eta A[8]$ is divisible by η^2, a contradiction. If $d^{10}(\eta A[50,2]M_1^3\bar{M}_2) = B[60]M_1$ then $A[50,2]M_1^3\bar{M}_2$ shows that $\eta A[50,2]M_1^2\bar{M}_2$ is homologous to $\eta A[52,2]M_1\bar{M}_2$ since $B[60] = d^8(\eta A[52,2]M_1\bar{M}_2)$. Then $r_{(1,1)}$ shows that $A[50,2]M_1^2$ has a representative with boundary $\eta A[52,2]$ which contradicts that $\nu A[50,2]$ is nonzero and not divisible by η.

We show that $\nu A[57] = 0$. Assume that $\nu A[57] = \eta A[59,2]$. Since $d^{14}(4C[44]M_1^7\bar{M}_2) = A[57]\bar{M}_2$ and $d^8(2C[44]M_1^7\bar{M}_2) = 2\sigma C[44]M_1^3\bar{M}_2$, there must be a hidden differential $d^r(4C[44]M_1^7\bar{M}_2) = X$ with X in either the 53 or 55 row. There are three possibilities for X: $\nu A[50,1]M_1^2M_2$, $A[8]D[45]M_1^2\bar{M}_2$ and

$\eta A[54,2]M_1\bar{M}_2$. Now $\nu A[50,1]M_1^2M_2$ can only bound from below the 36 row, and thus $X \neq \nu A[50,1]M_1^2M_2$. A representative of X will be homologous to a representative of $A[57]\bar{M}_2$. Thus, a representative of X will have boundary which represents $\eta A[59,2]M_1$. We know that this true if $X = \eta A[54,2]M_1\bar{M}_2$ because $A[54,2]\bar{M}_2$ has a representative with boundary A[59,2]. Moreover, we know that this is false if $X = A[8]D[45]M_1^2\bar{M}_2$: $d^8(A[8]M_1^2\bar{M}_2) = A[14]M_1$, $A[14]D[45] \neq A[59,1]$ (see the derivation of π_{60}^S), $A[59,2] = A[14]A[45,2]$ and we can thus choose a representative of D[45] such that $A[14]D[45] = 0$. Therefore, $d^{12}(4C[44]M_1^7M_2) = \eta A[54,2]M_1\bar{M}_2$. Let $\mathfrak{C}$ denote the mapping cone of $\eta A[54,2]$ with $\rho: S \longrightarrow \mathfrak{C}$ the canonical map. In the Atiyah-Hirzebruch spectral sequence for $\mathfrak{C}_*BP$, $d^{14}(\rho(4C[44])M_1^7M_2) = \rho(A[57])\bar{M}_2$ and $d^4(\rho(A[57])\bar{M}_2) = \rho(\eta A[59,2])M_1$. Thus, $\rho(\eta A[59,2]) = 0$ and $\eta A[59,2] \in \eta A[54,2]\cdot\pi_5^S = 0$, a contradiction. Therefore, $\nu A[57] \neq \eta A[59,2]$ and

$$\nu A[57] = 0. \qquad [7.32]$$

Since $B[60] = d^8(\eta A[52,2]M_1\bar{M}_2)$, we see from Theorem 2.4.5(a) that

$$B[60] \in \langle\eta, \eta A[52,2], \eta, \nu\rangle. \qquad [7.33]$$

Thus, $2B[60] \in 2\langle\eta,\eta A[52,2],\eta,\nu\rangle \subset \langle\langle 2,\eta,\eta A[52,2]\rangle,\eta,\nu\rangle = \langle\eta A[54,2],\eta,\nu\rangle$ because $\nu \nmid 2B[60]$, $\pi_{55}^S = Z_2\eta A[54,2] \oplus Z_2\gamma_6$ and $\langle 2\gamma_6,\eta,\nu\rangle = \gamma_6\langle 2,\eta,\nu\rangle = 0$. Thus, $\eta A[54,2]M_2$ has a representative with boundary 2B[60] and $d^6(\eta A[54,2]M_1\bar{M}_2) = 2B[60]M_1$. Note that $\eta A[54,2]M_1\bar{M}_2$ must be nonzero in E^6. Thus, $d^4(A[52,1]M_1^3M_2)$ can not equal $\eta A[54,2]M_1\bar{M}_2$. Therefore, $\nu A[52,1]$ equals 0 and not $\eta A[54,2]$. Since $d^{12}(B[34]M_1^6) = 2D[45]$, $d^{12}(2B[34]M_1^5\bar{M}_2\bar{M}_3)$ $= 4D[45](M_1^6M_2+M_1^2M_3)$. We will show in the derivation of π_{62}^S that $\sigma A[32,1]M_1^6M_2^2$ $= d^{22}(4C[18]M_1^7\bar{M}_2M_2^2\bar{M}_3)$. Since $B[60]M_1$ could only bound from below the 53 row, $\nu A[50,1]M_1^2M_2$, $A[8]D[45]M_1^2\bar{M}_2$ and $\eta A[54,2]M_1\bar{M}_2$ transgress. Since $\gamma_2M_1^{20}$ = $r_{\Delta_1}(\gamma_2M_1^{21})$ and there is no possibility for a hidden differential on $\gamma_2M_1^{21}$, $d^{38}(\gamma_2M_1^{20})$ can not equal $B[60]M_1$. Thus, $\gamma_2M_1^{20}$ transgresses. We shall see in

Section 8.3 that θ_5 exists and that the only possibility is for $\eta^2\sigma M_1^{21}M_2^2$ to transgress to θ_5. Assume that $d^{42}(4\beta_2 M_1^{19}\bar{M}_2) = B[60]M_1$. Recall that $A[54,1] = d^{36}(\beta_2 M_1^{18})$. By Theorem 2.2.7(b), $\langle A[54,1],4,\eta,\nu\rangle$ is defined. By Theorem 2.4.6 (d), $B[60] \in \langle A[54,1],4,\eta,\nu\rangle \supset \langle 2A[54,1],2,\eta,\nu\rangle$ $= \lambda\langle \eta A[8]D[45],2,\eta,\nu\rangle \supset \lambda\eta D[45]\langle A[8],2,\eta,\nu\rangle = 0$. By Theorem 2.3.1(b), $B[60] \in \mathrm{Indet}\ \langle A[54,1],4,\eta,\nu\rangle = \langle \eta A[54,2],\eta,\nu\rangle + \langle A[54,1],\eta^2,\nu\rangle =$ $d^6(\eta A[54,2]M_2)$ modulo (η,ν). Thus $B[60] = 0$ in E^8, a contradiction. Therefore $4\beta_2 M_1^{19}\bar{M}_2$ transgresses. Let $A[61] = d^6(\nu^2 A[50,2]M_1^3)$. By Theorem 2.4.4(c),

$$A[61] \in \langle \eta,\nu,\nu^2 A[50,2]\rangle. \qquad [7.34]$$

Then $2A[61] \in 2\langle \eta,\nu,\nu^2 A[50,2]\rangle = \langle 2,\eta,\nu\rangle\nu^2 A[50,2] = 0$. Note that $\eta\sigma A[54,1] = 0$ and $\eta\sigma A[54,2] = \sigma(\nu A[52,2]) = 0$. We shall see that $\eta^2 B[60]$ is nonzero. Moreover, $A[61]$ projects to $h_0(A+A')$ in the Adams spectral sequence, an element of filtration degree 9 which can not be divisible by σ. Therefore,

$$\sigma A[54,1] = \sigma A[54,2] = 0. \qquad [7.35]$$

We have thus proved the following theorem.

THEOREM 7.5.1 $\pi_{61}^S = Z_2 A[61] \oplus Z_2 \eta B[60]$

The computations of Section 6 show that we have the following leaders.

Row	Degree	Leader	Row	Degree	Leader
9	63	$\eta^2\sigma M_1^{21}M_2^2$	46	66	$\eta^2 C[44]M_1^7\bar{M}_2$
11	65	$\beta_1 M_1^{20}\bar{M}_3$	47	65	$\nu C[44]M_1^3\langle M_2^2\rangle$
18	64	$4C[18]M_1^7\bar{M}_2 M_2^2\bar{M}_3$	51	63	$\eta A[50,2]M_1^3\bar{M}_2$
19	63	$4\beta_2 M_1^{19}\bar{M}_2$	51	65	$\sigma C[44]M_1^4\bar{M}_2$
23	63	$\gamma_2 M_1^{20}$	52	64	$A[52,1]M_1^3 M_2$
32	66	$A[32,1]M_1^2\langle M_4\rangle$	53	63	$\nu A[50,1]M_1^2 M_2$, $A[8]D[45]M_1^2\bar{M}_2$
34	64	$2B[34]M_1^5\bar{M}_2\bar{M}_3$	54	66	$\eta A[8]D[45]M_1^3\bar{M}_2$
39	63	$\sigma A[32,1]M_1^6 M_2^2$	56	66	$A[56]M_1^2\bar{M}_2$, $\nu^2 A[50,2]M_1^2 M_2$

40	66	$\eta A[39,3]M_1^3\bar{M}_2<M_3>$	59	65	$A[59,1]M_2$, $A[59,1]\bar{M}_2$,
40	64	$\eta\sigma A[32,1]M_1^5<M_3>$			$A[59,2]\bar{M}_2$
44	66	$2C[44]M_1^2M_2^3$	60	66	$B[60]\bar{M}_2$
45	65	$2D[45]<M_1^4><M_2^2>$	61	63	$A[61]M_1$, $\eta B[60]M_1$
45	63	$4D[45]M_1^6M_2$			

FIGURE 7.5.1: Leaders from Rows 1 to 61 of Degree at Least 63

From Figure 7.5.1, we see that there are eleven leaders of degree 63 and four leaders of degree 64. We observed in the derivation of π_{61}^S that $d^{12}(2B[34]M_1^5\bar{M}_2\bar{M}_3) = 4D[45](M_1^6M_2+M_1^2M_3)$. If $4C[18]M_1^7\bar{M}_2M_2^2\bar{M}_3$ survives to E^{28} then $d^{28}(4C[18]M_1^7\bar{M}_2M_2^2\bar{M}_3) = 2D[45]M_1^2M_3$ since $d^{28}(4C[18]M_1^{11}M_2) = D[45]$ and $2D[45](M_1^6M_2+M_2^3)$ is a d^{12}-boundary. However, $d^4(2D[45]M_1^2M_3) = 2\nu D[45]M_3 \neq 0$. Thus, $4C[18]M_1^7\bar{M}_2M_2^2\bar{M}_3$ must hit an element below the 45 row. The only possibility is $d^{22}(4C[18]M_1^7\bar{M}_2M_2^2\bar{M}_3) = \sigma A[32,1]M_1^6M_2^2$.

We show that $\eta\sigma A[32,1]M_1^5<M_3>$ transgresses and that $d^{10}(A[52,1]M_1^3M_2) = A[61]M_1$. Since $\nu\cdot\pi_{55}^S = 0$, $A[8]A[50,2] \in <\eta,\nu,2\nu>A[50,2] = \eta<\nu,2\nu,A[50,2]>$ $\subset <\nu,2\nu,\eta A[50,2]> = <\nu,2\nu,\nu^2A[45,1]> = <\nu,2\nu,\nu^2>A[45,1] = 0$. Thus, $A[8]A[50,2] = 0$. Observe that $\eta A[61] \in \eta<\nu^2A[50,2],\nu,\eta> \subset \eta<A[50,2],\nu^3,\eta>$ $= \eta<A[50,2],\eta A[8],\eta> \supset \eta<A[50,2],A[8],\eta^2> \subset <A[50,2],A[8],\eta^3>$ $= <A[50,2],A[8],4\nu> \supset 2\nu<A[50,2],A[8],2> \subset 2\nu\cdot\pi_{59}^S = 0$. Thus, $\eta A[61] \in (\eta^2,\eta A[50,2])$, $\eta A[61] = 0$ and $A[61]M_1$ must be a boundary. Since $\eta\sigma A[32,1]M_1^5<M_3> = r_{\Delta_1}(\eta A[39,3]M_1^3\bar{M}_2<M_3>)$ (see the derivation of π_{61}^S) and there is no possibility for a hidden differential on $\eta A[39,3]M_1^3\bar{M}_2<M_3>$, $\eta\sigma A[32,1]M_1^5<M_3>$ can bound neither $A[8]D[45]M_1^2\bar{M}_2$, $\eta A[50,2]M_1^3\bar{M}_2$, $A[61]M_1$ nor $\eta B[60]M_1$. Thus, $\eta\sigma A[32,1]M_1^5<M_3>$ transgresses. The only remaining possibility is $d^{10}(A[52,1]M_1^3M_2) = A[61]M_1$.

Now $B[62] = d^{40}(\gamma_2M_1^{20})$, $A[62,1] = d^{44}(\eta^2\sigma M_1^{21}M_2^2)$, $A[62,2] = d^{44}(4\beta_2M_1^{19}\bar{M}_2)$, $A[62,3] = d^{10}(\nu A[50,1]M_1^2M_2)$, $A[62,4] = d^{12}(\eta A[50,2]M_1^3\bar{M}_2)$,

$A'[62] = d^{10}(A[8]D[45]M_1^2\bar{M}_2)$ and $\eta^2B[60]$ are nonzero. We shall see in Section 8.3 that $A[62,1] = \theta_5$ and that $2\theta_5 = 0$. By Theorem 2.4.5(b),

$$A[62,3] \in \langle\eta,\nu,\nu A[50,1],\nu\rangle. \qquad [7.36]$$

Thus, $2A[62,3] \in 2\langle\eta,\nu,\nu A[50,1],\nu\rangle \subset \langle\langle 2,\eta,\nu\rangle,\nu A[50,1],\nu\rangle = \langle 0,\nu A[50,1],\nu\rangle$ $= \nu\cdot\pi_{59}^S$. Now $\nu A[59,2] = \eta^2B[60]$ and $A[59,1]M_1^2$ bounds. Thus, $\nu A[59,1] \in (\eta)$ $= Z_2\eta^2B[60]$, and there is a choice of $A[59,1]$ such that

$$\nu A[59,1] = 0. \qquad [7.37]$$

As observed by Mark Mahowald, $A[62,3]$ is represented by by h_5n in the Adams spectral sequence from which it follows that

$$A[62,3] \in \langle A[30],2,A[31]\rangle. \qquad [7.38]$$

Then $2A[62,3] \in 2\langle A[30],2,A[31]\rangle = \langle 2,A[30],2\rangle A[31] = \eta A[30]A[31] = 0$. Note that $\nu\langle 2,\eta,\eta A[14]\rangle = 2\langle\eta,\eta A[14],\nu\rangle = 2(2C[20]) = \nu(\nu A[14])$ and $\eta\langle 2,\eta,\eta A[14]\rangle = \langle\eta,2,\eta\rangle\eta A[14] = 2\nu(\eta A[14]) = 0$. Thus,

$$\langle 2,\eta,\eta A[14]\rangle = \nu A[14]. \qquad [7.39]$$

Since $d^8(A[8]D[45]M_1^2\bar{M}_2) = D[45]d^8(A[8]M_1^2\bar{M}_2) = D[45](\eta A[14]M_1)$, $A'[62] \in \langle\eta,\eta A[14],D[45]\rangle$. Thus, $2A'[62] \in 2\langle\eta,\eta A[14],D[45]\rangle = \langle 2,\eta,\eta A[14]\rangle D[45]$ $= \nu A[14]D[45] = 0$. We shall see in the derivation of π_{63}^S that $2B[62] = A'[62]$. Thus,

$$2B[62] \in \langle\eta,\eta A[14],D[45]\rangle. \qquad [7.40]$$

Since $A[62,2] = d^{44}(4\beta_2M_1^{19}\bar{M}_2)$ and $A[56] = d^{38}(2\beta_2M_1^{16}\bar{M}_2)$, Theorem 2.4.6(d) implies that

$$A[62,2] \in \langle\nu,\eta,2,A[56]\rangle. \qquad [7.41]$$

Now $2A[62,2] \in 2\langle A[56],2,\eta,\nu\rangle \subset \langle\langle 2,A[56],2\rangle,\eta,\nu\rangle \subset \langle\eta A[56],\eta,\nu\rangle$ $= \langle\nu A[54,1],\eta,\nu\rangle \supset A[54,1]\langle\nu,\eta,\nu\rangle = A[8]A[54,1] = \langle\eta,\nu,2\nu\rangle A[54,1]$ $= \eta\langle\nu,2\nu,A[54,1]\rangle$. Thus, $2A[62,2] \in (\eta,\nu) = Z_2\eta^2B[60]$. Assume that $2A[62,2]$ $= \eta^2B[60]$. Let $\mathfrak{C}$ denote the mapping cone of $A[56]$ with $\rho: S \longrightarrow \mathfrak{C}$ the canonical map and $\partial: \pi_*\mathfrak{C} \longrightarrow \pi_{*-57}^S$ the connecting homomorphism in the long exact sequence induced by multiplication by $A[56]$. In the Atiyah-Hirzebruch spectral sequence for $\mathfrak{C}_*BP$, $\rho(A[62,2])M_1$ is nonzero in E^6 because

$\nu\partial^{-1}(\eta^2) = \rho(A[62,2])$ would imply $A[62,2] \in \langle\nu, A[56], \eta^2\rangle$ and $2A[62,2] \in 2\langle\eta^2, A[56], \nu\rangle = \langle 2, \eta^2, A[56]\rangle\nu$, $A[59,2] \in \langle 2, \eta^2, A[56]\rangle$ and $\eta A[59,2] \in \eta\langle 2, \eta^2, A[56]\rangle = \langle\eta, 2, \eta^2\rangle A[56] = 0$, a contradiction. By 7.41, $d^6(\partial^{-1}(2)M_1M_2) = \rho(A[62,2])M_1$ and $\rho(A[62,2]) \in \langle\nu, \eta, \partial^{-1}(2)\rangle$. Then $2\rho(A[62,2]) \in 2\langle\nu, \eta, \partial^{-1}(2)\rangle \subset \langle 2\nu, \eta, \partial^{-1}(2)\rangle \supset \nu\langle 2, \eta, \partial^{-1}(2)\rangle$. Thus, $\rho(A[59,2])$ is an element of $\langle 2, \eta, \partial^{-1}(2)\rangle$. Then $\rho(\eta A[59,2]) \in \eta\langle 2, \eta, \partial^{-1}(2)\rangle$ $= \langle\eta, 2, \eta\rangle\partial^{-1}(2) = 2\nu\partial^{-1}(2)$ and $\rho(B[60])$ is divisible by ν. Thus, $\rho(\eta B[60]) = 0$ and $\eta B[60]$ is divisible by $A[56]$, a contradiction. Therefore, $2A[62,2] \neq \eta^2 B[60]$ and $2A[62,2] = 0$. We shall see that $\eta A[62,4] \neq 0$. Therefore, $A[62,4]$ is not divisible by 2. We have thus proved the following theorem.

THEOREM 7.5.2 π^S_{62} has a composition series

$$Z_2A[62,4],\ Z_2A[62,1] \oplus Z_2A[62,2] \oplus Z_2A[62,3] \oplus Z_4B[62] \oplus Z_2\eta^2B[60]$$

where $2A[62,4] \in Z_2\eta^2B[60]$.

The computations of Section 6 show that we have the following leaders.

Row	Degree	Leader	Row	Degree	Leader
11	65	$\beta_1M_1^{20}\bar{M}_3$	52	66	$A[52,1]M_1^7$
23	65	$2\gamma_2M_1^{18}\bar{M}_2$	54	66	$\eta A[8]D[45]M_1^3\bar{M}_2$
32	66	$A[32,1]M_1^2\langle M_4\rangle$	56	66	$A[56]M_1^2\bar{M}_2$, $\nu^2A[50,2]M_1^2M_2$
40	64	$\eta\sigma A[32,1]M_1^5\langle M_3\rangle$	59	65	$A[59,1]M_2$, $A[59,1]\bar{M}_2$,
40	66	$\eta A[39,3]M_1^3\bar{M}_2\langle M_3\rangle$			$A[59,2]\bar{M}_2$
44	66	$2C[44]M_1^2M_2^3$	60	66	$B[60]\bar{M}_2$
45	65	$2D[45]\langle M_1^4\rangle\langle M_2^2\rangle$	62	64	$A[62,1]M_1$, $A[62,2]M_1$,
46	66	$\eta^2C[44]M_1^7\bar{M}_2$			$A[62,4]M_1$, $2B[62]M_1$, $\eta^2B[60]M_1$
47	65	$\nu C[44]M_1^3\langle M_2^2\rangle$	62	66	$A[62,3]M_1^2$
51	65	$\sigma C[44]M_1^4\bar{M}_2]$			

FIGURE 7.5.2: Leaders from Rows 1 to 62 of Degree at Least 64

From Figure 7.5.2 we see that there are six leaders of degree 64 and eight leaders of degree 65. Clearly $A[59,1]M_2$ transgresses. Since $\nu A[59,1] = 0$, $A[59,1]\bar{M}_2$ transgresses. Since $\nu A[59,2] = \eta^2 B[60]$, $d^4(A[59,2]\bar{M}_2) = \eta^2 B[60]M_1$. Since $2D[45]\langle M_1^4\rangle\langle M_2^2\rangle$ is twice a boundary, it must be a boundary. Observe that a representative of $8M_1^{26}M_2^2+M_1^{32}$ has boundary $2\eta^2\sigma M_1^{21}\bar{M}_2^2$ union $\beta_1 M_1^{20}M_2^2$ modulo $(2\beta_1)$ and elements of filtration degree 50. Since $r_{\Delta_1}(\beta_1 M_1^{20}\bar{M}_3)$ $= \beta_1 M_1^{20}M_2^2$, $\beta_1 M_1^{20}\bar{M}_3$ has a representative $\mathcal{R}$ with boundary $(A[62,1]2 \wedge \mu_1) \cup (A[62,1] \wedge B_{2\eta})$. Then $\mathcal{R} \cup (B_{A[62,1]2} \wedge \mu_1)$ represents $\beta_1 M_1^{20}\bar{M}_3$ and has boundary $(B_{A[62,1]2} \wedge \eta) \cup (A[62,1] \wedge B_{2\eta})$. Thus, $\beta_1 M_1^{20}\bar{M}_3$ transgresses to an element $B[64,1]$ of $\langle A[62,1],2,\eta\rangle$ and $2B[64,1] \in 2\langle\eta,2,A[62,1]\rangle = \langle 2,\eta,2\rangle A[62,1] = \eta^2 A[62,1]$.

We show that $\eta A'[62] = 0$ and that $A'[62] = 2B[62]$. By 7.40, $A'[62] \in \langle\eta^2,A[14],D[45]\rangle$ and $\eta A'[62] \in \eta\langle\eta^2,A[14],D[45]\rangle \subset \langle\eta^3,A[14],D[45]\rangle =$ $\langle 4\nu,A[14],D[45]\rangle \subset \langle 2\nu,0,D[45]\rangle = 2\nu\cdot\pi_{60}^S + D[45]\cdot\pi_{18}^S = \{C[18]D[45],\eta\alpha_2 D[45]\}$. Now $C[18]D[45] \in D[45]\langle\sigma,2\sigma,\nu\rangle = \langle D[45],\sigma,2\sigma\rangle\nu \subset \nu\cdot\pi_{60}^S = 0$. Since $\eta\cdot\pi_{53}^S =$ $Z_2\eta A[8]D[45]$, $\eta\alpha_2 D[45] \in \eta D[45]\langle\sigma,16,\alpha_1\rangle = \eta\alpha_1\langle D[45],\sigma,16\rangle = k\eta\alpha_1 A[8]D[45]$ $= k\alpha_1\nu^3 D[45] = 0$. Thus, $\eta A'[62] = 0$ and $A'[62]M_1$ must be a boundary. Assume that $d^{12}(\sigma C[44]M_1^4\bar{M}_2) = A'[62]M_1$. Then a representative of $A[40,1]M_1^2\bar{M}_2\langle M_3\rangle$ would show that $\sigma C[44]M_1^6$ is homologous to $A[8]D[45]M_1^2\bar{M}_2$. It would follow that $d^{14}(\eta A[40,1]M_1^3\bar{M}_2\langle M_3\rangle) = \eta A[8]D[45]M_1^3\bar{M}_2$. However, this would contradict the fact that $\eta A[40,1]M_1^3\bar{M}_2\langle M_3\rangle$ is a d^8-boundary. Thus, $\sigma C[44]M_1^4\bar{M}_2$ transgresses. Similarly, if $d^{16}(\nu C[44]M_1^3\langle M_2^2\rangle) = A'[62]M_1$ then a representative of $A[40,1]M_1^6\langle M_2^2\rangle$ would show that $\nu C[44]M_1^2\langle M_2^2\rangle$ is homologous to $A[8]D[45]M_1^2\bar{M}_2$. Then $d^{14}(\eta A[40,1]M_1^7\langle M_2^2\rangle) = \eta A[8]D[45]M_1^3\bar{M}_2$ which contradicts the fact that $\eta A[40,1]M_1^7\langle M_2^2\rangle$ is a d^8-boundary. The only remaining possibility is $d^{40}(2\gamma_2 M_1^{21}) = A'[62]M_1$. Since $2\gamma_2 M_1^{21} = (2M_1)(\gamma_2 M_1^{20})$, $d^{40}(2\gamma_2 M_1^{21}) = 2B[62]M_1$ and $A'[62] = 2B[62]$.

We show that $A[62,4]M_1$ can not be a boundary. Assume that $d^{16}(\nu C[44]M_1^3<M_2^2>) = A[62,4]M_1$. Let $\mathfrak{C}$ denote the mapping cone of $\eta C[44]$ with $\rho: S \longrightarrow \mathfrak{C}$ the canonical map. Let $\partial: \pi_*^S \longrightarrow \pi_{*-46}^S$ denote the connecting homomorphism in the long exact sequence induced by multiplication by $\eta C[44]$. The following observations show that $\rho(A[62,4])M_1$ is nonzero in E^{16} of the Atiyah-Hirzebruch spectral sequence for $\mathfrak{C}_*BP$.

1) η does not divide $A[62,4]$.

2) $d^2((\partial^{-1}(\gamma_1)M_1) = \partial^{-1}(\eta\gamma_1)$ and $\partial^{-1}(\eta A[14])M_1 = d^8(\partial^{-1}(A[8])M_1^2\bar{M}_2)$.

3) The only elements from $Z_8\partial^{-1}(\beta_1) \otimes H_6BP$ which surivive to E^6 are d^{12}-boundaries.

4) The only elments from $[Z_2\partial^{-1}(\eta A[8]) \oplus Z_2\partial^{-1}(\eta^2\sigma) \oplus Z_2\partial^{-1}(\alpha_1)] \otimes H_8BP$ which survive to E^8 are d^{10}-boundaries.

5) The only element from $Z_8\partial^{-1}(2\sigma) \otimes H_{10}BP$ which survives to E^{10} is $\partial^{-1}(2\sigma)M_1^5$, and $d^{10}(\partial^{-1}(2\sigma)M_1^5) = \partial^{-1}(A[16])$.

6) The only elements from $Z_8\partial^{-1}(\nu) \otimes H_{16}BP$ which survive to E^{14} are $Z_2\partial^{-1}(\nu)\{M_1^8, M_1<M_3>\}$. Assume that one of these elements $\partial^{-1}(\nu)X$ bounds $\rho(A[62,4])M_1$. Let $\mathfrak{C}'$ denote the mapping cone of η with $\rho': S \longrightarrow \mathfrak{C}'$ the canonical map and ∂' the connecting homomorphism in the long exact sequence induced by multiplication by η. In the Atiyah-Hirzebruch spectral sequence for $\mathfrak{C}'_*BP$, $\rho'(A[62,4])M_1 = d^{12}(C[44]\partial'^{-1}(\nu)X) = C[44]d^{12}(\partial'^{-1}(\nu)X)$. Thus, $\rho'(A[62,4])$ is divisible by $C[44]$. The only possibility is $\rho'(A[62,4]) \in \{C[44]\rho'(C[18]), C[44]\rho'(\eta\alpha_2)\}$ and $A[62,4] \in \{C[44]C[18], C[44]\eta\alpha_2\} + Z_2\eta^2B[60]$. Now $C[44]C[18] = d^{12}(2\sigma C[44]M_1^6) = 0$ in E^{12} and $A[62,4]$ is nonzero in E^{12}. Thus, $A[62,4]$ is divisible by η, a contradiction. Therefore, neither element of $Z_2\partial^{-1}(\nu)\{M_1^8, M_1<M_3>\}$ can bound $\rho(A[62,4])M_1$.

Now $d^{16}(\rho(\nu C[44])M_1^3<M_2^2>) = \rho(A[62,4])M_1$. Applying r_{Δ_1}, we see that $\rho(C[44])M_1M_2<M_2^2>$ shows that $\rho(\nu C[44])M_1^2<M_2^2>$ is homologous to $\rho(\sigma C[44])M_1^3M_2$ not

$\rho(\eta A[50,2])M_1^3\bar{M}_2$, a contradiction. Thus, $d^{16}(\nu C[44]M_1^3\langle M_2^2\rangle)$ is not $A[62,4]M_1$. There is no other way for $A[62,4]M_1$ to bound from below the 51 row, and $A[62,4]M_1$ is not a boundary.

Now $A[62,1]M_1$, $A[62,2]M_1$ and $\eta\sigma A[32,1]M_1^5\langle M_3\rangle$ can only bound from below the 40 row. There are no such leaders of degree 65 remaining which could bound any of these elements. Thus, $A[63] = d^{24}(\eta\sigma A[32,1]M_1^5\langle M_3\rangle)$, $\eta A[62,1]$, $\eta A[62,2]$ and $\eta A[62,4]$ are nonzero. Since $\eta^2 A[62,1]$ and $\eta^2 A[62,4]$ will be seen to be nonzero, $2A[63] \in Z_2\eta A[62,2]$. We have thus proved the following theorem.

THEOREM 7.5.3. π_{63}^S has a composition series:

$$Z_2A[63],\ Z_2\eta A[62,1] \oplus Z_2\eta A[62,2] \oplus Z_2\eta A[62,4] \oplus Z_{128}\gamma_7$$

where $2A[63] \subset Z_2\eta A[62,2]$.

The computations of Section 6 show that we have the following leaders.

Row	Degree	Leader	Row	Degree	Leader
11	65	$\beta_1 M_1^{20}\bar{M}_3$	54	66	$\eta A[8]D[45]M_1^3\bar{M}_2$
32	66	$A[32,1]M_1^2\langle M_4\rangle$	56	66	$A[56]M_1^2\bar{M}_2$, $\nu^2 A[50,2]M_1^2M_2$
40	66	$\eta A[39,3]M_1^3\bar{M}_2\langle M_3\rangle$	59	65	$A[59,1]\bar{M}_2$, $A[59,1]M_2$
44	66	$2C[44]M_1^2M_2^3$	60	66	$B[60]\bar{M}_2$
45	65	$2D[45]\langle M_1^4\rangle\langle M_2^2\rangle$	62	66	$A[62,1]M_1^2$, $A[62,2]M_1^2$, $A[62,3]M_1^2$,
46	66	$\eta^2 C[44]M_1^7\bar{M}_2$			$2B[62]M_1^2$, $A[62,4]M_1^2$
47	65	$\nu C[44]M_1^3\langle M_2^2\rangle$	63	65	$A[63]M_1$, $\eta A[62,1]M_1$,
51	65	$\sigma C[44]M_1^4\bar{M}_2$			$\eta A[62,2]M_1$, $\eta A[62,4]M_1$
52	66	$A[52,1]M_1^7$			

FIGURE 7.5.3: Leaders from Rows 1 to 63 of Degree at Least 65

There are ten leaders of degree 65 and fourteen leaders of degree 66. In the derivation of π^S_{63} we showed that $2D[45]\langle M_1^4\rangle\langle M_2^2\rangle$ must bound. Since $2D[45]\langle M_1^4\rangle\langle M_2^2\rangle$ can only bound from below the 34 row, the only possibility is $d^{14}(A[32,1]M_1^2\langle M_4\rangle) = 2D[45]\langle M_1^4\rangle\langle M_2^2\rangle$. Clearly $A[62,1]M_1^2$, $A[62,2]M_1^2$, $A[62,3]M_1^2$, $2B[62]M_1^2$ and $A[62,4]M_1^2$ transgress. Since $\nu B[60] = \nu C[20]^3 = 0$, $B[60]\bar{M}_2$ transgresses. Note that $d^{38}(2\beta_2 M_1^{20}\bar{M}_2) = A[56]M_1^4$, $r_{\Delta_1}(2\beta_2 M_1^{20}\bar{M}_2) = 2\beta_2 M_1^{22}$ and $d^{44}(2\beta_2 M_1^{22}) = A[62,2]$. Thus, $A[56]M_1^4$ is homologous to $A[62,2]M_1$, $d^8(A[56]M_1^2\bar{M}_2) = \eta A[62,2]M_1$ and

$$\sigma A[56] = \eta A[62,2] \qquad [7.42]$$

In addition, $\nu^2 A[50,2]M_1^2M_2$ can not hit $\eta A[62,1]M_1$, $\eta A[62,4]M_1$ or $A[63]M_1$ because $\nu^2 A[50,2]M_1^2M_2$ is in Image r_{Δ_1} and there is no possibility for a hidden differential on $\nu^2 A[50,2]M_1^3M_2$. Thus, $\nu^2 A[50,2]M_1^2M_2$ also transgresses. Clearly $\eta A[8]D[45]M_1^3\bar{M}_2$ survives to E^{10} and $d^{10}(\eta A[8]D[45]M_1^3\bar{M}_2) = \eta d^{10}(A[8]D[45]M_1^2\bar{M}_2)M_1 = \eta(2B[62])M_1 = 0$. Therefore, $\eta A[8]D[45]M_1^3\bar{M}_2$ must transgress. Since $\nu A[37] = \eta\sigma A[32,1] + \eta A[39,3]$, $\eta A[39,3]M_1^5\langle M_3\rangle = \eta\sigma A[32,1]M_1^5\langle M_3\rangle$ in E^6. Applying r_{Δ_1}, we see that $d^{24}(\eta A[39,3]M_1^3\bar{M}_2\langle M_3\rangle) = A[63]M_1$. Since $\beta_1 M_1^{20}\bar{M}_3$, $\nu C[44]M_1^3\langle M_2^2\rangle$, $\sigma C[44]M_1^4\bar{M}_2$ and $A[59,1]M_2$ can only bound from below the 40 row, none of these elements can be a boundary.

The remaining leaders of degree 66 are $\eta^2 C[44]M_1^7\bar{M}_2$, $2C[44]M_1^4M_3$ and $A[52,1]M_1^7$. Note that $A[59,1]M_2$, $A[59,1]\bar{M}_2$ can only bound from below the 38 row, 52 row, respectively. If $d^{14}(\eta^2 C[44]M_1^7\bar{M}_2) = A[59,1]\bar{M}_2$ then $\eta^2 C[44]M_1^7$ has a representative with boundary $A[59,1]$ and $d^{14}(\eta^2 C[44]M_1^5\bar{M}_2) = A[59,1]M_1$ which contradicts that $\eta^2 C[44]M_1^5\bar{M}_2$ is a d^8-boundary and that $A[59,1]M_1$ bounds from the 38 row. Thus, the only possible differentials are $d^{18}(\eta^2 C[44]M_1^7\bar{M}_2) = \eta A[62,4]M_1$, $d^{16}(2C[44]M_1^2M_2^3) = A[59,1]\bar{M}_2$, $d^{20}(2C[44]M_1^2M_2^3) = \eta A[62,4]M_1$ or $d^{12}(A[52,1]M_1^7) = \eta A[62,4]M_1$. However, we see from the Adams spectral sequence that the order of π^S_{64} is 2^8. Thus, $\eta^2 C[44]M_1^7\bar{M}_2$, $2C[44]M_1^2M_2^3$ and $A[52,1]M_1^7$ must transgress.

Let $A[64,1] = d^{14}(\sigma C[44]M_1^4\bar{M}_2)$, $A[64,2] = d^{6}(A[59,1]M_1^3)$, $A[64,3] = d^{6}(A[59,1]M_2)$, $B[64,1] = d^{54}(\beta_1 M_1^{20}\bar{M}_3)$ and $B[64,2] = d^{18}(\nu C[44]M_1^3<M_2^2>)$. Then $A[64,1]$, $A[64,2]$, $A[64,3]$, $B[64,1]$, $B[64,2]$, $\eta^2A[62,1]$ and $\eta^2A[62,4]$ must be nonzero. By Theorem 2.4.6(c),

$$A[64,1] \in <\eta,\nu,\sigma C[44],\sigma>. \qquad [7.43]$$

Thus modulo $2\cdot$Indet $<\eta,\nu,\sigma C[44],\sigma>$, $2A[64,1] \in 2<\eta,\nu,\sigma C[44],\sigma>$ $= 2<\eta,\nu,\sigma,\sigma C[44]> = <<2,\eta,\nu>,\sigma,\sigma C[44]> = <0,\sigma,\sigma C[44]> = \sigma C[44]\cdot\pi_{13}^S = 0$. Then $2A[64,1] \in 2\cdot$Indet $<\eta,\nu,\sigma C[44],\sigma> = 2<\eta,\nu,X> \cup 2<\eta,Y,\sigma> \cup 2<Z,\sigma C[44],\sigma>$ where $X \in \pi_{59}^S$ with $\nu X = 0$, $Y \in \pi_{55}^S$ with $\eta Y = \sigma Y = 0$ and $Z \in \pi_5^S = 0$. Thus, $2A[64,1] \in <2,\eta,\nu>X \cup <2,\eta,\eta A[54,2]>\sigma \cup 2\sigma\cdot\pi_{57}^S = \sigma<2,\eta,\eta A[54,2]>$. Now $\eta<2,\eta,\eta A[54,2]> = <\eta,2,\eta>A[54,2] = 2\nu A[54,2] = 0$. Thus, $2A[64,1] = \sigma(\eta A[56])$ $= 0$. $\nu A[61] \in \nu<\eta,\nu,\nu^2A[50,2]> = <\nu,\eta,\nu>\nu^2A[50,2] = A[8]\nu^2A[50,2] = 0$, and

$$\nu A[61] = 0. \qquad [7.44]$$

By Theorem 2.4.4(a),

$$A[64,2] \in <\eta,A[59,1],\nu>. \qquad [7.45]$$

Thus, $2A[64,2] \in 2<\eta,A[59,1],\nu> = <2,\eta,A[59,1]>\nu \subset \nu\cdot\pi_{61}^S = 0$. By Theorem 2.4.4(b),

$$A[64,3] \in <A[59,1],\eta,\nu>. \qquad [7.46]$$

Thus, $2A[64,3] \in 2<A[59,1],\eta,\nu> = <2,A[59,1],\eta>\nu \subset \nu\cdot\pi_{61}^S = 0$. Recall that in the derivation of π_{63}^S we showed that $2B[64,1] = \eta^2A[62,1]$ and

$$B[64,1] \in <\eta,2,A[62,1]>. \qquad [7.47]$$

If $2A[62,4] = 0$ then, by Lemma 3.3.14, $\eta^2A[62,4]$ must be divisble by 2. If $2A[62,4] = \eta^2B[60]$ then $\eta^2A[62,4] \in <2,\eta,2>A[62,4] \subset <2,\eta,2A[62,4]>$ $= <2,\eta,\eta^2B[60]> = <2,\eta,\nu A[59,2]> \equiv <2,\eta,\nu>A[59,2] = 0$ modulo (2). Thus, in both cases $\eta^2A[62,4]$ is divisible by 2. The only possiblity is $2B[64,2] = \eta^2A[62,4]$. We have thus proved the following theorem.

THEOREM 7.5.4 $\pi_{64}^S = Z_2A[64,1] \oplus Z_2A[64,2] \oplus Z_2A[64,3] \oplus Z_4B[64,1] \oplus Z_4B[64,2] \oplus Z_2\eta\gamma_7$

where $2B[64,1] = \eta^2A[62,1]$ and $2B[64,2] = \eta^2A[62,4]$.

In the Adams spectral sequence $gg_2 = C[20]C[44]$ and $h_3Q_2 = \sigma A[57]$ are nonzero infinite cylces distinct from B[64,1], 2B[64,1], B[64,2] and 2B[64,2]. Since both C[20]C[44] and $\sigma A[57]$ are zero in E^{14}, $\{C[20]C[44], \sigma A[57]\}$ $= \{A[64,2], A[64,3]\}$. Note that $\sigma A[57]M_1^2 = d^8(A[57]\langle M_2^2\rangle) = 0$ in E^8 because $A[57]\langle M_2^2\rangle$ is a d^{14}-boundary. Since $A[64,2]M_1^2$ is a d^6-boundary and $A[64,3]M_1^2$ is nonzero in E^8,

$$\sigma A[57] = A[64,2] \text{ and } C[20]C[44] = A[64,3]. \qquad [7.48]$$

6. Tentative Differentials

In this section we give the tentative differentials determined by the differentials on leaders of degree greater than or equal to 47 which were determined in this chapter. We continue the computations made in Chapter 6, Section 4 and use the same format and notation. These computations are complete through degree 68.

<u>DEGREE 9:</u> $\eta^2\sigma$ and α_1

The only differential in degrees less than 70 is $d^{30}(\eta^2\sigma M_1^{21}M_2^2) = A[62,1]$.

<u>DEGREE 11:</u> β_1

The leading differential $d^{40}(4\beta_1(M_1^7\overline{M}_2^3\overline{M}_3 + M_1^{10}\overline{M}_2^2\overline{M}_3 + M_1^{14}\overline{M}_2^3)) = A[50,1]M_2$ determines tentative differentials whose kernel in degrees less than 70 is given below. These differentials are computed by making the following assignments to monomials in $Z_8\beta_1 \otimes H_{46}BP$: $M_1^7M_2^3M_3$ is assigned 1 and all other monomials are assigned 0.

DEGREE	GROUP	GENERATOR	DEGREE	GROUP	GENERATOR
(54,11)	Z_2	20 0 1 0	(58,11)	Z_2	4/ 7 0 1 1
					4/ 13 3 1 0
					6/ 4 6 1 0
					6/ 10 4 1 0
					1/ 20 3 0 0
					5/ 22 0 1 0
					1/ 26 1 0 0

The leading differential $d^{52}(\beta_1 M_1^{20}\overline{M}_3) = B[64,1]$ determines tentative differentials by making the following assignments to monomials in $Z_8\beta_1 \otimes H_{54}BP$: $M_1\overline{M}_2^4\overline{M}_3^2$, $M_1^3\overline{M}_2\overline{M}_3^3$, $M_1^3\overline{M}_2^3\overline{M}_4$, $M_1^5\overline{M}_3\overline{M}_4$, $M_1^9\overline{M}_2^6$, $M_1^{13}\overline{M}_3^2$, $M_1^{17}\overline{M}_2\overline{M}_3$, $M_1^{21}\overline{M}_2^2$ are assigned 1 and all other monomials are assigned 0. We have the tentative differential $d^{52}(\beta_1 M_1^{22}\overline{M}_3) = B[64,1]M_1^2$. There are no remaining elements in degrees less than 70.

<u>DEGREE 17:</u> $\eta^2\gamma_1$ and α_2

The leading differential $d^{32}(\eta^2\gamma_1 M_1^{19}) = \eta A[47]M_1^3$ is computed by assigning 1 to the monomials $\eta^2\gamma_1 M_1\overline{M}_2^6$, $\eta^2\gamma_1 M_1^7\overline{M}_2^4$, $\eta^2\gamma_1 M_1^{13}\overline{M}_2^2$, $\eta^2\gamma_1 M_1^{19}$ and assigning 0 to all the other monomials of degree 55. The tentative differentials have kernel in degrees less than 70 given by the table below, and the new leader is $\alpha_1 M_1^{14}\overline{M}_2$.

DEGREE	BASIS	DEGREE	BASIS	DEGREE	BASIS
(34,17)	14 1 0 0	(36,17)	15 1 0 0	(42,17)	14 0 1 0
	15 2 0 0	(44,17)	15 0 1 0	(46,17)	13 1 1 0 23 0 0 0
	14 3 0 0	(48,17)	14 1 1 0		15 3 0 0
(50,17)	12 2 1 0		15 1 1 0	(52,17)	13 2 1 0 19 0 1 0

The leading differential $d^{32}(\alpha_2 M_1^{14}\overline{M}_2) = A[50,1]$ is computed by assigning 1 to the monomials $\eta^2\gamma_1 M_1^5\overline{M}_2^4$, $\eta^2\gamma_1 M_1^{11}\overline{M}_2^2$, $\eta^2\gamma_1 M_1^{17}$, $\alpha_2 M_1^2\overline{M}_2^5$, $\alpha_2 M_1^8\overline{M}_2^3$, $\alpha_2 M_1^{14}\overline{M}_2$ and 0 to all other monomials of degree 51. These tentative differentials have the following kernel in degrees less than 70:

DEGREE	BASIS	DEGREE	BASIS
(42,17)	15 2 0 0	(46,17)	13 1 1 0 23 0 0 0

The leading differential $d^{38}(\eta^2\gamma_1 M_1^{15}\overline{M}_2^2) = A[52,1]M_1^3$ determines tentative differentials which are a monomorphism on the remaining elements in degrees less than 70. Thus, there are no remaining elements.

DEGREE 18: C[18]

The differential $d^{22}(4C[18]M_1^7\overline{M}_2 M_2^2\overline{M}_3) = \sigma A[32,1]M_1^6M_2^2$ leaves no remaining elements in degrees less than 69.

DEGREE 19: β_2

The leading differential $d^{36}(\beta_2 M_1^{18}) = A[54,1]$ determines tentative differentials whose kernel in degrees less than 70 is given below. These differentials are computed by making the following assignments to monomials of $Z_8\beta_2 \otimes H_{36}BP$: M_1^{18} is assigned 1, and all other monomials are assigned 0. The new β_2-leader is $2\beta_2 M_1^{16}\overline{M}_2$.

DEGREE	GROUP	GENERATOR	DEGREE	GROUP	GENERATOR	DEGREE	GROUP	GENERATOR
(38,19)	Z_2	2/ 16 1 0 0	(42,19)	Z_2	14 0 1 0		Z_2	2/ 18 1 0 0
(44,19)	Z_4	2/ 19 1 0 0 6/ 22 0 0 0	(46,19)	Z_2	1/ 14 3 0 0 6/ 20 1 0 0		Z_2	2/ 20 1 0 0
(48,19)	Z_4	14 1 1 0	(50,19)	Z_2	1/ 15 1 1 0 7/ 18 0 1 0 1/ 22 1 0 0		Z_2 Z_2	2/ 16 3 0 0 2/ 22 1 0 0

The leading differentials $d^{38}(2\beta_2 M_1^{16}\overline{M}_2) = A[56]$ determines tentative differentials by making the following assignments to monomials of $Z_8\beta_2 \otimes H_{38}BP$: M_1^{19} is assigned 1, $M_1^{13}M_2^2$ is assigned 6 and all other monomials are assigned 0. The kernel of these differentials in degrees less than 70 is given by the table below and the new β_2-leader is $\beta_2 M_1^{14}\bar{M}_3$.

DEGREE	GROUP	GENERATOR	DEGREE	GROUP	GENERATOR	DEGREE	GROUP	GENERATOR
(42,19)	Z_2	1/ 14 0 1 0 6/ 18 1 0 0	(44,19)	Z_2	4/ 19 1 0 0	(46,19)	Z_2	1/ 14 3 0 0 6/ 20 1 0 0
(48,19)	Z_4	14 1 1 0	(50,19)	Z_2	1/ 15 1 1 0 7/ 18 0 1 0 7/ 22 1 0 0			

The leading differential $d^{40}(\beta_2 M_1^{14}\bar{M}_3) = \eta A[57]M_1$ determines tentative differentials which are computed by assigning 1 to $\beta_2 M_1^{15}M_2^2$ and 0 to all other monomials of $Z_8\beta_2 \otimes H_{42}BP$. The kernel of these differentials in degrees less than 70 is given by the table below.

DEGREE	GROUP	GENERATOR	DEGREE	GROUP	GENERATOR
(44,19)	Z_2	4/ 19 1 0 0	(48,19)	Z_2	2/ 14 1 1 0

The leading differential $d^{44}(4\beta_2 M_1^{19}\bar{M}_2) = A[62,2]$ determines tentative differentials which are computed by making the following assignments to monomials of $Z_8\beta_2 \otimes H_{44}BP$: $M_1^{19}M_2$ is assigned 1, $M_1^{16}M_2^2$ and M_1^{22} are assigned 2, $M_1^{13}M_2^3$ is assigned 6, $M_1^{15}M_3$ and $M_1^{12}M_2M_3$ are assigned 4 while all other monomials are assigned 0. There are no tentative differentials in degrees less than 70. The only remaining element is $2\beta_2 M_1^{14}\bar{M}_2\bar{M}_3$.

DEGREE 21: $\nu C[18]$

The leading differential $d^{24}(\nu C[18]M_1^6\bar{M}_2\bar{M}_3) = (\eta A[39,3]+\eta\sigma A[32,1])M_1^3\bar{M}_2$ determines tentative differentials which are a monomorphism on the remaining elements of $Z_2(\nu C[18]) \otimes H_*BP$ in degrees less than 68.

DEGREE 22: $\nu A[19]$

The leading differential $d^{30}(\nu A[19]M_1^7M_2^2M_3) = \sigma C[44]M_1^2\bar{M}_2$ leaves no remaining elements in degrees less than 69.

DEGREE 23: γ_2

The leading differential $d^{40}(\gamma_2 M_1^{20}) = B[62]$ determines tentative differentials by assigning 1 to $\gamma_2 M_1^{20}$ and 0 to all other monomials of $Z_{16}\gamma_2 \otimes H_{40}BP$. The kernel in degrees less than 70 is given by the table below, and the new γ_2-leader is $2\gamma_2 M_1^{18}\overline{M}_2$.

DEGREE	GROUP	GENERATOR	DEGREE	GROUP	GENERATOR	DEGREE	GROUP	GENERATOR
(42,23)	Z_2	2/ 18 1 0 0	(44,23)	Z_4	2/ 22 0 0 0	(46,23)	Z_2	2/ 23 0 0 0
								6/ 20 1 0 0
	Z_8	2/ 20 1 0 0						

The leading differential $d^{40}(2\gamma_2 M_1^{18}\overline{M}_2) = 2B[62]M_1$ determines tentative differentials by assigning 1 to $\gamma_2 M_1^{21}$ and 0 to all other monomials of $Z_{16}\gamma_2 \otimes H_{42}BP$. The kernel in degrees less than 70 is given by the table below, and the new γ_2-leader is $2\gamma_2 M_1^{22}$.

DEGREE	GROUP	GENERATOR	DEGREE	GROUP	GENERATOR
(44,23)	Z_4	2/ 22 0 0 0	(46,23)	Z_4	4/ 20 1 0 0

DEGREE 24: $\eta A[23]$

The leading differential $d^{22}(\eta A[23]M_1^{15}\overline{M}_2) = 2D[45]M_1^4 M_2$ determines tentative differentials which are a monomorphism on the remaining elements of degree less than 69.

DEGREE 30: $A[30]$

The leading differential $d^{16}(A[30]\langle M_4\rangle) = \sigma A[32,1]M_1^4 M_2^2$ determines tentative differentials which are a monomorphism on

$Z_2 A[30]\langle M_4\rangle \otimes Z_2[\langle M_1^4\rangle^2, \langle M_2^2\rangle^2, \langle M_3\rangle^2, \langle M_4\rangle^2, \{M_5\}, \ldots, \{M_n\}, \ldots]$. There are no remaining elements.

DEGREE 32: $A[32,1]$

The leading differential $d^8(A[32,1](M_1^2M_3+M_2^3)) = A[39,1]M_1^2M_2$ determines tentative differentials with kernel in degrees less than 69 given by $Z_2A[32,1]\{M_1^8\bar{M}_3, M_1^2\bar{M}_4+M_1^2\bar{M}_2^5+M_1^8\bar{M}_2^3+M_1^{14}\bar{M}_2\}$. The new $A[32,1]$-leader is $A[32,1]M_1^8\bar{M}_3$. The leading differential $d^{14}(A[32,1]M_1^8\bar{M}_3) = 2D[45]M_1\langle M_3\rangle$ determines no tentative differentials by assigning 1 to $A[32,1]M_1^8M_3$ and 0 to all other monomials of $Z_2A[32,1] \otimes H_{30}BP$. The only remaining element in degree less than 69 is $A[32,1](M_1^2\bar{M}_4+M_1^2\bar{M}_2^5+M_1^8\bar{M}_2^3+M_1^{14}\bar{M}_2)$. Moreover,

$$d^{32}(A[32,1](M_1^2\bar{M}_4+M_1^2\bar{M}_2^5+M_1^8\bar{M}_2^3+M_1^{14}\bar{M}_2)) = 2D[45]\langle M_1^4\rangle\langle M_2^2\rangle.$$

DEGREE 34: $B[34]$

The leading differential $d^8(B[34]M_1^4M_2) = \eta A[40,1]M_1^3$ determines tentative differentials by making the following assignments to monomials of $Z_2\otimes [Z_4B[34] \otimes H_{14}BP]$: $M_1^4M_2$ and M_1^7 are assigned 1 and all other monomials are assigned 0. The kernel of these tentative differentials in degrees less than 69 is given by the table below.

DEGREE	GENERATOR	DEGREE	GENERATOR	DEGREE	GENERATOR
(12,34)	6 0 0 0	(14,34)	0 0 1 0	(18,34)	2 0 1 0
			4 1 0 0		6 1 0 0
(20,34)	4 2 0 0	(22,34)	4 0 1 0	(24,34)	6 2 0 0
			8 1 0 0		
(26,34)	0 2 1 0		6 0 1 0	(28,34)	14 0 0 0
	4 3 0 0		10 1 0 0		
(30,34)	8 0 1 0		2 2 1 0	(34,34)	4 2 1 0
	12 1 0 0		6 3 0 0		8 3 0 0
					10 0 1 0
					14 1 0 0

The leading differential $d^6(B[34]M_1^6) = 2D[45]$ determines tentative differentials by assigning 1 to $B[34]M_1^6$ and 0 to all other monomoials of

$Z_2 \otimes (Z_4 B[34] \otimes H_{12} BP)$. These tentative differentials are a monomorphism on the elements of the table above. There are no elements of $Z_2 \otimes [Z_4 B[34] \otimes H_{14} BP]$ remaining in degrees less than 69.

The leading differential $d^{12}(2B[34]M_1^5 \bar{M}_2 \bar{M}_3) = 4D[45]M_1^6 M_2$ determines the tentative differential $d^{12}(2B[34]M_1 \bar{M}_2^3 \bar{M}_3) = 2D[45]M_1^5<M_2^2>$. There are no remaining elements in degrees less than 69.

<u>DEGREE 36:</u> A[36]

The leading differential $d^{10}(A[36]M_1^2 \bar{M}_3) = 4D[45]M_1 M_2$ determines tentative differentials which have kernel in degrees less than 69 equal to $Z_2(A[36]M_1^6 \bar{M}_2^2)$. Moreover, $d^{16}(A[36]M_1^6 \bar{M}_2^2) = \nu A[50,1]M_1^3$.

<u>DEGREE 38:</u> $\eta\sigma A[30] = \eta A[37]$

The leading differential $d^{10}(\eta\sigma A[30]M_1^3 \bar{M}_2) = \eta^2 D[45]M_1$ determines tentative differentials which are a monomorphsim on $Z_2 \eta\sigma A[30]M_1^3 \bar{M}_2 \otimes B<4>$. There are no remaining elements.

<u>DEGREE 38:</u> B[38]

The relation $\sigma B[38] = 4D[45]$ determines tentative d^8-differentials in degrees less than 69 with kernel $Z_2(B[38]M_1^2 \bar{M}_2 \bar{M}_3)$. Moreover, $d^{23}(B[38]M_1^2 \bar{M}_2 \bar{M}_3)$ $= A[59,1]M_1$.

<u>DEGREE 39:</u> $\eta B[38]$

The leading differential $d^{12}(\eta B[38]M_1^3 \bar{M}_2) = \eta A[47]M_1$ determines tentative differentials which are a monomorphism on $Z_2 \eta B[38]M_1^3 \bar{M}_2 \otimes B<4>$. There are no remaining elements.

DEGREE 39: $A[39,1] = \sigma A[32,3]$

The leading differential $d^8(A[39,1]M_1^5) = \eta^2 C[44]M_1$ determines tentative differentials which in degrees less than 70 have kernel

$Z_2A[39,1]\{M_1^2M_2, M_1^3M_2, M_1^6M_2+M_1^3M_2^2, M_1M_2^3, M_1M_2M_3, M_1^2M_2M_3, M_1^{10}M_2, M_1^3M_2M_3+M_1^4M_2^3, M_1^{11}M_2, M_1^6\overline{M}_2^3\}$

Thus, the new $A[39,1]$-leader is $A[39,1]M_1^2M_2$.

The leading differential $d^8(A[32,1](M_1^2M_3+M_2^3)) = A[39,1]M_1^2M_2$ determines tentative differentials which in degrees less than 70 have cokernel $Z_2A[39,1]\{M_1^3M_2, M_1M_2^3, M_1M_2M_3, M_1^2M_2M_3, M_1^3M_2M_3+M_1^4M_2^3, M_1^{11}M_2, M_1^6\overline{M}_2^3\}$. Thus, the new $A[39,1]$-leader is $A[39,1]M_1^3M_2$.

The leading differential $d^{10}(A[39,1]M_1^3M_2) = A[50,2]$ determines tentative differentials which are a monomorphism on $Z_2(A[39,1]M_1^3M_2) \otimes B\langle 4\rangle$. The only remaining elements in degrees less than 70 are $A[39,1]\{M_1M_2M_3, M_1^6\overline{M}_2^3\}$. Moreover, $d^{14}(A[39,1]M_1M_2M_3) = A[52,1]M_1M_2$.

DEGREE 39: $A[39,3]$

The leading differential $d^8(A[39,3]M_1^2\overline{M}_2) = \eta A[45,1]M_1$ determines tentative differentials which are a monomorphism on $Z_2(A[39,3]M_1^2\overline{M}_2) \otimes B\langle 4\rangle$. There are no remaining elements.

DEGREE 39: $\sigma A[32,1]$

The leading differential $d^{10}(\sigma A[32,1]M_1^4\overline{M}_2) = \nu A[45,1]M_1^2$ determines tentative differentials with kernel in degrees less than 68 equal to $Z_2\sigma A[32,1]\{M_1^4M_2^2, M_1^6M_2^2\}$. Moreover, $d^{16}(A[30]\langle M_4\rangle) = \sigma A[32,1]M_1^4M_2^2$ and $d^{22}(4C[18]M_1^7\overline{M}_2M_2^2\overline{M}_3) = \sigma A[32,1]M_1^6M_2^2$.

DEGREE 40: $C[20]^2$

The leading differential $d^8(2C[20]^2M_1\bar{M}_2) = B[47]$ determines tentative differentials which are a monomorphism on $Z_2(2C[20]^2)\{M_1\bar{M}_2, M_1^3\bar{M}_2\} \otimes B\langle 4\rangle$. There are no remaining elements.

DEGREE 40: $A[40,1]$

The leading differential $d^8(A[40,1]M_1^6) = \nu C[44]M_1^2$ determines tentative differentials with kernel in degrees less than 69 equal to $Z_2A[40,1]\{\bar{M}_2\bar{M}_3, M_1^2\bar{M}_2\bar{M}_3\}$. Thus, the new $A[40,1]$-leader is $A[40,1](\bar{M}_2\bar{M}_3+M_1^2M_2^2)$.

The leading differential $d^{12}(A[40,1]\bar{M}_2\bar{M}_3) = \sigma C[44]M_1^4$ determines the tentative differential $d^{12}(A[40,1]M_1^2\bar{M}_2\bar{M}_3) = \sigma C[44]M_1^6$. There are no remaining elements.

DEGREE 40: $A[40,2]$

The leading differential $d^8(A[40,2]M_1^6\bar{M}_2) = A[47]M_1^2\bar{M}_2$ determines only the tentative differential $d^8(A[40,2]M_1^6\bar{M}_3) = A[47]M_1^6\bar{M}_2$ in degrees less than 69. There are no remaining elements.

DEGREE 40: $\eta A[39,3]$

The leading differential $d^{20}(\nu C[18]M_1^6\bar{M}_2\bar{M}_3) = (\eta A[39,3]+\eta\sigma A[32,1])M_1^3\bar{M}_2$ determines tentative differentials with image $Z_2(\eta A[39,3]M_1^3\bar{M}_2+\eta\sigma A[32,1]M_1^3\bar{M}_2)\{1, \langle M_1^4\rangle, \langle M_2^2\rangle, \langle M_1^4\rangle^2\}$ in degrees less than 69. The only remaining element is $\eta A[39,3]M_1^3\bar{M}_2\langle M_3\rangle$.

DEGREE 40: $\eta\sigma A[32,1]$

The leading differential $d^{16}(\eta\sigma A[32,1]M_1^5\langle M_3\rangle) = A[63]$ determines the tentative differential $d^{16}(\eta\sigma A[32,1]M_1<M_2^2><M_3>) = A[63]M_1^2$. There are no remaining elements in degrees less than 69.

<u>DEGREE 41:</u> $\eta A[40,1]$

The leading differential $d^8(B[34]M_1^4M_2) = \eta A[40,1]M_1^3$ determines tentative differentials which are a monomorphsim on the remaining elements in degrees less than 68.

<u>DEGREE 42:</u> $C[42]$

The leading differential $d^6(2C[42]M_1^3) = A[47]$ determines tentative differentials which are a monomorphism on $Z_2(2C[42])\{M_1^3, M_1^2M_2, M_1^3M_2\} \otimes B<4>$. There are no remaining elements in degrees less than 69.

<u>DEGREE 42:</u> $\eta^2C[20]^2$

The leading differential $d^6(\eta^2C[20]^2M_1\bar{M}_2) = \eta^2A[45,2]M_1$ determines tentative differentials which are a monomorphism on $Z_2\eta^2C[20]^2\{M_1\bar{M}_2, M_1^3\bar{M}_2\} \otimes B<4>$. There are no remaining elements.

<u>DEGREE 44:</u> $C[44]$

The leading differential $d^4(C[44]M_1^2) = \nu C[44]$ determines tentative differentials which are a monomorphism on $Z_2 \otimes (Z_8C[44]\{M_1^2, \bar{M}_2, M_1^2\bar{M}_2\} \otimes B<4>)$. The remaining elements from $Z_2 \otimes (Z_8C[44] \otimes H_*BP)$ in degrees less than 69 are $Z_2C[44]\{<M_1^4>, <M_2^2>, <M_3>, <M_1^4><M_2^2>, <M_1^4><M_3>, <M_1^4>^3\}$. Thus, the new $C[44]$-leader is $C[44]<M_1^4>$.

The leading differential $d^8(C[44]<M_1^4>) = \sigma C[44]$ determines tentative differentials which are a monomorphism on

$Z_2C[44]\{\langle M_1^4\rangle, \langle M_2^2\rangle, \langle M_3\rangle, \langle M_1^4\rangle\langle M_2^2\rangle, \langle M_1^4\rangle\langle M_3\rangle, \langle M_1^4\rangle^3\}$. There are no remaining elements from $Z_2 \otimes (Z_8C[44] \otimes H_*BP)$ in degrees less than 69.

The leading differential $d^{12}(2C[44]M_1^6) = 2\sigma C[44]M_1^2$ determines tentative differentials with kernel on $Z_2 \otimes [Z_4(2C[44]) \otimes H_*BP]$ in degrees less than 69 equal to $Z_2(2C[44])\{M_1^2M_2^3, M_1^5M_3+M_1^6M_2^2+M_1^3M_2^3\}$.

The leading differential $d^{14}(4C[44]M_1^7) = A[57]$ determines tentative differentials which are a monomorphism on $Z_2(4C[44])\{M_1^7, M_1^6\bar{M}_2, M_1^7\bar{M}_2, M_1^4\bar{M}_3\}$. The only remaing element in degrees less than 69 is $4C[44]M_1^3M_2^3$.

<u>DEGREE 45:</u> D[45]

The leading differential $d^{12}(B[34]M_1^6) = 2D[45]$ determines tentative differentials with cokernel in $Z_2 \otimes (Z_8(2D[45]) \otimes H_*BP)$ in degrees less than 68 given by the table below. The 2D[45]-leader is $2D[45]M_1^2$.

DEGREE	GENERATOR	DEGREE	GENERATOR	DEGREE	GENERATOR
(4,45)	2 0 0 0	(6,45)	3 0 0 0	(8,45)	1 1 0 0
(10,45)	2 1 0 0	(12,45)	3 1 0 0		6 0 0 0
(14,45)	1 2 0 0		7 0 0 0	(16,45)	1 0 1 0
	2 2 0 0		5 1 0 0	(18,45)	2 0 1 0
	3 2 0 0		6 1 0 0	(20,45)	3 0 1 0
	0 1 1 0		1 3 0 0		4 2 0 0
	7 1 0 0		10 0 0 0	(22,45)	1 1 1 0
	4 0 1 0		2 3 0 0		5 2 0 0
	11 0 0 0				

The d^8-differentials determined by the relation $\sigma B[38] = 4D[45]$ have cokernel on the elements from $Z_4(4D[45]) \otimes H_*BP$ in degrees less than 68 equal to

$(Z_2(4D[45])\{M_1M_2, M_1^3M_2\} \otimes B\langle 4\rangle) \oplus Z_2 4D[45]M_1^6\overline{M}_2$. Thus, the new 4D[45]-leader is $4D[45]M_1M_2$, and the new 8D[45]-leader is 0.

The leading differential $d^4(D[45]M_1^2) = \nu D[45]$ determines leading differentials which are a monomorphism on $[Z_4 \otimes (Z_{16}D[45]\{M_1^2, \overline{M}_2, M_1^2\overline{M}_2\} \otimes B\langle 4\rangle)]$ $\oplus [Z_2 \otimes ((Z_8(2D[45])\{M_1\overline{M}_2, M_1^3\overline{M}_2\} \otimes B\langle 4\rangle)]$. The remaining elements from $Z_4 \otimes (Z_{16}D[45] \otimes H_*BP)$ in degrees less than 68 are $Z_2(2D[45])\{M_1\langle M_2^2\rangle, M_1\langle M_3\rangle, M_2\langle M_3\rangle, \langle M_1^4\rangle\langle M_2^2\rangle, \langle M_1^4\rangle M_3, M_1^5\langle M_2^2\rangle\}$. Thus, the new D[45]-leader is 0, and the new 2D[45]-leader is $2D[45]M_2\langle M_1^4\rangle$.

The leading differential $d^{22}(\eta A[23]M_1^{15}\overline{M}_2) = 2D[45]M_1^4M_2$ determines tentative differentials with image $Z_2(2D[45])\{M_1^4M_2, M_2\langle M_3\rangle, M_3\langle M_1^4\rangle\}$ in degrees less than 68. The remaining elements from $Z_2 \otimes [Z_8(2D[45]) \otimes H_*BP]$ are $Z_2(2D[45])\{M_1\langle M_3\rangle, \langle M_1^4\rangle\langle M_2^2\rangle, M_1^5\langle M_2^2\rangle\}$, and the new 2D[45]-leader is $2D[45]M_1\langle M_3\rangle$.

The leading differentials $d^{14}(A[32,1]M_1^8\overline{M}_3) = 2D[45]M_1\langle M_3\rangle$ and $d^{14}(A[32,1]M_1^2\langle M_4\rangle) = 2D[45]\langle M_1^4\rangle\langle M_2^2\rangle$ determine no tentative differentials in degrees less than 68. The only remaining element from $Z_2 \otimes [Z_8(2D[45]) \otimes H_*BP]$ is $2D[45]M_1^5\langle M_2^2\rangle$.

The leading differential $d^{10}(A[36]M_1^2\overline{M}_3) = 4D[45]M_1M_2$ determines tentative differentials with cokernel on the remaining elements from $Z_4(4D[45]) \otimes H_*BP$ in degrees less than 68 equal to $Z_2(4D[45])\{M_1^3M_2, M_1^6\overline{M}_2, M_1^3M_2\langle M_1^4\rangle\}$. Thus, the new 4D[45]-leader is $4D[45]M_1^3M_2$.

The leading differential $d^{12}(4D[45]M_1^3M_2) = A[52,1]M_1^2$ determines tentative differentials which are a monomorphism on $Z_2(4D[45])\{M_1^3M_2, M_1^3M_2\langle M_1^4\rangle\}$. Thus, the only remaining element from $Z_4(4D[45]) \otimes H_*BP$ in degrees less than 68 is $4D[45]M_1^6M_2$.

The leading differential $d^{12}(2B[34]M_1^5\overline{M}_2\overline{M}_3) = 4D[45]M_1^6M_2$ determines the tentative differential $d^{12}(2B[34]M_1\overline{M}_2^3\overline{M}_3) = 2D[45]M_1^5\langle M_2^2\rangle$. There are no remaining elements from $Z_8C[44] \otimes H_*BP$ in degrees less than 69.

DEGREE 45: A[45,1]

The leading differential $d^4(A[45,1]M_1^2) = \nu A[45,1]$ determines tentative differentials which are a monomorphism on $Z_2A[45,1]\{M_1^2, \overline{M}_2, M_1^2\overline{M}_2\} \otimes B\langle 4\rangle$. There are no remaining elements.

DEGREE 45: A[45,2]

The leading differential $d^4(A[45,2]\overline{M}_2) = \eta B[47]M_1$ determines tentative differentials which are a monomorphism on $Z_2A[45,2]\{\overline{M}_2, M_1^2\overline{M}_2\} \otimes B\langle 4\rangle$. There are no remaining elements.

DEGREE 46: ηD[45] and ηA[45,2]

Both $\eta^2D[45]$ and $\eta^2A[45,2]$ are nonzero. Thus, the only element of E^4 with a representative in $(Z_2\eta D[45] \oplus Z_2\eta A[45,2]) \otimes H_*BP$ is 0.

DEGREE 46: $\eta^2C[44]$

The leading differential $d^2(\eta C[44]M_1) = \eta^2C[44]$ determines tentative differentials with image $Z_2(\eta^2C[44]) \otimes B\langle 2\rangle$ and cokernel $Z_2(\eta^2C[44]M_1) \otimes B\langle 2\rangle$. The $\eta^2C[44]$-leader is $\eta^2C[44]M_1$.

The leading differential $d^8(A[39,1]M_1^5) = \eta^2C[44]M_1$ determines tentative differentials with cokernel in degrees less than 69 equal to $Z_2\eta^2C[44]\{M_1^7\overline{M}_2, M_1^5\overline{M}_2^2\}$. Moreover, $d^{14}(\eta^2C[44]M_1^7\overline{M}_2) = A[59,1]\overline{M}_2$.

<u>DEGREE 46:</u> $\eta A[45,1]$

The leading differential $d^2(A[45,1]M_1) = \eta A[45,1]$ determines tentative differentials with image $Z_2(\eta A[45,1]) \otimes B<2>$ and cokernel $Z_2(\eta A[45,1]M_1) \otimes B<2>$. The $\eta A[45,1]$-leader is $\eta A[45,1]M_1$.

The leading differential $d^8(A[39,3]M_1^2\bar{M}_2) = \eta A[45,1]M_1$ determines tentative differentials with image $Z_2\eta A[45,1]M_1 \otimes B<4>$. The remaining elements are $Z_2\eta A[45,1]\{M_1^3, M_1\bar{M}_2, M_1^3\bar{M}_2\} \otimes B<4>$, and the new $\eta A[45,1]$-leader is $\eta A[45,1]M_1^3$.

The leading differential $d^6(\eta A[45,1]M_1^3) = 2\sigma C[44]$ determines tentative differentials which are a monomorphism on $Z_2\eta A[45,1]\{M_1^3, M_1\bar{M}_2, M_1^3\bar{M}_2\} \otimes B<4>$. There are no remaining elements.

<u>DEGREE 47:</u> $\eta^2 A[45,2] = 2B[47] = \eta A[14]A[32,2]$

The leading differential $d^2(\eta A[45,2]M_1) = \eta^2 A[45,2]$ determines tentative differentials with image $Z_2(\eta^2 A[45,2]) \otimes B<2>$ and cokernel $Z_2(\eta^2 A[45,2]M_1) \otimes B<2>$. The $\eta^2 A[45,2]$-leader is $\eta^2 A[45,2]M_1$.

The leading differential $d^6(\eta^2 C[20]^2 M_1\bar{M}_2) = \eta^2 A[45,2]M_1$ determines tentative differentials with image $Z_2\eta^2 A[45,2]\{M_1, M_1^3\} \otimes B<4>$. The remaining elements are $Z_2\eta^2 A[45,2]\{M_1\bar{M}_2, M_1^3\bar{M}_2\} \otimes B<4>$, and the new $\eta^2 A[45,2]$-leader is $\eta^2 A[45,2]M_1\bar{M}_2$.

The leading differential $d^8(\eta^2 A[45,2]M_1M_2) = A[54,2]$ determines tentative differentials which are a monomorphism on $Z_2\eta^2 A[45,2]\{M_1\bar{M}_2, M_1^3\bar{M}_2\} \otimes B<4>$. Thus, there are no remaining elements.

<u>DEGREE 47:</u> $\eta^2 D[45]$

The leading differential $d^2(\eta D[45]M_1) = \eta^2 D[45]$ determines tentative differentials with image $Z_2(\eta^2D[45]) \otimes B\langle 2\rangle$ and cokernel $Z_2(\eta^2D[45]M_1) \otimes B\langle 2\rangle$. The $\eta^2D[45]$-leader is $\eta^2D[45]M_1$.

The tentative differentials determined by the leading differential $d^{10}(\eta\sigma A[30]M_1^3\bar{M}_2) = \eta^2D[45]M_1$ have image $Z_2\eta^2D[45]M_1 \otimes B\langle 4\rangle$. The remaining elements are $Z_2(\eta^2D[45])\{M_1^3, M_1\bar{M}_2, M_1^3\bar{M}_2\} \otimes B\langle 4\rangle$, and the new $\eta^2D[45]$-leader is $\eta^2D[45]M_1^3$.

The leading differential $d^6(\eta^2D[45]M_1^3) = A[52,1]$ determines tentative differentials which are a monomorphism on $Z_2(\eta^2D[45])\{M_1^3, M_1\bar{M}_2, M_1^3\bar{M}_2\} \otimes B\langle 4\rangle$. There are no remaining elements.

<u>DEGREE 47:</u> $\nu C[44]$

The leading differential $d^4(C[44]M_1^2) = \nu C[44]$ determines tentative differentials with image $Z_2\nu C[44]\{1, M_1, M_2\} \otimes B\langle 4\rangle$ and cokernel $Z_2\nu C[44]\{M_1^2, M_1^3, M_1M_2, M_1^2M_2, M_1^3M_2\} \otimes B\langle 4\rangle$. The $\nu C[44]$-leader is $\nu C[44]M_1^2$.

The leading differential $d^8(A[40,1]M_1^6) = \nu C[44]M_1^2$ determines tentative differentials whose image in degrees less than 68 is all of $Z_2\nu C[44]\{M_1^2, M_1^3, M_1^2M_2\} \otimes B\langle 4\rangle$ except for $Z_2\nu C[44]M_1^3\langle M_2^2\rangle$. The remaining elements in degrees less than 68 are $Z_2\nu C[44]M_1^3\langle M_2^2\rangle \oplus Z_2\nu C[44]\{M_1M_2, M_1^3M_2\} \otimes B\langle 4\rangle$, and the new $\nu C[44]$-leader is $\nu C[44]M_1M_2$.

The leading differential $d^8(\nu C[44]M_1M_2) = \eta\sigma C[44]M_1$ determines tentative differentials which are a monomorphism on $Z_2\nu C[44]\{M_1M_2, M_1^3M_2\} \otimes B\langle 4\rangle$. The only remaining element in degrees less than 68 is $\nu C[44]M_1^3\langle M_2^2\rangle$. Moreover, $d^{18}(\nu C[44]M_1^3\langle M_2^2\rangle) = B[64,2]$.

<u>DEGREE 47:</u> $A[47] = \sigma A[40,2]$

The leading differential $d^6(2C[42]M_1^3) = A[47]$ determines leading differentials with image $Z_2A[47]\{1,M_1^2,\bar{M}_2\} \otimes B\langle 4\rangle$. Since $\eta A[47] \neq 0$, the remaining elements are $Z_2A[47]M_1^2\bar{M}_2 \otimes B\langle 4\rangle$, and the new A[47]-leader is $A[47]M_1^2\bar{M}_2$.

The leading differential $d^8(A[40,2]M_1^6\bar{M}_2) = A[47]M_1^2\bar{M}_2$ determines tentative differentials with image in degrees less than 68 equal to $Z_2A[47]M_1^2\bar{M}_2 \otimes B\langle 4\rangle$. There are no remaining elements.

DEGREE 47: B[47]

The leading differential $d^6(2C[20]^2M_1\bar{M}_2) = B[47]$ determines tentative differentials with image $Z_2B[47]\{1,M_1^2\} \otimes B\langle 4\rangle$. Since $\eta B[47] \neq 0$, the remaining elements are $Z_2 \otimes (Z_4B[47]\{\bar{M}_2,M_1^2\bar{M}_2\} \otimes B\langle 4\rangle)$, and the new B[47]-leader is $B[47]\bar{M}_2$.

The leading differential $d^6(B[47]\bar{M}_2) = A[52,2]$ determines tentative differentials which are a monomorphism on $Z_2 \otimes (Z_4B[47]\{\bar{M}_2,M_1^2\bar{M}_2\} \otimes B\langle 4\rangle)$. There are no remaining elements.

DEGREE 48: $\eta A[47]$

The leading differential $d^2(A[47]M_1) = \eta A[47]$ determines tentative differentials with image $Z_2(\eta A[47]) \otimes B\langle 2\rangle$ and cokernel $Z_2(\eta A[47]M_1) \otimes B\langle 2\rangle$. The $\eta A[47]$-leader is $\eta A[47]M_1$.

The leading differential $d^4(\eta B[38]M_1^3\bar{M}_2) = \eta A[47]M_1$ determines tentative differentials with image $Z_2\eta A[47]M_1 \otimes B\langle 4\rangle$. The remaining elements are $Z_2\eta A[47]\{M_1^3,M_1\bar{M}_2,M_1^3\bar{M}_2\} \otimes B\langle 4\rangle$, and the new $\eta A[47]$-leader is $\eta A[47]M_1^3$.

The leading differential $d^{32}(\eta^2\gamma_1M_1^{19}) = \eta A[47]M_1^3$ determines tentative differentials which are surjective in degrees less than 69. Thus, there are no remaining elements.

<u>DEGREE 48:</u> $\nu A[45,1]$

The leading differential $d^4(A[45,1]M_1^2) = \nu A[45,1]$ determines tentative differentials with cokernel $Z_2\nu A[45,1]\{M_1^2, M_1^3, M_1M_2, M_1^2M_2, M_1^3M_2\} \otimes B\langle 4\rangle$. The new $\nu A[45,1]$-leader is $\nu A[45,1]M_1^2$.

The leading differential $d^{10}(\sigma A[32,1]M_1^4\bar{M}_2) = \nu A[45,1]M_1^2$ determines tentative differentials with image in degrees less than 69 equal to $Z_2\nu A[45,1]\{M_1^2, M_1M_2\} \otimes B\langle 4\rangle$. The remaining elements in degrees less than 69 are $Z_2\nu A[45,1]\{M_1^3, M_1^2M_2, M_1^3M_2\} \otimes B\langle 4\rangle$, and the new $\nu A[45,1]$-leader is $\nu A[45,1]M_1^3$.

The leading differential $d^4(\nu A[45,1]M_1^3) = \nu^2A[45,1]M_1 = \eta A[50,2]M_1$ determines tentative differentials which are a monomorphism on $Z_2\nu A[45,1]\{M_1^3, M_1^2M_2, M_1^3M_2\} \otimes B\langle 4\rangle$. There are no remaining elements in degrees less than 69.

<u>DEGREE 48:</u> $\nu D[45]$

The leading differential $d^4(D[45]M_1^2) = \nu D[45]$ determines tentative differentials with image $[Z_4(\nu D[45])\{1, M_1, M_2\} \otimes B\langle 4\rangle] \oplus [Z_2(2\nu D[45])\{M_1^2, M_1M_2\} \otimes B\langle 4\rangle]$. The remaining elements are $[Z_4(\nu D[45])\{M_1^3, M_1^2M_2, M_1^3M_2\} \otimes B\langle 4\rangle] \oplus [Z_2(\nu D[45])\{M_1^2, M_1M_2\} \otimes B\langle 4\rangle]$. Thus, the $\nu D[45]$-leader is $\nu D[45]M_1^2$.

The leading differential $d^4(\nu D[45]M_1^2) = \nu^2D[45]$ determines tentative differentials which are a monomorphism on $Z_2 \otimes (Z_4\nu D[45]\{M_1^2, M_1^3, M_1M_2, M_1^2M_2, M_1^3M_2\} \otimes B\langle 4\rangle)$. The remaining elements are $Z_2(2\nu D[45])\{M_1^3, M_1^2M_2, M_1^3M_2\}$, and the new $\nu D[45]$-leader is $2\nu D[45]M_1^3$.

The leading differential $d^6(2\nu D[45]M_1^3) = A[8]D[45]$ determines tentative differentials which are a monomorphism on $Z_2(2\nu D[45])\{M_1^3, M_1^2M_2, M_1^3M_2\}$. Thus, there are no remaining elements.

DEGREE 48: $\eta B[47]$

The leading differential $d^2(B[47]M_1) = \eta B[47]$ determines tentative differentials with image $Z_2(\eta B[47]) \otimes B\langle 2\rangle$ and cokernel $Z_2(\eta B[47]M_1) \otimes B\langle 2\rangle$. The $\eta B[47]$-leader is $\eta B[47]M_1$.

The leading differential $d^4(A[45,2]\overline{M}_2) = \eta B[47]M_1$ determines tentative differentials with image $Z_2\eta B[47]\{M_1, M_1^3\} \otimes B\langle 4\rangle$. The remaining elements are $Z_2\eta B[47]\{M_1\overline{M}_2, M_1^3\overline{M}_2\} \otimes B\langle 4\rangle$, and the new $\eta B[47]$-leader is $\eta B[47]M_1\overline{M}_2$.
The leading differential $d^6(\eta B[47]M_1\overline{M}_2) = \eta A[52,2]M_1$ determines tentative differentials which are a monomorphism on $Z_2\eta B[47]\{M_1\overline{M}_2, M_1^3\overline{M}_2\} \otimes B\langle 4\rangle$. Thus, there are no remaining elements.

DEGREE 50: $A[50,1]$

The leading differential $d^{34}(\eta^2\gamma_1M_1^{17}) = A[50,1]$ determines tentative differentials with image in degrees less than 69 equal to $Z_2A[50,1]\{1, M_1\} \otimes B\langle 4\rangle$. Thus, the remaining elements are $Z_2A[50,1]\{M_1^2, M_1^3, M_2, M_1M_2, M_1^2M_2, M_1^3M_2\} \otimes B\langle 4\rangle$, and the new $A[50,1]$-leader is $A[50,1]M_1^2$.

The leading differential $d^4(A[50,1]M_1^2) = \nu A[50,1]$ determines tentative differentials which are a monomorphism on $Z_2A[50,1]\{M_1^2, M_1^3, M_1M_2, M_1^2M_2, M_1^3M_2\} \otimes B\langle 4\rangle$. The remaining elements in degrees less than 69 are $Z_2A[50,1]M_2 \otimes B\langle 4\rangle$, and the new $A[50,1]$-leader is $A[50,1]M_2$.

The leading differential $d^{40}(4\beta_1(M_1^7\overline{M}_2^3\overline{M}_3+M_1^{10}\overline{M}_2^2\overline{M}_3+M_1^{14}\overline{M}_2^3)) = A[50,1]M_2$ determines tentative differentials with image $Z_2A[50,1]M_2 \otimes B\langle 4\rangle$ in degrees less than 69. Thus, there are no remaining elements.

DEGREE 50: $A[50,2]$

The leading differential $d^{12}(A[39,1]M_1^3M_2) = A[50,2]$ determines tentative differentials with image equal to $Z_2A[50,2] \otimes B\langle 4\rangle$. Since $\eta A[50,2]$ is nonzero, the remaining elements are $Z_2A[50,2]\{M_1^2,\bar{M}_2,M_1^2\bar{M}_2\} \otimes B\langle 4\rangle$, and the $A[50,2]$-leader is $A[50,2]M_1^2$.

The leading differential $d^4(A[50,2]M_1^2) = \nu A[50,2]$ determines tentative differentials which are a monomorphism on $Z_2A[50,2]\{M_1^2,\bar{M}_2,M_1^2\bar{M}_2\} \otimes B\langle 4\rangle$. Thus, there are no remaining elements.

DEGREE 51: $\sigma C[44]$

The leading differential $d^8(C[44]\langle M_1^4\rangle) = \sigma C[44]$ determines tenatiative differentials with image $Z_2(\sigma C[44])\{1,M_1^2,\bar{M}_2,M_2^2,\bar{M}_3,M_1^8\}$ in degrees less than 68. Since $\eta\sigma C[44] \neq 0$, the remaining elements are $Z_2(\sigma C[44])\{M_1^4,M_1^2\bar{M}_2,M_1^6,M_1^4\bar{M}_2,M_1^2M_2^2\}$. Thus, the $\sigma C[44]$-leader is $\sigma C[44]M_1^4$.

The leading differential $d^{12}(A[40,1]\bar{M}_2\langle M_3\rangle) = \sigma C[44]M_1^4$ determines the tentative differential $d^{12}(A[40,1]M_1^2\bar{M}_2\langle M_3\rangle) = \sigma C[44]M_1^6$. The remaining elements in degrees less than 68 are $Z_2(\sigma C[44])\{M_1^2\bar{M}_2,M_1^4\bar{M}_2,M_1^2M_2^2\}$, and the new $\sigma C[44]$-leader is $\sigma C[44]M_1^2\bar{M}_2$.

The leading differential $d^{30}(\nu A[19]M_1^7M_2^2\langle M_3\rangle) = \sigma C[44]M_1^2\bar{M}_2$ determines no tentative differentials in degrees less than 68. The remaining elements in degrees less than 68 are $Z_2(\sigma C[44])\{M_1^4\bar{M}_2,M_1^2M_2^2\}$ and the new $\sigma C[44]$-leader is $\sigma C[44]M_1^4\bar{M}_2$. Moreover, $d^{14}(\sigma C[44]M_1^4\bar{M}_2) = A[64,1]$.

DEGREE 51: $2\sigma C[44]$

The leading differential $d^6(\eta A[45,1]M_1^3) = 2\sigma C[44]$ determines tentative differentials with image $Z_2(2\sigma C[44])\{1,M_1,M_2\} \otimes B\langle 4\rangle$. The remaining elements are $Z_2(2\sigma C[44])\{M_1^2,M_1^3,M_1M_2,M_1^2M_2,M_1^3M_2\} \otimes B\langle 4\rangle$, and the $2\sigma C[44]$-leader is $2\sigma C[44]M_1^2$.

The leading differential $d^8(2C[44]<M_2^2>) = 2\sigma C[44]M_1^2$ determines tentative differentials with image in degrees less than 68 equal to all the remaining elements listed above.

<u>DEGREE 51:</u> $\nu^2D[45]$

The leading differential $d^4(\nu D[45]M_1^2) = \nu^2D[45]$ determines tentative differentials with image $Z_2\nu^2D[45]\{1, M_1, M_1^2, M_2, M_1M_2\} \otimes B<4>$. The remaining elements are $Z_2\nu^2D[45]\{M_1^3, M_1^2M_2, M_1^3M_2\} \otimes B<4>$, and the $\nu^2D[45]$-leader is $\nu^2D[45]M_1^3$.

The leading differential $d^4(\nu^2D[45]M_1^3) = \eta A[8]D[45]M_1$ determines tentative differentials which are a monomorphism on $Z_2\nu^2D[45]\{M_1^3, M_1^2M_2, M_1^3M_2\} \otimes B<4>$. There are no remaining elements.

<u>DEGREE 51:</u> $\eta A[50,2]$

The leading differential $d^2(A[50,2]M_1) = \eta A[50,2]$ determines tentative differentials with image $Z_2\eta A[50,2] \otimes B<2>$ and cokernel $Z_2\eta A[50,2]M_1 \otimes B<2>$. Thus, the $\eta A[50,2]$-leader is $\eta A[50,2]M_1$.

The leading differential $d^4(\nu A[45,1]M_1^3) = \eta A[50,2]M_1$ determines tentative differentials with image $Z_2\eta A[50,2]\{M_1, M_1^3, M_1\bar{M}_2\} \otimes B<4>$. The remaining elements are $Z_2\eta A[50,2]M_1^3\bar{M}_2 \otimes B<4>$, and the new $\eta A[50,2]$-leader is $\eta A[50,2]M_1^3\bar{M}_2$. The leading differential $d^6(\eta A[50,2]M_1^3\bar{M}_2) = A[62,4]$ determines tentative differentials which are a monomorphism on $Z_2\eta A[50,2]M_1^3\bar{M}_2 \otimes B<4>$. There are no remaining elements.

<u>DEGREE 52:</u> $A[52,1]$

The leading differential $d^6(\eta^2D[45]M_1^3) = A[52,1]$ determines tentative differentials with image $Z_2A[52,1]\{1,M_1,M_2\} \otimes B<4>$. The remaining elements are $Z_2A[52,1]\{M_1^2,M_1^3,M_1M_2,M_1^2M_2,M_1^3M_2\} \otimes B<4>$, and the A[52,1]-leader is $A[52,1]M_1^2$.

The leading differential $d^8(4D[45]M_1^3M_2) = A[52,1]M_1^2$ determines tentative differentials with image $Z_2A[52,1]M_1^2 \otimes B<4>$. The remaining elements are $Z_2A[52,1]\{M_1^3,M_1M_2,M_1^2M_2,M_1^3M_2\} \otimes B<4>$, and the A[52,1]-leader is $A[52,1]M_1^3$.

The leading differential $d^{36}(\eta^2\gamma_1 M_1^{15}\bar{M}_2^2) = A[52,1]M_1^3$ determines tentative differentials with image in degrees less than 69 equal to $Z_2A[52,1]\{M_1^3,M_1^2M_2\}$. The remaining elements in degrees less than 69 are $Z_2A[52,1]\{M_1M_2,M_1^3M_2,M_1^7,M_1^5M_2\}$, and the new A[52,1]-leader is $A[52,1]M_1M_2$.

The leading differential $d^{14}(A[39,1]M_1M_2M_3) = A[52,1]M_1M_2$ determines no tentative differentials in degrees less than 69. The remaining elements in degrees less than 69 are $Z_2A[52,1]\{M_1^3M_2,M_1^7,M_1^5M_2\}$, and the new A[52,1]-leader is $A[52,1]M_1^3M_2$.

The leading differential $d^{10}(A[52,1]M_1^3M_2) = A[61]M_1$ determines tentative differentials which are a monomorphism on $Z_2A[52,1]M_1^3M_2 \otimes B<4>$. The only remianing elements in degrees less than 69 are $Z_2A[52,1]\{M_1^7,M_1^5M_2\}$.

<u>DEGREE 52:</u> A[52,2]

The leading differential $d^6(B[47]\bar{M}_2) = A[52,2]$ determines tentative differentials with image $Z_2A[52,2]\{1,M_1^2\} \otimes B<4>$. Since $\eta A[52,2]$ is nonzero, the remaining elements are $Z_2A[52,2]\{\bar{M}_2,M_1^2\bar{M}_2\} \otimes B<4>$, and the A[52,2]-leader is $A[52,2]\bar{M}_2$.

The leading differential $d^4(A[52,2]\bar{M}_2) = \eta A[54,2]M_1$ determines tentative differentials which are a monomorphism on $Z_2A[52,2]\{\bar{M}_2,M_1^2\bar{M}_2\} \otimes B<4>$. There are no remaining elements.

DEGREE 52: $\eta\sigma C[44]$

The leading differential $d^2(\sigma C[44]M_1) = \eta\sigma C[44]$ determines tentative differentials with cokernel $Z_2\eta\sigma C[44]M_1 \otimes B\langle 2\rangle$, and the $\eta\sigma C[44]$-leader is $\eta\sigma C[44]M_1$.

The leading differential $d^6(\nu C[44]M_1M_2) = \eta\sigma C[44]M_1$ determines tentative differentials with image $Z_2\eta\sigma C[44]\{M_1, M_1^3\} \otimes B\langle 4\rangle$. The remaining elements are $Z_2\eta\sigma C[44]\{M_1\bar{M}_2, M_1^3\bar{M}_2\} \otimes B\langle 4\rangle$, and the new $\eta\sigma C[44]$-leader is $\eta\sigma C[44]M_1\bar{M}_2$.

The leading differential $d^8(\eta\sigma C[44]M_1\bar{M}_2) = A[59,1]$ determines tentative differentials which are a monomorphism on $Z_2(\eta\sigma C[44])\{M_1\bar{M}_2, M_1^3\bar{M}_2\} \otimes B\langle 4\rangle$. There are no remaining elements.

DEGREE 53: $\eta A[52,2]$

The leading differential $d^2(A[52,2]M_1) = \eta A[52,2]$ determines tentative differentials with image $Z_2\eta A[52,2] \otimes B\langle 2\rangle$ and cokernel $Z_2\eta A[52,2]M_1 \otimes B\langle 2\rangle$. The $\eta A[52,2]$-leader is $\eta A[52,2]M_1$.

The leading differential $d^6(\eta B[47]M_1\bar{M}_2) = \eta A[52,2]M_1$ determines tentative differentials with image $Z_2\eta A[52,2]\{M_1, M_1^3\} \otimes B\langle 4\rangle$. The remaining elements are $Z_2\eta A[52,2]\{M_1\bar{M}_2, M_1^3\bar{M}_2\} \otimes B\langle 4\rangle$, and the new $\eta A[52,2]$-leader is $\eta A[52,2]M_1\bar{M}_2$.

The leading differential $d^8(\eta A[52,2]M_1\bar{M}_2) = B[60]$ determines tentative differentials which are a monomorphsim on $Z_2\eta A[52,2]\{M_1\bar{M}_2, M_1^3\bar{M}_2\} \otimes B\langle 4\rangle$. There are no remaining elements.

DEGREE 53: $A[8]D[45]$

The leading differential $d^6(2\nu D[45]M_1^3) = A[8]D[45]$ determines tentative differentials with image $Z_2A[8]D[45]\{1, M_1^2, \bar{M}_2\} \otimes B\langle 4\rangle$. Since $\eta A[8]D[45] \neq 0$, the remaining elements are $Z_2A[8]D[45]M_1^2\overline{M}_2 \otimes B<4>$, and the A[8]D[45]-leader is $A[8]D[45]M_1^2\overline{M}_2$.

The leading differential $d^8(A[8]D[45]M_1^2\overline{M}_2) = 2B[62]$ determines tentative differentials which are a monomorphism on $Z_2A[8]D[45]M_1^2\overline{M}_2 \otimes B<4>$. There are no remaining elements.

<u>DEGREE 53:</u> $\nu A[50,1]$

The leading differential $d^4(A[50,1]M_1^2) = \nu A[50,1]$ determines tentative differentials with image $Z_2\nu A[50,1]\{1, M_1, M_1^2, M_2, M_1M_2\} \otimes B<4>$. The remaining elements are $Z_2\nu A[50,1]\{M_1^3, M_1^2M_2, M_1^3M_2\} \otimes B<4>$, and the $\nu A[50,1]$-leader is $\nu A[50,1]M_1^3$.

The leading differential $d^{18}(A[36]M_1^6\overline{M}_2^2) = \nu A[50,1]M_1^3$ determines no tentative differentials in degrees less than 68. The remaining elements are $(Z_2\nu A[50,1]\{M_1^2M_2, M_1^3M_2\} \otimes B<4>) \oplus Z_2\nu A[50,1]M_1^7$, and the new $\nu A[50,1]$-leader is $\nu A[50,1]M_1^2M_2$.

The tentative differential $d^{10}(\nu A[50,1]M_1^2M_2) = A[62,3]$ determines tentative differentials which are a monomorphism on $Z_2\nu A[50,1]\{M_1^2M_2, M_1^3M_2\} \otimes B<4>$. The only remaining element in degrees less than 68 is $\nu A[50,1]M_1^7$.

<u>DEGREE 53:</u> $\nu A[50,2]$

The leading differential $d^4(A[50,2]M_1^2) = \nu A[50,2]$ determines tentative differentials with image $Z_2\nu A[50,2]\{1, M_1, M_2\} \otimes B<4>$. The remaining elements are $Z_2\nu A[50,2]\{M_1^2, M_1^3, M_1M_2, M_1^2M_2, M_1^3M_2\} \otimes B<4>$, and the $\nu A[50,2]$-leader is $\nu A[50,2]M_1^2$.

The leading differential $d^4(\nu A[50,2]M_1^2) = \nu^2A[50,2]$ determines tentative differentials which are a monomorphism on $Z_2\nu A[50,2]\{M_1^2, M_1^3, M_1M_2, M_1^2M_2, M_1^3M_2\} \otimes B<4>$. There are no remaining elements.

DEGREE 54: $A[54,1]$

The leading differential $d^{36}(\beta_2M_1^{18}) = A[54,1]$ determines tentative differentials with image $Z_2A[54,1]\{1, M_1, M_1^2, M_2, M_1M_2\} \otimes B<4>$ in degrees less than 69. The remaining elements in degrees less than 69 are $Z_2A[54,1]\{M_1^3, M_1^2M_2, M_1^3M_2\} \otimes B<4>$, and the $A[54,1]$-leader is $A[54,1]M_1^3$.

The leading differential $d^4(A[54,1]M_1^3) = \eta A[56]M_1$ determines tentative differentials which are a monomorphism on $Z_2A[54,1]\{M_1^3, M_1^2M_2, M_1^3M_2\} \otimes B<4>$. There are no remaining elements in degrees less than 69.

DEGREE 54: $A[54,2] = A[14]C[20]^2$

The leading differential $d^8(\eta^2A[45,2]M_1\bar{M}_2) = A[54,2]$ determines tentative differentials with image $Z_2A[54,2]\{1, M_1^2\} \otimes B<4>$. Since $\eta A[54,2] \neq 0$, the remaining elements are $Z_2A[54,2]\{\bar{M}_2, M_1^2\bar{M}_2\} \otimes B<4>$, and the $A[54,2]$-leader is $A[54,2]\bar{M}_2$.

The leading differential $d^6(A[54,2]\bar{M}_2) = A[59,2]$ determines tentative differentials which are a monomorphism on $Z_2A[54,2]\{\bar{M}_2, M_1^2\bar{M}_2\} \otimes B<4>$. There are no remaining elements.

DEGREE 54: $\eta A[8]D[45]$

The leading differential $d^2(A[8]D[45]M_1) = \eta A[8]D[45]$ determines tentative differentials with image $Z_2\eta A[8]D[45] \otimes B<2>$ and cokernel $Z_2\eta A[8]D[45]M_1 \otimes B<2>$. The $\eta A[8]D[45]$-leader is $\eta A[8]D[45]M_1$.

The leading differential $d^4(\nu^2 D[45]M_1^3) = \eta A[8]D[45]M_1$ determines tentative differentials with image equal to $Z_2(\eta A[8]D[45])\{M_1, M_1^3, M_1\bar{M}_2\} \otimes B\langle 4\rangle$. The remaining elements are $Z_2\eta A[8]D[45]M_1^3\bar{M}_2 \otimes B\langle 4\rangle$, and the new $\eta A[8]D[45]$-leader is $\eta A[8]D[45]M_1^3\bar{M}_2$.

<u>DEGREE 55:</u> $\eta A[54,2]$

The leading differential $d^2(A[54,2]M_1) = \eta A[54,2]$ determines tentative differentials with image $Z_2\eta A[54,2] \otimes B\langle 2\rangle$ and cokernel $Z_2\eta A[54,2]M_1 \otimes B\langle 2\rangle$. The $\eta A[54,2]$-leader is $\eta A[54,2]M_1$.

The leading differential $d^4(A[52,2]\bar{M}_2) = \eta A[54,2]M_1$ determines tentative differentials with image $Z_2\eta A[54,2]\{M_1, M_1^3\} \otimes B\langle 4\rangle$. The remaining elements are $Z_2\eta A[54,2]\{M_1\bar{M}_2, M_1^3\bar{M}_2\} \otimes B\langle 4\rangle$.

The leading differential $d^6(\eta A[54,2]M_1\bar{M}_2) = \eta A[59,2]M_1$ determines tentative differentials which are a monomorphism on $Z_2\eta A[54,2]\{M_1\bar{M}_2, M_1^3\bar{M}_2\} \otimes B\langle 4\rangle$. There are no remaining elements.

<u>DEGREE 56:</u> $A[56]$

The leading differential $d^{38}(2\beta_2 M_1^{16}\bar{M}_2) = A[56]$ determines tentative differentials with image in degrees less than 69 equal to $Z_2A[56]\{1, M_1^2, \bar{M}_2\} \otimes B\langle 4\rangle$. Since $\eta A[56] \neq 0$, the only remaining element is $A[56]M_1^2\bar{M}_2$.

<u>DEGREE 56:</u> $\nu^2 A[50,2]$

The leading differential $d^4(\nu A[50,2]M_1^2) = \nu^2 A[50,2]$ determines tentative differentials with image $Z_2(\nu^2 A[50,2])\{1, M_1, M_1^2, M_2, M_1M_2\} \otimes B\langle 4\rangle$. The remaining elements are $Z_2(\nu^2 A[50,2])\{M_1^3, M_1^2M_2, M_1^3M_2\} \otimes B\langle 4\rangle$, and the $\nu^2 A[50,2]$-leader is $\nu^2 A[50,2]M_1^3$.

The leading differential $d^6(\nu^2A[50,2]M_1^3)$ =A[61] determines tentative differentials which are a monomorphism on $Z_2(\nu^2A[50,2])\{M_1^3, M_1^2M_2, M_1^3M_2\}$ $\otimes$ B<4>. There are no remaining elements.

DEGREE 57: A[57]

The leading differential $d^{14}(4C[44]M_1^7)$ = A[57] determines tentative differentials with image $Z_2A[57]\{1, M_1^2, \bar{M}_2, M_1^4\}$ in degrees less than 68. Since $\eta A[57] \neq 0$, the only remaining element in degrees less than 68 is $A[57]M_1^2\bar{M}_2$.

DEGREE 57: $\eta A[56]$

The leading differential $d^2(A[56]M_1)$ = $\eta A[56]$ determines tentative differentials with image $Z_2\eta A[56]$ $\otimes$ B<2> and cokernel $Z_2\eta A[56]M_1$ $\otimes$ B<2>. The $\eta A[56]$-leader is $\eta A[56]M_1$.

The leading differential $d^4(A[54,1]M_1^3)$ = $\eta A[56,2]M_1$ determines tentative differentials with image $Z_2\eta A[56]\{M_1, M_1^3, M_1\bar{M}_2\}$ $\otimes$ B<4>. The remaining elements are $Z_2(\eta A[56]M_1^3\bar{M}_2)$ $\otimes$ B<4>, and the new $\eta A[56]$-leader is $\eta A[56]M_1^3\bar{M}_2$.

DEGREE 58: $\eta A[57]$

The leading differential $d^2(A[57]M_1)$ = $\eta A[57]$ determines tentative differentials with image $Z_2\eta A[57]$ $\otimes$ B<2> and cokernel $Z_2\eta A[57]M_1$ $\otimes$ B<2>. The $\eta A[57]$-leader is $\eta A[57]M_1$.

The leading differential $d^8(\beta_2 M_1^{14}\bar{M}_3) = \eta A[57]M_1$ determines tentative differentials with image in degrees less than 69 equal to $Z_2\eta A[57]\{M_1, M_1^3, M_1\bar{M}_2, M_1^5\}$. There are no remaining elements.

<u>DEGREE 59:</u> A[59,1]

The leading differential $d^8(\eta\sigma C[44]M_1\bar{M}_2) = A[59,1]$ determines tentative differentials with image $Z_2A[59,1]\{1, M_1^2\} \otimes B\langle 4\rangle$. The remaining elements are $Z_2A[59,1]\{M_1, M_1^3, M_2, M_1M_2, M_1^2M_2, M_1^3M_2\} \otimes B\langle 4\rangle$, and the A[59,1]-leader is $A[59,1]M_1$.

The leading differential $d^{22}(B[38]M_1^2\bar{M}_2\bar{M}_3) = A[59,1]M_1$ determines no tentative differentials in degrees less than 68. The remaining elements are $Z_2A[59,1]\{M_1^3, M_2, M_1M_2\}$, and the new A[59,1]-leaders are $A[59,1]M_1^3$ and $A[59,1]M_2$.

The leading differentials $d^6(A[59,1]M_1^3) = A[64,2]$ and $d^6(A[59,1]M_2) = A[64,3]$ determine tentative differentials which are a monomorphism on $Z_2A[59,1]\{M_1^3, M_2, M_1M_2, M_1^2M_2, M_1^3M_2\} \otimes B\langle 4\rangle$. (The first leading differential determines tentative differentials by assigning 1 to $A[59,1]M_1^3$ and 0 to $A[59,1]M_2$. The second leading differential determines tentative differentials by assigning 1 to $A[59,1]M_1^3$ and $A[59,1]M_2$.) There are no remaining elements in degrees less than 68.

<u>DEGREE 59:</u> A[59,2] = A[14]A[45,2]

The leading differential $d^6(A[54,2]\bar{M}_2) = A[59,2]$ determines tentative differentials with image $Z_2A[59,2]\{1, M_1^2\} \otimes B\langle 4\rangle$. Since $\eta A[59,2] \neq 0$, the remaining elements are $Z_2A[59,2]\{\bar{M}_2, M_1^2\bar{M}_2\} \otimes B\langle 4\rangle$, and the A[59,2]-leader is $A[59,2]\bar{M}_2$.

The leading differential $d^4(A[59,2]\bar{M}_2) = \eta^2 B[60]M_1$ determines tentative differentials which are a monomorphism on $Z_2A[59,2]\{\bar{M}_2, M_1^2\bar{M}_2\} \otimes B\langle 4\rangle$. There are no remaining elements.

<u>DEGREE 60:</u> $B[60] = C[20]^3$

The leading differential $d^8(\eta A[52,2]M_1\bar{M}_2) = B[60]$ determines tentative differentials with image equal to $Z_2B[60]\{1, M_1^2\} \otimes B\langle 4\rangle$. Since $\eta B[60] \neq 0$, the remaining elements from $Z_2 \otimes (Z_4B[60] \otimes H_*BP)$ are $Z_2B[60]\{\bar{M}_2, M_1^2\bar{M}_2\} \otimes B\langle 4\rangle$, and the B[60]-leader is $B[60]\bar{M}_2$.

<u>DEGREE 60:</u> $2B[60] = \eta A[59,2]$

The leading differential $d^2(A[59,2]M_1) = \eta A[59,2]$ determines differentials with image $Z_2(\eta A[59,2]) \otimes B\langle 2\rangle$ and cokernel $Z_2(\eta A[59,2]M_1) \otimes B\langle 2\rangle$. The $\eta A[59,2]$-leader is $\eta A[59,2]M_1$.

The leading differential $d^6(\eta A[54,2]M_1\bar{M}_2) = \eta A[59,2]M_1$ determines tentative differentials with image $Z_2\eta A[59,2]\{M_1, M_1^3\} \otimes B\langle 4\rangle$. The remaining elements are $Z_2\eta A[59,2]\{M_1\bar{M}_2, M_1^3\bar{M}_2\} \otimes B\langle 4\rangle$, and the new $\eta A[59,2]$-leader is $\eta A[59,2]M_1\bar{M}_2$.

<u>DEGREE 61:</u> A[61]

The leading differential $d^6(\nu^2 A[50,2]M_1^3) = A[61]$ determines tentative differntials with image $Z_2A[61]\{1, M_1^2, \bar{M}_2\} \otimes B\langle 4\rangle$. The remaining elements are $Z_2A[61]\{M_1, M_1^3, M_1\bar{M}_2, M_1^2\bar{M}_2, M_1^3\bar{M}_2\} \otimes B\langle 4\rangle$, and the A[61]-leader is $A[61]M_1$.

The leading differential $d^{10}(A[52,1]M_1^3M_2) = A[61]M_1$ determines tentative differentials with image $Z_2A[61]M_1^3M_2 \otimes B\langle 4\rangle$. The remaining elements are $Z_2A[61]\{M_1^3, M_1\bar{M}_2, M_1^2\bar{M}_2, M_1^3\bar{M}_2\} \otimes B\langle 4\rangle$, and the new A[61]-leader is $A[61]M_1^3$.

DEGREE 61: $\eta B[60]$

Since $\eta^2 B[60] \neq 0$, the only element of E^4 with a representative in $Z_2\eta B[60] \otimes H_*BP$ is zero.

DEGREE 62: $A[62,1] = \theta_5$

The leading differential $d^{54}(\eta^2\sigma M_1^{21}M_2^2) = A[62,1]$ determines no tentative differentials in degrees less than 69. Since $\eta A[62,1] \neq 0$, the remaining elements in degrees less than 69 are $Z_2A[62,1]\{M_1^2,\bar{M}_2\}$, and the $A[62,1]$-leader is $A[62,1]M_1^2$.

DEGREE 62: $A[62,2]$

The leading differential $d^{44}(4\beta_2M_1^{19}\bar{M}_2) = A[62,2]$ determines no tentative differentials in degrees less than 69. Since $\eta A[62,2] \neq 0$ the remaining elements are $Z_2A[62,2]\{M_1^2,\bar{M}_2\}$, and the $A[62,2]$-leader is $A[62,2]M_1^2$.

DEGREE 62: $A[62,3]$

The leading differential $d^{10}(\nu A[50,1]M_1^2M_2) = A[62,3]$ determines tentative differentials with image $Z_2A[62,3]\{1,M_1\} \otimes B\langle 4\rangle$. The remaining elements are $Z_2A[62,3]\{M_1^2,M_1^3,M_2,M_1M_2,M_1^2M_2,M_1^3M_2\} \otimes B\langle 4\rangle$, and the $A[62,3]$-leader is $A[62,3]M_1^2$.

DEGREE 62: $A[62,4]$

The leading differential $d^{12}(\eta A[50,2]M_1^3\bar{M}_2) = A[62,4]$ determines tentative differentials with image $Z_2A[62,4] \otimes B\langle 4\rangle$. Since $\eta A[62,4] \neq 0$, the remaining elements are $Z_2A[62,4]\{M_1^2,\bar{M}_2,M_1^2\bar{M}_2\} \otimes B\langle 4\rangle$, and the $A[62,4]$-leader is $A[62,4]M_1^2$.

DEGREE 62: $B[62]$

The leading differential $d^{40}(\gamma_2 M_1^{20}) = B[62]$ determines tentative differentials whose image in degrees less than 69 equals $Z_2 B[62]\{1, M_1, M_1^2, \bar{M}_2\}$. The only remaining element is $B[62]M_1^3$.

DEGREE 62: $2B[62]$

The leading differential $d^{10}(A[8]D[45]M_1^2\bar{M}_2) = 2B[62]$ determines tentative differentials with image $Z_2(2B[62]) \otimes B<4>$. The remaining elements are $Z_2(2B[62])\{M_1, M_1^2, M_1^3, M_2, M_1M_2, M_1^2M_2, M_1^3M_2\} \otimes B<4>$, and the $2B[62]$-leader is $2B[62]M_1$.

The leading differential $d^{40}(2\gamma_2 M_1^{18}\bar{M}_2) = 2B[62]M_1$ determines a tentative differential with image $2B[62]M_2$ in degrees less than 69. The only remaining elements are $Z_2(2B[62])\{M_1^2, M_1^3\}$.

DEGREE 62: $\eta^2 B[60]$

The leading differential $d^2(\eta B[60]M_1) = \eta^2 B[60]$ determines tentative differentials with image $Z_2\eta^2 B[60] \otimes B<2>$ and cokernel $Z_2\eta^2 B[60]M_1 \otimes B<2>$. The $\eta^2 B[60]$-leader is $\eta^2 B[60]M_1$.

The leading differential $d^4(A[59,2]\bar{M}_2) = \eta^2 B[60]M_1$ determines tentative differentials with image $Z_2(\eta^2 B[60])\{M_1, M_1^3\} \otimes B<4>$. There are no elements remaining in degrees less than 69.

DEGREE 63: $A[63]$

The leading differential $d^{24}(\eta\sigma A[32,1]M_1^5<M_3>) = A[63]$ determines tentative differentials with image $Z_2A[63]\{1,M_1^2\}$ in degrees less than 68. The only remaining element in degrees less than 68 is $A[63]M_1$ and $d^{24}(A[39,3]M_1^3\overline{M}_2) = A[63]M_1$.

DEGREE 63: $\eta A[62,1]$ and $\eta A[62,4]$

Since $\eta^2 A[62,1] \neq 0$ and $\eta^2 A[62,4] \neq 0$, the only element of E^4 with a representative in $(Z_2\eta A[62,1] \oplus Z_2\eta A[62,4]) \otimes H_*BP$ is zero.

DEGREE 63: $\eta A[62,2]$

The leading differential $d^2(A[62,2]M_1) = \eta A[62,2]$ determines tentative differentials with image $Z_2\eta A[62,2] \otimes B\langle 2\rangle$. The remaining elements are $Z_2\eta A[62,2]M_1 \otimes B\langle 2\rangle$, and the $\eta A[62,2]$-leader is $\eta A[62,2]M_1$.

The leading differential $d^8(A[56]M_1^2\overline{M}_2) = \eta A[62,2]M_1$ determines tentative differentials with image $Z_2\eta A[62,2]M_1 \otimes B\langle 4\rangle$. The remaining elements are $Z_2\eta A[62,2]\{M_1^2,M_1^3,M_2,M_1M_2,M_1^2M_2,M_1^3M_2\} \otimes B\langle 4\rangle$, and the new $\eta A[62,2]$-leader is $\eta A[62,2]M_1^2$.

CHAPTER 8: THE ELEMENTS OF ARF INVARIANT ONE

1. Introduction

One of the most important open problems in homotopy theory is whether or not there exist elements $\theta_N \in \pi^S_{2^{N+1}-2}$ of Arf invariant one. These elements arose in the work of Kervaire [26] and Kervaire and Milnor [27] as obstructions in surgery theory. Browder [14] showed that the nonvanishing of these obstructions is equivalent to the elements $h_N^2 \in E_2^{2,2^{N+1}-2,2} = \mathrm{Ext}^2_{\mathfrak{A}}(Z_2,Z_2)_{2^{N+1}-2}$ being infinite cylcles in the classical Adams spectral sequence for π^S_*. Thus, an element $\theta_N \in \pi^S_{2^{N+1}-2}$ has Arf invariant one if and only if the secondary cohomology operation Φ_N defined by the following Adem relation is nonzero in the mapping cone of θ_N:

$$(8.1.1) \qquad 0 = \sum_{i=0}^{N} Sq^{2^{N+1}-2^i} Sq^{2^i}.$$

The first three elements of Arf invariant one are merely η^2, $\nu^2$and σ^2. The next two elements of Arf invariant one, $\theta_4 \in \pi^S_{30}$ and $\theta_5 \in \pi^S_{62}$, have been shown to exist using the classical Adams spectral sequence [37], [11]. It is not known whether θ_N exists for $N \geq 6$. The reader can find a more detailed exposition of this problem in [12] and [13].

In Section 2 we show that the element $A[30] \in \pi^S_{30}$ has Arf invariant one by calculating that the secondary operation Φ_4 is nonzero in the mapping cone of A[30]. In Section 3 we identify θ_5 as A[62,1] by showing that $\theta_4^2 = 0$ using an argument of Mahowald based upon a generalization of [34A,Theorem 16]. The construction of Barratt, Jones and Mahowald [11] shows that θ_5 exists but does not determine the order of θ_5. The argument of Section 3 shows that there are choices of θ_5 of order two.

In [35] Mahowald showed that the elements $h_1h_N \in E_2^{2^N,2}$ of the Adams spectral sequence are infinite cycles which are represented by the elements $\eta_N \in \pi^S_{2^N}$. In Section 2 we identify η_5 as A[32,1]. In Section 3, we identify η_6 as B[64,1].

2. The Existence of θ_4

Recall from Theorem 5.3.10 that $\pi^S_{30} = Z_2A[30]$ and $A[30] = d^{12}(2\sigma M_1^{12})$. We will show that the secondary operation Φ_4 is nonzero in the mapping cone of A[30]. It follows that $A[30] = \theta_4$ has Arf invariant one. We will assume the definitions and basic properties of secondary cohomology operations and functional cohomology operations [47].

Let $f:S^{23} \longrightarrow BP^{(16)}$ be the attaching map of the cell represented in homology by $\langle M_1^4\rangle^3$. This map has image in the 16-skeleton becasause $\langle M_1^4\rangle^3$ survives to E^8. Let $i:BP^{(14)} \longrightarrow BP^{(16)}$ be the natural inclusion. Since $d^8(2\sigma\langle M_1^4\rangle^3) = 0$, there is a lifing F of $2\sigma f$ to $BP^{(14)}$:

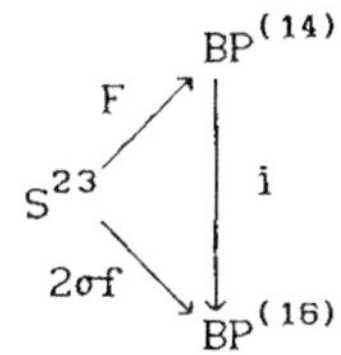

FIGURE 8.2.1: Definition of F

The next step is to define a map $G:\Sigma^7C_f \longrightarrow C_F$. We begin by defining $G_1 = G|\Sigma^7BP^{(8)}:\Sigma^7BP^{(8)} \longrightarrow BP^{(8)}$ as the composite of the projection map $P:\Sigma^7BP^{(8)} \longrightarrow \Sigma^7BP^{(8)}/\Sigma^7BP^{(6)} = S^{15}\vee S^{15}$ followed by $g \vee *$ where the first sphere is represented by $\langle M_1^4\rangle$ in homology, the second sphere is represented by $M_1\bar{M}_2$ in homology and g is the attaching map of the cell represented by $\langle M_1^4\rangle^2$ in homology. Since $\langle M_1^4\rangle^2$ survives to E^8 and $d^8(\langle M_1^4\rangle^2) = 2\sigma\langle M_1^4\rangle$, the following diagram commutes:

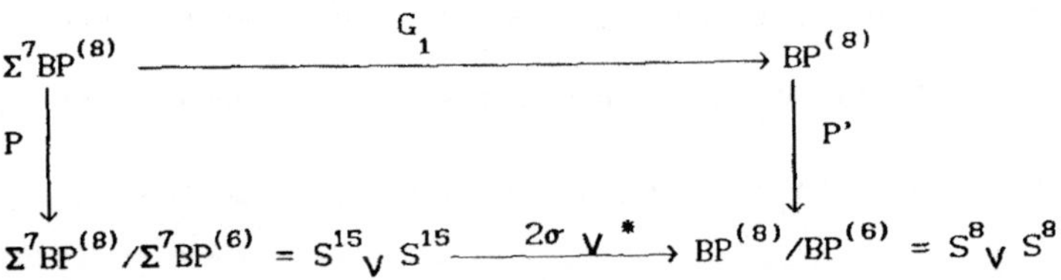

FIGURE 8.2.2: Definition of G_1

P' is the natural projection map above. Now define $G_2 : \Sigma^7 BP^{(16)} \longrightarrow BP^{(16)}$ as the composite of the projection map

$$P'' : \Sigma^7 BP^{(16)} \longrightarrow \Sigma^7 BP^{(16)} / \Sigma^7 BP^{(14)} = S^{23} \vee 3S^{23}$$

followed by $(g \wedge \alpha)' \vee 3*$. Here the first copy of S^{23} is represented by $\langle M_1^4 \rangle^3$ in homology and $\alpha : (D^8, S^7) \longrightarrow (BP^{(8)}, S)$ represents $\langle M_1^4 \rangle$. Also, $(g \wedge \alpha)' : S^{15} \wedge S^8 \longrightarrow BP^{(16)}$ is an extension of $S^{15} \wedge D^8 \xrightarrow{g \wedge \alpha} BP^{(8)} \wedge BP^{(8)} \xrightarrow{\varepsilon} BP^{(16)}$, thinking of D^8 as the upper hemisphere of S^8. This extension to S^{15} smash the bottom hemisphere exists as a map into $BP^{(8)}$ because $2\sigma^2 = 0$ in π_*^S. The top square in Figure 8.2.3 commutes because G_2 restricts to 2σ on Σ^7 of the cell C represented by $\langle M_1^4 \rangle^2$ in homology. Thus, the map G_3 must exist making the bottom square commute. Now G_3 maps all cells into $BP^{(14)}$ except for the cell $\Sigma^7 C$, and G_2 on this cell is $2\sigma \wedge 1$. In $\Sigma^7 C_f$, $C\Sigma^7 S^{23}$ is attached to this cell by $\Sigma^7 f$. Therefore G_3 on this cell is $2\sigma \wedge f$ which, as in Figure 8.2.1, lifts to $BP^{(14)}$. Thus G_3 lifts to a map G.

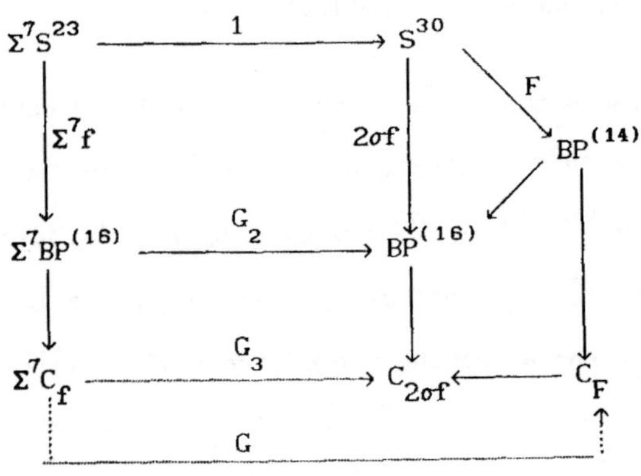

FIGURE 8.2.3: Definition of G

In $H^*(C_G;Z_2)$, let $u(X)$ denote the element dual to $X \in H_*(C_G;Z_2)$. Let $Y \in H_{16}(C_G;Z_2)$ denote the element determined by the first sphere in Figure 8.2.2. By the definition of G_1, Y represents a cell with the same attaching map as $\langle M_1^4\rangle^2$. Therefore, $Sq^{16}u(1) = u(Y)$. Thus, the functional secondary cohomology operation Sq_G^{16} is defined on $u(1) \in H^0(C_F;Z_2)$ and equals $S^7u\langle M^4\rangle \in H^{15}(\Sigma^7C\ ;Z\ )$. By the Peterson-Stein formula: $G^*\circ\Phi\ (u(1))$
$= Sq^{31}Sq_G^1(u(1))+Sq^{30}Sq_G^2(u(1))+Sq^{28}Sq_G^4(u(1))+Sq^{24}Sq_G^8(u(1))+Sq^{16}Sq_G^{16}(u(1))$.
Since $H^k(\Sigma^7C_f;Z_2) = 0$ for $k = 0,1,3$, we must have $Sq_G^1(u(1)) = 0$, $Sq_G^2(u(1)) = 0$ and $Sq_G^4(u(1)) = 0$. Since $G_1|\Sigma^7BP^{(6)} = *$, $Sq_G^8(u(1))$ must be 0 not $S^7u(1)$. Thus, $G^*\circ\Phi_4(u(1)) = Sq^{16}Sq_G^{16}(u(1)) = Sq^{16}(S^7u\langle M_1^4\rangle) = S^7u(\langle M_1^4\rangle^3) \neq 0$. Thus, $\Phi_4(u(1)) \neq 0$ in $H^{31}(C_F;Z_2)$. Note that there is a unique top dimensional cell of degree 31 in C_F which determines a nononzero element $\tau \in H^{31}(C_F;Z_2)$. Hence $\tau = \Phi_4(u(1)) \neq 0$ Since $d^{12}(2\sigma M_1^{12}) = A[30]$ and F represents the boundary of $2\sigma M_1^{12}$, the triangle in the following diagram must commute up to homotopy. Therefore, there is an induced map J making the square commute.

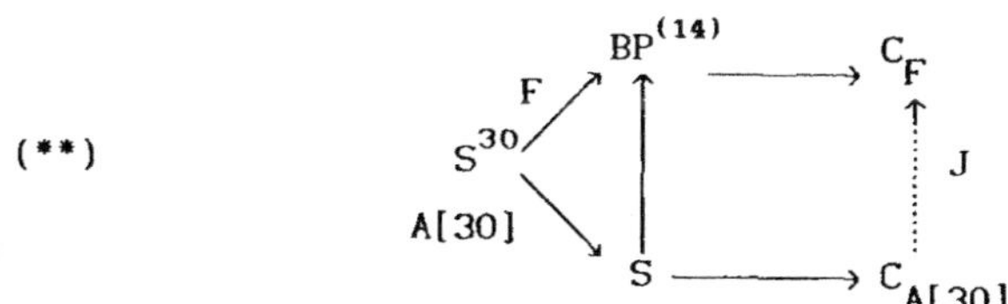

Now $\Phi_4(u(\iota)) = \Phi_4\circ J^*(u(1)) = J^*\circ\Phi_4(u(1)) = J^*(\tau) \neq 0$. Thus, A[30] must have Arf invariant one. We have thus proved the following theorem.

THEOREM 8.2.1 A[30] has Arf invariant one.

We derive several Toda brackets involving elements related to θ_4. The first Toda bracket below was proved by Hoffman [24]. We give a proof using our spectral sequence.

THEOREM 8.2.2 (a) $\theta_4 = A[30] \in \langle\sigma, 2\sigma, 2\sigma, \sigma\rangle$

(b) $\nu\theta_4 = \nu A[30] \in \langle C[18], \sigma, 2\sigma\rangle$

(c) $\theta_4 = A[30] \in \langle\sigma, 2\sigma, \sigma^2, 2\rangle = \langle\sigma^2, 2, \sigma^2, 2\rangle$

(d) $\eta\theta_4 = \eta A[30] \in \langle A[16], 2, \sigma^2\rangle$

PROOF. (a) Represent $\langle M_1^4\rangle^2$ by μ_8 such that $\partial(\mu_8) = (\sigma \wedge 2\mu_4) \cup (B_{\sigma 2\sigma})$. Since $2\cdot\pi_{22}^S = 0$, $\sigma A[14] = 0$ and $\sigma\gamma_1 = 0$, it follows that $\langle\sigma, 2\sigma, 2\sigma\rangle$ $= 2\langle\sigma, 2\sigma, \sigma\rangle = 0$. Thus, $2\sigma M_1^{12} \in E_{24,7}^{24}$ is represented by

$$\mathcal{M} = (\mu_4 \wedge \sigma \wedge 2\mu_8) \cup (B_{\sigma^2 2} \wedge \mu_8) \cup (\mu_4 \wedge B_{\sigma 2\sigma} \wedge 2\mu_4) \cup (B_{\langle\sigma, \sigma 2, \sigma 2\rangle} \wedge \mu_4) \cup (\mu_4 \wedge B_{\langle\sigma 2, \sigma 2, \sigma\rangle})$$

because $\partial\mathcal{M} = (B_{\sigma^2 2} \wedge B_{\sigma 2\sigma}) \cup (B_{\langle\sigma, \sigma 2, \sigma 2\rangle} \wedge \sigma) \cup (\sigma \wedge B_{\langle\sigma 2, \sigma 2, \sigma\rangle})$. Since $d^{24}(2\sigma M_1^{12}) = A[30]$, $\partial\mathcal{M}$ represents $A[30]$ and clearly $\partial\mathcal{M} \in \langle\sigma, \sigma 2, \sigma 2, \sigma\rangle$.

(b), (c) The four-fold Toda bracket $\langle\sigma, 2\sigma, \sigma^2, 2\rangle$ is defined by Theorem 2.2.7(a) because $\langle\sigma, 2\sigma, \sigma^2\rangle \in \pi_{29}^S = 0$ and $\langle 2\sigma, \sigma^2, 2\rangle = \sigma\langle 2, \sigma^2, 2\rangle + 2\cdot\pi_{22}^S$ $= \sigma(\eta\sigma^2) = 0$. Now $\nu A[30] \in \nu\langle\sigma, 2\sigma, 2\sigma, \sigma\rangle \subset \langle\langle\nu, \sigma, 2\sigma\rangle, 2\sigma, \sigma\rangle = \langle C[18], 2\sigma, \sigma\rangle$. Since $\mathrm{CokJ}_{26} = Z_2\nu^2 C[20]$, $\nu A[30] \in \langle C[18], \sigma, 2\sigma\rangle \subset \langle C[18], \sigma^2, 2\rangle$ $= \langle\langle\nu, \sigma, 2\sigma\rangle, \sigma^2, 2\rangle = \nu\langle\sigma, 2\sigma, \sigma^2, 2\rangle$. Thus, $\langle\sigma, 2\sigma, \sigma^2, 2\rangle$ contains $A[30]$. Note that $\langle\sigma^2, 2, \sigma^2, 2\rangle$ is defined by Theorem 2.2.7(a) because $\langle\sigma^2, 2, \sigma^2\rangle \in \pi_{29}^S = 0$ and $\langle 2, \sigma^2, 2\rangle = \eta\sigma^2 = 0$. Now $\langle\sigma^2, 2, \sigma^2, 2\rangle \subset \langle\sigma, 2\sigma, \sigma^2, 2\rangle = \{A[30]\}$.

(d) $\eta A[30] \in \eta\langle 2, \sigma^2, 2\sigma, \sigma\rangle \subset \langle\langle\eta, 2, \sigma^2\rangle, 2\sigma, \sigma\rangle = \langle A[16], 2\sigma, \sigma\rangle + \langle\eta\gamma_1, 2\sigma, \sigma\rangle$. Now $\langle\eta\gamma_1, 2\sigma, \sigma\rangle \supset \eta\langle\gamma_1, 2\sigma, \sigma\rangle$. Since $\nu\langle\gamma_1, 2\sigma, \sigma\rangle = \langle\nu, \gamma_1, 2\sigma\rangle\sigma = 0$, $\langle\gamma_1, 2\sigma, \sigma\rangle$ can not equal $A[30]$ and must therefore equal zero. It follows that $\langle\eta\gamma_1, 2\sigma, \sigma\rangle$ $= \eta\gamma_1\cdot\pi_{15}^S + \sigma\cdot\pi_{24}^S = \eta\xi$ where $\nu\xi = 0$. Thus, $\langle\eta\gamma_1, 2\sigma, \sigma\rangle = 0$ and $\eta A[30] \in \langle A[16], 2\sigma, \sigma\rangle$. ∎

We conclude this section by identifying the Mahowald element $\eta_5 \in \pi_{32}^S$ as $A[32, 1]$.

THEOREM 8.2.3 Let η_5 be <u>any</u> element of π_{32}^S which projects to h_1h_5 in $E_\infty^{32,2}$ of the Adams spectral sequence. Then η_5 projects to A[32,1] in $E^{24}_{0,32}$ of the Atiyah-Hirzebruch spectral sequence.

PROOF. From the computation of E_2 of the Adams spectral sequence by Tangora [59], it follows from the fact that h_1h_5 is an infinite cycle that $h_1^3h_5$ is a nonbounding infinite cycle. Thus, if η_5 is any element that projects to h_1h_5 then $\eta^2\eta_5 \neq 0$. Since $\eta^2\cdot\pi_{32}^S = Z_2\eta^2A[32,1]$ for any choice of A[32,1] modulo $Z_2A[32,2] \oplus Z_2A[32,3] \oplus Z_2\eta\gamma_3$, it follows that $\eta_5 \in A[32,1] + (Z_2A[32,2] \oplus Z_2A[32,3] \oplus Z_2\eta\gamma_3)$. Now the theorem follows from the observation that $Z_2A[32,2] \oplus Z_2A[32,3] \oplus Z_2\eta\gamma_3$ projects to zero in $E^{24}_{0,32}$ of the Atiyah-Hirzebruch spectral sequence.∎

3. The Existence of θ_5

In this section we show that A[62,1] has Arf invariant one and is thus entitled to be denoted as θ_5. We also identify the Mahowald element η_6 as B[64,1]. In addition, we derive a few miscellaneous results which are relevant to the Arf invariant problem. We begin with the following well known lemma which can be proved from a computation of $\mathrm{Ext}_{\mathfrak{A}}(Z_2,Z_2)$ as the homology of the Λ-algebra.

LEMMA 8.3.1 The following elements are nonzero in $\mathrm{Ext}_{\mathfrak{A}}(Z_2,Z_2)$:

(a) h_N and h_N^2 for $N \geq 0$;

(b) $h_0h_N^2$ and h_1h_N for $N \geq 3$;

(c) $h_1h_N^2$ for $N \geq 4$;

(d) $h_1^2h_N^2$ for $N \geq 5$.

Adams's proof [2] of the nonexistence of elements of Hopf invariant one in degrees 2^N-1, $N \geq 4$, is equivalent to the following differentials in the Adams spectral sequence. The elements listed in Lemma 8.3.1 and the differentials of Theorem 8.3.2 for $N \geq 4$ are depicted in Figure 8.3.1. Note that there are other elements in the bidegrees of that figure which are not depicted.

THEOREM 8.3.2 $d^2(h_N) = h_0 h_{N-1}^2$ for $N \geq 4$.

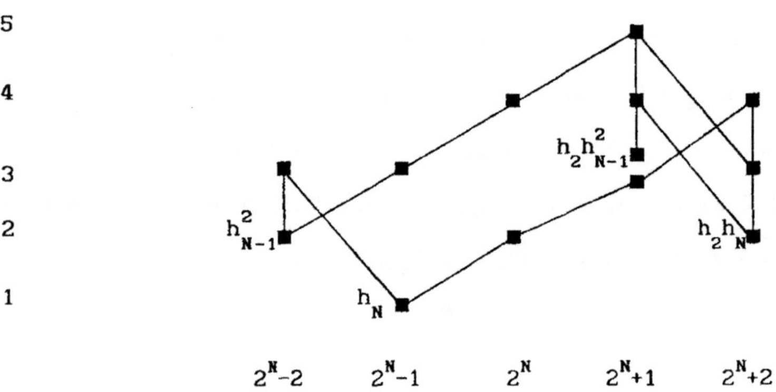

FIGURE 8.3.1: Part of E^2 of the Adams Spectral Sequence $(N \geq 6)$

The following lemmas will be used to identify θ_5 as A[62,1]. The entire argument is based upon ideas of Mahowald [34A] and is a rewording of a detailed proof which he sent to me.

LEMMA 8.3.3 $\langle\sigma^2, 2, A[30]\rangle \subset Z_2(\eta C[44]) \oplus Z_2(8D[45])$

PROOF. Note that $\eta^2\langle\sigma^2, 2, A[30]\rangle = \sigma^2\langle 2, A[30], \eta^2\rangle \subset \sigma^2\cdot\pi_{33}^S = 0$. Also, $\nu^2\langle\sigma^2, 2, A[30]\rangle = \sigma^2\langle 2, A[30], \nu^2\rangle \subset \sigma^2\cdot\pi_{37}^S = \sigma(4C[44]) = 0$. In addition, $2\langle\sigma^2, 2, A[30]\rangle = \sigma^2\langle 2, A[30], 2\rangle = 0$. The only elements of π_{45}^S which satisfy these three conditions are $Z_2(\eta C[44]) \oplus Z_2(8D[45])$. ∎

LEMMA 8.3.4 If $\xi \in \pi_*^S$ and $\xi A[36] = 0$ then

$$\xi C[44] \in \langle\eta\xi, \eta A[30], \nu, \sigma\rangle.$$

PROOF. By Theorem 2.4.6(a), if $\langle\eta,\eta A[30],\nu,\sigma\rangle$ were defined then it would contain C[44]. Now $\langle\eta A[30],\nu,\sigma\rangle \supset A[30]\langle\eta,\nu,\sigma\rangle = 0$, $\eta A[30]\cdot\pi^S_{11} = 0$ and $\sigma\cdot\pi^S_{35} = 0$. Thus, $\langle\eta A[30],\nu,\sigma\rangle = 0$. However, $A[36] \in \langle\eta,\eta A[30],\nu\rangle$. Since $\xi A[36] = 0$, $\langle\eta\xi,\eta A[30],\nu,\sigma\rangle$ is defined by Theorem 2.2.7(b). Thus, $\xi C[44] \in \langle\eta\xi,\eta A[30],\nu,\sigma\rangle$. ∎

LEMMA 8.3.5 (a) $A[16]A[36] = 0$.

(b) $\eta A[16]A[30] = 0$.

(c) $A[16]C[44] = 0$.

PROOF. (a) $A[16]A[36] \in A[36]\langle\eta,2,\sigma^2\rangle = \langle A[36],\eta,2\rangle\sigma^2 \in \sigma^2\cdot\pi^S_{38} = 0$.

(b) $\eta A[16]A[30] \in \eta A[30]\langle\eta,2,\sigma^2\rangle \subset \langle 0,2,\sigma^2\rangle = \sigma^2\cdot\pi^S_{33} = 0$.

(c) Since $A[16]A[36] = 0$, $A[16]C[44] \in \langle\eta A[16],\eta A[30],\nu,\sigma\rangle$ $\supset \langle\eta A[16]A[30],\eta,\nu,\sigma\rangle = \langle 0,\eta,\nu,\sigma\rangle$. (Note that $\langle\eta A[16]A[30],\eta,\nu,\sigma\rangle$ is defined by Theorem 2.2.7(b) because $0 \in \langle\eta A[16]A[30],\eta,\nu\rangle$ and $0 = \langle\eta,\nu,\sigma\rangle$.) Since $\sigma\cdot\pi^S_{53} = 0$, $A[16]C[44] \in \langle\pi^S_{49},\nu,\sigma\rangle + \eta A[16]\cdot\pi^S_{43} + \langle\eta A[16],\pi^S_{35},\sigma\rangle$

$= \langle\alpha_6,\nu,\sigma\rangle+\langle\eta^2\gamma_5,\nu,\sigma\rangle+\langle\eta A[16],\eta A[14]C[20],\sigma\rangle+\langle\eta A[16],\nu A[32,3],\sigma\rangle+\langle\eta A[16],\beta_4,\sigma\rangle$

$= \langle\alpha_6,\nu,\sigma\rangle+\eta\gamma_5\langle\eta,\nu,\sigma\rangle+\langle\eta A[16],\eta C[20],0\rangle+A[16]A[32,3]\langle\eta,\nu,\sigma\rangle+A[16]\langle\eta,\beta_4,\sigma\rangle$

$= \langle\alpha_6,\nu,\sigma\rangle$. By Theorem 4.2.3 and Figure 4.2.2, it follows that $\langle\alpha_6,\nu,\sigma\rangle$ projects to an element of filtration degree at least 26 in the Adams spectral sequence. The only such element is $h_0^2P^5g = d^2(h_0P^4k)$. Thus, $0 = \langle\alpha_6,\nu,\sigma\rangle = A[16]C[44]$. ∎

LEMMA 8.3.6 $A[30]^2 = 0$

PROOF. $A[30]^2 \in A[30]\langle 2,\sigma^2,2,\sigma^2\rangle \subset \langle\langle A[30],2,\sigma^2\rangle,2,\sigma^2\rangle$

$\subset \langle\eta C[44],2,\sigma^2\rangle + \langle 8D[45],2,\sigma^2\rangle \supset C[44]\langle\eta,2,\sigma^2\rangle + \langle 2\sigma B[38],2,\sigma^2\rangle$

$\supset C[44]A[16] + \sigma\langle 2B[38],2,\sigma^2\rangle \subset \sigma\cdot\pi^S_{53} = 0$. Since $\eta A[36] = 0$,

$A[30]^2 \in \eta C[44]\cdot\pi^S_{15} + 8D[45]\cdot\pi^S_{15} + \sigma^2\cdot\pi^S_{46} = \eta\gamma_1 C[44] \in \gamma_1\langle\eta^2,\eta A[30],\nu,\sigma\rangle$

$\subset \langle\eta^2,\eta A[30],\langle\nu,\sigma,\gamma_1\rangle\rangle \supset \langle\eta^2,A[30],\eta\langle\nu,\sigma,\gamma_1\rangle\rangle = \langle\eta^2,A[30],\langle\eta,\nu,\sigma\rangle\gamma_1\rangle$

$= \langle\eta^2, A[30], 0\rangle = \eta^2\cdot\pi^S_{58} = 0$. Thus, $A[30]^2 \in \langle\nu,\sigma,\gamma_1\rangle\cdot\pi^S_{34} \subset \{\nu^2 C[20], \eta\alpha_3\}\cdot\pi^S_{34}$ $= \eta\alpha_3 A[14]C[20] = \eta A[14]C[20]\langle 8\sigma, 2, \alpha_2\rangle = \eta A[14]\alpha_2\langle C[20], 8\sigma, 2\rangle \in (\eta\cdot\pi^S_{31})\cdot\pi^S_{28}$ $= (\eta\gamma_3)(A[8]C[20]) = 0$. ∎

THEOREM 8.3.7 $A[62,1] + \mathrm{Span}\,\{A[62,2], A[62,3], A[62,4], B[62], \eta^2 B[60]\}$ are all the elements of π^S_{62} of Arf invariant one. In particular, there are choices of θ_5 of order two.

PROOF. Since $\theta_4 = A[30]$ exists, $2\theta_4 = 0$ and $\theta_4^2 = 0$, it follows from [12, Theorem 2.1] that θ_5 exists and has order two. From Figure 8.3.1, we see that any element θ_5 of Arf invariant one satisfies $\eta^2\theta_5 \neq 0$. Since $\eta^2\cdot\pi^S_{62} = Z_2\eta^2 A[62,1] \oplus Z_2\eta^2 A[62,4]$, $\mathrm{Span}\,\{A[62,2], A[62,3], B[62], \eta^2 B[60]\}$ has Adams filtration at least three. Since $A[62,4] = d^{12}(\eta A[50,2]M_1^3\bar{M}_2)$ $= d^{12}(\nu^2 A[45,1]M_1^3\bar{M}_2)$ and $C[20] = d^{12}(\nu^3 M_1^3\bar{M}_2)$, $\nu A[62,4] = C[20]A[45,1]$. From Figure 8.3.1, $\nu\theta_5$ is nonzero and is represented in the Adams spectral sequence by $h_2h_5^2$ in filtration degree three while $C[20]A[45,1]$ has Adams filtration at least nine. Thus, $A[62,4]$ has Adams filtration at least three. Now all the elements of $\mathrm{Span}\,\{A[62,2], A[62,3], A[62,4], B[62], \eta^2 B[60]\}$ have Adams filtration at least three. Therefore, all the elements of $A[62,1] + \mathrm{Span}\,\{A[62,2], A[62,3], A[62,4], B[62], \eta^2 B[60]\}$ have Arf invariant one. ∎

Next we identify the Mahowald element η_6 in terms of the Atiyah-Hirzebruch spectral sequence. Recall that η_6 denotes any element of π^S_{64} which projects to h_1h_6 in $E_\infty^{64,2}$ of the Adams spectral sequence.

THEOREM 8.3.8 (a) Any choice of η_6 projects to B[64, 1] in $E^{54}_{0,64}$ of the Atiyah-Hirzebruch spectral sequence.

(b) All the choices of η_6 are

$B[64,1] + (Z_2A[64,1] \oplus Z_2A[64,2] \oplus Z_2A[64,3] \oplus Z_4B[64,2] \oplus Z_2\eta^2A[62,1] \oplus Z_2\eta\gamma_6)$.

(c) All the values of $2\eta_6$ are $\eta^2\theta_5 + Z_2\eta^2A[62,4]$, and $4\eta_6 = 0$.

(d) There are choices of η_5 and η_6 such that $\eta_5^2 = 2\eta_6$.

PROOF. Since A[64,1], A[64,2], A[64,3], B[64,2], $\eta^2A[62,1]$ and $\eta\gamma_7$ project to zero in $E_2^{64,2}$ of the Adams spectral sequence, all the choices for η_6 are

$\eta_6 = B[64,1] + pA[64,1] + qA[64,2] + rA[64,3] + sB[64,2] + t\eta^2A[62,1] + u\eta\gamma_7$.

All of these elements project to B[64,1] in $E^{54}_{0,64}$. Moreover, $2\eta_6 = 2B[64,1] + 2sB[64,2] = \eta^2A[62,1] + s\eta^2A[62,4] = \eta^2\theta_5 + s\eta^2A[62,4]$ and $4\eta_6 = 0$. Note that η_5^2 projects to $h_1^2h_5^2$ in the Adams spectral sequence. Thus, η_5^2 is not zero, and by Mahowald [36] there are choices of η_5 and η_6 such that $2\eta_6 = \eta_5^2$.∎

APPENDIX 1: THE STABLE STEMS

The following table summarizes the group structure of the first 64 stable homotopy groups of spheres. In dimensions 54, 62 and 63 we give composition series for π_N^S.

DEGREE	STABLE STEM	REFERENCE
0	Z	
1	$Z_2\eta$	3.2.1
2	$Z_2\eta^2$	3.2.1
3	$Z_8\nu$	3.3.1
4	0	3.3.6
5	0	3.3.6
6	$Z_2\nu^2$	3.3.7
7	$Z_{16}\sigma$	3.4.1
8	$Z_2\eta\sigma \oplus Z_2A[8]$	3.4.11
9	$Z_2\eta A[8] \oplus Z_2\eta^2\sigma \oplus Z_2\alpha_1$	5.2.1
10	$Z_2\eta\alpha_1$	5.2.2
11	$Z_8\beta_1$	5.2.2
12	0	5.2.2
13	0	5.2.2
14	$Z_2\sigma^2 \oplus Z_2A[14]$	5.2.2
15	$Z_2\eta A[14] \oplus Z_{32}\gamma_1$	5.2.3
16	$Z_2A[16] \oplus Z_2\eta\gamma_1$	5.2.4
17	$Z_2\eta A[16] \oplus Z_2\nu A[14] \oplus Z_2\eta^2\gamma_1 \oplus Z_2\alpha_2$	5.2.5
18	$Z_8C[18] \oplus Z_2\eta\alpha_2$	5.2.6
19	$Z_2A[19] \oplus Z_8\beta_2$	5.2.7
20	$Z_8C[20]$	5.3.1

21	$Z_2\eta C[20] \oplus Z_2\nu A[18]$	5.3.1
22	$Z_2\eta^2 C[20] \oplus Z_2\nu A[19]$	5.3.1
23	$Z_8\nu C[20] \oplus Z_2 A[23] \oplus Z_{16}\gamma_2$	5.3.2
24	$Z_2\eta A[23] \oplus Z_2\eta\gamma_2$	5.3.3
25	$Z_2\eta^2\gamma_2 \oplus Z_2\alpha_3$	5.3.4
26	$Z_2\nu^2 C[20] \oplus Z_2\eta\alpha_3$	5.3.6
27	$Z_8\beta_3$	5.3.7
28	$Z_2 A[8]C[20]$	5.3.7
29	0	5.3.9
30	$Z_2 A[30]$	5.3.10
31	$Z_2\eta A[30] \oplus Z_2 A[31] \oplus Z_{64}\gamma_3$	5.3.10
32	$Z_2 A[32,1] \oplus Z_2 A[32,2] \oplus Z_2 A[32,3] \oplus Z_2\eta\gamma_3$	6.2.2
33	$Z_2\nu A[30] \oplus Z_2\eta A[32,1] \oplus Z_2\eta A[32,2] \oplus Z_2\alpha_4 \oplus Z_2\eta^2\gamma_3$	6.2.3
34	$Z_4 B[34] \oplus Z_2 A[14]C[20] \oplus Z_2\nu A[31] \oplus Z_2\eta\alpha_4$	6.2.4
35	$Z_2\eta A[14]C[20] \oplus Z_2\nu A[32,3] \oplus Z_8\beta_4$	6.2.5
36	$Z_2 A[36]$	6.2.7
37	$Z_2 A[37] \oplus Z_2\sigma A[30]$	6.2.9
38	$Z_4 B[38] \oplus Z_2\eta\sigma A[30]$	6.2.11
39	$Z_2\sigma A[32,1] \oplus Z_2\eta B[38] \oplus Z_2 A[39,1] \oplus Z_2 A[39,2]$ $\oplus Z_2 A[39,3] \oplus Z_{16}\gamma_4$	6.3.1
40	$Z_4 C[20]^2 \oplus Z_2\eta A[39,3] \oplus Z_2\eta\sigma A[32,1] \oplus Z_2 A[40,1]$ $\oplus Z_2 A[40,2] \oplus Z_2\eta\gamma_4$	6.3.3
41	$Z_2\eta C[20]^2 \oplus Z_2\eta A[40,1] \oplus Z_2\eta A[40,2] \oplus Z_2\alpha_5 \oplus Z_2\eta^2\gamma_4$	6.3.4
42	$Z_8 C[42] \oplus Z_2\eta^2 C[20]^2 \oplus Z_2\eta\alpha_5$	6.3.5
43	$Z_8\beta_5$	6.3.6
44	$Z_8 C[44]$	6.3.7
45	$Z_{16} D[45] \oplus Z_2\eta C[44] \oplus Z_2 A[45,1] \oplus Z_2 A[45,2]$	6.3.8
46	$Z_2\eta D[45] \oplus Z_2\eta A[45,1] \oplus Z_2\eta A[45,2] \oplus Z_2\eta^2 C[44]$	7.2.1

47	$Z_4B[47] \oplus Z_2A[47] \oplus Z_2\eta^2D[45] \oplus Z_2\nu C[44] \oplus Z_{32}\gamma_5$	7.2.3
48	$Z_4\nu D[45] \oplus Z_2\nu A[45,1] \oplus Z_2\eta A[47] \oplus Z_2\eta B[47] \oplus Z_2\eta\gamma_5$	7.2.4
49	$Z_2\alpha_6 \oplus Z_2\eta^2\gamma_5$	7.2.5
50	$Z_2A[50,1] \oplus Z_2A[50,2] \oplus Z_2\eta\alpha_6$	7.2.6
51	$Z_4\sigma C[44] \oplus Z_2\eta A[50,2] \oplus Z_2\nu^2D[45] \oplus Z_8\beta_6$	7.3.1
52	$Z_2A[52,1] \oplus Z_2A[52,2] \oplus Z_2\eta\sigma C[44]$	7.3.2
53	$Z_2\eta A[52,2] \oplus Z_2\nu A[50,1] \oplus Z_2\nu A[50,2] \oplus Z_2A[8]D[45]$	7.3.3
*54	$Z_2A[54,1]$, $Z_2A[54,2] \oplus Z_2\eta A[8]D[45]$	7.3.4
55	$Z_2\eta A[54,2] \oplus Z_{16}\gamma_6$	7.3.5
56	$Z_2A[56] \oplus Z_2\nu^2A[50,2] \oplus Z_2\eta\gamma_6$	7.4.2
57	$Z_2A[57] \oplus Z_2\eta A[56] \oplus Z_2\alpha_7 \oplus Z_2\eta^2\gamma_6$	7.4.4
58	$Z_2\eta A[57] \oplus Z_2\eta\alpha_7$	7.4.5
59	$Z_2A[59,1] \oplus Z_2A[59,2] \oplus Z_8\beta_7$	7.4.6
60	$Z_4B[60]$	7.4.8
61	$Z_2A[61] \oplus Z_2\eta B[60]$	7.5.1
*62	$Z_2A[62,4]$, $Z_2A[62,1] \oplus Z_2A[62,2] \oplus Z_2A[62,3]$	
	$\oplus Z_4B[62] \oplus Z_2\eta^2B[60]$	7.5.2
*63	$Z_2A[63]$, $Z_2\eta A[62,1] \oplus Z_2\eta A[62,2] \oplus Z_2\eta A[62,4] \oplus Z_{128}\gamma_7$	7.5.3
64	$Z_2A[64,1] \oplus Z_2A[64,2] \oplus Z_2A[64,3] \oplus Z_4B[64,1]$	
	$\oplus Z_4B[64,2] \oplus Z_2\eta\gamma_7$	7.5.4

* See the referenced theorems for the additive extensions which remain undetermined.

APPENDIX 2: MULTIPLICATIVE RELATIONS

The following table gives the structure of π_*^S as a module over (η, ν, σ). The column labeled "ADAMS" gives the projection of the element into E_∞ of the Adams spectral sequence. We include a column "DEC" in which we enter D if the element is decomposable and I if the element is indecomposable by Lemma 1.3.10. (If the element is indecomposable for some other reason we give the explanation in a footnote.) The four entries of ?? in the table indicates that we have not determined the appropriate entry for that position.

DEG.	X	ADAMS	ηX	νX	σX	DEC	REFERENCES
1	η	h_1	η^2	0	$\eta\sigma$	I	3.2.1, 3.3.6, 3.4.11
2	η^2	h_1^2	4ν	0	$\eta^2\sigma$	D	3.3, 3.3.6, 5.2.1
3	ν	h_2	0	ν^2	0	I	3.3.6, 3.3.7, 3.4.5
6	ν^2	h_2^2	0	$\eta A[8]$	0	D	3.3.6, 3.3.15, 3.4.5
7	σ	h_3	$\eta\sigma$	0	σ^2	I	3.4.11, 3.4.5, 5.22
8	A[8]	h_1h_3 [1]	$\eta A[8]$	0	0	I	5.2.1, 3.5, 5.1
	$\eta\sigma$	h_1h_3	$\eta^2\sigma$	0	0	D	5.2.1, 3.3.6, 3.4
9	$\eta A[8]$	h_1c_0	0	0	0	D	5.2.1, 3.3.6, 5.1
14	A[14]	d_0	$\eta A[14]$	$\nu A[14]$	0	I	5.2.3, 5.2.5, 5.3.5
	σ^2	h_3^2	0	0	$\nu C[18]$	D	5.2.2, 3.4.5, 5.3.5
15	$\eta A[14]$	h_1d_0	0	0	0	D	5.2.4, 3.3.6, 5.3.5
16	A[16]	h_1h_4	$\eta^2 A[16]$	0	A[23]	I	5.2.5, 5.2.7, 6.2.1
17	$\eta A[16]$	$h_1^2h_4$	4C[18]	0	$\eta A[23]$	D	5.2.6, 3.3.6, 6.2.1
	$\nu A[14]$	h_0e_0	0	4C[20]	0	D	3.3.6, 5.3.1, 3.4.5

[1] A[8] + $\eta\sigma$ projects to c_0 .

18	C18]	h_2h_4	0	$\nu C[18]$	0	I	5.2.7, 5.3.1, 5.3.4
19	A[19]	c_1	0	$\nu A[19]$	0	I	5.3.1, 5.3.1, 5.3.6
20	C[20]	g	$\eta C[20]$	$\nu C[20]$	0	I	5.3.1, 5.3.2, 5.3.7
21	$\eta C[20]$	h_1g	$\eta^2C[20]$	0	0	D	5.3.1, 3.3.6, 5.3.7
	$\nu C[18]$	$h_2^2h_4$	0	0	0	D	3.3.6, 5.4, 3.4.5
22	$\nu A[19]$	h_2c_1	0	0	0	D	3.3.6, 5.3.4, 3.4.5
	$\eta^2C[20]$	Pd_0	$4\nu C[20]$	0	0	D	5.3.2, 3.3.6, 5.3.7
23	A[23]	h_4c_0	$\eta A[23]$	0	0	D	5.3.3, 6.2.1, 5.3.10
	$\nu C[20]$	h_2g	0	$\nu^2C[20]$	0	D	3.3.6, 5.3.6, 3.4.5
24	$\eta A[23]$	$h_1h_4c_0$	0	0	0	D	5.3.4, 3.3.6, 5.3.10
26	$\nu^2C[20]$	h_2^2g	0	0	0	D	3.3.6, 5.3.9, 3.4.5
28	A[8]C[20]	Pg	0	0	0	D	5.3.8, 3.5, 5.3.7
30	A[30]	h_4^2	$\eta A[30]$	$\nu A[30]$	$\sigma A[30]$	I	5.3.10, 6.2.3, 6.2.9
31	A[31]	n	0	$\nu A[31]$	0	I[2]	6.2.2, 6.2.4, 6.2.11
	$\eta A[30]$	$h_1h_4^2$	0	0	$\eta\sigma A[30]$	D	6.2.1, 3.3.6, 6.2.11
32	A[32,1]	h_1h_5	$\eta A[32,1]$	0	$\eta\sigma A[32,1]$	I	6.2.3, 6.2.5, 6.3.1
	A[32,2]	q	$\eta A[32,2]$	$\eta A[14]C[20]$	0	I	6.2.3, 6.3, Adams ss
	A[32,3]	d_1	0	$\nu A[32,3]$	A[39,1]	I[3]	6.2.3, 6.2.5, 6.12
33	$\eta A[32,1]$	$h_1^2h_5$	2B[34]	0	$\eta\sigma A[32,1]$	D	6.2.4, 3.3.6, 6.3.3
	$\eta A[32,2]$	h_1q	0	0	0	D	6.2.4,3.3.6, Adams ss
	$\nu A[30]$	p	0	0	0	D	3.3.6, 6.2.6, 3.4.5
34	B[34]	$h_0h_2h_5$	0	0	$\eta A[40,1]$	I	6.2.5, 6.2.9, 7.1
	$\nu A[31]$	h_2n	0	0	0	D	3.3.6, 6.2.8, 3.4.5
	A[14]C[20]	e_0^2	$\eta A[14]C[20]$	0	0	D	6.2.5, 6.2.9, 5.3.7

[2] Use $\nu A[31] \neq 0$.

[3] Use $\nu A[32,3] \neq 0$ and $\sigma A[32,3] \neq 0$.

35	$\nu A[32,3]$	h_2d_1	0	$\eta\sigma A[30]$	0	D	3.3.6, 6.2.10, 3.4.5
	$\eta A[14]C[20]$	$h_1e_0^2$	0	0	0	D	5.2.4, 3.3.6, 5.3.7
36	$A[36]$	t	0	$\eta B[38]$	0	I[4]	6.2.8, 6.3.2, 6.3.6
37	$A[37]$	$h_2^2h_5$	$\eta\sigma A[30]$	$\eta A[39,3]$ $+\eta\sigma A[32,1]$	0	I[5]	6.2.11, 7.39, 6.2.10
	$\sigma A[30]$	x	$\eta\sigma A[30]$	0	$4C[44]$	D	6.2.11, 3.4.5, 6.3.7
38	$B[38]$	$h_0^2h_3h_5$	$\eta B[38]$	0	$4D[45]$	I	6.3.1, 6.3.4, 7.6
	$\eta\sigma A[30]$	h_1x	0	0	0	D	6.3.1, 3.3.6, 3.4.5
39	$A[39,1]$	h_1e_1	0	0	$\eta^2C[44]$	D	6.3.3, 6.17, 7.4
	$A[39,2]$	u	$2C[20]^2$	$\eta^2C[20]^2$	??	I[6]	6.3.3, 6.13
	$A[39,3]$	h_5c_0	$\eta A[39,3]$	0	0	I[7]	6.3.3, 6.17, 7.2.2
	$\eta B[38]$	c_1g	0	0	0	D	6.3.3, 3.3.6, 7.6
	$\sigma A[32,1]$	$h_1h_3h_5$	$\eta\sigma A[32,1]$	0	0	D	6.3.3, 3.4.5, 7.2.2
40	$C[20]^2$	g^2	$\eta C[20]^2$	0	0	D	6.3.4, 6.3.6, 5.3.7
	$A[40,1]$	f_1	$\eta A[40,1]$	0	$\nu C[44]$	I[8]	6.3.4, 6.3.6, 7.9
	$A[40,2]$	Ph_1h_5	$\eta A[40,2]$	0	$A[47]$	I	6.3.4, 6.3.6, 7.4.1
	$\eta A[39,3]$	$h_1h_5c_0$	0	0	0	D	6.3.4, 3.3.6, 7.2.2
	$\eta\sigma A[32,1]$	$h_1^2h_3h_5$	0	0	0	D	6.3.4, 3.3.6, 3.4.5
41	$\eta C[20]^2$	z	$\eta^2C[20]^2$	0	0	D	6.3.5, 3.3.6, 5.3.7
	$\eta A[40,1]$	h_1f_1	0	0	0	D	6.3.5, 3.3.6, 7.9
	$\eta A[40,2]$	$h_1Ph_1h_5$	$4C[42]$	0	$\eta A[47]$	D	6.3.5, 3.3.6, 7.4.1

[4] Use $\nu A[36] \neq 0$.

[5] Use $\eta A[37] \neq 0$ and $\nu A[37] \neq 0$.

[6] Use $\eta A[39,2] \neq 0$ and $\nu A[39,2] \neq 0$.

[7] Use the Adams spectral sequence.

[8] Use $\eta A[40,1] \neq 0$ and $\sigma A[40,1] \neq 0$.

42	$C[42]$	Ph_2h_5	0	$8D[45]$	0	I	6.3.6, 6.3.8, 7.2.5
	$\eta^2C[20]^2$	Pe_0^2	0	0	0	D	6.3.6, 3.3.6, 5.3.7
44	$C[44]$	g_2	$\eta C[44]$	$\nu C[44]$	$\sigma C[44]$	I	6.3.8, 7.2.3, 7.3.1
45	$D[45]$	h_4^3	$\eta D[45]$	$\nu D[45]$	0	I	7.2.1, 7.2.4, 7.14
	$A[45,1]$	h_5d_0	$\eta A[45,1]$	$\nu A[45,1]$	0	I[9]	7.2.1, 7.2.4, 7.14
	$A[45,2]$	w	$\eta A[45,2]$	$\eta B[47]$	0	I[10]	7.2.1, 7.7, 7.14
	$\eta C[44]$	h_1g_2	$\eta^2C[44]$	0	$\eta\sigma C[44]$	D	7.2.1, 3.3.6, 7.3.1
46	$\eta D[45]$	B_1	$\eta^2D[45]$	0	0	D	7.2.3, 3.3.6, 7.14
	$\eta A[45,1]$	$h_1h_5d_0$	0	0	0	D	7.2.3, 3.3.6, 7.14
	$\eta A[45,2]$	gj	$2B[47]$	0	0	D	7.2.3, 3.3.6, 7.14
	$\eta^2C[44]$	N	0	0	0	D	7.2.3, 3.3.6, 7.3.3
47	$B[47]$	e_0r	$\eta B[47]$	0	0	I[11]	7.2.4, 7.2.6, 7.19
	$A[47]$	Ph_5c_0	$\eta A[47]$	0	0	D	7.2.4, 7.2.6, 7.17
	$\eta^2D[45]$	h_1B_1	0	0	0	D	7.2.4, 3.3.6, 7.14
	$\nu C[44]$	h_2g_2	0	0	0	D	3.3.6, 7.2.6, 3.4.5
48	$\eta A[47]$	h_1Ph_5c	0	0	0	D	7.2.5, 3.3.6, 7.17
	$\eta B[47]$	Pg^2	0	0	0	D	7.2.5, 3.3.6, 7.19
	$\nu A[45,1]$	$h_0h_5e_0$	0	$\eta A[50,2]$	0	D	3.3.6, 7.13, 3.4.5
	$\nu D[45]$	B_2	0	$\nu^2D[45]$	0	D	3.3.6, 7.3.1, 3.4.5

[9] Use $\eta A[45,1] \neq 0$ and $\eta A[45,2] \neq 0$.

[10] Use $\eta A[45,2] \neq 0$ and $\nu A[45,2] \neq 0$.

[11] Use $2B[47] \neq 0$ and $\eta B[47] \neq 0$.

50	A[50,1]	h_5c_1 [12]	0	νA[50,1]	0[13]	I	7.3.1, 7.3.3
	A[50,2]	C	ηA[50,2]	νA[50,2]	??	I[14]	7.3.1, 7.3.3
51	ηA[50,2]	gn	0	0	0	D	7.3.2, 3.3.6, 7.23
	σC[44]	h_5g	$\eta\sigma$C[44]	0	0	D	7.3.2, 7.3.3, 7.30
	ν^2D[45]	h_2B_2	0	ηA[8]D[45]	0	D	3.3.6, 3.3.15, 3.4.5
52	A[52,1]	d_1g	0	0	0	I[15]	7.3.3, 7.20, 7.31
	A[52,2]	$g\ell$	ηA[52,2]	ηA[54,2]	0	D	7.3.3, 7.20, 7.31
	$\eta\sigma$C[44]	h_1h_5g	0	0	0	D	7.3.3, 3.3.6, 5.2.2
53	ηA[52,2]	Pw	0	0	0	D	7.3.4, 3.3.6, 7.31
	νA[50,1]	$h_2h_5c_1$	0	0	0	D	3.3.6, 7.4.1, 3.4.5
	νA[50,2]	h_2C	0	ν^2A[50,2]	0	D	3.3.6, 7.4.2, 3.4.5
	A[8]D[45]	x'	ηA[8]D[45]	0	0	D	7.3.4, 3.5, 5.1
54	A[54,1]	h_0h_5i	0	ηA[56]	0	I	7.3.5, 7.25, 7.35
	A[54,2]	e_0^2g	ηA[54,2]	0	0	D	7.3.5, 7.25, 7.35
	ηA[8]D[45]	h_1x'	0	0	0	D	5.2.1, 3.3.6, 5.1
55	ηA[54,2]	Pe_0r	0	0	0	D	7.4.2, 3.3.6, 7.35
56	A[56]	Ph_5e_0	ηA[56]	0	ηA[62,2]	I	7.4.4, 7.29, 7.42
	ν^2A[50,2]	gt	0	0	0	D	3.3.6, 7.29, 3.4.5
57	A[57]	Q_2	ηA[57]	0	A[64,2]	I[16]	7.4.5, 7.32, 7.48
	ηA[56]	$h_1Ph_5e_0$	0	0	0	D	7.4.5, 3.3.6, 7.42

[12] $h_5c_1 \in \langle A[19], 2, A[30]\rangle$ and $\nu^2(h_5n) \subset \nu^2\langle A[19], 2, A[30]\rangle = \langle \nu^2, A[19], 2\rangle A[30]$ $= \nu^2 C[20]A[30] = 0$, and $\nu^2 A[50,2] \neq 0$.

[13] $\sigma(h_5c_1) \in \sigma\langle A[19], 2, A[30]\rangle = \langle\sigma, A[19], 2\rangle A[30] = 0$.

[14] Use $\eta A[50,2] \neq 0$ and $\nu^2 A[50,2] \neq 0$.

[15] Use the Adams spectral sequence.

[16] Use $\eta A[57] \neq 0$ and $\sigma A[57] \neq 0$. Also note that in E_2 of the Adams spectral sequence $h_3C \neq Q_2$ since $h_0C = 0$ and $h_0Q_2 \neq 0$.

58	ηA[57]	h_1Q_2	0	0	D	7.4.6, 3.3.6
59	A[59,1]	B_{21}	0	0	??	7.4.7, 7.36
	A[59,2]	d_0w	ηA[59,2]	η^2B[60]	D	7.4.7, 7.4.7
60	B[60]	g^3	ηB[60]	0	D	7.5.1, 6.3.6
61	A[61]	$h_0(A+A')$	0	0	I[17]	7.5.2, 7.44
	ηB[60]	gz	η^2B[60]	0	D	7.5.2, 3.3.6
62	A[62,1]	h_5^2	ηA[62,1]		I	7.5.3
	A[62,2]	E_1	ηA[62,2]		I	7.5.3
	A[62,3]	h_5n	0		??	7.5.3
	A[62,4]	H_1	ηA[62,4]		I[18]	7.5.3
	B[62]	B_{22}	0		I	7.5.3
63	A[63]	$\varepsilon X_2+(1-\varepsilon)C'$	0		I[19]	7.5.4
	ηA[62,2]	h_1E_1	0		D	7.5.4
	ηA[62,1]	$h_1h_5^2$	2B[64,1]		D	7.5.4
	ηA[62,4]	h_1H_1	2B[64,2]		D	7.5.4

[17] Use the Adams spectral sequence noting $h_0(A+A') \neq 0$ in E_2.

[18] Use η^2A[62,4] $\neq$ 0.

[19] Use the Adams spectral sequence and note that A[19]C[44] = C[18]D[45] = 0.

APPENDIX 3: TODA BRACKETS

	REFERENCE
$\eta^2 \in \langle 2,\eta,2\rangle$	3.2.1
$2\nu \in \langle \eta,2,\eta\rangle$	3.1
$\nu^2 \in \langle \eta,\nu,\eta\rangle$	3.2
$A[8] \in \langle \eta,\nu,2\nu\rangle$	3.3
$A[8] = \langle \nu,\eta,\nu\rangle$	3.4.12
$A[8] \in \langle \eta^2,\nu,\eta,2\rangle$	7.28
$A[14] \in \langle \nu,A[8],2,\eta\rangle$	3.6
$A[14] \in \langle 2,A[8],\nu,\eta\rangle$	6.5
$\sigma^2 \in \langle \nu,\sigma,\nu\rangle$	3.7
$A[16] \in \langle \eta,2,\sigma^2\rangle = \langle \eta,\sigma^2,2\rangle$	3.8
$\nu A[14] \in \langle 2,\eta,\eta A[14]\rangle$	7.35
$C[18] \in \langle \nu,\sigma,2\sigma\rangle$	3.9
$C[18] \in \langle \sigma,\nu,\sigma\rangle$	5.2
$\nu^3 \in \langle \eta,\nu^2,\eta\rangle$	3.4
$\eta A[8] \in \langle 2,A[8],2\rangle$ modulo $\eta^2\sigma$	5.3.8
$\eta A[14] \in \langle 2,A[14],2\rangle$	5.3.8
$\nu A[14] \in \langle 2,\eta,\eta A[14]\rangle$	7.36
$C[18] \in \langle \nu,\sigma,2\sigma\rangle$	3.9
$C[18] \in \langle \sigma,\nu,\sigma\rangle$	5.5
$A[19] \in \langle \eta,\sigma^2,\nu\rangle = \langle \nu,\eta,\sigma^2\rangle$	5.2
$C[20] \in \langle \nu,\eta,2,A[14]\rangle$	5.3.8
$2C[20] \in \langle \eta A[14],\eta,\nu\rangle$	5.3.8
$\eta C[20] \in \langle \nu^2,2,A[14]\rangle$	5.3.8
$A[23] \in \langle \eta,\nu,2C[18]\rangle$	5.3
$A[30] \in \langle \sigma,2\sigma,2\sigma,\sigma\rangle$	8.2.2

$A[30] \in \langle \sigma, 2\sigma, \sigma^2, 2 \rangle$ 8.2.2

$A[30] \in \langle \sigma^2, 2, \sigma^2, 2 \rangle$ 8.2.2

$\eta A[30] \in \langle A[16], 2, \sigma^2 \rangle$ 8.2.2

$A[31] \in \langle \eta, \nu, \nu A[19], \nu \rangle$ 5.6

$A[32,1] \in \langle \eta, 2, A[30] \rangle$ 7.24

$A[32,3] \in \langle A[19], \sigma, \nu, \eta \rangle$ 7.5

$\nu A[30] \in \langle \sigma, C[18], \sigma \rangle = \langle C[18], \sigma, 2\sigma \rangle$ 6.2

$B[34] \in \langle \eta, 2, A[32,1] \rangle$ 6.1

$A[36] \in \langle \nu, \eta, A[31] \rangle$ 6.4

$A[37] \in \langle \nu, \eta, A[32,3] \rangle$ 6.6

$A[37] \in \langle \eta, \nu, A[32,1] \rangle$ 6.10

$A[39,1] \in \langle \eta, \nu, \nu A[31] \rangle$ 6.8

$A[39,2] \in \langle \eta, \nu, A[14]C[20] \rangle$ 6.9

$A[39,3] \in \langle \eta, \nu, B[34] \rangle$ 6.11

$A[40,1] \in \langle \eta, (\eta A[30], \nu A[32,3]), (\sigma, \nu)^T \rangle$ 6.14

$A[40,2] \in \langle \eta, 2, 2B[38] \rangle$ 6.15

$\eta A[40,1] \in \langle \nu, \eta, A[36] \rangle$ 6.16

$C[42] \in \langle 2, \eta, \eta\sigma A[32,1] \rangle$ 6.19

$2C[42] \in \langle \eta, 2, A[40,2] \rangle$ 6.20

$C[44] \in \left\langle \sigma, \begin{bmatrix} A[31], \nu \end{bmatrix}, \begin{bmatrix} \eta & 0 \\ 0 & \eta \end{bmatrix}, \begin{bmatrix} \nu \\ \eta A[30] \end{bmatrix} \right\rangle$ 7.8

$2C[44] \in \langle \nu, \nu A[30], \sigma \rangle = \langle \nu A[30], \nu, \sigma \rangle$ 6.18

$A[45,2] \in \langle \eta, \nu, C[20]^2 \rangle$ 6.21

$2D[45] \in \langle \nu(B[34], A[40,1]), (\sigma, \eta)^T \rangle$ 6.22

$4D[45] \in \langle \eta, \eta\sigma A[32,1], \nu \rangle$ 6.23

$A[47] \in \langle \eta, \nu, C[42] \rangle$ 7.3

$B[47] \in \langle \eta, 2, A[45,2] \rangle$ 7.2

$B[47] \in \langle \nu, \eta, C[20]^2, \eta \rangle$ 7.18

$2\sigma C[44] \in \langle \eta, \eta A[45,1], \nu \rangle$ 7.10

$A[52,1] \in \langle \eta, \eta^2 D[45], \nu \rangle$ 7.11

$A[52,2] \in \langle \eta, \nu, B[47] \rangle$ 7.12

$A[54,2] \in \langle \nu, \eta, \eta^2 A[45,2], \eta \rangle$ 7.15

$A[54,2] \in \langle \eta, 2, A[52,2] \rangle$ 7.16

$A[56] \in \langle \eta, (2, \eta A[8]), (A[54,1], \lambda D[45])^T \rangle$ 7.21

$A[57] \in \langle \sigma, 4C[44], \nu, \eta \rangle$ 7.22

$A[59,1] \in \langle \eta, \eta\sigma C[44], \eta, \nu \rangle$ 7.27

$A[59,2] \in \langle \eta, \nu, A[54,2] \rangle$ 7.26

$B[60] \in \langle \eta, \eta A[52,2], \eta, \nu \rangle$ 7.33

$A[61] \in \langle \eta, \nu, \nu^2 A[50,2] \rangle$ 7.34

$A[62,2] \in \langle \nu, \eta, 2, A[56] \rangle$ 7.41

$A[62,3] \in \langle \eta, \nu, \nu A[50,1], \nu \rangle$ 7.36

$A[62,3] \in \langle A[30], 2, A[31] \rangle$ 7.38

$2B[62] \in \langle \eta, \eta A[14], D[45] \rangle$ 7.40

$A[64,1] \in \langle \eta, \nu, \sigma C[44], \sigma \rangle$ 7.43

$A[64,2] \in \langle \eta, A[59,1], \nu \rangle$ 7.45

$A[64,3] \in \langle \nu, \eta, A[59,1] \rangle$ 7.46

$B[64,1] \in \langle \eta, 2, A[62,1] \rangle$ 7.47

APPENDIX 4: LEADERS

1 2 3 4 6 7 8 9 10

$\eta \leftarrow M_1$ $\quad \nu \longleftarrow M_1^2$ $\quad \nu^2 \longleftarrow \nu M_1^2$ $\quad \eta\sigma \longleftarrow \sigma M_1$

$\eta^2 \longleftarrow \eta M_1$ $\quad \sigma \longleftarrow \langle M_1^4 \rangle$ $\quad \eta A[8] \longleftarrow A[8]M_1$

$\eta^3 \longleftarrow \eta^2 M_1$ $\quad A[8] \longleftarrow 2\nu M_1^3$

11 12 14 15 16 17 18

$\nu^3 M_1 \longleftarrow \nu^2 M_1^3$ $\quad \sigma^2 \longleftarrow \sigma\langle M_1^4 \rangle$ $\quad A[16] \longleftarrow 2\sigma M_1^5$ $\quad 4C[18] \longleftarrow$

$A[14] \longleftarrow 4\nu M_1^3 \bar{M}_2$ $\quad C[18] \longleftarrow$

$\eta A[14] \longleftarrow A[14]M_1$ $\quad \nu A[14] \longleftarrow A[14]M_1^2$

$\eta A[16] \longleftarrow A[16]M_1$

$\eta A[14]M_1 \longleftarrow A[8]M_1^2 \bar{M}_2$

19 20 21 22 23 24

$\longleftarrow \eta A[16]M_1$ $\quad C[20] \longleftarrow \nu^3 M_1^3 \bar{M}_2$ $\quad \nu A[19] \longleftarrow A[19]M_1^2$ $\quad \eta A[23] \longleftarrow$

$\longleftarrow 2\sigma M_1^6$ $\quad 2C[20] \longleftarrow \eta A[14]M_1^3$ $\quad \eta^2 C[20] \longleftarrow \eta C[20]M_1$

$A[19] \longleftarrow \sigma^2 M_1^3$ $\quad \nu C[18] \longleftarrow C[18]M_1^2$ $\quad 4\nu C[20] \longleftarrow \eta^2 C[20]M_1$

$4C[20] \longleftarrow \nu A[14]M_1^2$ $\quad \nu C[20] \longleftarrow C[20]M_1^2$

$\eta C[20] \longleftarrow C[20]M_1$ $\quad A[23] \longleftarrow 2C[18]\bar{M}_2$

$4C[18]M_1 \longleftarrow 8\sigma M_1^7$

25 26 27 28 29 30 31

$\longleftarrow A[23]M_1$ $\quad \nu C[18]\bar{M}_2 \longleftarrow \sigma^2 M_1^4 \bar{M}_2$ $\quad A[30] \longleftarrow 2\sigma M_1^{12}$

$\nu^2 C[20] \longleftarrow \nu C[20]M_1^2$ $\quad A[8]C[20] \longleftarrow 2\nu C[20]M_1^3$ $\quad \eta A[30] \longleftarrow$

$\eta A[23]M_1 \longleftarrow \eta^2 \sigma M_1^9$ $\quad \nu A[19]M_1^3 \longleftarrow \eta^2 \sigma M_1^7 \bar{M}_2$ $\quad A[31] \longleftarrow$

$A[8]C[20]M_1 \longleftarrow \beta_1 M_1^{10}$

32 | 33 | 34 | 35 | 36

$\nu^2C[20]M_1^3 \longleftarrow \beta_1M_1^{11}$ $\quad A[32,3]M_1 \longleftarrow \eta^2\sigma M_1^7\bar{M}_2^2$ $\quad 2B[34]M_1 \longleftarrow$

$\longleftarrow A[30]M_1$ $\quad B[34] \longleftarrow \beta_1M_1^6\bar{M}_2^2$ $\quad A[36] \longleftarrow$

$\longleftarrow \nu A[19]M_1^2\bar{M}_2$ $\quad 2B[34] \longleftarrow \eta A[32,1]M_1$

$A[23]M_1^2\bar{M}_2 \longleftarrow A[16]M_1^6\bar{M}_2$ $\quad \eta A[32,2]M_1 \longleftarrow A[8]C[20]M_1\bar{M}_2$

$\nu A[30] \longleftarrow A[30]M_1^2$ $\quad \nu A[32,3] \longleftarrow A[32,3]M_1^2$

$\eta A[32,1] \longleftarrow A[32,1]M_1$ $\quad \eta A[14]C[20] \longleftarrow A[14]C[20]M_1$

$\eta A[32,2] \longleftarrow A[32,2]M_1$

$\eta A[30]M_1 \longleftarrow \sigma^2M_1^4\bar{M}_2^2$

$A[32,1] \longleftarrow \eta^2\sigma M_1^5\bar{M}_3$ $\quad A[14]C[20] \longleftarrow 4\nu C[20]M_1^3\bar{M}_2$

$A[32,2] \longleftarrow 2\beta_1M_1^8\bar{M}_2$ $\quad \nu A[31] \longleftarrow A[31]M_1^2$

$A[32,3] \longleftarrow \nu C[18]M_1^3\bar{M}_2$

37 | 38 | 39 | 40

$\longleftarrow \gamma_1M_1^8\bar{M}_2$ $\quad B[38] \longleftarrow \gamma_1M_1^{12}$ $\quad C[20]^2 \longleftarrow$

$\longleftarrow A[31]M_2$ $\quad A[36]M_1 \longleftarrow \eta A[30]M_1\bar{M}_2$ $\quad 2C[20]^2 \longleftarrow$

$\eta A[14]C[20]M_1 \longleftarrow A[32,2]\bar{M}_2$ $\quad (\sigma A[30]+A[37])M_1 \longleftarrow 2C[18](M_1^4\bar{M}_3+3M_1^8\bar{M}_2)$

$\nu A[30]M_1^2 \longleftarrow C[18](M_1^4\bar{M}_2^2+2M_1^7\bar{M}_2)$ $\quad A[40,1] \longleftarrow$

$A[8]C[20]M_1^2\bar{M}_2 \longleftarrow 2\beta_1M_1^{11}M_2$ $\quad A[40,2] \longleftarrow$

$A[14]C[20]M_1^2 \longleftarrow 4\beta_1M_1^{11}\bar{M}_2$ $\quad \eta A[39,3] \longleftarrow$

$\eta\sigma A[30] \longleftarrow \sigma A[30]M_1$ $\quad \eta\sigma A[32,1] \longleftarrow$

$\sigma A[30] \longleftarrow A[30]M_1^4$ $\quad A[36]M_1^2 \longleftarrow$

$A[37] \longleftarrow A[32,3]M_2$ $\quad \eta B[38] \longleftarrow B[38]M_1$

$A[39,1] \longleftarrow \nu A[31]M_1^3$

$A[39,2] \longleftarrow A[14]C[20]\bar{M}_2$

$A[39,3] \longleftarrow B[34]M_1^3$

$\sigma A[32,1] \longleftarrow A[32,1]M_1^4$

$\eta\sigma A[30]M_1 \longleftarrow$

41	42	43	44
$\longleftarrow \eta A[32,2]M_1\bar{M}_2$	$\eta^2C[20]^2 \longleftarrow$	$\eta C[20]^2M_1$	$C[44] \longleftarrow$
$\longleftarrow A[39,2]M_1$	$\eta A[39,3]M_1 \longleftarrow$ $+\eta\sigma A[32,1]M_1$	$A[37]\bar{M}_2$	$2C[44] \longleftarrow$
$\eta B[38]M_1 \longleftarrow$	$A[36]M_1^3$		$4C[44] \longleftarrow$
$\longleftarrow \eta A[30]M_1^5$	$C[42] \longleftarrow$	$4\gamma_1(M_1^{11}\bar{M}_2+2M_1^{14})$	$4C[42]M_1 \longleftarrow$
$\longleftarrow 4\gamma_1M_1^{10}\bar{M}_2$	$\eta\sigma A[32,1]M_1 \longleftarrow$	$2\gamma_1(M_1^{11}\bar{M}_2+10M_1^{14})$	$\eta^2C[20]^2M_1 \longleftarrow$
$\longleftarrow A[39,3]M_1$	$2C[20]^2M_1 \longleftarrow$	$\eta A[14]C[20]M_1\bar{M}_2$	
$\longleftarrow \sigma A[32,1]M_1$	$4C[42] \longleftarrow$	$\eta A[40,2]M_1$	
$\longleftarrow 2\beta_1M_1^6\bar{M}_2^3$	$2B[34]M_1\bar{M}_2 \longleftarrow$	$2\beta_1M_1^7\bar{M}_2^3$	
$A[37]M_1^2 \longleftarrow$	$A[32,1]M_1^2\bar{M}_2$	$\eta A[40,1]M_1 \longleftarrow$	$A[36]M_1M_2$
$A[39,1]M_1 \longleftarrow$	$A[32,3]M_1^5$		
$\eta A[40,1] \longleftarrow$	$A[40,1]M_1$		
$\eta A[40,2] \longleftarrow$	$A[40,2]M_1$		
$\eta C[20]^2 \longleftarrow$	$C[20]^2M_1$		
$\longleftarrow \nu A[32,3]M_1^3$			

45 $\qquad$ 46 $\qquad$ 47 $\qquad$ 48 $\qquad$ 49

$\leftarrow \eta A[30]M_1M_2^2 \qquad \eta A[45,1] \longleftarrow A[45,1]M_1 \qquad \eta^2C[44]M_1 \longleftarrow A[39,1]M_1^5$

$\leftarrow \nu A[30]M_1^6 \qquad \eta A[45,2] \longleftarrow A[45,2]M_1 \qquad \eta A[45,1]M_1 \longleftarrow A[39,3]M_1^2\bar{M}_2$

$\leftarrow \sigma A[30]M_1^4 \qquad \eta D[45] \longleftarrow D[45]M_1 \qquad \eta A[47] \longleftarrow A[47]M_1$

$\leftarrow 16\gamma_1M_1^8\langle M_3\rangle \qquad \eta^2C[44] \longleftarrow \eta C[44]M_1 \qquad \eta B[47] \longleftarrow B[47]M_1$

$\leftarrow A[39,2]\bar{M}_2 \qquad \eta A[40,1]M_1^3 \longleftarrow B[34]M_1^4M_2 \qquad A[39,1]M_1^2M_2 \longleftarrow$

$D[45] \longleftarrow 4C[18](M_1^7\bar{M}_3+M_1^{11}\bar{M}_2) \qquad \nu A[45,1] \longleftarrow A[45,1]M_1^2$

$2D[45] \longleftarrow B[34]M_1^6 \qquad A[47] \longleftarrow 2C[42]M_1^3 \qquad 4D[45]M_1^2 \longleftarrow$

$4D[45] \longleftarrow \eta\sigma A[32,1]M_1^3 \qquad B[47] \longleftarrow 2C[20]^2M_1\bar{M}_2 \qquad \eta^2D[45]M_1 \longleftarrow$

$8D[45] \longleftarrow C[42]M_1^2 \qquad 2B[47] \longleftarrow \eta A[45,2]M_1 \qquad \eta^2A[45,2]M_1 \longleftarrow$

$\eta C[44] \longleftarrow C[44]M_1 \qquad \eta^2D[45] \longleftarrow \eta D[45]M_1$

$A[45,1] \longleftarrow 2B[34]M_1^3\bar{M}_2 \qquad \nu C[44] \longleftarrow C[44]M_1^2$

$A[45,2] \longleftarrow C[20]^2\bar{M}_2 \qquad \nu D[45] \longleftarrow D[45]M_1^2$

$2\nu D[45] \longleftarrow 2D[45]M_1^2$

50 $\qquad$ 51 $\qquad$ 52 $\qquad$ 53 $\qquad$ 54

$\eta A[47]M_1 \longleftarrow \eta B[38]M_1^3\bar{M}_2 \qquad \nu A[45,1]M_1^2 \longleftarrow \sigma A[32,1]M_1^4\bar{M}_2 \qquad \eta A[47]M_1^3 \longleftarrow$

$\eta B[47]M_1 \longleftarrow A[45,2]\bar{M}_2 \qquad \begin{bmatrix}\eta A[39,3]+\\ \eta\sigma A[32,1]\end{bmatrix}M_1^3\bar{M}_2 \longleftarrow \nu C[18]M_1^6\bar{M}_2\bar{M}_3 \qquad \eta\sigma C[44]M_1 \longleftarrow$

$A[50,1] \longleftarrow \eta^2\gamma_1M_1^{17} \qquad A[52,1] \longleftarrow \eta^2D[45]M_1^3 \qquad A[54,1] \longleftarrow$

$A[50,2] \longleftarrow A[39,1]M_1^3M_2 \qquad A[52,2] \longleftarrow B[47]\bar{M}_2 \qquad A[54,2] \longleftarrow$

$\leftarrow A[32,1](M_1^2M_3+M_2^3) \qquad \eta\sigma C[44] \longleftarrow \sigma C[44]M_1 \qquad \eta A[8]D[45] \longleftarrow$

$\nu^2D[45] \longleftarrow \nu D[45]M_1^2 \qquad 4D[45]M_1M_2 \longleftarrow A[36]M_1^2\bar{M}_3$

$\leftarrow B[38]M_1^6 \qquad 8D[45]M_1^3 \longleftarrow 2B[38]M_1^4\bar{M}_2 \qquad \eta A[50,2]M_1 \longleftarrow \nu A[45,1]M_1^3$

$\leftarrow \eta A[37]M_1^3\bar{M}_2 \qquad \nu C[44]M_1^2 \longleftarrow A[40,1]M_1^6 \qquad A[8]D[45] \longleftarrow 2\nu D[45]M_1^3$

$\leftarrow \eta^2C[20]^2M_1\bar{M}_2 \qquad \sigma C[44] \longleftarrow C[44]M_1^4 \qquad \eta A[52,2] \longleftarrow A[52,2]M_1$

$2\sigma C[44] \longleftarrow \eta A[45,1]M_1^3 \qquad \nu A[50,1] \longleftarrow A[50,1]M_1^2$

$\eta A[50,2] \longleftarrow A[50,2]M_1 \qquad \nu A[50,2] \longleftarrow A[50,2]M_1^2$

55		56		57		58		59	
$\leftarrow \eta^2\gamma_1 M_1^{19}$		$A[50,1]M_2$	$\longleftarrow$	$4\beta_1 M_1^7\bar{M}_2^3\bar{M}_3$		$A[52,1]M_1^3$	$\leftarrow$	$\eta^2\gamma_1 M_1^{15}\bar{M}_2^2$	
$\leftarrow \nu C[44]M_1M_2$		$\eta A[8]D[45]M_1$	$\leftarrow$	$\nu^2 D[45]M_1^3$				$\sigma A[32,1]M_1^4M_2^2$	$\longleftarrow$
$\leftarrow \beta_2 M_1^{18}$		$A[56]$	$\longleftarrow$	$2\beta_2 M_1^{16}\bar{M}_2$		$\eta A[57]$	$\longleftarrow$	$A[57]M_1$	
$\leftarrow \eta^2 A[45,2]M_1\bar{M}_2$		$A[52,1]M_1^2$	$\longleftarrow$	$4D[45]M_1^3M_2$				$2D[45]M_1^4M_2$	$\longleftarrow$
$\leftarrow A[8]D[45]M_1$				$A[47]M_1^2\bar{M}_2$	$\longleftarrow$	$A[40,2]M_1^6\bar{M}_2$		$\sigma C[44]M_1^4$	$\longleftarrow$
		$\nu^2 A[50,2]$	$\longleftarrow$	$\nu A[50,2]M_1^2$				$\nu A[50,1]M_1^3$	$\longleftarrow$
$2\sigma C[44]M_1^2$	$\longleftarrow$	$2C[44]M_1^6$		$\eta A[54,2]M_1$	$\longleftarrow$	$A[52,2]\bar{M}_2$		$\eta A[56]M_1$	$\longleftarrow$
$\eta A[52,2]M_1$	$\longleftarrow$	$\eta B[47]M_1\bar{M}_2$		$A[57]$	$\longleftarrow$	$4C[44]M_1^7$		$A[59,2]$	$\longleftarrow$
$\eta A[54,2]$	$\longleftarrow$	$A[54,2]M_1$		$\eta A[56]$	$\longleftarrow$	$A[56]M_1$		$A[59,1]$	$\longleftarrow$

60		61		62		63	
$\eta A[57]M_1$	$\longleftarrow$	$\beta_2 M_1^{14}\bar{M}_3$		$A[62,1]$	$\longleftarrow$	$\eta^2\sigma M_1^{21}M_2^2$	
$\leftarrow A[30]\langle M_4\rangle$				$A[62,2]$	$\longleftarrow$	$4\beta_2 M_1^{19}\bar{M}_2$	
$A[52,1]M_1M_2$	$\longleftarrow$	$A[39,1]M_1M_2M_3$		$B[62]$	$\longleftarrow$	$\gamma_2 M_1^{20}$	
$\leftarrow \eta A[23]M_1^{15}\bar{M}_2$		$A[59,1]M_1$	$\longleftarrow$	$B[38]M_1^2\bar{M}_2\bar{M}_3$		$A[63]$	$\longleftarrow$
$\leftarrow A[40,1]\bar{M}_2\langle M_3\rangle$		$\sigma C[44]M_1^2\bar{M}_2$	$\longleftarrow$	$\nu A[19]M_1^7M_2^2\langle M_3\rangle$		$A[61]M_1$	$\longleftarrow$
$\leftarrow A[36]M_1^6\bar{M}_2^2$		$A[61]$	$\longleftarrow$	$\nu^2 A[50,2]M_1^3$		$\eta A[62,4]$	$\longleftarrow$
$\leftarrow A[54,1]M_1^3$				$A[62,4]$	$\longleftarrow$	$\eta A[50,2]M_1^3\bar{M}_2$	
$\leftarrow A[54,2]\bar{M}_2$		$\eta B[60]$	$\longleftarrow$	$B[60]M_1$		$4D[45]M_1^6M_2$	$\longleftarrow$
$\leftarrow \eta\sigma C[44]M_1\bar{M}_2$		$2D[45]M_1\langle M_3\rangle$	$\longleftarrow$	$A[32,1]M_1^8\bar{M}_3$		$\sigma A[32,1]M_1^6M_2^2$	$\longleftarrow$
$\eta A[59,2]$	$\longleftarrow$	$A[59,2]M_1$				$\eta A[62,1]$	$\longleftarrow$
$B[60]$	$\longleftarrow$	$\eta A[52,2]M_1\bar{M}_2$		$2B[62]$	$\longleftarrow$	$A[8]D[45]M_1^2\bar{M}_2$	
				$\eta A[59,2]M_1$	$\longleftarrow$	$\eta A[54,2]M_1\bar{M}_2$	
				$A[62,3]$	$\longleftarrow$	$\nu A[50,1]M_1^2M_2$	
				$\eta^2 B[60]$	$\longleftarrow$	$\eta B[60]M_1$	
						$\eta A[62,2]$	$\longleftarrow$

<u>64</u>		<u>65</u>		<u>66</u>
$B[64,1]$	$\longleftarrow$	$\beta_1 M_1^{20}\bar{M}_3$		$A[62,1]M_1^2$
$2B[62]M_1$	$\longleftarrow$	$2\gamma_2 M_1^{18}\bar{M}_2$		$B[60]\bar{M}_2$
$A[64,1]$	$\longleftarrow$	$\sigma C[44]M_1^4\bar{M}_2$		$A[62,2]M_1^2$
$\leftarrow\ \eta\sigma A[32,1]M_1^5\langle M_3\rangle$		$2D[45]\langle M_1^4\rangle\langle M_2^2\rangle$	$\longleftarrow$	$A[32,1]M_1^2\langle M_4\rangle$
$\leftarrow\ A[52,1]M_1^3M_2$		$A[63]M_1$	$\longleftarrow$	$\eta A[39,3]M_1^3\bar{M}_2\langle M_3\rangle$
$\leftarrow\ A[62,4]M_1$				$A[52,1]M_1^7$
$A[64,2]$	$\longleftarrow$	$A[59,1]M_1^3$		$\eta^2C[44]M_1^7\bar{M}_2$
$\leftarrow\ 2B[34]M_1^5\bar{M}_2\bar{M}_3$		$\eta A[62,2]M_1$	$\longleftarrow$	$A[56]M_1^2\bar{M}_2$
$\leftarrow\ 4C[18]M_1^7\bar{M}_2M_2^2\bar{M}_3$				$\nu^2A[50,2]M_1^2M_2$
$\leftarrow\ A[62,1]M_1$				$2C[44]M_1^2M_2^3$
$B[64,2]$	$\longleftarrow$	$\nu C[44]M_1^3\langle M_2^2\rangle$		$\eta A[8]D[45]M_1^3\bar{M}_2$
$\eta^2B[60]M_1$	$\longleftarrow$	$A[59,2]\bar{M}_2$		$A[62,3]M_1^2$
$A[64,3]$	$\longleftarrow$	$A[59,1]M_2$		$2B[62]M_1^2$
$2B[64,1]$	$\longleftarrow$	$\eta A[62,1]M_1$		$A[62,4]M_1^2$
$\leftarrow\ A[62,2]M_1$				
$2B[64,2]$	$\longleftarrow$	$\eta A[62,4]M_1$		

APPENDIX 5: THE COMPUTER PROGRAMS

All of the computer programs used to make the computations of this paper are written in FORTRAN 77 [9]. They were compiled and linked by Ryan-McFarland RM/FORTRAN version 2.00 on an IBM PC/AT microcomputer running DOS version 3.0. There are eight component programs which are linked into five programs. The scheme for linking them is given in the left columns of Figures 1 and 2. The files S70.FOR, MODB70.FOR and MODS70.FOR contain subroutines for manipulating the arrays which represent monomials, polynomials and bases of poloynomials. The boxes in the right columns represent files which store data. The arrows indicate the data which is required as input for each program and the data which each program generates. For each of the five programs, we describe what the program computes and briefly indicate how the program carries out the computations. The 100 pages of complete program listings are available from the author.

PROGRAM I. This program uses formula 1.2.3 to compute the Hazewinkel generators V_i, $1 \le i \le 5$, as polynomials in the M_N and to compute the M_i, $1 \le i \le 5$, as polynomials in the V_N. This information is stored in the file HAZEWINK. In the second part of this program we consider all monomials M_E in the M_N, degree $M_E \le 70$, all Quillen operation r_I, degree $r_I \le$ degree M_E, and all $t \le (\text{degree } M_E)/2$. Let $U_N = V_N/2$ denote polynomial generators of H_*BP. We determine the coefficient $C(E,I,t)$ of U_1^t in $r_I(M_E)$ written as a polynomial in the U_N. This is accomplished by first computing $r_I(M_E)$ as a polynomial in the M_N using the Cartan formula and the fact that

$$r_I(M_s) = \begin{cases} M_k & \text{if } s \ge k \text{ and } I = 2^k\Delta_{s-k} \\ 0 & \text{otherwise} \end{cases}.$$

Then we use the observation that when M_N is written as a polynomial in the U_k, the coefficient of $U_1^{2^N-1}$ is 2^{2^N-N-1}. The values of the $C(E,I,t)$ are used in the second program to compute the d^{2t}-differentials which originiate on the 0 row. The C(E,I,t) are stored in the files STRDMON1,...,STRDMON7.

PROGRAM II. This program computes the differentials on the 0 row in all degrees 2t less than or equal to 70. Elements of $E^{2r}_{2t,0}$ are written as polynomials in the U_N, and elements of $E^{2r}_{2t-2r,2r-1}$ are written as polynomials in the $\bar{M}_N$. The d^{2r}-differentials are computed by converting an element X in $E^{2r}_{2t,0}$ to a polynomial in the M_N using the information in the file HAZEWINK. The coefficient of M_I in $d^{2r}(X)$, written as a polynomial in the M_N, is $2^{k(r)-r}$ times the coefficient C(X;I,r) of U_1^r in $r_I(X)$, written as a polynomial in the U_N. Here k(r) equals 1, 2, $\mathcal{E}((N-8)/8)$ if r is congruent modulo 4 to 1, 2, 4, respectively. The C(X;I,r) are determined from the C(E,I,r) which were stored in the files STRDMON1,...,STRDMON7 in Program I. Now $d^{2r}(X)$ has been determined as a polynomial in the M_N and is converted to a polynomial in the $\bar{M}_N$. When r = 4s+1, we have only determined the summand of $d^{8s+2}(X)$ in $Z_2\alpha_s \otimes B\langle 2\rangle$. If $s \geq 1$, we determine the summand of $d^{8s+2}(X)$ in $Z_2\eta^2\gamma_{s-1}M_1 \otimes B\langle 2\rangle$ by carrying out the above procedure for finding "$d^{8s+4}(X)$" using k(4s+2) = 3. When r = 4s+2 we have only determined $d^{8s+4}(X)$ in $Z_4 \otimes (Z_4\beta_s \oplus Z_8\beta_s M_1) \otimes B\langle 2\rangle$. To complete the determination of the coefficients in $Z_8\beta_s M_1 \otimes B\langle 2\rangle$, we carry out the above procedure for finding "$d^{8s+6}(X)$" using k(4s+3) = 3. We use elementary row and column operations to diagonalize the matrix D of d^{2r} keeping a record of the elements of $E^{2r}_{2t,0}$ represented by each row of the matrix and the elements of $E^{2r}_{2t-2r,2r-1}$ represented by each column of the matrix. The basis of Kernel d^{2r} are the appropriate powers of two times the row representatives of D and Cokernel d^{2r} is the direct sum of the cyclic groups generated by the column representatives of D corresponding to the diagonal entries $d \neq 1$. The cokernels of these differentials are stored in the files STRJCOK1,...,STRJCOK4 as polynomials in the $\bar{M}_N$. The $E^{2r}_{2t,0}$, $0 < t \leq 35$, are stored in the files STRBSGP1,...,STRCSGP6 as polynomials in the U_N.

PROGRAM III. This program reorganizes the data stored in STRJCOK1,...,STRJCOK4 in the way that it will be used in Program V to compute differentials originating on the rows of CokJ. That is, Cok $[d^{2r}:E^{2r}_{2t,0} \longrightarrow E^{2r}_{2t-2r,2r-1}]$ in STRJCOK1,...,STRJCOK4 is stored in the lexicographical order of (t,r). The output of this program stores these cokernels in INFILE*R* where $R = 2r-1$. We only need store this information for $R \leq 23$ because the cokernels turn out to be zero in degrees less than 70 when $R > 23$.

PROGRAM IV. This program computes images of differentials in the bidegree (N,t) of a leader X on which a nonzero differential originates. In order to use Quillen operations to compute the tentative differentials determined by this differential on X in bidegrees (N',t), $N' \geq N$, we need to know all elements of $E^2_{N,t}$ which are homologous to zero in $E^{2s}_{N,t}$ for $1 \leq s < t$. This program is used to determine these elements. The output of this program is very short and is produced on the monitor. The input file INPIPE contains the $E^{2s}_{N+2s,t-2s+1}$ on which the d^{2s} are to be computed. Any of the files INFILE3,...,INFILE23 or any of the files OUTDOM1,...,OUTDOM9, OUTRANG1,...,OUTRANG9 produced by Program V can be renamed INPIPE and used as input for this program. When the d^{2s} originate on the 0 row, the required information can be obtained from Program II.

PROGRAM V. This program is the analogue of Prgram II for computing the cokernels of differentials d^{2r} which originate on the t row where $t > 0$. The input file INPIPE contains the $E^{2r}_{*,t}$. Any of the files INFILE3,...,INFILE23 or any of the files OUTDOM1,...,OUTDOM9, OUTRANG1,...,OUTRANG9 produced by a previous running of program V can be renamed INPIPE and used as input for this program. The $E^{2r+2}_{*,t}$ are stored in OUTDOM1 and the $E^{2r+2}_{*,t+2r-1}$ are stored in

OUTRANG1. A sequence of differentials of this sort $d^{2r(k)}$, $1 \le k \le 9$, can be computed with one run of the program where t is fixed and $r(1) \le \cdots \le r(9)$. The $E_{*,t}^{2r(k)+2}$ are stored in OUTDOM*k* and the $E_{*,t+2r(k)-1}^{2r(k)+2}$ are stored in OUTRANG*k*.

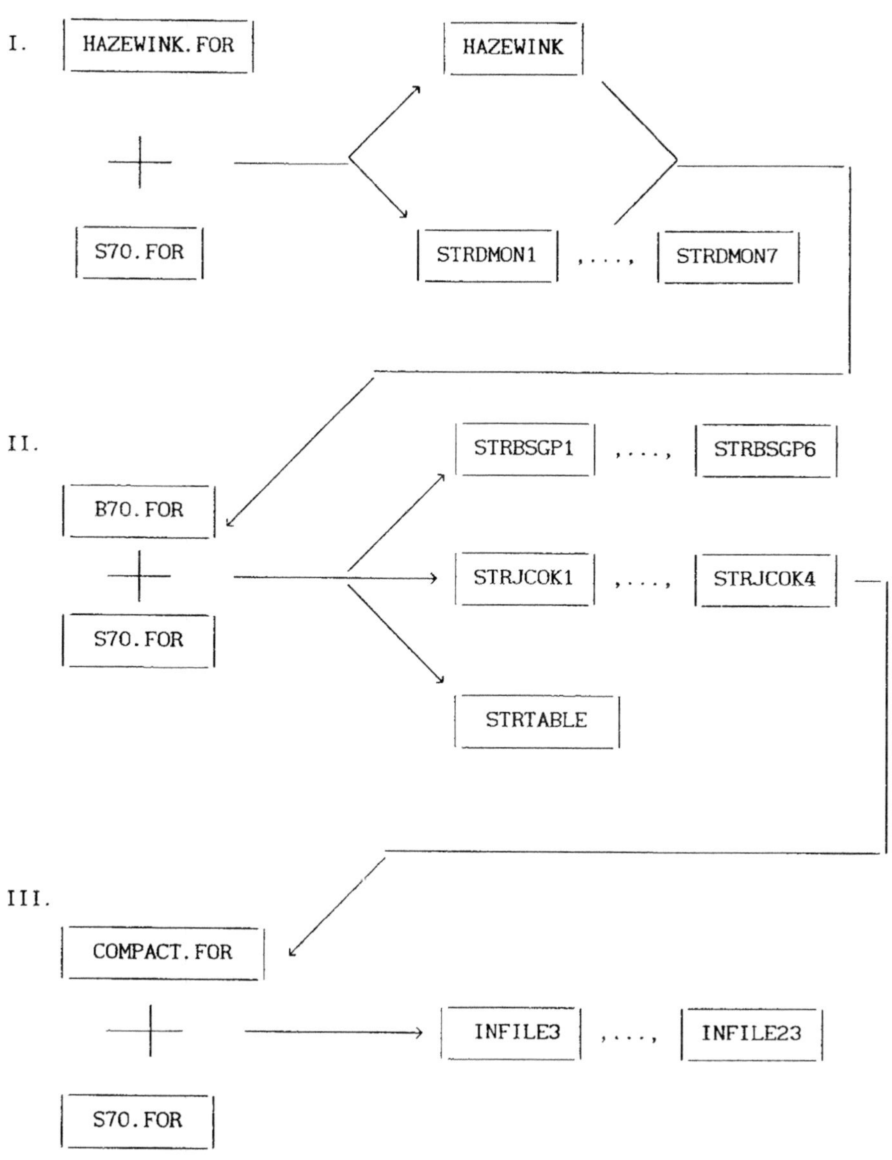

FIGURE 1: COMPUTING DIFFERENTIALS ORIGINATING ON THE 0 ROW

IV.

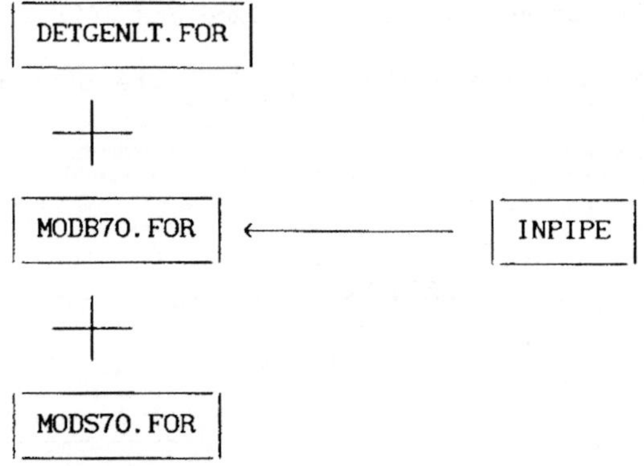

V.

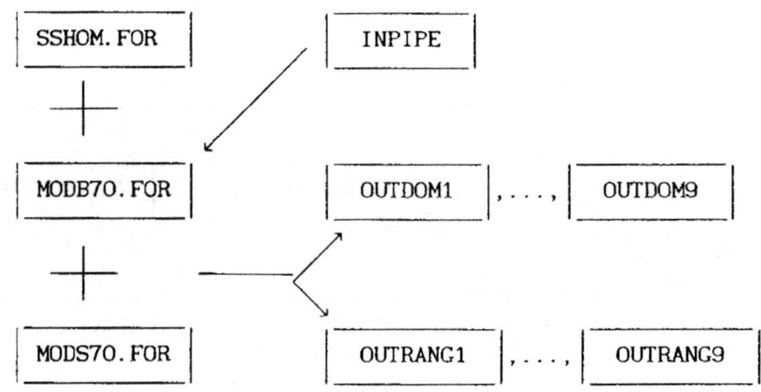

FIGURE 2: COMPUTING DIFFERENTIALS ABOVE THE 0 ROW

APPENDIX 6: THE ADAMS SPECTRAL SEQUENCE

The tables below depict the E_2-term of the classical mod 2 Adams spectral sequence for π_*^S. The notation is standard: vertical lines represent multiplication by h_0, lines of positive slope represent multiplication by h_1 and lines of negative slope represent nonzero differentials. If solid vertical lines from both A and B land on C, this indicates that $C = h_0A = h_0B$. If these lines are dotted, this indicates that $C = h_0A + h_0B$. Nontrivial extensions given by multiplication by 2 or η are denoted by dotted lines. To make the tables readable we do not label any of the elements or include lines indicating multiplication by h_2. Instead, each table is followed by a list of labels indexed by bidegree. In each beidegree the elements are labeled from left to right. For each infinite cycle we use the symbol $X \longleftarrow \xi$ to indicate the name of the element ξ in the Atiyah-Hirzebruch spectral sequence which projects to X. We also include tables giving products with h_2.

These tables are based upon the tables of E_2 of the Adams spectral sequence of Mahowald [55] and Tangora [59]. The differentials in degrees less than or equal to 45 were computed by Mahowald and Tangora [37], Barratt, Mahowald and Tangora [10] and Bruner [16]. The differentials in degrees 46 through 59 confirm the tentative differentials given by Mahowald in [55]. The differentials in degrees greater than 59 are tentative in the following sense:

(1) some are consequences of differentials in lower degrees;

(2) some are consequences of the computation of π_n^S, $n \le 64$, in this paper;

(3) some are the most reasonable choices among several possibilities which agree with the computations of this paper.

The use of these tables in Sections 7.5, 8.3 and Appendix 2 do not rely on any of the choices described in (3).

The tables below include the following entries which were accidentally omitted from the tables in [55] and [59]:

$d^3(h_5 i) = h_0 x'$ in (53,11), $h_0 R_1 = S_1$ in (54,11), $d^4(gm) = Pgj$ in (54,15), $h_0^2 R = Ph_5 i$ in (62,12), $h_0 P^4 r = P^4 s$ in (62,23), $h_1 h_3 Q_2 = h_0^2 h_3 D_2$ in (65,9) and several products which are marked with asterisks in the tables below.

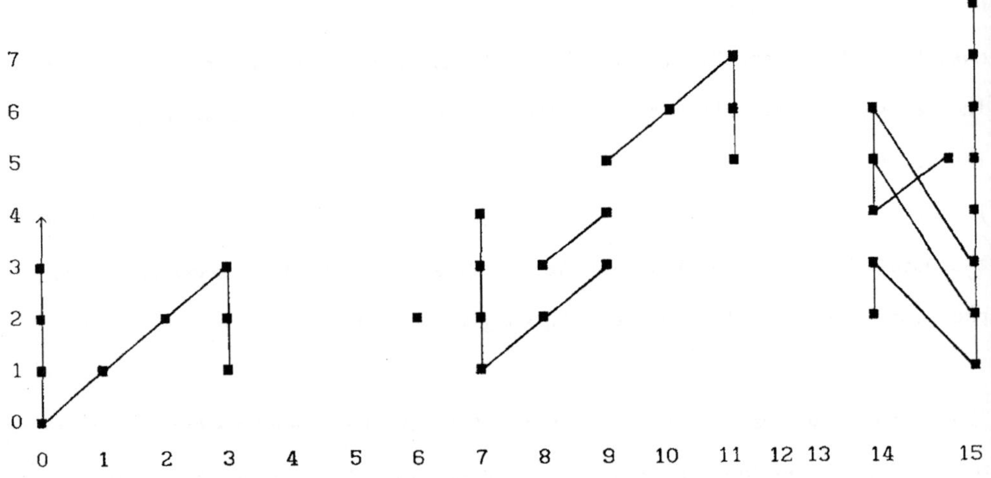

Notation:

(0,1) $h_0 \leftarrow 2$ (1,1) $h_1 \leftarrow \eta$ (3,1) $h_2 \leftarrow \nu$

(6,2) $h_2^2 \leftarrow \nu^2$ (7,1) $h_3 \leftarrow \sigma$ (8,3) $c_0 \leftarrow A[8]+\eta\sigma$

(9,5) $Ph_1 \leftarrow \alpha_1$ (11,5) $Ph_2 \leftarrow \beta_1$ (14,2) $h_3^2 \leftarrow \sigma^2$

(14,4) $d_0 \leftarrow A[14]$ (15,1) h_4 (15,4) $h_0^3 h_4 \leftarrow \gamma_1$

Multiplication by h_2:

(9,3) $h_1^2 h_3 = h_2^3$ (14,6) $h_2 Ph_2 = h_0^2 d_0$

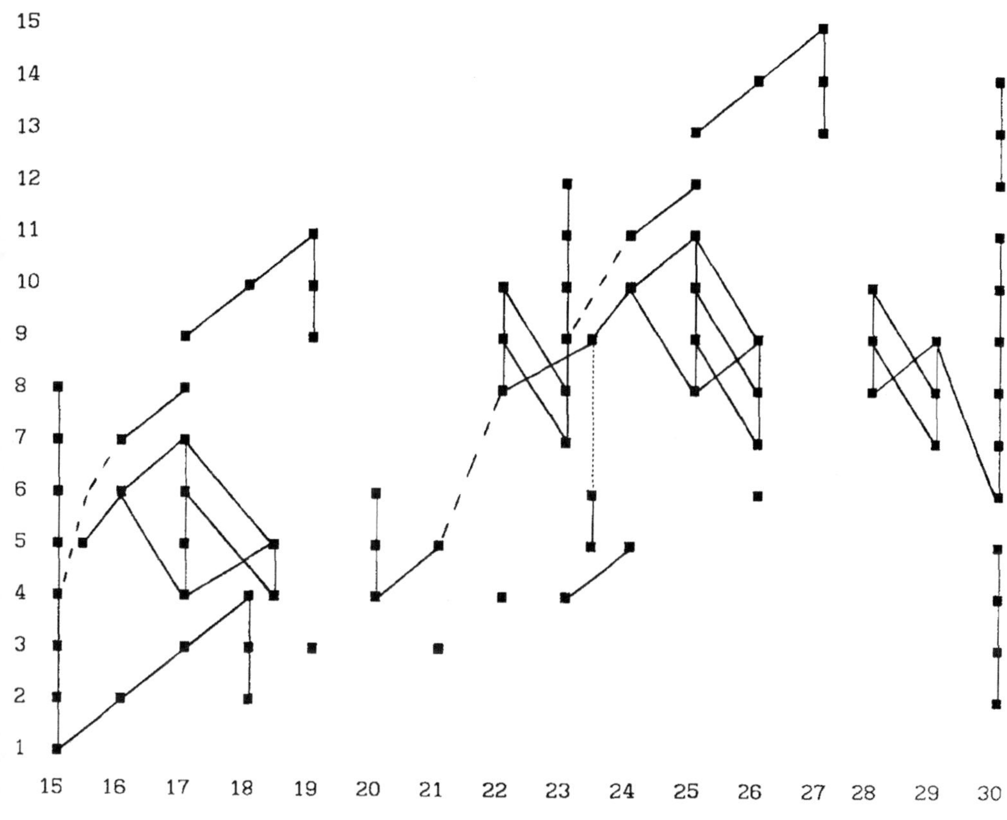

Notation:

(16,2) $h_1h_4 \leftarrow A[16]$ (16,7) $Pc_0 \leftarrow \eta\gamma_1$ (17,4) e_0

(17,9) $P^2h_1 \leftarrow \alpha_2$ (18,2) $h_2h_4 \leftarrow C[18]$ (18,4) f_0

(19,3) $c_1 \leftarrow A[19]$ (19,9) $P^2h_2 \leftarrow \beta_2$ (20,4) $g \leftarrow C[20]$

(21,3) $h_2^2h_4 \leftarrow \nu C[18]$ (22,4) $h_2c_1 \leftarrow \nu A[19]$ (22,8) $Pd_0 \leftarrow \eta^2C[20]$

(23,4) $h_4c_0 \leftarrow A[23]$ (23,5) $h_2g \leftarrow \nu C[20]$ (23,7) i

(23,9) $h_0^2i \leftarrow \gamma_2$ (24,11) $P^2c_0 \leftarrow \eta\gamma_2$ (25,8) Pe_0

(25,13) $P^3h_1 \leftarrow \alpha_3$ (26,6) $h_2^2g \leftarrow \nu^2C[20]$ (26,7) j

(27,13) $P^3h_2 \leftarrow \beta_3$ (28,8) $Pg \leftarrow A[8]C[20]$ (29,7) k

(30,2) $h_4^2 \leftarrow A[30]$ (30,6) r (30,12) P^2d_0

Multiplication by h_2:

(17,5) $h_2d_0 = h_0e_0$ (20,5) $h_2e_0 = h_0g$ (22,10) $h_2P^2h_2 = h_0^2Pd_0$

(25,9) $h_2Pd_0 = h_0Pe_0$ (26,8) $h_2i = h_0j$ (28,9) $h_2Pe_0 = h_0Pg$

(29,8) $h_2j = h_0k$

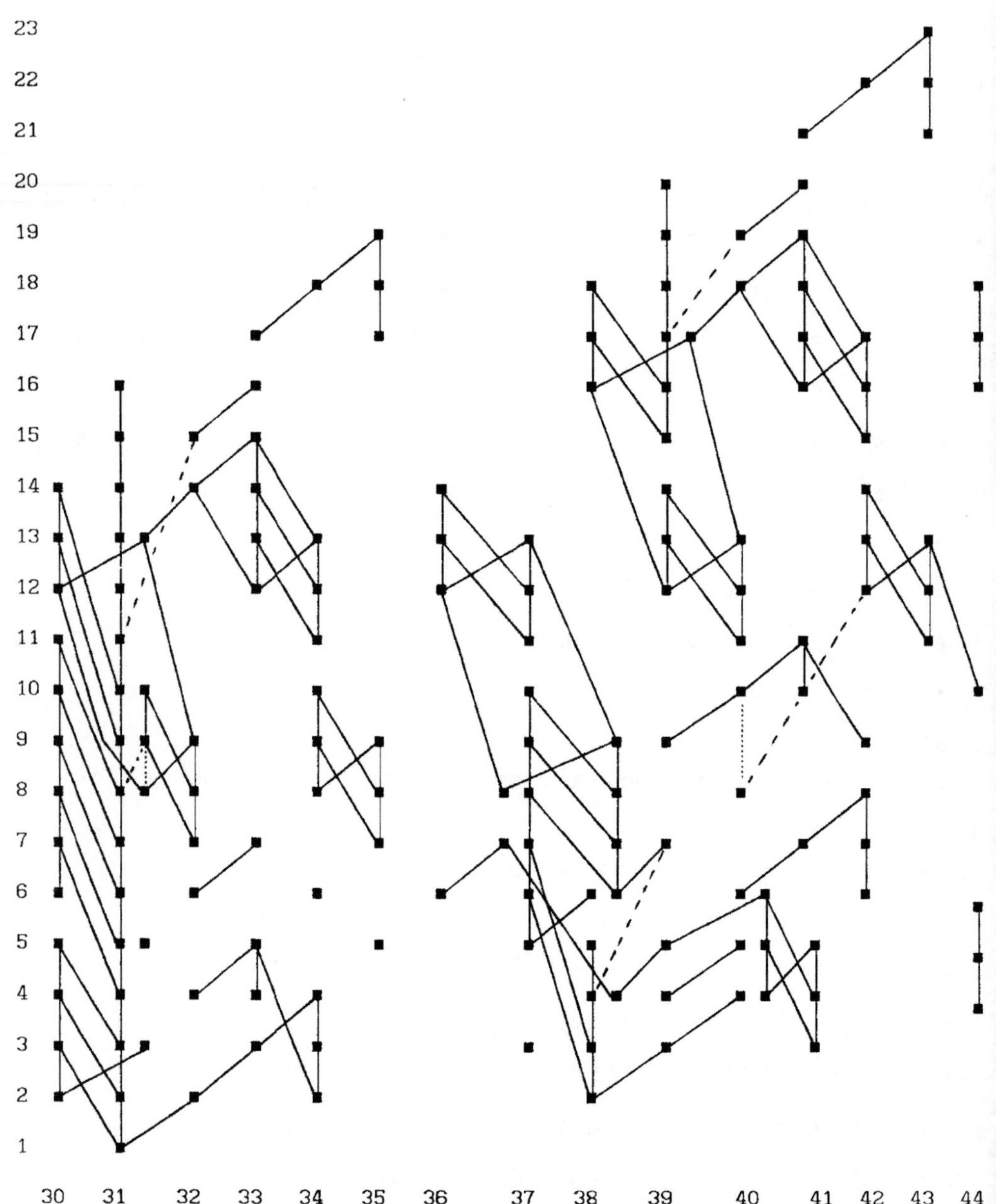
23
22
21
20
19
18
17
16
15
14
13
12
11
10
9
8
7
6
5
4
3
2
1
30 31 32 33 34 35 36 37 38 39 40 41 42 43 44

Notation:

(31,1)	h_5	(31,5)	$n \leftarrow A[31]$	(31,8)	d_0e_0
(31,11)	$h_0^{10}h_5 \leftarrow \gamma_3$	(32,2)	$h_1h_5 \leftarrow A[32,1]$	(32,4)	$d_1 \leftarrow A[32,3]$
(32,6)	$q \leftarrow A[32,2]$	(32,7)	ℓ	(32,15)	$P^3c_0 \leftarrow \eta\gamma_3$
(33,4)	$p \leftarrow \nu A[30]$	(33,12)	P^2e_0	(33,17)	$P^4h_1 \leftarrow \alpha_4$
(34,2)	h_2h_5	(34,3)	$h_0h_2h_5 \leftarrow B[34]$	(34,6)	$h_2n \leftarrow \nu A[31]$
(34,8)	$e_0^2 \leftarrow A[14]C[20]$	(34,11)	Pj	(35,5)	$h_2d_1 \leftarrow \nu A[32,3]$
(35,7)	m	(35,17)	$P^4h_2 \leftarrow \beta_4$	(36,6)	$t \leftarrow A[36]$
(36,12)	P^2g	(37,3)	$h_2^2h_5 \leftarrow A[37]$	(37,5)	$x \leftarrow \sigma A[30]$
(37,8)	e_0g	(37,11)	Pk	(38,2)	h_3h_5
(38,4)	$h_0^2h_3h_5 \leftarrow B[38]$	(38,4)	e_1	(38,6)	y
(38,16)	P^3d_0	(39,3)	$h_1h_3h_5 \leftarrow A[39,3]$	(39,4)	$h_5c_0 \leftarrow \sigma A[32,1]$
(39,5)	$h_1e_1 \leftarrow A[39,1]$	(39,7)	$c_1g \leftarrow \eta B[38]$	(39,9)	$u \leftarrow A[39,2]$
(39,12)	Pd_0e_0	(39,15)	P^2i	(39,17)	$h_0^2P^2i \leftarrow \gamma_4$
(40,4)	$f_1 \leftarrow A[40,1]$	(40,6)	$Ph_1h_5 \leftarrow A[40,2]$	(40,8)	$g^2 \leftarrow C[20]^2$
(40,11)	d_0j	(40,19)	$P^4c_0 \leftarrow \eta\gamma_4$	(41,3)	c_2
(41,10)	$z \leftarrow \eta C[20]^2$	(41,16)	P^3e_0	(41,21)	$P^5h_1 \leftarrow \alpha_5$
(42,6)	$Ph_2h_5 \leftarrow C[42]$	(42,9)	v	(42,12)	$Pe_0^2 \leftarrow \eta^2C[20]^2$
(42,15)	P^2j	(43,11)	Pm	(43,21)	$P^5h_2 \leftarrow \beta_5$
(44,4)	$g_2 \leftarrow C[44]$	(44,10)	d_0r	(44,16)	P^3g

Multiplication by h_2:

(31,9)	$h_2Pg = h_0d_0e_0$	(32,8)	$h_3k = h_0\ell$	(33,13)	$h_2P^2d_0 = h_0P^2e_0$
(34,9)	$h_2d_0e_0 = h_0e_0^2$	(35,8)	$h_2\ell = h_0m$	(36,13)	$h_2P^2e_0 = h_0P^2g$
(37,12)	$h_2Pj = h_0Pk$	(38,6)	$h_2^2d_1 = h_1x$	(38,8)	$h_2m = h_0^2y$
*(38,9)	$h_2h_0m = h_1e_0g$	(38,18)	$h_2P^4h_2 = h_0^2P^3d_0$	*(39,7)	$h_2t = c_1g = h_1y$
(39,13)	$h_2P^2g = h_0Pd_0e_0$	(40,4)	$h_2^3h_5 = h_1^2h_3h_5$	(40,12)	$h_2Pk = h_0d_0j$
(41,17)	$h_2P^3d_0 = h_0P^3e_0$	(42,13)	$h_0Pd_0e_0 = h_0Pe_0^2$	(42,16)	$h_2P^2i = h_0P^2j$
(43,12)	$h_2d_0j = h_0Pm$	(44,17)	$h_2P^3e_0 = h_0P^3g$		

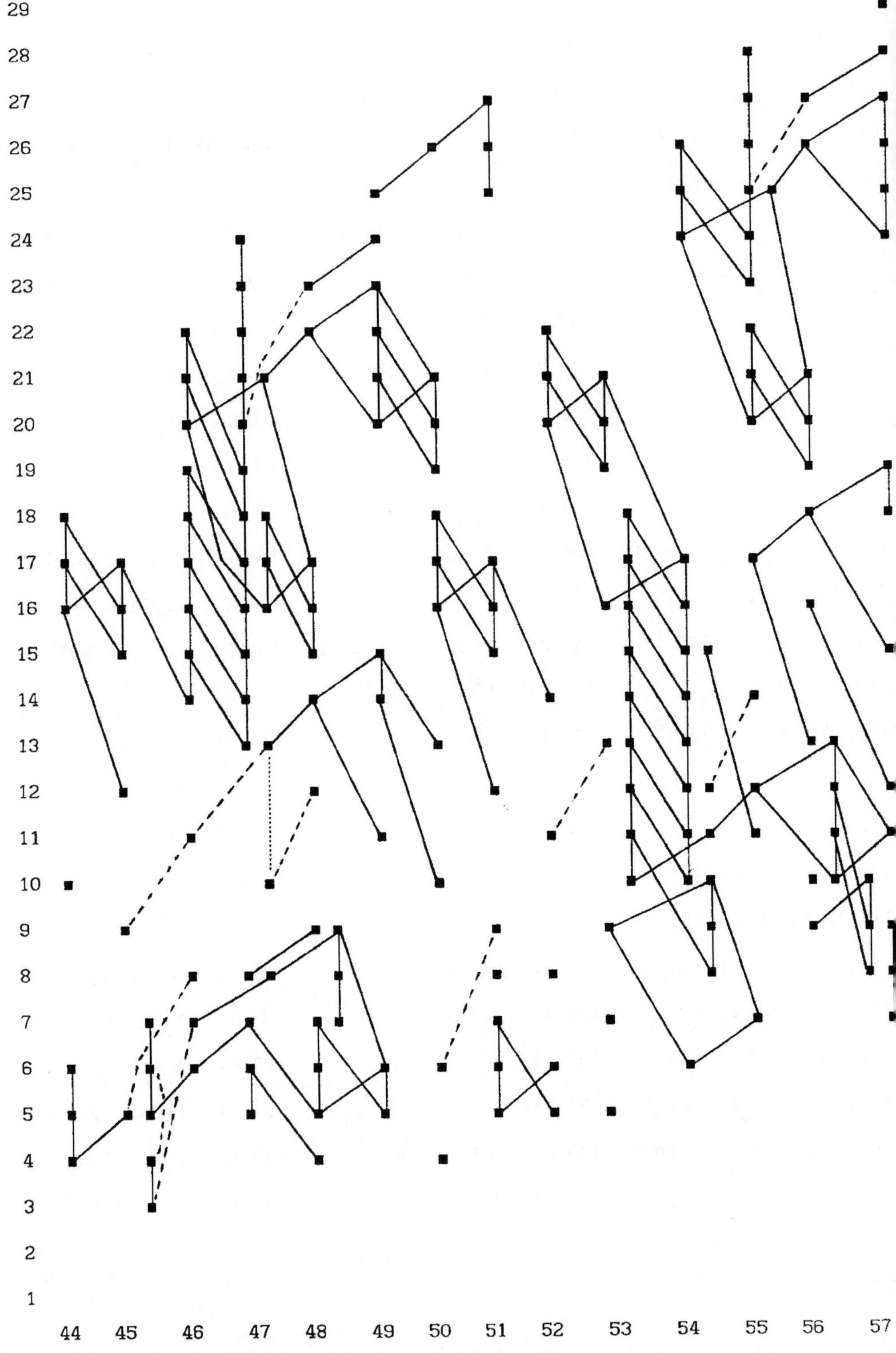
29
28
27
26
25
24
23
22
21
20
19
18
17
16
15
14
13
12
11
10
9
8
7
6
5
4
3
2
1
44
45
46
47
48
49
50
51
52
53
54
55
56
57

Notation:

(45,3)	$h_4^3 \leftarrow D[45]$	(45,5)	$h_5 d_0 \leftarrow A[45,1]$	(45,9)	$w \leftarrow A[45,2]$
(45,12)	$Pe_0 g$	(45,15)	$P^2 k$	(46,7)	$B_1 \leftarrow \eta D[45]$
(46,8)	$N \leftarrow \eta^2 C[44]$	(46,11)	$gj \leftarrow \eta A[45,2]$	(46,14)	$P^2 r$
(46,20)	$P^4 d_0$	(47,5)	$h_2 g_2 \leftarrow \nu C[44]$	(47,8)	$Ph_5 c_0 \leftarrow A[47]$
(47,10)	$e_0 r \leftarrow B[47]$	(47,13)	Q'	(47,13)	$Pu \leftarrow \eta^2 A[45,2]$
(47,16)	$P^2 e_0 d_0$	(47,20)	$h_0^7 Q' \leftarrow \gamma_5$	(48,4)	$h_3 c_2$
(48,5)	$h_5 e_0$	(48,6)	$h_0 h_5 e_0 \leftarrow \nu A[45,1]$	(48,7)	$B_2 \leftarrow \nu D[45]$
(48,12)	$Pg^2 \leftarrow \eta B[47]$	(48,15)	$P^2 \ell$	(48,23)	$P^5 c_0 \leftarrow \eta \gamma_5$
(49,5)	$h_5 f_0$	(49,11)	gk	(49,14)	Pz
(49,20)	$P^4 e_0$	(49,25)	$P^6 h_1 \leftarrow \alpha_6$	(50,4)	$h_5 c_1 \leftarrow A[50,1]$
(50,6)	$C \leftarrow A[50,2]$	(50,10)	gr	(50,13)	Pv
(50,16)	$P^2 e_0^2$	(50,19)	$P^3 j$	(51,5)	$h_5 g \leftarrow \sigma C[44]$
(51,8)	$h_2 B_2 \leftarrow \nu^2 D[45]$	(51,9)	$gn \leftarrow \eta A[50,2]$	(51,12)	e_0^3
(51,15)	$P^2 m$	(51,25)	$P^6 h_2 \leftarrow \beta_6$	(52,5)	D_1
(52,8)	$d_1 g \leftarrow A[52,1]$	(52,11)	$g\ell \leftarrow A[52,2]$	(52,14)	$Pd_0 r$
(52,20)	$P^4 g$	(53,5)	$h_2 h_5 c_1 \leftarrow \nu A[50,1]$	(53,7)	$h_2 C \leftarrow \nu A[50,2]$
(53,9)	$Ph_5 d_0$	(53,10)	$x' \leftarrow A[8]D[45]$	(53,13)	$Pw \leftarrow \eta A[52,2]$
(53,16)	$P^2 e_0 g$	(53,19)	$P^3 k$	(54,6)	G
(54,8)	$h_5 i$	(54,9)	$h_0 h_5 i \leftarrow A[54,1]$	(54,10)	R_1
(54,12)	$e_0^2 g \leftarrow A[54,2]$	(54,15)	Pgj	*(54,17)	$h_1 P^2 e_0 g = h_0^7 R_1$
(54,24)	$P^5 d_0$	(55,11)	gm	(55,14)	$Pe_0 r \leftarrow \eta A[54,2]$
(55,17)	$P^2 u$	(55,20)	$P^3 e_0 d_0$	(55,23)	$P^4 i$
(55,25)	$h_0^2 P^4 i \leftarrow \gamma_6$	(56,9)	$Ph_5 e_0 \leftarrow A[56]$	(56,10)	$gt \leftarrow \nu^2 A[50,2]$
(56,10)	R'	(56,13)	$d_0 v$	(56,16)	$P^2 g_2$
(56,19)	$P^3 \ell$	(56,27)	$P^6 c_0 \leftarrow \eta \gamma_6$	(57,7)	$Q_2 \leftarrow A[57]$
(57,8)	$h_5 j$	(57,12)	$e_0 g^2$	(57,15)	Pgk
(57,18)	$P^2 z$	(57,24)	$P^5 e_0$	(57,29)	$P^7 h_1 \leftarrow \alpha_7$

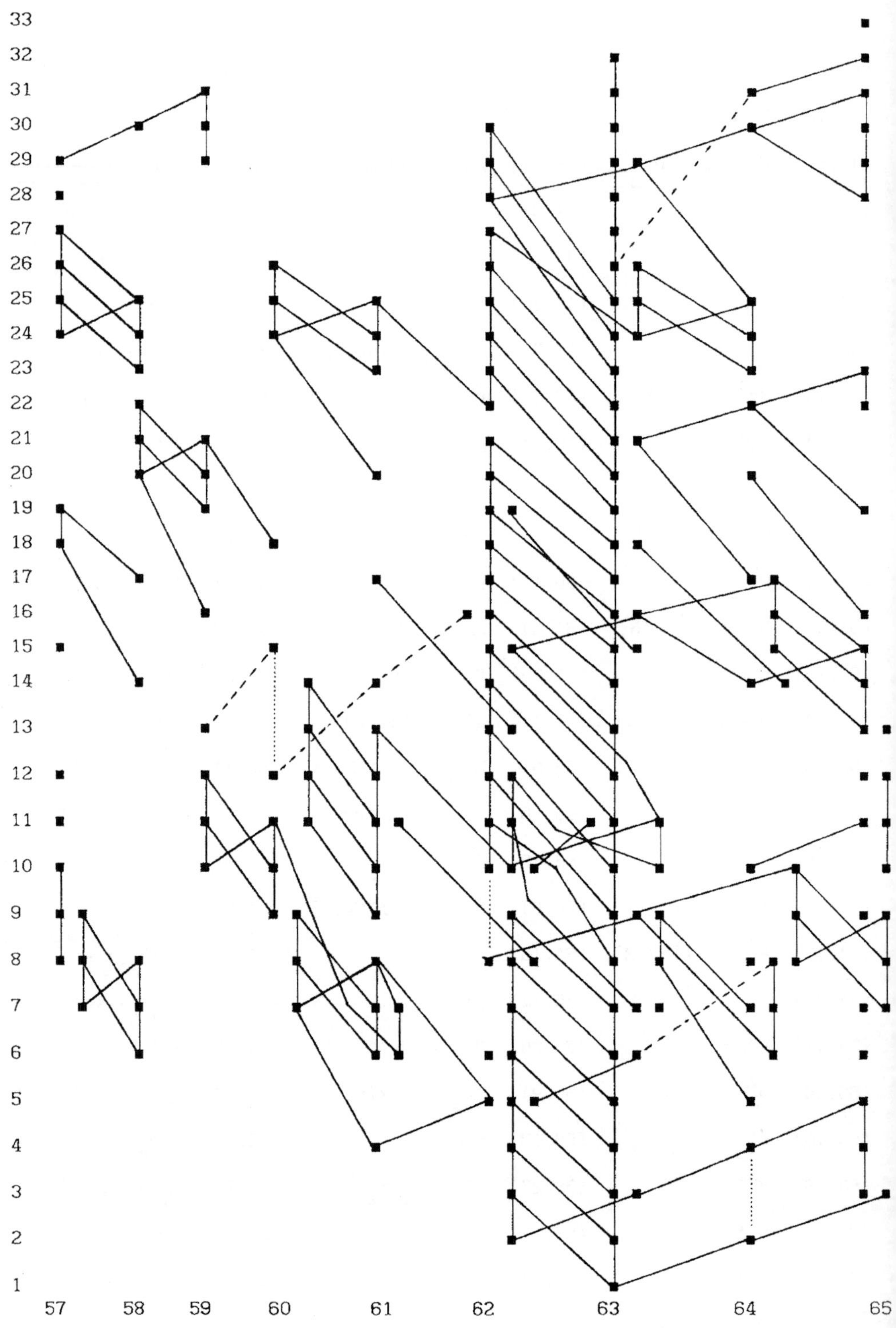

33
32
31
30
29
28
27
26
25
24
23
22
21
20
19
18
17
16
15
14
13
12
11
10
9
8
7
6
5
4
3
2
1
57
58
59
60
61
62
63
64
65

Notation:

(58,6)	D_2	(58,14)	Pgr	(58,17)	P^2v
(58,20)	$P^3e_0^2$	(58,23)	P^4j	(59,10)	$B_{21} \leftarrow A[59,1]$
(59,13)	$d_0w \leftarrow A[59,2]$	(59,16)	Pe_0^3	(59,19)	P^3m
(59,29)	$P^7h_2 \leftarrow \beta_7$	(60,7)	B_3	(60,9)	B_4
(60,11)	g_2'	(60,12)	$g^3 \leftarrow B[60]$	(60,15)	$Pg\ell \leftarrow \eta A[59,2]$
(60,18)	P^2d_0r	(60,24)	P^5g	(61,4)	D_3
(61,6)	A	(61,6)	$A+A'$	(61,7)	$h_0(A+A') \leftarrow A[61]$
(61,9)	X_1	(61,11)	rn	(61,14)	$gz \leftarrow \eta B[60]$
(61,17)	P^2w	(61,20)	P^3e_0g	(61,23)	P^4k
(62,2)	$h_5^2 \leftarrow A[62,1]$	(62,5)	$H_1 \leftarrow A[62,4]$	(62,6)	$h_5n \leftarrow A[62,3]$
(62,8)	$E_1 \leftarrow B[62]$	(62,8)	C_0	(62,10)	$R \leftarrow 2B[62]$
(62,10)	B_{22}	(62,10)	$PG \leftarrow A[62,2]$	(62,13)	gv
(62,15)	P^2B_1	(62,16)	$Pe_0^2g \leftarrow \eta^2B[60]$	(62,19)	P^2gj
(62,22)	P^4r	(62,28)	P^6d_0	(63,1)	h_6
(63,7)	$\varepsilon C'+(1-\varepsilon)X_2$	(63,7)	$(1-\varepsilon)C'+\varepsilon X_2 \leftarrow A[63]$	(63,8)	h_2B_3
(63,10)	h_2B_4	(63,15)	Pgm	(63,18)	P^2e_0r
(63,21)	P^3u	(63,24)	$P^4d_0e_0$	(63,26)	$h_0^{25}h_6 \leftarrow \gamma_7$
(64,2)	$h_1h_6 \leftarrow B[64,1]$	(64,5)	h_2D_3	(64,6)	A''
(64,7)	$h_2A = h_2A'$	(64,7)	$h_0A'' \leftarrow B[64,2]$	(64,8)	$gg_2 \leftarrow A[64,3]$
(64,8)	$h_3Q_2 \leftarrow A[64,2]$	(64,10)	$q_1 \leftarrow A[64,1]$	(64,14)	PQ_1
(64,14)	d_0gr	(64,15)	P^2B_2	(64,17)	Pd_0v
(64,20)	P^3g_2	(64,23)	$P^4\ell$	(64,31)	$P^7c_0 \leftarrow \eta\gamma_7$
(65,3)	$h_2h_5^2$	(65,6)	h_2H_1	(65,7)	h_2h_5n
(65,7)	h_3D_2	(65,9)	h_2C_0	(65,10)	B_{23}
(65,12)	Ph_5j	(65,13)	R_2	(65,13)	gw
(65,16)	Pe_0g^2	(65,19)	P^2gk	(65,22)	P^3z
(65,28)	P^6e_0	(65,33)	$P^8h_1 \leftarrow \alpha_8$		

Multiplication by h_2:

*(45,7)	$h_2Ph_2h_5 = h_0^2h_5d_0$	(45,16)	$h_2P^2j = h_0P^2k$	(47,17)	$h_2P^3g = h_0P^2e_0d_0$
(48,6)	$h_2h_5d_0 = h_0h_5e_0$	(48,16)	$h_2P^2k = h_0P^2\ell$	(49,21)	$h_2P^4d_0 = h_0P^4e_0$
(52,21)	$h_2P^4e_0 = h_0P^4g$	*(53,20)	$h_2P^3j = h_0P^3k$	(54,26)	$h_2P^6h_2 = h_0^2P^5d_0$
(55,21)	$h_2P^4g = h_0P^3e_0d_0$	(56,11)	$h_2x' = h_0R'$	(56,20)	$h_2P^3k = h_0P^3\ell$
*(56,21)	$h_2h_0P^3k = h_0^2P^3\ell$	*(57,9)	$h_2h_5i = h_0h_5j$	(57,11)	$h_2R_1 = h_1R'$
(57,25)	$h_2P^5d_0 = h_0P^5e_0$	(58,21)	$h_2P^3e_0d_0 = h_0P^3e_0^2$	(58,24)	$h_2P^4i = h_0P^4j$
*(59,11)	$h_2R' = h_0B_{21}$	*(59,20)	$h_2P^3\ell = h_0P^3m$	*(60,8)	$h_2Q_2 = h_0B_3$
(60,25)	$h_2P^5e_0 = h_0P^5g$	*(61,7)	$h_2D_2 = h_0A$	(61,24)	$h_2P^4j = h_0P^4k$
(62,11)	$h_2B_{21} = h_0B_{22}$	(62,29)	$h_2P^7h_2 = h_0^2P^6d_0$	(63,25)	$h_2P^5g = h_0P^4e_0d_0$
(64,24)	$h_2P^4k = h_0P^4\ell$	(65,11)	$h_2B_{22} = h_0B_{23}$	(65,29)	$h_2P^6d_0 = h_0P^6e_0$

APPENDIX 7: REPRESENTING MAPS

ELEMENT	REPRESENTATIVE	BOUNDARY
	B_{XY}	$X \wedge Y$
	$B_{\langle X,Y,Z\rangle}$	$(X \wedge B_{YZ}) \cup (B_{XY} \wedge Z)$
M_1	μ_1	η
M_1^2	μ_2	ν
$\langle M_1^4\rangle$	μ_4	σ
M_2	μ_{01}	$(\eta \wedge \mu_2) \cup B_{\eta\nu}$
$\bar{M}_2$	$\bar{\mu}_{01}$	$(\nu \wedge \mu_1) \cup B_{\nu\eta}$
M_2^2	μ_{02}	$(\nu \wedge \mu_4) \cup B_{\nu\sigma}$
$\langle M_2^2\rangle$	$\langle\mu_{02}\rangle$	$(\sigma \wedge \mu_2) \cup B_{\sigma\nu}$
M_3	μ_{001}	$(\eta \wedge \mu_{01}) \cup (B_{\eta\nu} \wedge \mu_4) \cup B_{\langle\eta,\nu,\sigma\rangle}$
$\langle M_3\rangle$	$\langle\mu_{001}\rangle$	$(\sigma \wedge \bar{\mu}_{01}) \cup (B_{\sigma\nu} \wedge \mu_1) \cup B_{\langle\sigma,\nu,\eta\rangle}$

BIBLIOGRAPHY

[1] J. F. Adams, *On the structure and applications of the Steenrod algebra*, Comm. Math. Helv. 32 (1958), 180-214.

[2] J. F. Adams, *On the non-existence of elements of Hopf invariant one*, Annals of Math. 72 (1960), 20-104.

[3] J. F. Adams, *On the groups J(X) - I*, Topology 2 (1963), 181-195.

[4] J. F. Adams, *On the groups J(X) - II*, Topology 3 (1965), 127-171.

[5] J. F. Adams, *On the groups J(X) - III*, Topology 3 (1965), 193-222.

[6] J. F. Adams, *On the groups J(X) - IV*, Topology 5 (1966), 21-71.

[7] J. F. Adams, *Stable homotopy and generalised homology*, Chicago Lecture Notes in Math., U. of Chicago Press, Chicago, Illinois, 1974.

[8] J. F. Adams, *A periodicity theorem in homological algebra*, Proc. Cambridge Phil. Soc. 62 (1966), 365-377.

[9] American National Standards Committee, *American National Standard Programming Language FORTRAN (FORTRAN 77)*, Document X3J3/90, X3 Secretariat, CBEMA/Standards, 1828 L Street, N.W., Washington, D.C., 20036.

[10] M. G. Barratt, M. E. Mahowald and M. C. Tangora, *Some differentials in the Adams spectral sequence - II*, Topology 9 (1970), 309-316.

[11] M. G. Barratt, J. D. S. Jones and M. E. Mahowald, *Relations amongst Toda brackets and the Kervaire invariant in dimension 62*, J. London Math. Soc. (2) 30 (1984), 533-550.

[12] M. G. Barratt, J. D. S. Jones and M. E. Mahowald, *The Kervaire invariant problem*, Contemporary Math., vol. 19, 9-22.

[13] M. G. Barratt, J. D. S. Jones and M. E. Mahowald, *The Kervaire invariant and the Hopf invariant*, Algebraic Topology, Springer Lecture Notes in Math., No. 1286, 1987, Berlin, 135-173.

[14] W. Browder, *The Kervaire invariant of framed manifolds and its generalizations*, Annals of Math. 90 (1969), 157-186.

[15] E. H. Brown and F. P. Peterson, *A spectrum whose Z_p-cohomology is the algebra of reduced p-th powers*, Topology 5 (1966), 149-154.

[16] R. Bruner, *A new differential in the Adams spectral sequence*, Topology 23 (1984), 271-276.

[17] H. Cartan, Algèebre d'Eilenberg-MacLane et homotopie, Séminaire Henri Cartan, 7e année: 1954/1955, 2e édition, Secrétariat mathématique, Paris, 1956.[18] J. M. Cohen, *The decomposition of stable homotopy*, Annals of Math. 87 (1968), 305-320.

[19] J. M. Cohen, *Stable homotopy*, Springer Lecture Notes in Math., vol. 165, Berlin, 1970.

[20] E. Dyer, *Cohomology theories*, Math. Lecture Notes Series, W. A. Benjamin Inc., New York, 1969.

[21] H. H. Gershenson, *Higher composition products*, J. Math. Kyoto Univ. 5 (1965), 1-37.

[22] H. Hazewinkel, *A universal formal group and complex cobordism*, Bull. Amer. Math. Soc. 81 (1975), 930-933.

[23] H. Hazewinkel, *Constructing formal groups III, applications to complex cobordism and Brown-Peterson cohomology,* J. Pure App. Algebra 10 (1977/78), 1-18.

[24] P. Hoffman, *Adams operations and homotopy composition,* Quaterly J. of Math. 19 (1968), 351-361.

[25] D. S. Kahn, *Cup-i products and the Adams spectral sequence,* Topology 9 (1970), 1-9.

[26] M. Kervaire, *A manifold which does not admit any differentiable structure,* Comm. Math. Helv. 34 (1966), 256-270.

[27] M. Kervaire and J. Milnor, *Groups of homotopy spheres I*, Annals of Math. 77 (1963), 504-537.

[28] S. O. Kochman, *A chain functor for bordism,* Trans. Amer. Math. Soc. 239 (1978), 167-196.

[29] S. O. Kochman, *Uniqueness of Massey products on the stable homotopy of spheres,* Can. J. of Math. 32 (1980), 576-589.

[30] S. O. Kochman, *Integral Cohomology Operations,* CMS Conference Proceedings, vol. 2, Part 1, 1982, 437-478.

[31] S. O. Kochman, *The symplectic cobordism ring III,* (to appear).

[32] S. O. Kochman and V. P. Snaith, *On the stable homotopy of symplectic classifying and Thom spaces,* Springer Lecture Notes in Math., vol. 741, 1979, Berlin, 394-448.

[33] A. Lawrence, *Higher order compositions in the Adams spectral sequence,* Bull. Amer. Math. Soc. 76 (1970), 874-877.

[34A] M. E. Mahowald, *Some remarks on the Kervaire invariant problem from the homotopy point of view,* Algebraic Topology, Proc. of Symposia in Pure Math., vol. XXII, Amer. Math. Soc., 1971.

[34] M. E. Mahowald, *Descriptive homotopy of the elements in the Image of the J-homomorphism,* Manifolds - Tokyo 1973, Univ. of Tokyo Press, Tokyo, 1975, 255-264.

[35] M.E. Mahowald, *A new ifinite family in* ${}_2\pi_*^S$, Topology 16 (1977), 249-256.

[36] M. E. Mahowald, *Some homotopy classes generated by* η_j, Algebraic Topology Aarhus 1978, Springer Lecture Notes in Math. No. 763, 1979, Berlin, 23-37.

[37] M. E. Mahowald and M. C. Tangora, *Some differentials in the Adams spectral sequence,* Topology 6 (1967), 349-369.

[38] M. E. Mahowald and M. C. Tangora, *An Infinite Subalgebra of* $Ext_A(Z_2,Z_2)$, Trans. Amer. Math.Soc. 132 (1968), 263-274.

[39] J.P. May, *The cohomology of restricted Lie algebras and of Hopf algebras; applications to the the Steenrod algebra,* Thesis, Princeton University, 1964.

[40] J. P. May, *Matric Massey products,* J. of Algebra, 12 (1969), 533-568.

[41] J. P. May, E_∞ *ring spaces and* E_∞ *ring spectra,* Lecture Notes in Math. No. 577, Springer Verlag, Berlin, 1977.

[42] H. R. Miller, D. C. Ravenel and W. S. Wilson, *Periodic phenomena in the Adams-Novikov spectral sequence,* Annals of Math. 106 (1977), 469-516.

[43] J. Milnor, *On the cobordism ring* Ω^* *and a complex analogue,* Amer. J. Math. 82 (1960), 505-521.

[44] M. Mimura, *On the generalized Hopf homomorphism and the higher composition, Part I, Part II,* J. Math. Kyoto Univ. 4 (1964/65), 171-190, 301-326.

[45] M. Mimura and H. Toda, *The (n+20)th homotopy groups of n-spheres,* J. Math. Kyoto Univ. 3 (1963), 37-58.

[46] M. Mimura, M. Mori and N. Oda, *Determination of 2-componenets of the 23- and 24-stems in homotopy groups of spheres,* Mem. Fac. Sci. Kyushu Univ. 29 (1975), 1-42.

[47] R. E. Mosher and M. C. Tangora, *Cohomology operations and applications in homotopy theory,* Harper's Series in Modern Math., New York, 1968.

[48] R. M. F. Moss, *Secondary compositions and the Adams spectral sequence,* Math. Zeit. 115, (1970), 283-310.

[49] S. P. Novikov, *The methods of algebraic topology from the viewpoint of cobordism theories,* Izv. Akad. Nauk SSSR Ser. Mat. 31 (1967), 855-951; translation, Math. USSR - Izv. (1967), 827-913.

[50] N. Oda, *Unstable homotopy groups of spheres,* Bull. of the Inst. for Advanced Research of Fukuoka Univ. 44 (1979), 49 - 152.

[51] K. Oguchi, *A generalization of secondary composition and applications,* J. Fac. Sci. Tokyo Univ. 10 (1963), 29-79.

[52] G. J. Porter, *Higher products,* Trans. Amer. Math. Soc. 148 (1970), 314-345.

[53] D. Quillen, *The Adams conjecture*, Topology 10 (1971), 1-10.

[54] D. Quillen, *On the formal group laws of unoriented and complex cobordism theory,* Bull. Amer. Math. Soc. 75 (1969), 1293-1298.

[55] D. C. Ravenel, *Complex cobordism and stable homotopy groups of spheres,* Pure and Applied Math., vol. 121, Academic Press, Orlando, Florida, 1986.

[56] N. Ray, *The symplectic bordism ring*, Proc. Camb. Phil. Soc. 71 (1972), 271-282.

[57] J. P. Serre, *Groupes d'homotopie et classes de groupes abelien,* Annals of Math. 58 (1953), 258-294.

[58] E. Spanier, *Secondary operations on mappings and cohomology,* Annals of Math. 75 (1962), 260-282.

[59] M. C. Tangora, *On the cohomology of the Steenrod algebra,* Math. Zeit. 116 (1970), 18-64.

[60] H. Toda, *Composition methods in homotopy groups of spheres*, Annals of Math. Studies No. 49, Princeton Univ. Press, Princeton, N.J., 1962.

[61] W. S. Wilson, *Brown-Peterson homology: an introduction and sampler,* Regional Conference Series in Math., No.48, Amer. Math. Soc., Providence, Rhode Island, 1980.

[62] R. Zahler, *The Adams-Novikov spectral sequence for the spheres,* Annals of Math. 96 (1972), 480-504.

Vol. 1259: F. Cano Torres, Desingularization Strategies for Three-Dimensional Vector Fields. IX, 189 pages. 1987.

Vol. 1260: N.H. Pavel, Nonlinear Evolution Operators and Semigroups. VI, 285 pages. 1987.

Vol. 1261: H. Abels, Finite Presentability of S-Arithmetic Groups. Compact Presentability of Solvable Groups. VI, 178 pages. 1987.

Vol. 1262: E. Hlawka (Hrsg.), Zahlentheoretische Analysis II. Seminar, 1984–86. V, 158 Seiten. 1987.

Vol. 1263: V.L. Hansen (Ed.), Differential Geometry. Proceedings, 1985. XI, 288 pages. 1987.

Vol. 1264: Wu Wen-tsün, Rational Homotopy Type. VIII, 219 pages. 1987.

Vol. 1265: W. Van Assche, Asymptotics for Orthogonal Polynomials. VI, 201 pages. 1987.

Vol. 1266: F. Ghione, C. Peskine, E. Sernesi (Eds.), Space Curves. Proceedings, 1985. VI, 272 pages. 1987.

Vol. 1267: J. Lindenstrauss, V.D. Milman (Eds.), Geometrical Aspects of Functional Analysis. Seminar. VII, 212 pages. 1987.

Vol. 1268: S.G. Krantz (Ed.), Complex Analysis. Seminar, 1986. VII, 195 pages. 1987.

Vol. 1269: M. Shiota, Nash Manifolds. VI, 223 pages. 1987.

Vol. 1270: C. Carasso, P.-A. Raviart, D. Serre (Eds.), Nonlinear Hyperbolic Problems. Proceedings, 1986. XV, 341 pages. 1987.

Vol. 1271: A.M. Cohen, W.H. Hesselink, W.L.J. van der Kallen, J.R. Strooker (Eds.), Algebraic Groups Utrecht 1986. Proceedings. XII, 284 pages. 1987.

Vol. 1272: M.S. Livšic, L.L. Waksman, Commuting Nonselfadjoint Operators in Hilbert Space. III, 115 pages. 1987.

Vol. 1273: G.-M. Greuel, G. Trautmann (Eds.), Singularities, Representation of Algebras, and Vector Bundles. Proceedings, 1985. XIV, 383 pages. 1987.

Vol. 1274: N. C. Phillips, Equivariant K-Theory and Freeness of Group Actions on C*-Algebras. VIII, 371 pages. 1987.

Vol. 1275: C.A. Berenstein (Ed.), Complex Analysis I. Proceedings, 1985–86. XV, 331 pages. 1987.

Vol. 1276: C.A. Berenstein (Ed.), Complex Analysis II. Proceedings, 1985–86. IX, 320 pages. 1987.

Vol. 1277: C.A. Berenstein (Ed.), Complex Analysis III. Proceedings, 1985–86. X, 350 pages. 1987.

Vol. 1278: S.S. Koh (Ed.), Invariant Theory. Proceedings, 1985. V, 102 pages. 1987.

Vol. 1279: D. Ieşan, Saint-Venant's Problem. VIII, 162 Seiten. 1987.

Vol. 1280: E. Neher, Jordan Triple Systems by the Grid Approach. XII, 193 pages. 1987.

Vol. 1281: O.H. Kegel, F. Menegazzo, G. Zacher (Eds.), Group Theory. Proceedings, 1986. VII, 179 pages. 1987.

Vol. 1282: D.E. Handelman, Positive Polynomials, Convex Integral Polytopes, and a Random Walk Problem. XI, 136 pages. 1987.

Vol. 1283: S. Mardešić, J. Segal (Eds.), Geometric Topology and Shape Theory. Proceedings, 1986. V, 261 pages. 1987.

Vol. 1284: B.H. Matzat, Konstruktive Galoistheorie. X, 286 pages. 1987.

Vol. 1285: I.W. Knowles, Y. Saitō (Eds.), Differential Equations and Mathematical Physics. Proceedings, 1986. XVI, 499 pages. 1987.

Vol. 1286: H.R. Miller, D.C. Ravenel (Eds.), Algebraic Topology. Proceedings, 1986. VII, 341 pages. 1987.

Vol. 1287: E.B. Saff (Ed.), Approximation Theory, Tampa. Proceedings, 1985–1986. V, 228 pages. 1987.

Vol. 1288: Yu. L. Rodin, Generalized Analytic Functions on Riemann Surfaces. V, 128 pages, 1987.

Vol. 1289: Yu. I. Manin (Ed.), K-Theory, Arithmetic and Geometry. Seminar, 1984–1986. V, 399 pages. 1987.

Vol. 1290: G. Wüstholz (Ed.), Diophantine Approximation and Transcendence Theory. Seminar, 1985. V, 243 pages. 1987.

Vol. 1291: C. Mœglin, M.-F. Vignéras, J.-L. Waldspurger, Correspondances de Howe sur un Corps p-adique. VII, 163 pages. 1987

Vol. 1292: J.T. Baldwin (Ed.), Classification Theory. Proceedings, 1985. VI, 500 pages. 1987.

Vol. 1293: W. Ebeling, The Monodromy Groups of Isolated Singularities of Complete Intersections. XIV, 153 pages. 1987.

Vol. 1294: M. Queffélec, Substitution Dynamical Systems – Spectral Analysis. XIII, 240 pages. 1987.

Vol. 1295: P. Lelong, P. Dolbeault, H. Skoda (Réd.), Séminaire d'Analyse P. Lelong – P. Dolbeault – H. Skoda. Seminar, 1985/1986. VII, 283 pages. 1987.

Vol. 1296: M.-P. Malliavin (Ed.), Séminaire d'Algèbre Paul Dubreil et Marie-Paule Malliavin. Proceedings, 1986. IV, 324 pages. 1987.

Vol. 1297: Zhu Y.-l., Guo B.-y. (Eds.), Numerical Methods for Partial Differential Equations. Proceedings. XI, 244 pages. 1987.

Vol. 1298: J. Aguadé, R. Kane (Eds.), Algebraic Topology, Barcelona 1986. Proceedings. X, 255 pages. 1987.

Vol. 1299: S. Watanabe, Yu.V. Prokhorov (Eds.), Probability Theory and Mathematical Statistics. Proceedings, 1986. VIII, 589 pages. 1988.

Vol. 1300: G.B. Seligman, Constructions of Lie Algebras and their Modules. VI, 190 pages. 1988.

Vol. 1301: N. Schappacher, Periods of Hecke Characters. XV, 160 pages. 1988.

Vol. 1302: M. Cwikel, J. Peetre, Y. Sagher, H. Wallin (Eds.), Function Spaces and Applications. Proceedings, 1986. VI, 445 pages. 1988.

Vol. 1303: L. Accardi, W. von Waldenfels (Eds.), Quantum Probability and Applications III. Proceedings, 1987. VI, 373 pages. 1988.

Vol. 1304: F.Q. Gouvêa, Arithmetic of *p*-adic Modular Forms. VIII, 121 pages. 1988.

Vol. 1305: D.S. Lubinsky, E.B. Saff, Strong Asymptotics for Extremal Polynomials Associated with Weights on ℝ. VII, 153 pages. 1988.

Vol. 1306: S.S. Chern (Ed.), Partial Differential Equations. Proceedings, 1986. VI, 294 pages. 1988.

Vol. 1307: T. Murai, A Real Variable Method for the Cauchy Transform, and Analytic Capacity. VIII, 133 pages. 1988.

Vol. 1308: P. Imkeller, Two-Parameter Martingales and Their Quadratic Variation. IV, 177 pages. 1988.

Vol. 1309: B. Fiedler, Global Bifurcation of Periodic Solutions with Symmetry. VIII, 144 pages. 1988.

Vol. 1310: O.A. Laudal, G. Pfister, Local Moduli and Singularities. V, 117 pages. 1988.

Vol. 1311: A. Holme, R. Speiser (Eds.), Algebraic Geometry, Sundance 1986. Proceedings. VI, 320 pages. 1988.

Vol. 1312: N.A. Shirokov, Analytic Functions Smooth up to the Boundary. III, 213 pages. 1988.

Vol. 1313: F. Colonius, Optimal Periodic Control. VI, 177 pages. 1988.

Vol. 1314: A. Futaki, Kähler-Einstein Metrics and Integral Invariants. IV, 140 pages. 1988.

Vol. 1315: R.A. McCoy, I. Ntantu, Topological Properties of Spaces of Continuous Functions. IV, 124 pages. 1988.

Vol. 1316: H. Korezlioglu, A.S. Ustunel (Eds.), Stochastic Analysis and Related Topics. Proceedings, 1986. V, 371 pages. 1988.

Vol. 1317: J. Lindenstrauss, V.D. Milman (Eds.), Geometric Aspects of Functional Analysis. Seminar, 1986–87. VII, 289 pages. 1988.

Vol. 1318: Y. Felix (Ed.), Algebraic Topology – Rational Homotopy. Proceedings, 1986. VIII, 245 pages. 1988

Vol. 1319: M. Vuorinen, Conformal Geometry and Quasiregular Mappings. XIX, 209 pages. 1988.

Vol. 1320: H. Jürgensen, G. Lallement, H.J. Weinert (Eds.), Semigroups, Theory and Applications. Proceedings, 1986. X, 416 pages. 1988.

Vol. 1321: J. Azéma, P.A. Meyer, M. Yor (Eds.), Séminaire de Probabilités XXII. Proceedings. IV, 600 pages. 1988.

Vol. 1322: M. Métivier, S. Watanabe (Eds.), Stochastic Analysis. Proceedings, 1987. VII, 197 pages. 1988.

Vol. 1323: D.R. Anderson, H.J. Munkholm, Boundedly Controlled Topology. XII, 309 pages. 1988.

Vol. 1324: F. Cardoso, D.G. de Figueiredo, R. Iório, O. Lopes (Eds.), Partial Differential Equations. Proceedings, 1986. VIII, 433 pages. 1988.

Vol. 1325: A. Truman, I.M. Davies (Eds.), Stochastic Mechanics and Stochastic Processes. Proceedings, 1986. V, 220 pages. 1988.

Vol. 1326: P.S. Landweber (Ed.), Elliptic Curves and Modular Forms in Algebraic Topology. Proceedings, 1986. V, 224 pages. 1988.

Vol. 1327: W. Bruns, U. Vetter, Determinantal Rings. VII,236 pages. 1988.

Vol. 1328: J.L. Bueso, P. Jara, B. Torrecillas (Eds.), Ring Theory. Proceedings, 1986. IX, 331 pages. 1988.

Vol. 1329: M. Alfaro, J.S. Dehesa, F.J. Marcellan, J.L. Rubio de Francia, J. Vinuesa (Eds.): Orthogonal Polynomials and their Applications. Proceedings, 1986. XV, 334 pages. 1988.

Vol. 1330: A. Ambrosetti, F. Gori, R. Lucchetti (Eds.), Mathematical Economics. Montecatini Terme 1986. Seminar. VII, 137 pages. 1988.

Vol. 1331: R. Bamón, R. Labarca, J. Palis Jr. (Eds.), Dynamical Systems, Valparaiso 1986. Proceedings. VI, 250 pages. 1988.

Vol. 1332: E. Odell, H. Rosenthal (Eds.), Functional Analysis. Proceedings, 1986–87. V, 202 pages. 1988.

Vol. 1333: A.S. Kechris, D.A. Martin, J.R. Steel (Eds.), Cabal Seminar 81–85. Proceedings, 1981–85. V, 224 pages. 1988.

Vol. 1334: Yu.G. Borisovich, Yu. E. Gliklikh (Eds.), Global Analysis – Studies and Applications III. V, 331 pages. 1988.

Vol. 1335: F. Guillén, V. Navarro Aznar, P. Pascual-Gainza, F. Puerta, Hyperrésolutions cubiques et descente cohomologique. XII, 192 pages. 1988.

Vol. 1336: B. Helffer, Semi-Classical Analysis for the Schrödinger Operator and Applications. V, 107 pages. 1988.

Vol. 1337: E. Sernesi (Ed.), Theory of Moduli. Seminar, 1985. VIII, 232 pages. 1988.

Vol. 1338: A.B. Mingarelli, S.G. Halvorsen, Non-Oscillation Domains of Differential Equations with Two Parameters. XI, 109 pages. 1988.

Vol. 1339: T. Sunada (Ed.), Geometry and Analysis of Manifolds. Procedings, 1987. IX, 277 pages. 1988.

Vol. 1340: S. Hildebrandt, D.S. Kinderlehrer, M. Miranda (Eds.), Calculus of Variations and Partial Differential Equations. Proceedings, 1986. IX, 301 pages. 1988.

Vol. 1341: M. Dauge, Elliptic Boundary Value Problems on Corner Domains. VIII, 259 pages. 1988.

Vol. 1342: J.C. Alexander (Ed.), Dynamical Systems. Proceedings, 1986–87. VIII, 726 pages. 1988.

Vol. 1343: H. Ulrich, Fixed Point Theory of Parametrized Equivariant Maps. VII, 147 pages. 1988.

Vol. 1344: J. Král, J. Lukeš, J. Netuka, J. Veselý (Eds.), Potential Theory – Surveys and Problems. Proceedings, 1987. VIII, 271 pages. 1988.

Vol. 1345: X. Gomez-Mont, J. Seade, A. Verjovski (Eds.), Holomorphic Dynamics. Proceedings, 1986. VII, 321 pages. 1988.

Vol. 1346: O. Ya. Viro (Ed.), Topology and Geometry – Rohlin Seminar. XI, 581 pages. 1988.

Vol. 1347: C. Preston, Iterates of Piecewise Monotone Mappings on an Interval. V, 166 pages. 1988.

Vol. 1348: F. Borceux (Ed.), Categorical Algebra and its Applications. Proceedings, 1987. VIII, 375 pages. 1988.

Vol. 1349: E. Novak, Deterministic and Stochastic Error Bounds in Numerical Analysis. V, 113 pages. 1988.

Vol. 1350: U. Koschorke (Ed.), Differential Topology. Proceedings, 1987. VI, 269 pages. 1988.

Vol. 1351: I. Laine, S. Rickman, T. Sorvali, (Eds.), Complex Analysis, Joensuu 1987. Proceedings. XV, 378 pages. 1988.

Vol. 1352: L.L. Avramov, K.B. Tchakerian (Eds.), Algebra – Some Current Trends. Proceedings, 1986. IX, 240 Seiten. 1988.

Vol. 1353: R.S. Palais, Ch.-l. Terng, Critical Point Theory and Submanifold Geometry. X, 272 pages. 1988.

Vol. 1354: A. Gómez, F. Guerra, M.A. Jiménez, G. López (Eds.), Approximation and Optimization. Proceedings, 1987. VI, 280 pages. 1988.

Vol. 1355: J. Bokowski, B. Sturmfels, Computational Synthetic Geometry. V, 168 pages. 1989.

Vol. 1356: H. Volkmer, Multiparameter Eigenvalue Problems and Expansion Theorems. VI, 157 pages. 1988.

Vol. 1357: S. Hildebrandt, R. Leis (Eds.), Partial Differential Equations and Calculus of Variations. VI, 423 pages. 1988.

Vol. 1358: D. Mumford, The Red Book of Varieties and Schemes. V, 309 pages. 1988.

Vol. 1359: P. Eymard, J.-P. Pier (Eds.), Harmonic Analysis. Proceedings, 1987. VIII, 287 pages. 1988.

Vol. 1360: G. Anderson, C. Greengard (Eds.), Vortex Methods. Proceedings, 1987. V, 141 pages. 1988.

Vol. 1361: T. tom Dieck (Ed.), Algebraic Topology and Transformation Groups. Proceedings, 1987. VI, 298 pages. 1988.

Vol. 1362: P. Diaconis, D. Elworthy, H. Föllmer, E. Nelson, G.C. Papanicolaou, S.R.S. Varadhan. École d'Été de Probabilités de Saint-Flour XV–XVII, 1985–87. Editor: P.L. Hennequin. V, 459 pages. 1988.

Vol. 1363: P.G. Casazza, T.J. Shura. Tsirelson's Space. VIII, 204 pages. 1988.

Vol. 1364: R.R. Phelps, Convex Functions, Monotone Operators and Differentiability. IX, 115 pages. 1989.

Vol. 1365: M. Giaquinta (Ed.), Topics in Calculus of Variations. Seminar, 1987. X, 196 pages. 1989.

Vol. 1366: N. Levitt, Grassmannians and Gauss Maps in PL-Topology. V, 203 pages. 1989.

Vol. 1367: M. Knebusch, Weakly Semialgebraic Spaces. XX, 376 pages. 1989.

Vol. 1368: R. Hübl, Traces of Differential Forms and Hochschild Homology. III, 111 pages. 1989.

Vol. 1369: B. Jiang, Ch.-K. Peng, Z. Hou (Eds.), Differential Geometry and Topology. Proceedings, 1986–87. VI, 366 pages. 1989.

Vol. 1370: G. Carlsson, R.L. Cohen, H.R. Miller, D.C. Ravenel (Eds.), Algebraic Topology. Proceedings, 1986. IX, 456 pages. 1989.

Vol. 1371: S. Glaz, Commutative Coherent Rings. XI, 347 pages. 1989.

Vol. 1372: J. Azéma, P.A. Meyer, M. Yor (Eds.), Séminaire de Probabilités XXIII. Proceedings. IV, 583 pages. 1989.

Vol. 1373: G. Benkart, J.M. Osborn (Eds.), Lie Algebras, Madison 1987. Proceedings. V, 145 pages. 1989.

Vol. 1374: R.C. Kirby, The Topology of 4-Manifolds. VI, 108 pages. 1989.

Vol. 1375: K. Kawakubo (Ed.), Transformation Groups. Proceedings, 1987. VIII, 394 pages, 1989.

Vol. 1376: J. Lindenstrauss, V.D. Milman (Eds.), Geometric Aspects of Functional Analysis. Seminar (GAFA) 1987–88. VII, 288 pages. 1989.

Vol. 1377: J.F. Pierce, Singularity Theory, Rod Theory, and Symmetry-Breaking Loads. IV, 177 pages. 1989.

Vol. 1378: R.S. Rumely, Capacity Theory on Algebraic Curves. III, 437 pages. 1989.

Vol. 1379: H. Heyer (Ed.), Probability Measures on Groups IX. Proceedings, 1988. VIII, 437 pages. 1989

GPSR Compliance

The European Union's (EU) General Product Safety Regulation (GPSR) is a set of rules that requires consumer products to be safe and our obligations to ensure this.

If you have any concerns about our products, you can contact us on ProductSafety@springernature.com

In case Publisher is established outside the EU, the EU authorized representative is:

Springer Nature Customer Service Center GmbH
Europaplatz 3
69115 Heidelberg, Germany

Batch number: 09624486

Printed by Printforce, the Netherlands